Angiogenesis in Adipose Tissue

Yihai Cao
Editor

Angiogenesis in Adipose Tissue

Springer

Editor
Yihai Cao
Department of Microbiology
Tumor and Cell Biology (MTC)
Karolinska Institute
Stockholm, Sweden

Department of Medicine
and Health Sciences
Linköping University
Linköping, Sweden

ISBN 978-1-4614-8068-6 ISBN 978-1-4614-8069-3 (eBook)
DOI 10.1007/978-1-4614-8069-3
Springer New York Heidelberg Dordrecht London

Library of Congress Control Number: 2013946570

Printed on acid-free paper

Springer is part of Springer Science+Business Media (www.springer.com)

Preface

The concept of therapeutic antiangiogenesis was proposed by Judah Folkman 40 years ago. Although this theory was initially hypothesized for cancer therapy by blocking tumor angiogenesis, the same principle may also apply for treatment of other angiogenesis-dependent nonmalignant diseases. Today, antiangiogenic drugs have successfully been used for treatment of various cancers and age-related macular disease in human patients. The clinical success of antiangiogenic therapy crystallizes more than 40-year extensive preclinical and clinical findings of angiogenesis research. It has also paved new avenues for therapeutic extensions to other areas including obesity and metabolic disorders.

Obesity and its related metabolic diseases pose the biggest threat to global human health. Unfortunately, pharmacological intervention of obesity by effective drugs is not available and development of such effective drugs is an inevitable urgent task for prevention and treatment of the fast growing obese population in both developed and developing countries. The original experimental evidence that expansion of the adipose tissue is dependent on angiogenesis was provided in mouse obese models, in which treatment of obese animals with antiangiogenic agents significantly prevented obese development. These findings resemble the antitumor effect of angiogenesis inhibitors. As angiogenesis is essentially required for almost all tissue growth and expansion, the preclinical findings of antiobesity effect by angiogenesis inhibitors are probably not surprising.

White adipose tissue (WAT) and brown adipose tissue (BAT) exhibit opposing metabolic effects, as the former stores excessive energy and latter flames the stored energy to become heat. Storage of excessive lipids in WAT and energy consumption in BAT seem to be dependent on appropriate vascular functions, which are tightly controlled by the number and structure of microvessels in the adipose tissues. Thus, blood vessels might play dual roles in modulation of adipose tissue growth or regression, depending on its relation to the metabolic status of the adipose tissue.

Emerging experimental evidence shows that the adipose vascular system is not merely a pipe-like structure that supplies oxygen and nutrients for adipocytes. Blood vessels in the adipose tissue may also serve as a reservoir of stem cells that under certain conditions differentiate into preadipocytes and adipocytes. In fact, both vascular endothelial cells and perivascular cells have been reported to have stem cell features, which allow them to trans-differentiate into adipocytes. Additionally, blood vessels in the adipose tissue provide other cell types including stromal cells and inflammatory cells that significantly modulate adipocyte functions. Similar to tumors, various cell types in the adipose tissue produce multiple angiogenic factors and cytokines that often collaboratively modulate adipogenesis. Emerging evidence suggests that a complex interplay exists between these angiogenic factors and cytokines.

This is the first book focusing on discussion of complex roles of the adipose vasculature in modulation of adipogenesis, obesity, and metabolism. Although most of these findings are based on preclinical animal models, the same mechanism most likely exists in humans. I express my deepest appreciation to all coauthors and other experts who have significantly contributed to the development of this exciting field and to this comprehensive book covering most aspects of the adipocyte-vascular interface. I also apologize that many other interesting aspects related to this topic may not be covered due to the space limit of this book.

Stockholm, Sweden Yihai Cao, M.D., Ph.D

Contents

Part I Historical Review of Angiogenesis

1 **Angiogenesis in Diseases and Therapy** 3
Yihai Cao

Part II Spatiotemporal Relation of Adipocytes and Vascular Cells During Development

2 **Adipose Stem Cells** 19
Carolyn Algire, Dasa Medrikova, and Stephan Herzig

3 **Stromal Vascular Cells** 41
Sahohime Matsumoto and Ichiro Manabe

4 **Vascular Adipose Complex** 53
J. Michael Sorrell

Part III Adipose Vasculature

5 **Blood Vessels in White and Brown Adipose Tissues** 77
Sharon Lim, Jennifer Honek, and Yihai Cao

6 **Lymphatic System in Adipose Tissues** 103
Bernhard Nausch, Sonia Rehal, and Pierre-Yves von der Weid

7 **Origin of Adipocyte Precursors from Adipose Vascular Endothelium** 131
Andrea Frontini, Silvia Corvera, and Saverio Cinti

8 **Vascular and Endothelial Regeneration** 157
Louis Casteilla, Patrick Laharrague, and V. Planat-Benard

Part IV Angiogenic Factors and Adipokines in Adipose Tissue

9 **Role of NPY in Brown Adipocytes and Obesity** 169
Sheng Bi

10 **Leptin, Adiponectin, and Other Adipokines in Regulation of Adipose Tissue Angiogenesis** 187
Ebba Brakenhielm and Yihai Cao

Part V Regulation of Adipose Angiogenesis by Inflammatory Cells

11 **Immune Cells in Adipose Tissue: Key Players in Metabolic Disorders** 231
Fanny Volat and Anne Bouloumié

12 **Adipose Tissue Hypoxia in Regulation of Angiogenesis and Obesity** 247
Zoi Michailidou and Jonathan R. Seckl

Part VI Mouse Models to Study Adipose Angiogenesis

13 **Adipose Angiogenesis Models in Animals and Methodology** 265
Jennifer Honek, Sharon Lim, and Yihai Cao

14 **Mouse Genetic Models in Studying Adipose Angiogenesis** 297
Ilse Scroyen and H. Roger Lijnen

Part VII Adipose Tissue Facilitates Tumor Angiogenesis and Growth

15 **Adipose Tissue-Derived Progenitor Cells and Cancer** 321
Chieh Tseng and Mikhail G. Kolonin

16 **Adipose-Derived Endothelial Precursor Cells Supporting Tumor Growth** 339
Patrizia Mancuso, Ines Martin-Padura, and Francesco Bertolini

Part VIII Engineering of Vascularized Adipose Tissue

17 **Successful Angiogenesis as Key to Engineered Adipose Tissue** 351
Paul Severin Wiggenhauser and Jan-Thorsten Schantz

Index 371

Part I
Historical Review of Angiogenesis

Chapter 1
Angiogenesis in Diseases and Therapy

Yihai Cao

Abstract Angiogenesis research has become an increasingly attractive field in biomedicine for academic and pharmaceutical scientists because of its broad involvement in various human diseases, including the most common and lethal diseases such as cancer, obesity, cardiovascular diseases, chronic inflammation, and ophthalmological diseases. Although the initiation of angiogenesis can be triggered by distinct and overlapping angiogenic factors and cytokines in different diseases, a common fundamental mechanism that underlies the angiogenic process exists under various pathological settings. Targeting pathological angiogenesis has provided an effective therapeutic approach of treatment of certain human diseases. For example, anti-vascular endothelial growth factor (anti-VEGF) drugs show markedly beneficial effects in vision improvement in patients with wet type of age-related macular degeneration (AMD). Similar anti-VEGF drugs in combination with chemotherapy have also successfully been developed for treatment of various cancers in human patients. Paradoxically, promoting neovascularization has also been therapeutically implied for treatment of ischemic diseases such as ischemic myocardium and chronic leg ischemia. However, proangiogenic therapy-based approaches have not been demonstrated to be beneficial in rigorous clinical trials. Would modulation of angiogenesis offer a new therapeutic approach for treatment of obesity and its related metabolic disorders? In some mouse disease models, administration of generic antiangiogenic agents has provided supportive evidence although such an approach needs validation in humans.

Keywords Angiogenesis • Disease therapy • Obesity • Diabetes • Cancer • Targeted therapy • Neovascularization

---

Y. Cao, M.D., Ph.D. (✉)
Department of Microbiology, Tumor and Cell Biology (MTC), Karolinska Institute, 171 77 Stockholm, Sweden

Department of Medicine and Health Sciences, Linköping University, 581 85 Linköping, Sweden
e-mail: yihai.cao@ki.se

Y. Cao (ed.), *Angiogenesis in Adipose Tissue*, DOI 10.1007/978-1-4614-8069-3_1,

1.1 Process of Angiogenesis

Blood vessel formation and growth are achieved via several processes including vasculogenesis, angiogenesis, and intussusception (Carmeliet and Jain 2000; Cao 2009a; Folkman 1996). While vasculogenesis is essential for the formation of the first vascular structure including primitive cardinal endothelium and aorta during early stages of embryogenesis (Risau and Flamme 1995), angiogenesis is the key process of neovascularization that employs vessel sprouting from the preexisting vasculature (Yancopoulos et al. 1998). It is believed that endothelial cells in quiescent vasculatures respond to angiogenic signals to proliferate, migrate, and eventually form tubular-like structures. In adults, vasculatures in most tissues remain quiescent and the dormant nature of these blood vessels suggests the existence of negative regulators that prevent vessel growth. In support of this notion, numerous endogenous angiogenesis inhibitors have been identified in both healthy and pathological tissues (Cao 2001; O'Reilly et al. 1994, 1997; Nyberg et al. 2005). The initiation of an angiogenic phenotype requires high levels of angiogenic stimuli to override negative regulators, tipping the balance toward vascular spouting (Cao 2004). Indeed, high levels of angiogenic factors including vascular endothelial growth factor (VEGF), fibroblast growth factor (FGF), platelet-derived growth factor (PDGF), angiopoietins, and hepatocyte growth factor (HGF) are often found in tissues that undergo neovascularization (Cao et al. 2009). These angiogenic driving factors are especially expressed at high levels in tumors and other pathological tissues that encounter tissue ischemia.

In response to angiogenic signals, endothelial cells have to break down the basement membrane to migrate and proliferate, leading to an organized and directed growth cone. This process is controlled by several signaling systems in which the interplay between VEGF and Notch signaling systems is primarily essential to ensure the formation of an optimal vascular network (Cao et al. 2009; Benedito et al. 2012; Sawamiphak et al. 2010; Tammela et al. 2008; Hellstrom et al. 2007; Siekmann and Lawson 2007; Ridgway et al. 2006; Noguera-Troise et al. 2006). In the presence of VEGF, inhibition of the Notch signaling pathway results in excessive formation of endothelial cell tips and vascular networks. The Dll4-triggered Notch signaling prevents the undirected tip formation in the angiogenic growth cone, leading to the directed growth of angiogenic vessels (Hellstrom et al. 2007; Siekmann and Lawson 2007; Ridgway et al. 2006; Noguera-Troise et al. 2006) (Fig. 1.1). Concomitant to the vascular sprouting process, angiogenic vessels recruit peri-vascular cells or mural cells, i.e., pericytes and vascular smooth muscle cells, to remodel and stabilize the nascent vasculature (Gerhardt et al. 2003). Peri-vascular coverage also prevents the excessive and undirected sprouting from blood vessels.

In pathological tissues, blood vessels exhibit several abnormalities that distinguish them from healthy vasculatures. For example, in tumor tissues vasculatures usually appear as disorganized and tortuous primitive vascular networks that are highly leaky (Cao 2005, 2009a) (Figs. 1.1 and 1.2) One of the key reasons why tumor vessels possess these pathological features is that VEGF expression levels are high in the tumor microenvironment. Genetic alterations in tumor cells often lead to

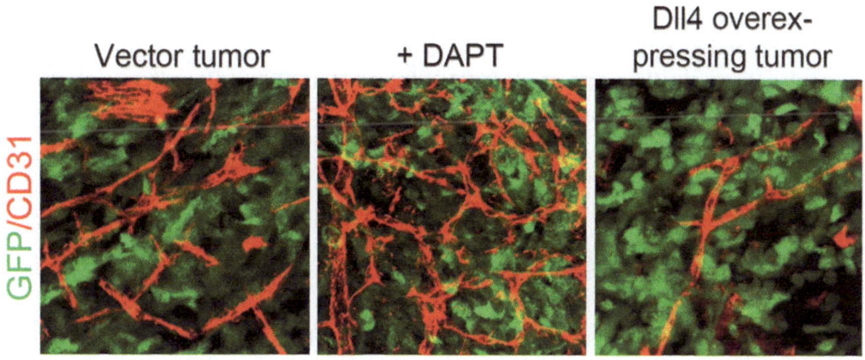

Fig. 1.1 Role of Notch signaling in the formation of a vascular network. Suppression of Notch signaling using DAPT (*middle*) results in the formation of a disorganized vascular network as compared to the vector tumor (*left*). Overexpression of Dll4 triggers Notch signaling and thereby prevents undirected tip formation resulting in a more normalized vascular structure (*right*; *green*: GFP-labelled tumor cells; *red*: CD31-positive endothelial cells)

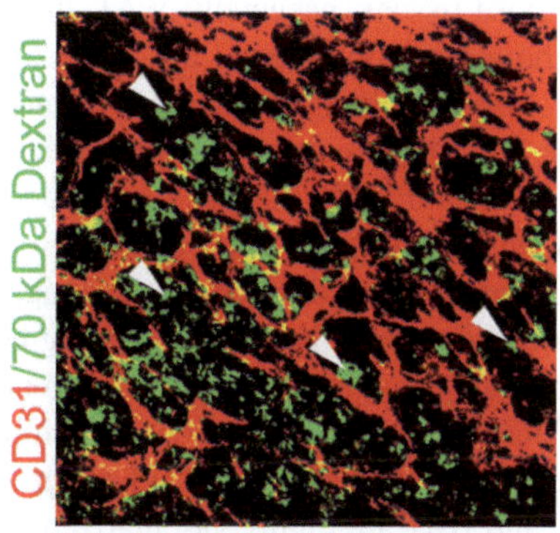

Fig. 1.2 Tumor tissue vasculatures are commonly disorganized and leaky. Tumor vessels were detected using the endothelial marker CD31 (*red color*). Perfusion with 70 kDa lysine-fixable rhodamine dextran (*green color*) results in extravasation of a large percentage of dextran indicating the high permeability and leakiness of the tumor vessels (*white arrowheads*: extravasated dextran)

elevated levels of VEGF expression (Konishi et al. 2000; Sager 1989). Additionally, infiltration of other cell types such as stromal fibroblasts and inflammatory cells also significantly contributes to VEGF production (Dong et al. 2004; Crawford et al. 2009; Inoue et al. 2002). It is known that the tumor cells and the tumor microenvironment are exposed to tissue hypoxia that substantially upregulates VEGF expression via transcriptional activation of the hypoxia-inducible factor 1α (HIF-1α) system (Makino et al. 2001; Maxwell et al. 1999; Carmeliet et al. 1998; Maltepe et al. 1997; Mukhopadhyay et al. 1995; Millauer et al. 1994; Shweiki et al. 1992). VEGF is also a potent vascular permeability factor that stimulates vascular leakiness (Dvorak et al. 1985; Senger et al. 1983). VEGF-induced angiogenesis and vascular permeability are mediated mainly via the VEGFR-2-triggered signaling pathway (Eriksson et al.

2003; Cao 2009b). Based on its broad and pivotal role in tumor angiogenesis, VEGF and its receptor signaling pathways have become attractive targets for anticancer drug development (Hurwitz et al. 2004; Motzer et al. 2007; Escudier et al. 2007). In fact, today's most commonly used antiangiogenic drugs used in cancer patients are based on inhibition of the VEGF signaling system (see below).

1.2 Adipose Angiogenesis

Among all tissues, adipose tissues are extraordinarily hypervascularized, and perhaps brown adipose tissue (BAT) is the most vascularized tissue in the body (Rupnick et al. 2002; Lim et al. 2012; Xue et al. 2010; Cao 2007, 2010a; Brakenhielm and Cao 2008; Brakenhielm et al. 2004). In both white adipose tissue (WAT) and BAT, each adipocyte is engulfed with microvessels that nourish and communicate with adipocytes (Rupnick et al. 2002; Lim et al. 2012; Xue et al. 2010; Cao 2007, 2010a; Brakenhielm and Cao 2008; Brakenhielm et al. 2004). Unlike most other tissues in the body, adipose tissues, especially WAT, undergo constant expansion and shrinkage during the entire adulthood (Cao 2007, 2010a). Adipose plasticity also demands adipose vascular changes that growth and regression relentlessly occur in order to cope with the adipose metabolic demand. Existence of high microvessel density in WAT and BAT suggests that adipocytes and probably other cell types in the adipose microenvironment produce angiogenic factors to maintain vascular homeostasis. In addition to production of a range of classical angiogenic factors, adipocytes produce several adipokines including leptin, adiponectin, resistin, and visfatin that modulate angiogenesis and vascular functions (Cao 2007). Adipose tissues also produce endogenous angiogenesis inhibitors that may involve in maintenance of vascular homeostasis and vascular regression (Cao 2007).

Although both WAT and BAT are highly vascularized tissues, the vasculature in these tissues may display opposing functions. In expanding WAT, the switch of an angiogenic phenotype could potentially facilitate energy deposition, whereas the same angiogenic phenotype may facilitate energy consumption in metabolically active BAT (Cao 2010a). In metabolically active adipose tissues, microvessels also play essential role in removing metabolic products. Bulky recent data demonstrate that vascular cells are the important source of supplying adipose precursor cells that can differentiate into adipocytes (Kahn 2008; Gupta et al. 2012; Tran et al. 2012). Thus, blood vessels in the adipose tissue do not merely provide nutrients and oxygen for adipocytes but also serve as a reservoir of stem cells.

1.3 Angiogenesis in Metabolic Diseases

Angiogenesis has been associated with development of metabolic diseases including type II diabetes and cancer (Folkman 1971; Smith et al. 1999; Gariano and Gardner 2005). Obesity-associated diabetic mellitus is a major worldwide metabolic disease and its prevalence is constantly increasing. A substantial number of diabetic patients eventually develop complications affecting multiple organs and tissues. Proliferative diabetic retinopathy (PDR) and diabetic macular edema (DME) are the two most common and severe retinal complications that are sight threatening in diabetic patients (Smith et al. 1999; Gariano and Gardner 2005). Interestingly, pathological angiogenesis and vascular dysfunctions are the primary cause for development of these diabetic complications (Hellstrom et al. 2007; Takeda et al. 2009; Ferrara and Alitalo 1999). In particular, the ischemia-induced VEGF expression has causally been linked to the development and progression of these retinal complications. As VEGF displays both angiogenic and vascular permeability effects on blood vessels, this factor is considered as a crucial target for therapeutic development of effective drugs. Early clinical assessments of anti-VEGF drugs in PDR patients have demonstrated some clinical benefits (Boras et al. 2011; Kakkassery et al. 2010; Nicholson and Schachat 2010).

In addition to retinal complications, diabetic patients often suffer from kidney complications in association with tissue hypoxia. Vascular dysfunction-induced hypoxia induces HIF-1α expression, which is modulated by prolylhydrylases (PHDs) (Miyata and de Strihou 2010). The vascular changes in diabetic kidney include reduction of the total number of peritubular capillaries by mechanisms of reduction of angiotensin II and nitric oxide, anemia, and high oxygen consumption (Carmeliet and Jain 2011; Manalo et al. 2011; Yu et al. 2012; Schodel et al. 2009; Nangaku 2009). Targeting PHDs have been implied for the treatment of diabetic nephropathy although clinical benefits based on this type of approached need future validation.

Opposing to hypervascularization in retinopathy and nephropathy, impaired angiogenesis often presents in legs of diabetic patients, leading to development of chronic ulcers. In this case, promoting, but not inhibiting, angiogenesis offers an alternative attractive approach for therapy. In fact, topical delivery of proangiogenic factors such as PDGF-BB has currently been used for treatment of chronic diabetic ulcers (Mulder et al. 2009; Steed 2006). Thus, targeting angiogenesis based on opposing principles has been used for treatment of various diabetes-associated clinical complications.

1.4 Antiangiogenic Cancer Therapy

The original concept that tumor growth is dependent on angiogenesis and inhibition of tumor angiogenesis would be a novel approach for cancer therapy was proposed by Dr. Judah Folkman 41 years ago (Folkman 1971). In most preclinical experimental settings, inhibition of tumor angiogenesis has produced remarkable effects on suppression of tumor growth (Cao et al. 2011; Cao and Langer 2010). Based on these exciting findings in preclinical animal models, the antiangiogenic principle has become an important and attractive approach for both pharmaceutical industry and academic scientists for development of cancer drugs.

There are two classes of angiogenesis inhibitors, broad-spectrum endogenous angiogenesis inhibitors and angiogenic factor antagonists (Cao 2009a; Folkman 2006, 2007). Endogenous angiogenesis inhibitors are those biological molecules that naturally exist in our body. For example, angiostatin, endostatin, and thrombospondin-1 are all endogenous angiogenesis inhibitors that display physiological and pathological functions by regulation of angiogenesis and vascular homeostasis (O'Reilly et al. 1994, 1997; Dameron et al. 1994). It has been proposed that endogenous angiogenesis inhibitors are often expressed at high levels to prevent vessel growth. In tumors, this endogenous angiogenesis inhibitor-dominant situation can be altered by overproduction of angiogenic factors, tipping toward an angiogenic phenotype. Endogenous angiogenesis inhibitors and many generic antiangiogenic agents usually block endothelial cell proliferation and migration via undefined and broad signaling pathways (Cao 2001; Folkman 2007). They often inhibit endothelial cell growth independent from the source of angiogenic stimuli. Angiogenic factor antagonists and angiogenic signaling pathway inhibitors usually specifically block one or several signaling systems required for endothelial cell growth or vascular functions. For example, bevacizumab and ramucirumab specifically block VEGF- and VEGFR-2-triggered signaling systems (Hurwitz et al. 2004; Garon et al. 2012; Krupitskaya and Wakelee 2009), respectively, whereas sunitinib, sorafenib, and pazopanib target a broad spectrum of tyrosine kinases including VEGF receptors (Motzer et al. 2007; Escudier et al. 2007; Xu et al. 2011).

The first marketed antiangiogenic drug, bevacizumab, for the treatment of cancer disease was approved by the US FDA in 2004 (Hurwitz et al. 2004). Bevacizumab as a neutralizing antibody specifically blocks VEGF and thus it is a monospecific drug. Initially, bevacizumab was approved for treatment of advanced and metastatic colorectal cancers in combination with the standard chemotherapy (Hurwitz et al. 2004). Unlike in mouse tumor models, administration of bevacizumab alone in cancer patients has not yielded statistical survival improvement. Why would anti-VEGF monotherapy be effective in mouse tumor models, but ineffective in human patients? What are the differences between mouse tumor models and human cancer patients? These critical issues remain currently unresolved (Cao 2011). Antiangiogenic drug-based combination therapy containing chemotherapeutics produced therapeutic benefits compared with chemotherapy alone although the survival benefits in most cancer types remain modest (Kerbel 2008). Antiangiogenic therapy is currently being used as one of the key components in the frontline and standard therapy for treatment of several cancer types including colorectal cancer, lung cancer, renal cell carcinoma, and gastrointestinal cancers (Hurwitz et al. 2004; Friedman et al. 2009; Ignoffo 2004; Miklos 2012; Mountzios and Pentheroudakis 2012; Sculier et al. 2007; Sharieff 2004; Sonpavde 2003; Tol et al. 2008).

Several mechanistic rationales have been proposed to explain the therapeutic benefit based on combination therapy. Vascular normalization is one of the attractive hypotheses that explain the possible mechanism underlying antiangiogenic therapy (Jain 2005). Tumor vasculatures often appear as disorganized primitive vascular networks that are highly leaky. The key reason why these tumor vessels are highly disorganized, tortuous, and leaky is that tumors often produce VEGF at high levels and VEGF induces vascular leakage and disorganization. Anti-VEGF drugs can significantly improve the structure of tumor vessels, leading to a normalized phenotype. The anti-VEGF-induced vascular normalization significantly improves the leaky features and blood perfusion that, in the presence of chemotherapeutics, improve drug delivery (Batchelor et al. 2007). Thus, anti-VEGF drug-induced vascular normalization may significantly modulate chemotherapy-based therapeutic efficacy. However, a recent study in human cancer patients does not support these preclinical findings (Van der Veldt et al. 2012).

Another interesting hypothesis of explicating the mechanism underlying the combination of antiangiogenic drugs with chemotherapy is that antiangiogenic agents might increase tolerance of chemotoxicity. It has been shown, in mouse tumor models, that tumor-derived VEGF significantly impairs the bone marrow function by suppression of hematopoiesis (Xue et al. 2008; Zhang et al. 2011). It is known that most chemotherapeutics also inhibit hematopoiesis in the bone marrow. Chemotherapy treatment of high VEGF-producing tumors resulted in severe suppression of bone marrow hematopoiesis, leading to early death of tumor-bearing mice (Zhang et al. 2011). However, treatment of these tumor-bearing mice with anti-VEGF drugs prior to chemotherapy led to a significant increase of survival by improving bone marrow hematopoiesis (Zhang et al. 2011). Given the fact that a substantial number of cancer patients die of chemotherapy-related toxicity, these findings imply that increasing tolerance of chemotoxicity could potentially provide an attractive mechanism explaining combination therapy.

Cancer is not a local disease, but a systemic disorder that affects functions of multiple tissues and organs. In addition to metastasis, cancer patients, especially at the advanced stage of diseases, often develop various degrees of systemic symptoms, manifesting cancer cachexia and paraneoplastic syndrome. In a mouse tumor model, it has been shown that tumor-derived VEGF enters the circulation and affects functions of multiple tissues and organs (Xue et al. 2008). Interestingly, anti-VEGF drugs at a low dose can significantly improve survivals of these tumor-bearing mice without affecting tumor growth (Xue et al. 2008). These findings demonstrate that off-tumor targets could potentially be beneficial sites of antiangiogenic drugs. These preclinical findings warrant clinical validation.

Clinical experiences with antiangiogenic drugs in cancer patients show that these drugs also produce broad adverse effects by affecting multiple tissue and organ systems (des Guetz et al. 2011; Hapani et al. 2009; Launay-Vacher and Deray 2009; Randall and Monk 2010; Zhu et al. 2007). Commonly seen clinical adverse effects include hypertension, proteinuria, hemorrhages, and gastrointestinal perforation (des Guetz et al. 2011; Hapani et al. 2009; Launay-Vacher and Deray 2009; Randall and Monk 2010; Zhu et al. 2007). In some rare cases, severe cardiovascular complications can occur and cause sudden death (des Guetz et al. 2011). Since these drugs are given to cancer patients via the systemic delivery route and VEGF displays broad physiological functions, the systemic impact of these drugs is not totally unexpected. One of the key and urgently unresolved issues related to clinical practice of antiangiogenic drugs is to define reliable biomarkers that distinguish nonresponsive patient population from the likely responsive population (Cao et al. 2011;

Cao and Langer 2010). Unfortunately, such a reliable biomarker is currently lacking during clinical practice. Noticeably, cancer patients also develop resistance to antiangiogenic drugs (Cao et al. 2009). Although the mechanism underlying the resistance is not completely understood, it is believed that multiple mechanisms are involved in development of antiangiogenic drug resistance.

While antiangiogenic drugs during clinical practice have encountered many unpredictable obstacles, survival improvements by currently available antiangiogenic drugs in combination with chemotherapy seen in various cancers are encouraging for development of future more effective drugs. Many unresolved challenging issues including understanding the fundamental mechanism of their actions, defining reliable biomarkers, overcoming drug resistance, minimizing adverse effects, optimizing combinations with chemotherapeutics, and long-lasting therapy have also offered tremendous opportunities for future intensive research in this important area.

1.5 Antiangiogenic Therapy in Ophthalmological Diseases

Several angiogenesis-related ophthalmological diseases including the wet type age-related macular degeneration (AMD) and diabetic retinopathy are the most common reasons of causing blindness in adult humans (Gariano and Gardner 2005; Boras et al. 2011; Kakkassery et al. 2010; Nicholson and Schachat 2010; Miyata and de Strihou 2010). The onset and progression of these retinal disorders are tightly associated with vascular abnormalities, manifesting excessive neovascularization and vascular leakiness as two major pathological features. VEGF is the key angiogenic factor that induces angiogenesis and vascular permeability under these pathological conditions. Thus, inhibition of VEGF and its signaling pathways offers an attractive and novel therapeutic approach for treatment of these ocular diseases.

In 2004, the US FDA approved the first anti-VEGF drug, pegaptanib (macugen, an anti-VEGF aptamer), for treatment of wet AMD in human patients (Gragoudas et al. 2004). Subsequently, the FDA authority approved ranibizumab (Stone 2006; Brown et al. 2006; Rosenfeld et al. 2006; Steinbrook 2006) (lucentis, a Fab fragment of anti-VEGF monoclonal antibody) for treatment of wet type of AMD. Unlike anti-VEGF drugs in oncology applications, anti-VEGF drugs have been used as monotherapy for treatment of AMD and have shown very robust effects in improvement of vision. It has been noticed that anti-vascular edema is one of the major mechanisms underlying clinical benefits. Bevacizumab, as a parental molecule of ranibizumab, has also been used as an off-label anti-VEGF drug for treatment of AMD (Martin and Maguire 2011). Today, anti-VEGF drugs are the most commonly used effective drugs for treatment of the wet type of AMD. Additionally, anti-VEGF drugs as monotherapy have also entered clinical trials for treatment of diabetic retinopathy and early clinical phase evaluation shows beneficial effects for this indication (Zechmeister-Koss and Huic 2012). Thus, anti-VEGF drugs have revolutionized clinical management of these otherwise untreatable eye diseases.

1.6 Cardiovascular Diseases

In paradox to antiangiogenic therapy for treatment of cancer and ophthalmological disorders, promoting angiogenesis has also been implemented for treatment of diseases (Powell et al. 2008; Khurana et al. 2005; Simons 2005). It was originally hypothesized that stimulation of angiogenesis would be clinically beneficial for treatment of ischemia-related diseases including ischemic myocardium and chronic leg ischemia. This concept remains undisputed and has been proved in various preclinical models by delivery of proangiogenic factors to the ischemic tissues. However, clinical translation of this concept and preclinical findings remain largely disappointing and unfulfilled promises (Powell et al. 2008; Khurana et al. 2005; Simons 2005). Several rigorous clinical trials using various proangiogenic factors including VEGF and FGF have not demonstrated clinical benefits for treatment of these ischemic disorders (Powell et al. 2008; Khurana et al. 2005; Simons 2005). It is not known why human patients responded differently from animal models. One of the possibilities to explain clinical failure is that these trials have been performed as monotherapy by delivering a single proangiogenic factor to the pathological site. For improvement of vascular functions in ischemic muscle tissues, it is desirable to establish collateral vascular networks rather than primitive vessels without appropriate functions. The arterial vessel wall consists of two cell types: endothelial cells and vascular smooth muscle cells. These cellular composition features suggest that a combination of arteriogenic and angiogenic factors should be considered for therapeutic development (Cao 2010b). In supporting this view, a combination of arteriogenic factor such as PDGF-BB and angiogenic factor FGF or VEGF has produced improved therapeutic benefits relative to their monotherapy (Zhang et al. 2009; Cao et al. 2003, 2008; Nissen et al. 2007; Lu et al. 2007; Cao and Liu 2007; Richardson et al. 2001). To date, combination therapy-based approaches have not been assessed in human trials.

1.7 Conclusions and Perspectives

Angiogenesis is essentially involved in development of most common and lethal human diseases. Understanding the fundamental mechanism of the angiogenic process and functions of key regulators under pathological conditions provides tremendous opportunities for therapeutic intervention. Clinical successes in oncology and ophthalmology areas have laid the cornerstone for future drug development for treatment of various angiogenesis-dependent diseases. Although the intensive basic research on angiogenesis has provided invaluable clues for clinical implications, clinical practice with antiangiogenic drugs has raised many unexpected issues that do not seem to match or even be against principles seen in various animal models. This is particularly true in the area of antiangiogenic cancer therapy. The only way to resolve these clinically related issues is to understand the fundamental mechanisms of vascular functions in modulating tumor growth and invasion in human patients. This task demands intimate collaborations between basic scientists and clinical oncologists. Improvement of therapeutic efficacy of current existing antiangiogenic drugs and future development of more effective drugs remain challenging. It is anticipated that antiangiogenic therapy will be further expanded in treatment of many more human diseases and will become one of the most important therapeutic approaches for improvement of quality of life and survival.

Acknowledgements The author thanks Sharon Lim, Patrik Andersson for the artwork and Dr. Jeannette Söderberg for critical reading. The author's laboratory is supported through research grants from the Swedish Research Council, the Swedish Cancer Foundation, the Karolinska Institute Foundation, the Karolinska Institute distinguished professor award, the Tianjin Natural Science Foundation (CMM-Tianjin, No. 09ZCZDSF04400) for international collaboration between Tianjin Medical University and Karolinska Institutet, the Torsten Söderbergs foundation, the European Union Integrated Project of Metoxia (Project no. 222741), and the European Research Council (ERC) advanced grant ANGIOFAT (Project no 250021).

References

Batchelor TT, Sorensen AG, di Tomaso E, et al. Azd2171, a pan-vegf receptor tyrosine kinase inhibitor, normalizes tumor vasculature and alleviates edema in glioblastoma patients. Cancer Cell. 2007;11:83–95.

Benedito R, Rocha SF, Woeste M, et al. Notch-dependent vegfr3 upregulation allows angiogenesis without vegf-vegfr2 signalling. Nature. 2012;484:110–4.

Boras I, Lazic R, Gabric N, et al. Anti-vegf in treatment of diabetic macular edema. Coll Antropol. 2011;35 Suppl 2:15–8.

Brakenhielm E, Cao Y. Angiogenesis in adipose tissue. Methods Mol Biol. 2008;456:65–81.

Brakenhielm E, Cao R, Gao B, et al. Angiogenesis inhibitor, tnp-470, prevents diet-induced and genetic obesity in mice. Circ Res. 2004;94:1579–88.

Brown DM, Kaiser PK, Michels M, et al. Ranibizumab versus verteporfin for neovascular age-related macular degeneration. N Engl J Med. 2006;355:1432–44.

Cao Y. Endogenous angiogenesis inhibitors and their therapeutic implications. Int J Biochem Cell Biol. 2001;33:357–69.

Cao Y. Antiangiogenic cancer therapy. Semin Cancer Biol. 2004;14:139–45.

Cao Y. Tumor angiogenesis and therapy. Biomed Pharmacother. 2005;59 Suppl 2:S340–3.

Cao Y. Angiogenesis modulates adipogenesis and obesity. J Clin Invest. 2007;117:2362–8.

Cao Y. Tumor angiogenesis and molecular targets for therapy. Front Biosci. 2009a;14:3962–73.

Cao Y. Positive and negative modulation of angiogenesis by vegfr1 ligands. Sci Signal. 2009b;2:re1.

Cao Y. Adipose tissue angiogenesis as a therapeutic target for obesity and metabolic diseases. Nat Rev Drug Discov. 2010a;9:107–15.

Cao Y. Therapeutic angiogenesis for ischemic disorders: what is missing for clinical benefits? Discov Med. 2010b;9:179–84.

Cao Y. Antiangiogenic cancer therapy: why do mouse and human patients respond in a different way to the same drug? Int J Dev Biol. 2011;55:557–62.

Cao Y, Langer R. Optimizing the delivery of cancer drugs that block angiogenesis. Sci Transl Med. 2010;2:15ps13.

Cao Y, Liu Q. Therapeutic targets of multiple angiogenic factors for the treatment of cancer and metastasis. Adv Cancer Res. 2007;97:203–24.

Cao R, Brakenhielm E, Pawliuk R, et al. Angiogenic synergism, vascular stability and improvement of hind-limb ischemia by a combination of pdgf-bb and fgf-2. Nat Med. 2003;9:604–13.

Cao Y, Cao R, Hedlund EM. R regulation of tumor angiogenesis and metastasis by fgf and pdgf signaling pathways. J Mol Med (Berl). 2008;86:785–9.
Cao Y, Zhong W, Sun Y. Improvement of antiangiogenic cancer therapy by understanding the mechanisms of angiogenic factor interplay and drug resistance. Semin Cancer Biol. 2009; 19:338–43.
Cao Y, Arbiser J, D'Amato RJ, et al. Forty-year journey of angiogenesis translational research. Sci Transl Med. 2011;3:114rv113.
Carmeliet P, Jain RK. Angiogenesis in cancer and other diseases. Nature. 2000;407:249–57.
Carmeliet P, Jain RK. Molecular mechanisms and clinical applications of angiogenesis. Nature. 2011;473:298–307.
Carmeliet P, Dor Y, Herbert JM, et al. Role of hif-1alpha in hypoxia-mediated apoptosis, cell proliferation and tumour angiogenesis. Nature. 1998;394:485–90.
Crawford Y, Kasman I, Yu L, et al. Pdgf-c mediates the angiogenic and tumorigenic properties of fibroblasts associated with tumors refractory to anti-vegf treatment. Cancer Cell. 2009; 15:21–34.
Dameron KM, Volpert OV, Tainsky MA, et al. Control of angiogenesis in fibroblasts by p53 regulation of thrombospondin-1. Science. 1994;265:1582–4.
des Guetz G, Uzzan B, Chouahnia K, et al. Cardiovascular toxicity of anti-angiogenic drugs. Target Oncol. 2011;6:197–202.
Dong J, Grunstein J, Tejada M, et al. Vegf-null cells require pdgfr alpha signaling-mediated stromal fibroblast recruitment for tumorigenesis. EMBO J. 2004;23:2800–10.
Dvorak HF, Senger DR, Dvorak AM, et al. Regulation of extravascular coagulation by microvascular permeability. Science. 1985;227:1059–61.
Eriksson A, Cao R, Roy J, et al. Small gtp-binding protein rac is an essential mediator of vascular endothelial growth factor-induced endothelial fenestrations and vascular permeability. Circulation. 2003;107:1532–8.
Escudier B, Eisen T, Stadler WM, et al. Sorafenib in advanced clear-cell renal-cell carcinoma. N Engl J Med. 2007;356:125–34.
Ferrara N, Alitalo K. Clinical applications of angiogenic growth factors and their inhibitors. Nat Med. 1999;5:1359–64.
Folkman J. Tumor angiogenesis: therapeutic implications. N Engl J Med. 1971;285:1182–6.
Folkman J. Tumor angiogenesis and tissue factor. Nat Med. 1996;2:167–8.
Folkman J. Angiogenesis. Annu Rev Med. 2006;57:1–18.
Folkman J. Angiogenesis: an organizing principle for drug discovery? Nat Rev Drug Discov. 2007;6:273–86.
Friedman HS, Prados MD, Wen PY, et al. Bevacizumab alone and in combination with irinotecan in recurrent glioblastoma. J Clin Oncol. 2009;27:4733–40.
Gariano RF, Gardner TW. Retinal angiogenesis in development and disease. Nature. 2005; 438:960–6.
Garon EB, Cao D, Alexandris E, et al. A randomized, double-blind, phase iii study of docetaxel and ramucirumab versus docetaxel and placebo in the treatment of stage iv non-small-cell lung cancer after disease progression after 1 previous platinum-based therapy (revel): treatment rationale and study design. Clin Lung Cancer. 2012;13:505–9.
Gerhardt H, Golding M, Fruttiger M, et al. Vegf guides angiogenic sprouting utilizing endothelial tip cell filopodia. J Cell Biol. 2003;161:1163–77.
Gragoudas ES, Adamis AP, Cunningham Jr ET, et al. Pegaptanib for neovascular age-related macular degeneration. N Engl J Med. 2004;351:2805–16.
Gupta RK, Mepani RJ, Kleiner S, et al. Zfp423 expression identifies committed preadipocytes and localizes to adipose endothelial and perivascular cells. Cell Metab. 2012;15:230–9.
Hapani S, Chu D, Wu S. Risk of gastrointestinal perforation in patients with cancer treated with bevacizumab: a meta-analysis. Lancet Oncol. 2009;10:559–68.
Hellstrom M, Phng LK, Hofmann JJ, et al. Dll4 signalling through notch1 regulates formation of tip cells during angiogenesis. Nature. 2007;445:776–80.
Hurwitz H, Fehrenbacher L, Novotny W, et al. Bevacizumab plus irinotecan, fluorouracil, and leucovorin for metastatic colorectal cancer. N Engl J Med. 2004;350:2335–42.

Ignoffo RJ. Overview of bevacizumab: a new cancer therapeutic strategy targeting vascular endothelial growth factor. Am J Health Syst Pharm. 2004;61:S21–6.

Inoue M, Hager JH, Ferrara N, et al. Vegf-a has a critical, nonredundant role in angiogenic switching and pancreatic beta cell carcinogenesis. Cancer Cell. 2002;1:193–202.

Jain RK. Normalization of tumor vasculature: an emerging concept in antiangiogenic therapy. Science. 2005;307:58–62.

Kahn CR. Medicine. Can we nip obesity in its vascular bud? Science. 2008;322:542–3.

Kakkassery V, Winterhalter S, Joussen AM. [Anti-vegf inhibitors and their role in the treatment of diabetic macular oedema]. Klin Monbl Augenheilkd. 2010;227:701–11.

Kerbel RS. Tumor angiogenesis. N Engl J Med. 2008;358:2039–49.

Khurana R, Simons M, Martin JF, et al. Role of angiogenesis in cardiovascular disease: a critical appraisal. Circulation. 2005;112:1813–24.

Konishi T, Huang CL, Adachi M, et al. The k-ras gene regulates vascular endothelial growth factor gene expression in non-small cell lung cancers. Int J Oncol. 2000;16:501–11.

Krupitskaya Y, Wakelee HA. Ramucirumab, a fully human mab to the transmembrane signaling tyrosine kinase vegfr-2 for the potential treatment of cancer. Curr Opin Investig Drugs. 2009;10:597–605.

Launay-Vacher V, Deray G. Hypertension and proteinuria: a class-effect of antiangiogenic therapies. Anticancer Drugs. 2009;20:81–2.

Lim S, Honek J, Xue Y, et al. Cold-induced activation of brown adipose tissue and adipose angiogenesis in mice. Nat Protoc. 2012;7:606–15.

Lu H, Xu X, Zhang M, et al. Combinatorial protein therapy of angiogenic and arteriogenic factors remarkably improves collaterogenesis and cardiac function in pigs. Proc Natl Acad Sci U S A. 2007;104:12140–5.

Makino Y, Cao R, Svensson K, et al. Inhibitory pas domain protein is a negative regulator of hypoxia-inducible gene expression. Nature. 2001;414:550–4.

Maltepe E, Schmidt JV, Baunoch D, et al. Abnormal angiogenesis and responses to glucose and oxygen deprivation in mice lacking the protein arnt. Nature. 1997;386:403–7.

Manalo DJ, Baek JH, Buehler PW, et al. Inactivation of prolyl hydroxylase domain (phd) protein by epigallocatechin (egcg) stabilizes hypoxia-inducible factor (hif-1alpha) and induces hepcidin (hamp) in rat kidney. Biochem Biophys Res Commun. 2011;416:421–6.

Martin DF, Maguire MG. Ying GS, et al Ranibizumab and bevacizumab for neovascular age-related macular degeneration. N Engl J Med. 2011;364:1897–908.

Maxwell PH, Wiesener MS, Chang GW, et al. The tumour suppressor protein vhl targets hypoxia-inducible factors for oxygen-dependent proteolysis. Nature. 1999;399:271–5.

Miklos GL. Bevacizumab in neoadjuvant treatment for breast cancer. N Engl J Med. 2012;366:1638; author reply 1639–40.

Millauer B, Shawver LK, Plate KH, et al. Glioblastoma growth inhibited in vivo by a dominant-negative flk-1 mutant. Nature. 1994;367:576–9.

Miyata T, de Strihou CY. Diabetic nephropathy: a disorder of oxygen metabolism? Nat Rev Nephrol. 2010;6:83–95.

Motzer RJ, Hutson TE, Tomczak P, et al. Sunitinib versus interferon alfa in metastatic renal-cell carcinoma. N Engl J Med. 2007;356:115–24.

Mountzios G, Pentheroudakis G. Bevacizumab in ovarian cancer. N Engl J Med. 2012;366:1257; author reply 1258.

Mukhopadhyay D, Tsiokas L, Zhou XM, et al. Hypoxic induction of human vascular endothelial growth factor expression through c-src activation. Nature. 1995;375:577–81.

Mulder G, Tallis AJ, Marshall VT, et al. Treatment of nonhealing diabetic foot ulcers with a platelet-derived growth factor gene-activated matrix (gam501): results of a phase 1/2 trial. Wound Repair Regen. 2009;17:772–9.

Nangaku M. Novel therapeutic approach targeting the hif-hre system in the kidney. Adv Exp Med Biol. 2009;645:81–6.

Nicholson BP, Schachat AP. A review of clinical trials of anti-vegf agents for diabetic retinopathy. Graefes Arch Clin Exp Ophthalmol. 2010;248:915–30.

Nissen LJ, Cao R, Hedlund EM, et al. Angiogenic factors fgf2 and pdgf-bb synergistically promote murine tumor neovascularization and metastasis. J Clin Invest. 2007;117:2766–77.
Noguera-Troise I, Daly C, Papadopoulos NJ, et al. Blockade of dll4 inhibits tumour growth by promoting non-productive angiogenesis. Nature. 2006;444:1032–7.
Nyberg P, Xie L, Kalluri R. Endogenous inhibitors of angiogenesis. Cancer Res. 2005; 65:3967–79.
O'Reilly MS, Holmgren L, Shing Y, et al. Angiostatin: a novel angiogenesis inhibitor that mediates the suppression of metastases by a lewis lung carcinoma. Cell. 1994;79:315–28.
O'Reilly MS, Boehm T, Shing Y, et al. Endostatin: an endogenous inhibitor of angiogenesis and tumor growth. Cell. 1997;88:277–85.
Powell RJ, Simons M, Mendelsohn FO, et al. Results of a double-blind, placebo-controlled study to assess the safety of intramuscular injection of hepatocyte growth factor plasmid to improve limb perfusion in patients with critical limb ischemia. Circulation. 2008;118:58–65.
Randall LM, Monk BJ. Bevacizumab toxicities and their management in ovarian cancer. Gynecol Oncol. 2010;117:497–504.
Richardson TP, Peters MC, Ennett AB, et al. Pol ymeric system for dual growth factor delivery. Nat Biotechnol. 2001;19:1029–34.
Ridgway J, Zhang G, Wu Y, et al. Inhibition of dll4 signalling inhibits tumour growth by deregulating angiogenesis. Nature. 2006;444:1083–7.
Risau W, Flamme I. Vasculogenesis. Annu Rev Cell Dev Biol. 1995;11:73–91.
Rosenfeld PJ, Brown DM, Heier JS, et al. Ranibizumab for neovascular age-related macular degeneration. N Engl J Med. 2006;355:1419–31.
Rupnick MA, Panigrahy D, Zhang CY, et al. Adipose tissue mass can be regulated through the vasculature. Proc Natl Acad Sci U S A. 2002;99:10730–5.
Sager R. Tumor suppressor genes: the puzzle and the promise. Science. 1989;246:1406–12.
Sawamiphak S, Seidel S, Essmann CL, et al. Ephrin-b2 regulates vegfr2 function in developmental and tumour angiogenesis. Nature. 2010;465:487–91.
Schodel J, Klanke B, Weidemann A, et al. Hif-prolyl hydroxylases in the rat kidney: physiologic expression patterns and regulation in acute kidney injury. Am J Pathol. 2009;174:1663–74.
Sculier JP, Meert AP, Paesmans M. Bevacizumab for non-small-cell lung cancer. N Engl J Med. 2007;356:1373–4; author reply 1374–5.
Senger DR, Galli SJ, Dvorak AM, et al. Tumor cells secrete a vascular permeability factor that promotes accumulation of ascites fluid. Science. 1983;219:983–5.
Sharieff W. Bevacizumab in colorectal cancer. N Engl J Med. 2004;351:1690–1; author reply 1690–1.
Shweiki D, Itin A, Soffer D, et al. Vascular endothelial growth factor induced by hypoxia may mediate hypoxia-initiated angiogenesis. Nature. 1992;359:843–5.
Siekmann AF, Lawson ND. Notch signalling limits angiogenic cell behaviour in developing zebrafish arteries. Nature. 2007;445:781–4.
Simons M. Angiogenesis: where do we stand now? Circulation. 2005;111:1556–66.
Smith LE, Shen W, Perruzzi C, et al. Regulation of vascular endothelial growth factor-dependent retinal neovascularization by insulin-like growth factor-1 receptor. Nat Med. 1999;5:1390–5.
Sonpavde G. Bevacizumab in renal-cell cancer. N Engl J Med. 2003;349:1674.
Steed DL. Clinical evaluation of recombinant human platelet-derived growth factor for the treatment of lower extremity ulcers. Plast Reconstr Surg. 2006;117:43S–9; discussion 150–1S.
Steinbrook R. The price of sight–ranibizumab, bevacizumab, and the treatment of macular degeneration. N Engl J Med. 2006;355:1409–12.
Stone EM. A very effective treatment for neovascular macular degeneration. N Engl J Med. 2006;355:1493–5.
Takeda A, Baffi JZ, Kleinman ME, et al. Ccr3 is a target for age-related macular degeneration diagnosis and therapy. Nature. 2009;460:225–30.
Tammela T, Zarkada G, Wallgard E, et al. Blocking vegfr-3 suppresses angiogenic sprouting and vascular network formation. Nature. 2008;454:656–60.

Tol J, Cats A, Mol L, et al. Gastrointestinal ulceration as a possible side effect of bevacizumab which may herald perforation. Invest New Drugs. 2008;26:393–7.

Tran KV, Gealekman O, Frontini A, et al. The vascular endothelium of the adipose tissue gives rise to both white and brown fat cells. Cell Metab. 2012;15:222–9.

Van der Veldt AA, Lubberink M, Bahce I, et al. Rapid decrease in delivery of chemotherapy to tumors after anti-vegf therapy: implications for scheduling of anti-angiogenic drugs. Cancer Cell. 2012;21:82–91.

Xu CF, Bing NX, Ball HA, et al. Pazopanib efficacy in renal cell carcinoma: evidence for predictive genetic markers in angiogenesis-related and exposure-related genes. J Clin Oncol. 2011;29:2557–64.

Xue Y, Religa P, Cao R, et al. Anti-vegf agents confer survival advantages to tumor-bearing mice by improving cancer-associated systemic syndrome. Proc Natl Acad Sci U S A. 2008; 105:18513–8.

Xue Y, Lim S, Brakenhielm E, et al. Adipose angiogenesis: quantitative methods to study microvessel growth, regression and remodeling in vivo. Nat Protoc. 2010;5:912–20.

Yancopoulos GD, Klagsbrun M, Folkman J. Vasculogenesis, angiogenesis, and growth factors: ephrins enter the fray at the border. Cell. 1998;93:661–4.

Yu X, Fang Y, Ding X, et al. Transient hypoxia-inducible factor activation in rat renal ablation and reduced fibrosis with l-mimosine. Nephrology (Carlton). 2012;17:58–67.

Zechmeister-Koss I, Huic M. Vascular endothelial growth factor inhibitors (anti-vegf) in the management of diabetic macular oedema: a systematic review. Br J Ophthalmol. 2012;96:167–78.

Zhang J, Cao R, Zhang Y, et al. Differential roles of pdgfr-alpha and pdgfr-beta in angiogenesis and vessel stability. FASEB J. 2009;23:153–63.

Zhang D, Hedlund EM, Lim S, et al. Antiangiogenic agents significantly improve survival in tumor-bearing mice by increasing tolerance to chemotherapy-induced toxicity. Proc Natl Acad Sci U S A. 2011;108:4117–22.

Zhu X, Wu S, Dahut WL, et al. Risks of proteinuria and hypertension with bevacizumab, an antibody against vascular endothelial growth factor: systematic review and meta-analysis. Am J Kidney Dis. 2007;49:186–93.

Part II
Spatiotemporal Relation of Adipocytes and Vascular Cells During Development

Chapter 2
Adipose Stem Cells

Carolyn Algire, Dasa Medrikova, and Stephan Herzig

Abstract Once considered an inert mass of stored energy, the past 2 decades have seen a surge in interest in the complexity of adipose tissue and its role in disease. In addition to serving as a site for energy storage, adipocytes secrete proteins involved in inflammation, appetite regulation, blood pressure control, and energy balance. Adipocytes are unique in their ability to store large quantities of lipids that can be rapidly released and used for energy by other organs, when necessary; however, excessive adipose tissue, particularly in the visceral adipose tissue depot, is associated with increased risk of insulin resistance, cardiovascular disease, and cancer. As such, adipose tissue is capable of extensive expansion or retraction depending on the energy balance or disease state of the host, a plasticity that is unparalleled in other organs. Expansion of adipose tissue is driven by both hypertrophy and hyperplasia of adipocytes, which can renew frequently to compensate for cell death, suggesting the necessity of adipocyte progenitor cells within the adipose tissue depot, that are capable of differentiating into mature and functional adipocytes.

Epidemiological studies estimate that more than one billion people worldwide are overweight and at least 400 million clinically obese. Therefore, in order to combat the global obesity pandemic, the origin and the molecular mechanisms controlling the development and expansion of adipocytes must be fully understood so that novel approaches to prevention and therapy can be developed.

Keywords Adipose tissue • Stem cells • White adipose tissue • Brown adipose tissue • Obesity • Cell differentiation • Signaling • Transcription factors

---

The authors Carolyn Algire and Dasa Medrikova have equally contributed.

C. Algire • D. Medrikova • S. Herzig (✉)
Joint Research Division Molecular Metabolic Control, German Cancer Research Center (DKFZ), Center for Molecular Biology (ZMBH) University of Heidelberg, Network Aging Research, University Hospital Heidelberg, Im Neuenheimer Feld 280, 69120 Heidelberg, Germany
e-mail: s.herzig@dkfz.de

Y. Cao (ed.), *Angiogenesis in Adipose Tissue*, DOI 10.1007/978-1-4614-8069-3_2,

2.1 Developmental Origin of Adipose Tissue

Formation of adipose tissue commences shortly after birth in most rodents and during mid-gestation in higher mammals, including humans.

The generation of mature adipocytes comprises two phases: determination and differentiation. In this process, multipotent stem cells become adipoblasts that can further differentiate into pre-adipocytes, cells already committed to become fat cells. Under appropriate stimulation, in the ultimate phase of differentiation, pre-adipocytes convert to mature, lipid-laden adipocytes.

The first phase, determination, involves the commitment of multipotent stem cells to the adipocyte lineage, at which point the cell is referred to as a pre-adipocyte. During the second phase, terminal differentiation, the pre-adipocyte undergoes multiple rounds of mitosis before exiting the cell cycle and differentiating into a mature adipocyte. In the progression from pre-adipocyte to adipocyte, the cell takes on the characteristics and metabolic capabilities of a mature adipocyte while under the control of a series of tightly regulated transcriptional and morphological changes. White adipose tissue (WAT) is readily available for study from human patient samples and experimental animal models; however, it is difficult to maintain in vitro and cannot be expanded. Terminal differentiation is more extensively characterized in immortalized cell lines, such as the mouse line 3T3-L1 (Green and Kehinde 1975; Russell and Ho 1976), which can undergo one or two rounds of cell division prior to differentiation and pre-adipocyte cell lines that differentiate without post-confluence mitosis (e.g., C3H10T1/2). Pre-adipocytes maintain the ability to divide and have been reported to have a turnover rate of up to 4.5 and 5 % per day for humans and mice, respectively. Mature adipocytes are widely believed to have lost the ability to divide following the completion of terminal differentiation, and adipocyte turnover has been reported to be up to 10 % per year (Spalding et al. 2008).

2.1.1 White Adipogenesis

2.1.1.1 Adipose Tissue-Resident Progenitors

Adipose tissue is characterized by striking plasticity, namely the capability of expansion and retraction, depending on changes in energy supply and demand, respectively. In adulthood, two means of adipose tissue expansion exist, hypertrophy and hyperplasia, which are complementary to each other. Hypertrophy is characterized by the enlargement of adipocytes, while hyperplasia leads to increased number of adipocytes (Sun et al. 2011). As mature adipocytes have lost the ability to divide, hyperplasia of adipose tissue must be achieved by differentiation of precursor cells. Adipocyte progenitors derive from pluripotent mesenchymal stem

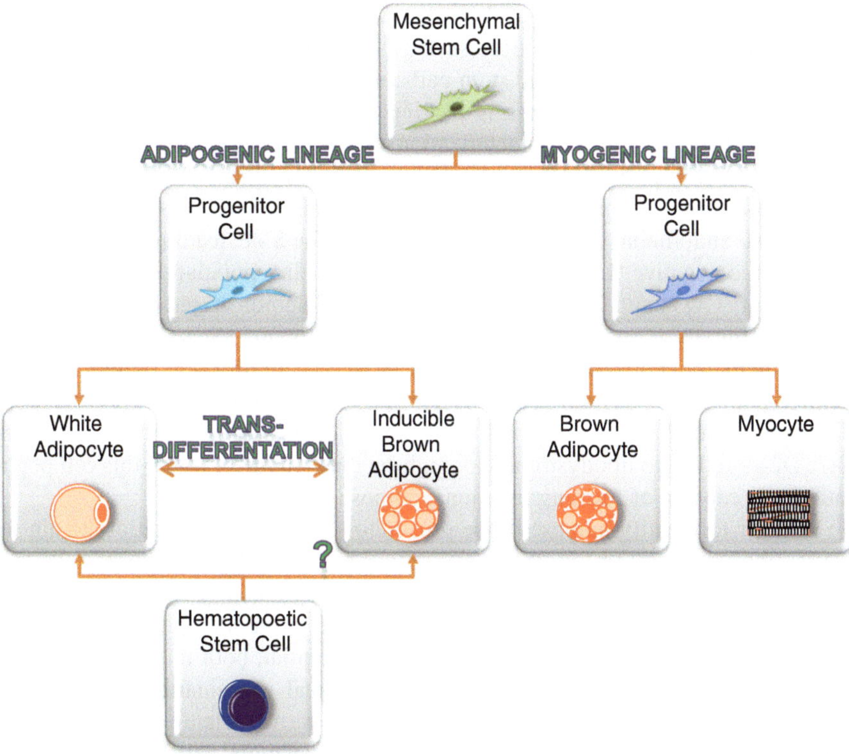

Fig. 2.1 Origin of adipocytes. Adipocytes (white, inducible brown, and brown) are mainly derived from mesenchymal stem cells (MSC) of mesodermal origin. MSC are present in almost every organ in the body and according to the localization give rise to different cell types, including cells of adipogenic and myogenic lineage. White adipocytes are derived from progenitors of adipogenic lineage or hematopoietic progenitors. Inducible brown adipocytes differentiate directly from adipocyte progenitors or may be produced through transdifferentiation from white adipocytes. Whether or not these cells can be derived from hematopoietic progenitors is not known. Finally, brown adipocytes originate from progenitors of myogenic lineage

cells (MSC) (Fig. 2.1), which are capable of differentiating into cells of both mesodermal (myocytes, adipocytes, chondrocytes, and osteoblasts) (Pittenger et al. 1999; Prockop 1997) and non-mesodermal (Ashjian et al. 2003; Brzoska et al. 2005; Ning et al. 2006) origin.

Adipocyte progenitors arise from the mesoderm and are localized to areas along blood capillaries (Poissonnet et al. 1984), indicating that adipose tissue develops in coordination with vasculature. MSC can be isolated from the stromal vascular fraction (SVF) of adipose tissue; however, the proliferative and adipogenic capacity of these cells is significantly affected by the location of the depot from which they were isolated and diverse conditions such as aging and obesity.

In contrast to hematopoietic stem cells, a clear definition of cells considered to be adipose tissue stem cells is lacking; however, Rodeheffer et al. (2008) largely contributed to the characterization of adipocyte precursors by applying fluorescence-activated cell sorting (FACS) on freshly isolated SVF to separate distinct cell subpopulations based on well-defined stem cell markers. Thus, hematopoietic and endothelial cells were sorted out from the SVF, leading to the creation of a lineage-negative (Lin−) population. From the Lin− population, cells positive for antigens CD34 and CD29 were selected as they demonstrated high adipogenic potential. Finally, based on staining for stem cell antigen-1 (Sca-1) and CD24, two surface markers known to be expressed on stem cells in different tissues, the progenitor population was defined as Lin−/CD29+/CD34+/Sca-1+/CD24+, and its adipogenic potential was confirmed in vivo by injecting Lin−/CD29+/CD34+/Sca-1+/CD24+ cells into lipodystrophic or high-fat diet-fed mice, where this cell population was capable of forming functional adipose tissue.

In order to identify adipose tissue progenitors and their localization within fat tissue, Tang et al. (2008) used lineage tracing based on the labeling of cells that express peroxisome proliferator-activated receptor gamma (PPARγ), the master regulator of adipogenesis. Labeled cells were found in the SVF and exhibited considerable proliferative capacity and adipogenic potential, both in culture and after transplantation to nude mice. Immunohistochemical examination revealed that PPARγ expressing cells were localized to the adipose vasculature and express the above mentioned Sca-1 and CD34, in addition to known mural cell markers (smooth-muscle actin (SMA), platelet-derived growth factor receptor beta (PDGFRβ) and chondroitin sulfate proteoglycan 4 known as NG2), indicating that adipocyte progenitors reside in the mural cell compartment of adipose depot. A close relationship between newly arising adipocytes and vasculature was also shown (Tran et al. 2012) using reporter genes expressed under the VE-cadherin promoter. VE-cadherin is indispensable for vasculature formation; however, it was also found to be expressed in pre-adipocytes. Together with morphological, immunohistochemical, and gene expression analyses, these results support the notion that pericytes (endothelial cells that surround capillaries) could serve as a source of adipocyte progenitors.

Studies of progenitor cells in humans are primarily based on FACS analysis and immunohistochemistry. Similar to mice, human adipose tissue-derived stem cells were shown to express stem cell markers Sca-1 and CD34 and were localized in close proximity to the vascular network (Lin et al. 2008). Furthermore, pericyte marker CD146 has been detected on adipose-derived progenitors and the adipogenic potential of those cells was confirmed after cell sorting (Zimmerlin et al. 2010). Under angiogenic conditions, human adipose tissue explants have angiogenic capabilities and can form new vessels, ex vivo, lined with cells containing lipid droplets (Tran et al. 2012). Overall, recent reports strongly support endothelial origin of human adipocytes, and propose a close relationship between adipogenic and angiogenic processes.

2.1.1.2 Adipose Tissue Nonresident Progenitors

Previously, the proposed source of progenitors for formation of new adipocytes has been restricted to adipose tissue-resident cells; however, a reason for reevaluation of this notion has emerged. It was shown that progenitors from other tissues could also contribute to the generation of new adipocytes (Crossno et al. 2006; Majka et al. 2010; Sera et al. 2009). Bone marrow containing a marked number of hematopoietic and mesenchymal progenitors seems to be a promising candidate (Fig. 2.1). After transplantation of green fluorescent protein (GFP)-labeled bone marrow-derived hematopoietic cell subpopulation into irradiated wild-type mice, GFP-expressing adipocytes appeared in the adipose tissue of recipient animals (Crossno et al. 2006). Selection of cells positive for a murine myeloid marker, CD11b, confirmed myeloid origin of GFP-labeled adipocytes. Furthermore, adipocytes originating from the myeloid lineage were characterized by increased expression of inflammatory and chemotactic genes. These cells accumulated preferentially in visceral adipose tissue, which is consistent with the increased inflammatory state of this depot, compared to subcutaneous adipose tissue (Majka et al. 2010).

2.1.2 Brown Adipogenesis

There exist two distinct kinds of adipose tissue: WAT, specifically designed for energy storage, and brown adipose tissue (BAT), which derives from a muscle-like precursors and, when activated, burns energy under conditions of non-shivering thermogenesis. Through the expression of uncoupling protein 1 (UCP1) BAT uncouples oxidative phosphorylation from the synthesis of ATP, resulting in the generation of heat and serves to protect an organism from hypothermia. The role of BAT in human physiology has gone overlooked in the past; however, since the discovery of metabolically active BAT in adult humans (Nedergaard et al. 2007), it resumed a place in the spotlight. Given the advent of the obesity pandemic in "westernized" countries, much attention has been paid to understanding the signals involved in the differentiation of such "energy burning" brown adipocytes.

Although conventional brown and white adipocytes differ considerably in function and morphology, they have long been considered to share a common progenitor. Both adipose tissues originate from MSC; however, brown adipocytes, unlike white adipocytes, are derived from the myogenic lineage, characterized by Myf5 (myogenic regulatory factor expressed exclusively in myogenic precursors) expression (Fig. 2.1). After deletion of PR domain containing 16 (PRDM16), the transcriptional regulator of brown adipocyte determination, in brown adipocyte precursors, the resulting cells exhibited a skeletal muscle phenotype, as opposed to the expected white adipocyte phenotype (Seale et al. 2008). In addition, lineage tracing experiments in mice expressing fluorescently tagged Myf5 revealed that brown adipocyte progenitors originate from the myogenic lineage, and thus are more closely related to skeletal muscle than to WAT (Seale et al. 2008).

Given that BAT is highly vascularized, it is not surprising that similar to white adipocytes, brown adipocytes might also be derived from endothelial cells (Tran et al. 2012); however, in addition to expressing VE-cadherin, brown preadipocytes must be Myf5-positive.

Based on cellular morphology, human brown adipocyte precursor cells were identified to be in close association with vasculature (Zingaretti et al. 2009), which further supports the data obtained from animal studies; however, characterization and localization of brown adipocyte progenitors in the human body still need to be elucidated. To date, human adipose tissue-derived stem cells were shown to differentiate to brown adipocytes only after stimulation of PPARγ with its potent activator rosiglitazone (Pisani et al. 2011), but there is little information about the clinical relevance of these observations.

2.1.2.1 Inducible Brown Fat

Following stimulation of β3-adrenergic receptors, via cold exposure or pharmacological treatment, fat cells morphologically resembling brown adipocytes can be recruited in WAT. These "de novo"-recruited brown adipocytes are characterized by high mitochondrial content, brownish color, and are consistent with conventional brown adipocytes in that they express UCP1 and contribute to energy dissipation. Although morphologically similar, brown adipocytes and inducible brown adipocytes do not share a common precursor cell (Seale et al. 2008). Two possibilities for how inducible brown adipocytes could be recruited in WAT have been proposed. The first is referred to as transdifferentiation of already mature white adipocytes (Barbatelli et al. 2010) and the second involves differentiation from existing progenitor cells (Fig. 2.1). Recently it was reported that progenitors of inducible BAT express specific markers, namely CD137 and transmembrane protein 26 (TMEM26), and displayed increased UCP1 expression compared to cells negative for those markers (Wu et al. 2012). In addition, it was reported that the gene expression pattern of human BAT resembles inducible BAT, more than conventional BAT, and these findings were consistent in both mature adipocytes and progenitors (Wu et al. 2012). Progenitors of inducible BAT were identified after bromodeoxyuridine (BrdU) was injected into mice treated with β3-adrenergic receptor agonist and newly differentiated UCP1-positive cells were detected in white fat. These results were confirmed by lineage tracing studies using constitutive and inducible reporter systems, and inducible brown adipocyte progenitors were further defined as Sca-1-, CD34-, and PDGFRα-positive (Lee et al. 2012), in contrast to white progenitors, that express PDGFRβ and endothelial markers (Tang et al. 2008). Interestingly, PDGFRα-positive progenitors developed into brown adipocytes under β3-adrenergic stimulation, whereas, in vivo, under conditions of high-fat diet feeding, the cells differentiated into white adipocytes, which demonstrated the bidirectional potential of differentiation depending of external stimuli (Lee et al. 2012).

2.1.3 *Microenvironment*

Adipogenic potential of progenitor cells does depend not merely on surface markers expressed but also on the microenvironment in which they reside. Many studies have demonstrated that adipose tissue-derived stem cells are capable of differentiation into various tissues depending on the differentiation signals provided (Halvorsen et al. 2001; Zuk et al. 2001). For example, selected progenitors were able to reconstitute adipose tissue only in lipodystrophic and high-fat diet-fed mice, but not in wild-type chow-fed mice (Rodeheffer et al. 2008), and the fate of bipotential adipocyte progenitors was highly dependent on external nutritional or β3-adrenergic stimuli (Lee et al. 2012) which further supports the importance of the microenvironment. It is also probable that not only one subpopulation of adipose progenitors exists and contributes to the renewal and hyperplastic growth of adipose tissue. More likely, adipose progenitors from different sources could contribute to a pool of stem cells in a depot-specific manner (Majka et al. 2010).

2.2 Signaling Pathways in Adipogenesis

The complete list of signals involved in the commitment phase of adipogenesis remains to be described; however, several factors have been identified that can either promote or inhibit the commitment of multipotent MSC to the pre-adipocyte lineage, most notably bone morphogenic proteins (BMP-2 and BMP-4) and Wnt signaling proteins. Accumulated evidence suggests that external growth factors and signaling molecules converge on the promoters of lineage-specific transcription factors. These external signals tightly regulate the transcriptional program to promote one cell type, while inhibiting another. For example, PPARγ stimulates adipogenesis while inhibiting chondrogenesis and the formation of cartilage, whereas inhibition of PPARγ can promote osteogenesis and inhibition of adipogenesis (Isenmann et al. 2009; Xu et al. 2007) (Fig. 2.2).

2.2.1 *Classical Pro-adipogenic Signals*

2.2.1.1 Bone Morphogenic Proteins

Both BMP-2 and BMP-4 have been shown to play a role in the commitment of multipotent MSC to the adipocyte lineage. Exposure of the dividing MSC cell line C3H10T1/2 to either BMP-4 or BMP-2 will result in pre-adipocyte-like cells that can be differentiated into adipocytes when treated with differentiation inducers following growth arrest (Bowers and Lane 2007; Huang et al. 2009). BMPs signal through two cell surface receptors, BMPr1 and BMPr2, both with serine/threonine

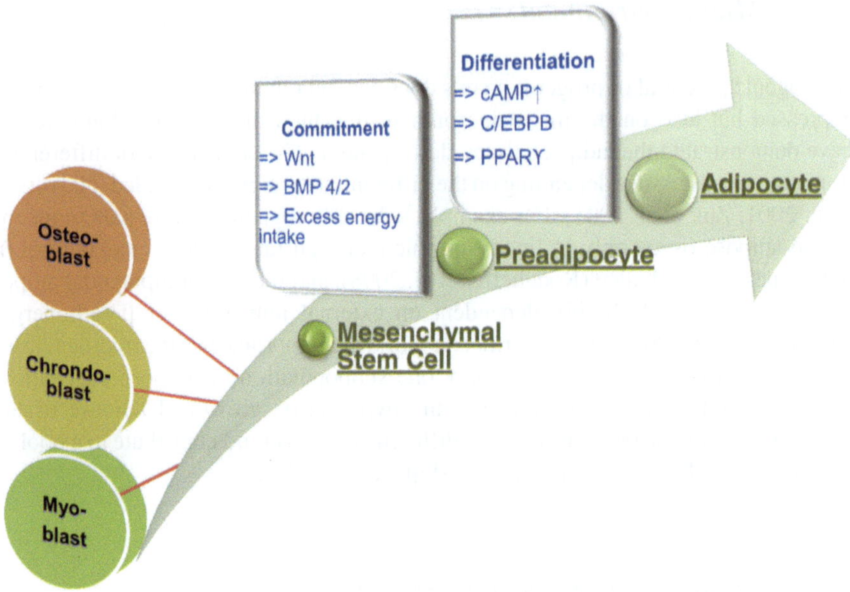

Fig. 2.2 Signaling pathways in adipogenesis. The proposed pathway from MSC to mature adipocyte. Following signals from the Wnt and BMP family of proteins, the MSC enter the pre-adipocyte lineage (Commitment). Later, differentiation signals promote the transition from pre-adipocyte to mature adipocyte (Differentiation)

kinase activities. Activation of these receptors leads to phosphorylation of BMPr1 kinase which then activates Smad-1, -5, and -8, leading to the formation of a complex with Smad-4 that translocates to the nucleus where it regulates gene expression. Expression of a constitutively active BMP receptor (CA-BMPr1A and CA-BMPr1B) induced commitment to the pre-adipocyte lineage in the absence of BMP-2 and BMP-4 (Huang et al. 2009). BMP signaling alters the expression of three cytoskeleton proteins, Lox, Tpt1, and αB-crystallin, that play important roles in adipocyte lineage commitment (Huang et al. 2011). Commitment to the adipocyte lineage can be eradicated by the knockdown of Lox1; however, it is only partially lost following the knockdown of Tpt1 and αB-crystallin. Given that pre-adipocytes differ dramatically in their morphology from MSC, BMPs may promote commitment to pre-adipocytes through the regulation of the cytoskeleton and consequent effects on cellular morphology.

2.2.1.2 Wnt Signaling

The Wnt family of signaling proteins exerts its effects through the frizzled receptor and the low-density lipoprotein-related protein 5/6 co-receptor, and can signal through both a "canonical" (Reya and Clevers 2005) and a "noncanonical" pathway (Christodoulides et al. 2009). Wnts play a role in both the lineage commitment and differentiation phase of adipogenesis, although with opposing effects (Kang et al. 2007; Ross et al. 2000). The canonical signaling pathway functions early during lineage commitment and involves the binding of Wnt to its receptors which promotes the dissociation of the "destruction complex," comprising adenomatous polyposis coli, axin, and glycogen synthase kinase (GSK)-3β, thereby permitting the accumulation of β-catenin, that was previously embedded in the complex (Reya and Clevers 2005). In the nucleus, β-catenin targets transcription factors that are differentially expressed in MSC (C3H10T1/2) vs. pre-adipocytes (A33, derived from C3H10T1/2), such as R-spondins-2 and -3 (Bowers and Lane 2007), which are dramatically upregulated in proliferating A33 cells and are known to activate canonical Wnt signaling, thus implicating Wnt signaling in the commitment phase of MSC to the adipocyte lineage. Later in the adipogenic program, Wnt inhibits adipocyte differentiation by inhibiting the expression of important adipogenic transcription factors, PPARγ and CCAAT/enhancer binding protein (C/EBP)α, and by doing so, maintains a balance among the myogenic, osteoblastogenic, and adipogenic cell fates (Ross et al. 2000; Tang and Lane 2012) Fig. 2.2.

2.2.2 *Differentiation of Adipocytes from Pre-adipocytes*

Cells isolated from the SVF of adipose tissue depots can be differentiated in culture when stimulated by specified contents of a "differentiation cocktail." In culture, pre-adipocytes will undergo multiple rounds of mitosis until they reach growth arrest, the G1 phase of the cell cycle. At this point, differentiation can be induced. Induction initiates a series of events that requires pre-adipocytes to reenter the cell cycle, undergo mitotic clonal expansion until they eventually exit the cell cycle, change their morphology, accumulate cytoplasmic triglycerides, and acquire the metabolic features and capabilities of mature adipocytes. The differentiation induction cocktail is composed of dexamethasone (a glucocorticoid), insulin, 3-isobutyl-1-methylxanthine (IBMX), and fetal bovine serum (FBS). The order in which the cells are exposed to these stimuli is crucial for the differentiation process as it was previously shown that dexamethasone can be added to the differentiation cocktail prior to IBMX, but the reverse will impede the differentiation of pre-adipocytes to mature adipocytes (Pantoja et al. 2008). Dexamethasone treatment is important in the initial phases of the induction of differentiation as it activates the transcription factor C/EBPβ. IBMX is a phosphodiesterase inhibitor that raises cyclic AMP (cAMP) levels in the cell, leading to the activation of the related transcription factor C/EBPδ. C/EBPβ and δ in turn induce transcription of C/EBPα and PPARγ (Rosen and MacDougald 2006; Rosen et al. 2000).

2.2.3 *Transcriptional Mediators of Adipogenesis*

2.2.3.1 C/EBPs

C/EBP family members are vital for the differentiation of adipocytes whereby early induction of C/EBPβ and C/EBPδ leads to the induction of C/EBPα and PPARγ. Almost immediately after the induction of differentiation, cAMP response element-binding protein (CREB) becomes phosphorylated and activates the expression of C/EBPβ (Zhang et al. 2004). Approximately 14–16 h after induction, C/EBPβ acquires DNA-binding capabilities as the pre-adipocytes reenter the cell cycle. C/EBPβ likely plays a role in mitotic clonal expansion, which is required for differentiation. The function of C/EBPβ and C/EBPδ may be redundant as knockout of C/EBPβ in mice had little effect on adipose tissue accumulation, whereas the double C/EBPβ–/–C/EBPδ–/– knockout showed markedly reduced adipose tissue mass, mostly due to decreased cell number (Tanaka et al. 1997). Early in the differentiation process, C/EBPβ is phosphorylated twice, by mitogen-activated protein kinase (MAPK) and GSK3β (Tang et al. 2005), which induces a conformational change resulting in the dimerization of two monomers of C/EBPβ and creates a DNA-binding pocket. Once C/EBPβ is capable of binding DNA, 14–16 h post-differentiation induction, it leads to the upregulation of C/EBPα and PPARγ which act together as pleiotropic transcriptional activators of the large group of genes that produce the adipocyte phenotype (Rosen and MacDougald 2006; Tang and Lane 2012).

2.2.3.2 PPARγ

PPARγ is the master regulator of adipogenesis, as it is both sufficient and necessary for adipogenesis (Rosen and MacDougald 2006; Rosen et al. 2000). It exists as three isoforms (PPARγ1, PPARγ2, and PPARγ3) that are differentially spliced from the same transcribed gene (Fajas et al. 1998; Zhu et al. 1995). In order to exert effects on peroxisome proliferator response elements, PPARs must form heterodimers with the retinoid X receptor (RXR). In primary adipocytes, PPARγ2 is the predominant isoform and is more efficient at promoting adipogenesis than the other isoforms (Rosen and MacDougald 2006). Efforts to describe endogenous PPARγ ligands have been largely unsuccessful; however, studies have found that the transcription factors sterol response element-binding protein-1c (SREBP-1c) and C/EBPβ can increase PPARγ ligand production although these findings did not lead to the identification of a PPARγ agonist (Hamm et al. 2001; Kim et al. 1998). Once expressed PPARγ and C/EBPα positively feedback on each other through their respective C/EBP regulatory elements; this action is presumed to perpetuate the adipocyte phenotype in mature adipocytes. In addition to its role in differentiation, PPARγ is crucial for the maintenance of mature adipocytes.

For example, expression of dominant negative PPARγ in differentiated 3T3-L1 adipocytes led to dedifferentiation and loss of lipid accumulation (Tamori et al. 2002).

2.2.4 Differentiation of Progenitor Cells into Brown Adipocytes

Brown pre-adipocyte recruitment is controlled by the TGF-β family of secreted proteins, such as BMP-7 (Tseng et al. 2008), while Wnt signaling represses the differentiation of brown pre-adipocytes into brown adipocytes. These inducible brown cells can arise from multipotent progenitors, or perhaps even from transdifferentiation of existing white adipocytes, in WAT depots when exposed to certain stimuli (Petrovic et al. 2010; Seale et al. 2008, 2011). The presence of inducible brown adipocytes can lead to weight loss, improved insulin sensitivity, and resistance to high-fat feeding in animal models (Vegiopoulos et al. 2010).

While the transcription factors encoded by C/EBP and PRDM16 have been shown to function as key regulators of BAT differentiation, defining the signaling pathways involved in the de novo recruitment of BAT remains an active area of research.

2.2.4.1 β-adrenergic Signaling

The sympathetic nervous system is the major regulator of BAT activity through β-adrenergic signaling (Rehnmark et al. 1990). In existing BAT depots, β1-adrenergic signaling promotes proliferation of BAT pre-adipocytes whereas in animal models β3-adrenergic receptor activation can promote the recruitment of inducible brown adipocytes in WAT depots (Bronnikov et al. 1999). In humans, β3-receptor agonists have not been successful in reducing adiposity, which could result from fewer β3-receptors or lack of inducible brown fat (Cannon and Nedergaard 2004). β3-receptor agonists lead to an increase in intracellular cAMP which activates the lipolytic pathway whereby triglycerides are hydrolyzed to free fatty acids (FA) and glycerol. The activation of this pathway is critical for BAT activity as it was recently shown that loss of desnutrin/ATGL, the first hydrolase in the triglyceride hydrolysis pathway, results in the conversion of BAT to a WAT-like tissue (Ahmadian et al. 2011).

2.2.4.2 COX-2

Cyclooxygenase-2 (COX-2) is a rate limiting enzyme in the synthesis of prostaglandins, and is downstream of the β-adrenergic signaling pathway. In Vegiopoulos et al. (2010), it was first reported that COX-2 overexpression and the synthesis of prostaglandins can shift the differentiation of adipocyte progenitors, isolated from the epididymal depot, towards a brown adipocyte phenotype in vitro. In vivo, COX-2 overexpressing mice and mice injected with a β3-adrenergic receptor agonist displayed de novo recruitment of brown adipocytes in the epididymal WAT. The appearance of inducible BAT was associated with increased energy expenditure and resistance to high-fat diet-induced weight gain.

In addition to the signals described above, recent advances in the field of brown adipocyte biology have suggested other factors can recruit inducible brown adipocytes. For example, Böstrom et al. (2012) recently reported that mice overexpressing PPARγ co-activator-1α (PGC-1α) in skeletal muscle have increased expression of fibronectin type III domain containing 5 (FNDC5), a membrane protein that is cleaved intracellularly and secreted as the peptide hormone irisin. This study reported that irisin acted on mature white adipocytes to increase expression of UCP1 and a broad program of brown adipocyte gene expression patterns. In vitro, it has been reported that treatment of murine white pre-adipocytes, isolated from the epididymal depot, with high concentrations of the PPARγ activator rosiglitazone promoted mitochondriogenesis, expression of PGC-1α and UCP1, an effect that could be augmented still following treatment with norepinephrine (Petrovic et al. 2010). These cells maintained genuine thermogenic capacity; however, they did not express the transcription factors known to be associated with classic brown adipocytes, such PRDM16, LIM homeobox protein 8 (Lhx8), and mesenchyme homeobox 2 (Meox2) (Walden et al. 2012). Therefore, these cells may represent an additional subset of adipocytes that maintain the transcriptional signature of white adipocytes while possessing the potential to express UCP1 and the brown adipocyte phenotype. The presence of inducible brown adipocytes in human adipose tissue, the signals required to recruit these cells in vivo, and their potential to curb diet-induced obesity remain under investigation.

2.3 Clinical Aspects

2.3.1 Adipose Tissue as a Source for MSC

Adipose tissue has a remarkable ability to undergo considerable changes in volume during the lifespan of an individual. Although relatively small increases in volume can be accommodated by adipocyte hypertrophy, larger changes are mediated by the generation of new adipocytes, e.g., hyperplasia. In fact, adipocyte hyperplasia has been found to rely on distinct populations of stem and progenitor cells residing within the so-called SVF of adipose tissue (see above).

As stated above, in vitro differentiation of adipocyte progenitors into mature adipocytes can be achieved by a hormonal cocktail, including insulin, glucocorticoid, and cAMP elevating agents that ultimately converge in the activation of nuclear receptor PPARγ. Interestingly, in contrast to murine adipocyte progenitors, differentiation of human pre-adipocytes requires the presence of a PPARγ agonist, highlighting intrinsic differences between murine and human progenitor populations which might become relevant considering therapeutic use of these cells (Jia et al. 2012).

In contrast to bone marrow-derived MSC, adipose tissue-derived stem/progenitor cells (ADSC) can be obtained in relatively large quantities with minimal risk and pain. In addition, studies have documented that the frequency of ADSC is in the order of 500-fold more than that found in bone marrow (Rodriguez et al. 2005).

Despite the differences in frequency and yield of ADSC and MSC, their biological properties are similar (Fraser et al. 2006). To this end, both cell populations express distinct cell surface markers, including CD105, SH3, Stro-1, CD90, CD44, and potentially CD34, but they do not express the hematopoietic marker CD45 nor the endothelial marker CD31 (Sengenes et al. 2005).

2.3.2 Application of MSC in Regenerative Medicine

Regenerative medicine can be defined as the process of generating living and functionally active tissue to repair or replace tissue or organ function lost due to disease, damage, congenital defects, or age. Indeed, regenerative medicine has been successfully used already for a number of decades to reconstitute the hematopoietic system of patients who have undergone chemotherapy or are suffering from genetic defects of the hematopoietic system through bone marrow transplantation protocols (Park et al. 2008).

Up to date, tissue repair and/or replacement has been achieved by two principle technological approaches:

(a) Replacement of tissue function with synthetic constructs (artificial organs), e.g., by seeding cells into a de-cellularized natural extracellular matrix or a polymer, metal, or ceramic scaffold (Hubbell 2004).
(b) Cellular therapies using stem cells or genetically modified cells to generate new tissues or organs by transplanting cell suspensions or aggregates into the affected locus (Daley and Scadden 2008).

Therapeutically, ADSC have thus far been employed for muscle cell-based therapies and adipose tissue reconstruction in mouse models (Rodriguez et al. 2005), the treatment of fistulas in human patients suffering from Crohn's disease (Garcia-Olmo et al. 2005), and are discussed as a source for autologous adult stem cells in the correction of soft tissue defects after trauma, tumor resection, or deep burns (Schaffler and Buchler 2007).

In contrast, no routine use has been made of ADSC for the regeneration of metabolic tissues in order to restore impaired local and systemic energy homeostasis and to counteract associated disorders, thereby establishing a novel "regenerative" arm in the therapy against metabolic dysfunction (Fig. 2.3).

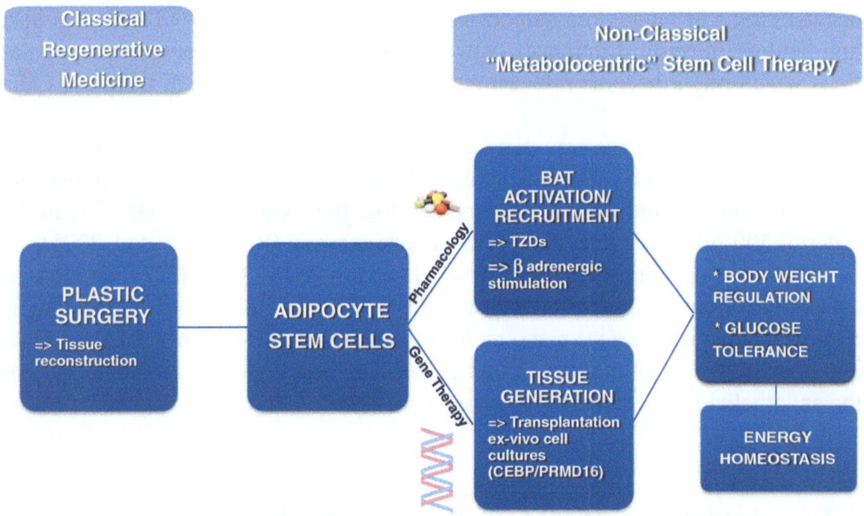

Fig. 2.3 Classical vs. nonclassical regenerative medicine. Adipocyte stem cells offer the future possibility to control energy homeostasis through either the pharmacological activation/recruitment of brown adipose tissue (BAT) in vivo or through the ex vivo generation of BAT via gene transfer of critical transcriptional regulators into progenitor cells

2.3.3 Adipose-Derived Stem Cells in Metabolic Control and Obesity-Related Disorders

2.3.3.1 The "Diabesity" Pandemic

The EU population is close to 500 million. More than half of the adult population are either overweight or obese while, among diabetic patients, ~90 % suffering from type 2 diabetes are obese.

Globally, the increase in the prevalence of overweight/obese people has reached a qualified epidemic stage with more than one billion overweight and at least 400 million clinically obese. Type 2 diabetes mellitus and the associated beta-cell failure is up to 90 % attributable to weight gain and has increased in parallel with the pandemic of obesity. Approximately 197 million people worldwide have impaired glucose tolerance, most commonly because of obesity and the associated Metabolic Syndrome, and this number is expected to increase to 420 million by 2025. This will lead to tremendous costs for health care systems, primarily because of the precipitation into end-stage type 2 diabetes and associated cardiovascular complications (Kopelman 2000).

In Western and emerging countries, metabolic and vascular diseases are the leading cause of disability and death with a major impact on health care costs. Metabolic overload is a paramount problem for people in industrial countries in which the caloric intake exceeds the dietary needs, due to associated risks of secondary disorders including obesity, diabetes, atherosclerosis, and hypertension (Dorrensteijn et al. 2012; Leroith et al. 2012). The WHO projects that by 2015, approximately 2.3 billion adults will be overweight and more than 700 million will be obese. Although the numbers of overweight and obese patients steadily increase worldwide, Europe as an example has one of the highest average body mass indices (BMI) of all WHO regions, totaling to almost 250 million overweight adults and 70–80 million severely obese patients in the European Community (Amos et al. 1997; Finucane et al. 2011).

Basic risk factors for obesity comprise excessive caloric intake, a sedentary lifestyle, as well as genetic predisposition, all of which precipitate in aberrant systemic energy/fat storage in the form of WAT (Saltiel 2001). In this setting, normal WAT function is severely damaged as it is compromised by tissue inflammation and aberrant metabolic properties. Excessive WAT depots can subsequently be made directly responsible for the manifestation of systemic and tissue-specific insulin resistance, hypertension, dyslipidemia, and hyperglycemia as hallmarks of the so-called Metabolic Syndrome, eventually culminating into end-stage diseases such as type 2 diabetes, atherosclerosis, cardiovascular failure, and even cancer (Calle and Kaaks 2004; Cohen and Leroith 2012).

2.3.4 Clinical Bottlenecks and Limitations

In the last 2 decades, the major scientific and financial contributions have focused on studies of genetic factors in response to an "obesogenic" environment, concentrating on counteracting excessive energy intake to tackle "diabesity."

This approach now appears rather disappointing in terms of Public Health and, until now, effective treatment strategies against obesity-related type 2 diabetes and the associated beta-cell failure induced by obesity are still missing.

This holds true for lifestyle intervention programs as well as for pharmacological therapies.

In lifestyle intervention programs, there is often an impressive short-term success in weight reduction, but this success is often blunted by weight regain after the program's end, which is in the most cases due to an insufficient patient compliance (Bray and Wilson 2008; Galani and Schneider 2007).

In particular, existing pharmacotherapy for the management of obesity, which is primarily aimed at weight loss, weight loss maintenance, and risk reduction, has only a modest efficacy reflected by a weight loss of 5–10 % of the initial body weight, followed by a weight loss plateau under further therapy (Ioannides-Demos et al. 2005). Therefore, in the past years thyroid hormone, amphetamines, phentermine, amfepramone, phenylpropanolamine, bupropion, naltrexone, and, more recently, sibutramine, orlistat, and rimonabant have been evaluated for obesity treatment regimes (Greenway et al. 2008). In addition to the missing impact on weight reduction, existing obesity therapies are often limited by safety concerns which have certainly increased after the recent market withdrawal of rimonabant in November 2008 (Soyka 2008).

Consequently, to date only a few antiobesity drugs have entered into clinical practice and are used at limited success rate. Clearly, pharmacological strategies have proven to be remarkably ineffective against the increasing problem of obesity. Besides the need for weight reduction, in recent years there is an increasing awareness for obesity-related diseases like dyslipidemia, high triglyceride-low HDL-syndromes, nonalcoholic fatty liver disease, and/or insulin resistance. However, similar to the therapeutic alternatives for weight reduction, treatments for those conditions are also underdeveloped and need urgent improvement.

2.3.5 BAT Recruitment and Activation as Novel Metabolocentric Therapy

From the perspective of regenerative medicine, the described studies and acquired knowledge on adipocyte progenitors, signaling pathways affecting their cellular and functional fate, and distinct properties of white vs. brown adipocytes open the intriguing possibility that even the regeneration and activation of small amounts of BAT in adult humans/patients would have a significant impact on systemic energy expenditure, thereby representing an unprecedented approach to the regulation of body weight and associated disorders, while simultaneously counteracting WAT tissue injury as imposed by energy overload and tissue-specific inflammation of WAT stores in the obese state.

Indeed, in humans, it is estimated that even as little as 50 g of stimulated and active BAT could account for a 20 % daily increase in energy expenditure (Rothwell and Stock 1983), and that up to 24 % of the increase in metabolism in lean men produced by ephedrine can be attributed to BAT activation (Astrup et al. 1985). Concomitant to the increase in energy expenditure, regeneration and/or activation of BAT can be envisaged to also beneficially influence other clinical parameters as associated with obesity, type 2 diabetes, and the Metabolic Syndrome.

While brown/BRITE adipocytes do not seem to display increased cellular insulin sensitivity per se (Mössenböck, K., Herzig, S. (2012). unpublished observation), both basal and insulin-stimulated glucose uptake are substantially elevated as compared with white adipocytes. Thus any WAT to BAT transformation or BAT recruitment will increase the systemic glucose clearance and potentially also improve glucose tolerance and insulin sensitivity. Indeed, a number of clinical studies have validated substantial cold-induced glucose uptake in human subjects as measured by PET studies using 18-fluordeoxyglucose. Depending on the experimental setting, cold-stimulated glucose clearance through BAT was enhanced between 4- and 15-fold, negatively correlating with BMI (Virtanen et al. 2009; van Marken Lichtenbelt et al. 2009).

Furthermore, insulin resistance is correlated to WAT inflammation, and studies on the differential metabolic impact of distinct WAT depots and their capacities to recruit BAT suggest that WAT to BAT transformation could reduce inflammation and improve insulin resistance (Frontini and Cinti 2010). In addition, prolonged treatment of a patient with levothyroxine (suppressive treatment for thyroid cancer) resulted in dramatic improvement in insulin sensitivity, paralleled by enhanced BAT volume and activity—in spite of extreme insulin resistance due to a mutation in the insulin receptor gene. Consequently, BAT activation could potentially mediate non-insulin-dependent glucose disposal (Skarulis et al. 2010). Indeed, the thyroid hormone receptor (TR) beta was found to be required for UCP1-dependent adaptive thermogenesis in mice (Ribeiro et al. 2010), potentially mediated largely via central rather than peripheral TR effects (Lopez et al. 2010).

In addition, fibroblast growth factor (FGF) 21 is produced by BAT and has been shown to promote beneficial effects on glucose metabolism (Hondares et al. 2010).

Finally, BAT activity controls triglyceride and/or FA clearance, thereby counteracting hypertriglyceridemia. In pathophysiological settings (mice fed a high-fat diet), cold exposure corrected hyperlipidemia and improved deleterious effects of insulin resistance, which could be attributed to enhanced BAT activation (Bartelt et al. 2011). In addition, tracer studies in humans also underlined the potential of BAT to efficiently take up FA and use these substrates in oxidative metabolism (Orava et al. 2011; Ouellet et al. 2012).

2.3.6 Therapeutic Outlook and Challenge

While the described studies have on the one side demonstrated the existence and feasibility of human adipocyte progenitors and on the other side provided clinical evidence for the functionality and metabolo-regulatory impact of human BAT, actual therapeutic use of this resource is currently still in its infancies. Indeed, clinical studies using β-adrenergic stimuli to activate BAT thermogenesis and counteract obesity showed limited efficacy and had to be abandoned due to cardiovascular complications as associated with enhanced beta sympathetic tone (Arch 2002). Along these lines, insulin sensitizers of the PPARγ ligand family that show a browning effect, at least in rodents, are concerned with substantial side effects in humans, including weight gain and cardiovascular events, eventually leading to the withdrawal of rosiglitazone from the market in 2010 (Nissen and Wolski 2010).

Despite these current obstacles, the increasing understanding of human BAT biology provides novel clues for innovative antiobesity therapeutic avenues in this direction.

While current therapeutic attempts had been centered on the increase of BAT content in obese subjects, the finding that virtually all younger humans possess relevant amounts of functionally intact BAT suggests that new approaches should focus on the maintenance of this cell pool and try to prevent the involution and inactivation of BAT during aging, particularly considering the proven sex differences in BAT activity in humans (Pfannenberg et al. 2010). To this end, the description of adipocyte white/BRITE progenitor isolation protocols for defined (Lin–/CD29+/CD34+/Sca-1+/CD24+) cell populations in mice (Vegiopoulos et al. 2010) may provide a promising starting point for clinical-therapeutic research in humans in this direction, which may lead to the identification of cellular pools that could generate/maintain proliferatively competent brown/BRITE adipocytes in humans during the aging process, thereby maintaining "young" levels of energy expenditure and obesity resistance throughout the lifetime of an individual.

Further clinical research in this direction is clearly validated in the future.

Acknowledgements We apologize to our colleagues whose contributions could not be cited due to space limitations. We thank members of our lab for critically reading the manuscript and discussions. In particular, we thank Tatiana Golea for administrative and graphical support. Our work is supported by grants from the Deutsche Forschungsgemeinschaft, the European Foundation for the Study of Diabetes, the FP7 DIABAT consortium, the German Cancer Aid, and the Network Aging Research.

References

Ahmadian M, Abbott MJ, Tang T, Hudak CS, Kim Y, Bruss M, Hellerstein MK, Lee HY, Samuel VT, Shulman GI, Wang Y, Duncan RE, Kang C, Sul HS. Desnutrin/ATGL is regulated by AMPK and is required for a brown adipose phenotype. Cell Metab. 2011;13:739–48.

Amos AF, McCarty DJ, Zimmet P. The rising global burden of diabetes and its complications: estimates and projections to the year 2010. Diabet Med. 1997;14 Suppl 5:S1–85.

Arch JR. beta(3)-Adrenoceptor agonists: potential, pitfalls and progress. Eur J Pharmacol. 2002;440:99–107.

Ashjian PH, Elbarbary AS, Edmonds B, DeUgarte D, Zhu M, Zuk PA, Lorenz HP, Benhaim P, Hedrick MH. In vitro differentiation of human processed lipoaspirate cells into early neural progenitors. Plast Reconstr Surg. 2003;111:1922–31.

Astrup A, Bulow J, Madsen J, Christensen NJ. Contribution of BAT and skeletal muscle to thermogenesis induced by ephedrine in man. Am J Physiol. 1985;248:E507–15.

Barbatelli G, Murano I, Madsen L, Hao Q, Jimenez M, Kristiansen K, Giacobino JP, De Matteis R, Cinti S. The emergence of cold-induced brown adipocytes in mouse white fat depots is determined predominantly by white to brown adipocyte transdifferentiation. Am J Physiol Endocrinol Metab. 2010;298:E1244–53.

Bartelt A, Bruns OT, Reimer R, Hohenberg H, Ittrich H, Peldschus K, Kaul MG, Tromsdorf UI, Weller H, Waurisch C, Eychmuller A, Gordts PL, Rinninger F, Bruegelmann K, Freund B, Nielsen P, Merkel M, Heeren J. Brown adipose tissue activity controls triglyceride clearance. Nat Med. 2011;17(2):200–5.

Bostrom P, Wu J, Jedrychowski MP, Korde A, Ye L, Lo JC, Rasbach KA, Bostrom EA, Choi JH, Long JZ, Kajimura S, Zingaretti MC, Vind BF, Tu H, Cinti S, Hojlund K, Gygi SP, Spiegelman BM. A PGC1-alpha-dependent myokine that drives brown-fat-like development of white fat and thermogenesis. Nature. 2012;481:463–8.

Bowers RR, Lane MD. A role for bone morphogenetic protein-4 in adipocyte development. Cell Cycle. 2007;6:385–9.

Bray GA, Wilson JF. In the clinic. Obesity. Ann Intern Med. 2008;149:ITC4-1-15; quiz ITC14-16.

Bronnikov G, Bengtsson T, Kramarova L, Golozoubova V, Cannon B, Nedergaard J. beta1 to beta3 switch in control of cyclic adenosine monophosphate during brown adipocyte development explains distinct beta-adrenoceptor subtype mediation of proliferation and differentiation. Endocrinology. 1999;140:4185–97.

Brzoska M, Geiger H, Gauer S, Baer P. Epithelial differentiation of human adipose tissue-derived adult stem cells. Biochem Biophys Res Commun. 2005;330:142–50.

Calle EE, Kaaks R. Overweight, obesity and cancer: epidemiological evidence and proposed mechanisms. Nat Rev Cancer. 2004;4:579–91.

Cannon B, Nedergaard J. Brown adipose tissue: function and physiological significance. Physiol Rev. 2004;84:277–359.
Christodoulides C, Lagathu C, Sethi JK, Vidal-Puig A. Adipogenesis and WNT signalling. Trends Endocrinol Metab. 2009;20:16–24.
Cohen DH, Leroith D. Obesity, type 2 diabetes and cancer: the insulin and insulin-like growth factor connection. Endocr Relat Cancer. 2012;19(5):F27–45.
Crossno Jr JT, Majka SM, Grazia T, Gill RG, Klemm DJ. Rosiglitazone promotes development of a novel adipocyte population from bone marrow-derived circulating progenitor cells. J Clin Invest. 2006;116:3220–8.
Daley GQ, Scadden DT. Prospects for stem cell-based therapy. Cell. 2008;132:544–8.
Dorresteijn JA, Visseren FL, Spiering W. Mechanisms linking obesity to hypertension. Obes Rev. 2012;13:17–26.
Fajas L, Fruchart JC, Auwerx J. PPARgamma3 mRNA: a distinct PPARgamma mRNA subtype transcribed from an independent promoter. FEBS Lett. 1998;438:55–60.
Finucane MM, Stevens GA, Cowan MJ, Danaei G, Lin JK, Paciorek CJ, Singh GM, Gutierrez HR, Lu Y, Bahalim AN, Farzadfar F, Riley LM, Ezzati M. National, regional, and global trends in body-mass index since 1980: systematic analysis of health examination surveys and epidemiological studies with 960 country-years and 9.1 million participants. Lancet. 2011;377:557–67.
Fraser JK, Wulur I, Alfonso Z, Hedrick MH. Fat tissue: an underappreciated source of stem cells for biotechnology. Trends Biotechnol. 2006;24:150–4.
Frontini A, Cinti S. Distribution and development of brown adipocytes in the murine and human adipose organ. Cell Metab. 2010;11:253–6.
Galani C, Schneider H. Prevention and treatment of obesity with lifestyle interventions: review and meta-analysis. Int J Public Health. 2007;52:348–59.
Garcia-Olmo D, Garcia-Arranz M, Herreros D, Pascual I, Peiro C, Rodriguez-Montes JA. A phase I clinical trial of the treatment of Crohn's fistula by adipose mesenchymal stem cell transplantation. Dis Colon Rectum. 2005;48:1416–23.
Green H, Kehinde O. An established preadipose cell line and its differentiation in culture. II. Factors affecting the adipose conversion. Cell. 1975;5:19–27.
Greenway FL, Whitehouse MJ, Guttadauria M, Anderson JW, Atkinson RL, Fujioka K, Gadde KM, Gupta AK, O'Neil P, Schumacher D, Smith D, Dunayevich E, Tollefson GD, Weber E, Cowley MA. Rational design of a combination medication for the treatment of obesity. Obesity (Silver Spring). 2008;17(1):30–9.
Halvorsen YD, Franklin D, Bond AL, Hitt DC, Auchter C, Boskey AL, Paschalis EP, Wilkison WO, Gimble JM. Extracellular matrix mineralization and osteoblast gene expression by human adipose tissue-derived stromal cells. Tissue Eng. 2001;7:729–41.
Hamm JK, Park BH, Farmer SR. A role for C/EBPbeta in regulating peroxisome proliferator-activated receptor gamma activity during adipogenesis in 3T3-L1 preadipocytes. J Biol Chem. 2001;276:18464–71.
Hondares E, Rosell M, Gonzalez FJ, Giralt M, Iglesias R, Villarroya F. Hepatic FGF21 expression is induced at birth via PPARalpha in response to milk intake and contributes to thermogenic activation of neonatal brown fat. Cell Metab. 2010;11:206–12.
Huang H, Song TJ, Li X, Hu L, He Q, Liu M, Lane MD, Tang QQ. BMP signaling pathway is required for commitment of C3H10T1/2 pluripotent stem cells to the adipocyte lineage. Proc Natl Acad Sci U S A. 2009;106:12670–5.
Huang HY, Hu LL, Song TJ, Li X, He Q, Sun X, Li YM, Lu HJ, Yang PY, Tang QQ. Involvement of cytoskeleton-associated proteins in the commitment of C3H10T1/2 pluripotent stem cells to adipocyte lineage induced by BMP2/4. Mol Cell Proteomics. 2011;10(M110):002691.
Hubbell JA. Biomaterials science and high-throughput screening. Nat Biotechnol. 2004;22: 828–9.
Ioannides-Demos LL, Proietto J, McNeil JJ. Pharmacotherapy for obesity. Drugs. 2005;65: 1391–418.
Isenmann S, Arthur A, Zannettino AC, Turner JL, Shi S, Glackin CA, Gronthos S. TWIST family of basic helix-loop-helix transcription factors mediate human mesenchymal stem cell growth and commitment. Stem Cells. 2009;27:2457–68.

Jia B, Madsen L, Petersen RK, Techer N, Kopperud R, Ma T, Doskeland SO, Ailhaud G, Wang J, Amri EZ, Kristiansen K. Activation of protein kinase A and exchange protein directly activated by cAMP promotes adipocyte differentiation of human mesenchymal stem cells. PLoS One. 2012;7:e34114.

Kang S, Bennett CN, Gerin I, Rapp LA, Hankenson KD, Macdougald OA. Wnt signaling stimulates osteoblastogenesis of mesenchymal precursors by suppressing CCAAT/enhancer-binding protein alpha and peroxisome proliferator-activated receptor gamma. J Biol Chem. 2007;282:14515–24.

Kim JB, Wright HM, Wright M, Spiegelman BM. ADD1/SREBP1 activates PPARgamma through the production of endogenous ligand. Proc Natl Acad Sci U S A. 1998;95:4333–7.

Kopelman PG. Obesity as a medical problem. Nature. 2000;404:635–43.

Lee YH, Petkova AP, Mottillo EP, Granneman JG. In vivo identification of bipotential adipocyte progenitors recruited by beta3-adrenoceptor activation and high-fat feeding. Cell Metab. 2012; 15:480–91.

Leroith D. Pathophysiology of the metabolic syndrome: implications for the cardiometabolic risks associated with type 2 diabetes. Am J Med Sci. 2012;343:13–6.

Lin G, Garcia M, Ning H, Banie L, Guo YL, Lue TF, Lin CS. Defining stem and progenitor cells within adipose tissue. Stem Cells Dev. 2008;17:1053–63.

Lopez M, Varela L, Vazquez MJ, Rodriguez-Cuenca S, Gonzalez CR, Velagapudi VR, Morgan DA, Schoenmakers E, Agassandian K, Lage R, Martinez de Morentin PB, Tovar S, Nogueiras R, Carling D, Lelliott C, Gallego R, Oresic M, Chatterjee K, Saha AK, Rahmouni K, Dieguez C, Vidal-Puig A. Hypothalamic AMPK and fatty acid metabolism mediate thyroid regulation of energy balance. Nat Med. 2010;16:1001–8.

Majka SM, Fox KE, Psilas JC, Helm KM, Childs CR, Acosta AS, Janssen RC, Friedman JE, Woessner BT, Shade TR, Varella-Garcia M, Klemm DJ. De novo generation of white adipocytes from the myeloid lineage via mesenchymal intermediates is age, adipose depot, and gender specific. Proc Natl Acad Sci U S A. 2010;107:14781–6.

Nedergaard J, Bengtsson T, Cannon B. Unexpected evidence for active brown adipose tissue in adult humans. Am J Physiol Endocrinol Metab. 2007;293:E444–52.

Ning H, Lin G, Lue TF, Lin CS. Neuron-like differentiation of adipose tissue-derived stromal cells and vascular smooth muscle cells. Differentiation. 2006;74:510–8.

Nissen SE, Wolski K. Rosiglitazone revisited: an updated meta-analysis of risk for myocardial infarction and cardiovascular mortality. Arch Intern Med. 2010;170:1191–201.

Orava J, Nuutila P, Lidell ME, Oikonen V, Noponen T, Viljanen T, Scheinin M, Taittonen M, Niemi T, Enerback S, Virtanen KA. Different metabolic responses of human brown adipose tissue to activation by cold and insulin. Cell Metab. 2011;14:272–9.

Ouellet V, Labbe SM, Blondin DP, Phoenix S, Guerin B, Haman F, Turcotte EE, Richard D, Carpentier AC. Brown adipose tissue oxidative metabolism contributes to energy expenditure during acute cold exposure in humans. J Clin Invest. 2012;122:545–52.

Pantoja C, Huff JT, Yamamoto KR. Glucocorticoid signaling defines a novel commitment state during adipogenesis in vitro. Mol Biol Cell. 2008;19:4032–41.

Park DH, Borlongan CV, Eve DJ, Sanberg PR. The emerging field of cell and tissue engineering. Med Sci Monit. 2008;14:RA206–20.

Petrovic N, Walden TB, Shabalina IG, Timmons JA, Cannon B, Nedergaard J. Chronic peroxisome proliferator-activated receptor gamma (PPARgamma) activation of epididymally derived white adipocyte cultures reveals a population of thermogenically competent, UCP1-containing adipocytes molecularly distinct from classic brown adipocytes. J Biol Chem. 2010;285:7153–64.

Pfannenberg C, Werner MK, Ripkens S, Stef I, Deckert A, Schmadl M, Reimold M, Haring HU, Claussen CD, Stefan N. Impact of age on the relationships of brown adipose tissue with sex and adiposity in humans. Diabetes. 2010;59:1789–93.

Pisani DF, Djedaini M, Beranger GE, Elabd C, Scheideler M, Ailhaud G, Amri EZ. Differentiation of human adipose-derived stem cells into "brite" (brown-in-white) adipocytes. Front Endocrinol (Lausanne). 2011;2:87.

Pittenger MF, Mackay AM, Beck SC, Jaiswal RK, Douglas R, Mosca JD, Moorman MA, Simonetti DW, Craig S, Marshak DR. Multilineage potential of adult human mesenchymal stem cells. Science. 1999;284:143–7.

Poissonnet CM, Burdi AR, Garn SM. The chronology of adipose tissue appearance and distribution in the human fetus. Early Hum Dev. 1984;10:1–11.

Prockop DJ. Marrow stromal cells as stem cells for nonhematopoietic tissues. Science. 1997; 276:71–4.

Rehnmark S, Nechad M, Herron D, Cannon B, Nedergaard J. Alpha- and beta-adrenergic induction of the expression of the uncoupling protein thermogenin in brown adipocytes differentiated in culture. J Biol Chem. 1990;265:16464–71.

Reya T, Clevers H. Wnt signalling in stem cells and cancer. Nature. 2005;434:843–50.

Ribeiro MO, Bianco SD, Kaneshige M, Schultz JJ, Cheng SY, Bianco AC, Brent GA. Expression of uncoupling protein 1 in mouse brown adipose tissue is thyroid hormone receptor-beta isoform specific and required for adaptive thermogenesis. Endocrinology. 2010;151:432–40.

Rodeheffer MS, Birsoy K, Friedman JM. Identification of white adipocyte progenitor cells in vivo. Cell. 2008;135:240–9.

Rodriguez AM, Elabd C, Amri EZ, Ailhaud G, Dani C. The human adipose tissue is a source of multipotent stem cells. Biochimie. 2005;87:125–8.

Rosen ED, MacDougald OA. Adipocyte differentiation from the inside out. Nat Rev Mol Cell Biol. 2006;7:885–96.

Rosen ED, Walkey CJ, Puigserver P, Spiegelman BM. Transcriptional regulation of adipogenesis. Genes Dev. 2000;14:1293–307.

Ross SE, Hemati N, Longo KA, Bennett CN, Lucas PC, Erickson RL, MacDougald OA. Inhibition of adipogenesis by Wnt signaling. Science. 2000;289:950–3.

Rothwell NJ, Stock MJ. Luxuskonsumption, diet-induced thermogenesis and brown fat: the case in favour. Clin Sci (Lond). 1983;64:19–23.

Russell TR, Ho R. Conversion of 3T3 fibroblasts into adipose cells: triggering of differentiation by prostaglandin F2alpha and 1-methyl-3-isobutyl xanthine. Proc Natl Acad Sci U S A. 1976;73:4516–20.

Saltiel AR. New perspectives into the molecular pathogenesis and treatment of type 2 diabetes. Cell. 2001;104:517–29.

Schaffler A, Buchler C. Concise review: adipose tissue-derived stromal cells–basic and clinical implications for novel cell-based therapies. Stem Cells. 2007;25:818–27.

Seale P, Bjork B, Yang W, Kajimura S, Chin S, Kuang S, Scime A, Devarakonda S, Conroe HM, Erdjument-Bromage H, Tempst P, Rudnicki MA, Beier DR, Spiegelman BM. PRDM16 controls a brown fat/skeletal muscle switch. Nature. 2008;454:961–7.

Seale P, Conroe HM, Estall J, Kajimura S, Frontini A, Ishibashi J, Cohen P, Cinti S, Spiegelman BM. Prdm16 determines the thermogenic program of subcutaneous white adipose tissue in mice. J Clin Invest. 2011;121:96–105.

Sengenes C, Lolmede K, Zakaroff-Girard A, Busse R, Bouloumie A. Preadipocytes in the human subcutaneous adipose tissue display distinct features from the adult mesenchymal and hematopoietic stem cells. J Cell Physiol. 2005;205:114–22.

Sera Y, LaRue AC, Moussa O, Mehrotra M, Duncan JD, Williams CR, Nishimoto E, Schulte BA, Watson PM, Watson DK, Ogawa M. Hematopoietic stem cell origin of adipocytes. Exp Hematol. 2009;37:1108–20. 1120 e1101–1104.

Skarulis MC, Celi FS, Mueller E, Zemskova M, Malek R, Hugendubler L, Cochran C, Solomon J, Chen C, Gorden P. Thyroid hormone induced brown adipose tissue and amelioration of diabetes in a patient with extreme insulin resistance. J Clin Endocrinol Metab. 2010;95:256–62.

Soyka M. Rimonabant and depression. Pharmacopsychiatry. 2008;41:204–5.

Spalding KL, Arner E, Westermark PO, Bernard S, Buchholz BA, Bergmann O, Blomqvist L, Hoffstedt J, Naslund E, Britton T, Concha H, Hassan M, Ryden M, Frisen J, Arner P. Dynamics of fat cell turnover in humans. Nature. 2008;453:783–7.

Sun K, Kusminski CM, Scherer PE. Adipose tissue remodeling and obesity. J Clin Invest. 2011;121:2094–101.

Tamori Y, Masugi J, Nishino N, Kasuga M. Role of peroxisome proliferator-activated receptor-gamma in maintenance of the characteristics of mature 3T3-L1 adipocytes. Diabetes. 2002; 51:2045–55.

Tanaka T, Yoshida N, Kishimoto T, Akira S. Defective adipocyte differentiation in mice lacking the C/EBPbeta and/or C/EBPdelta gene. EMBO J. 1997;16:7432–43.

Tang QQ, Gronborg M, Huang H, Kim JW, Otto TC, Pandey A, Lane MD. Sequential phosphorylation of CCAAT enhancer-binding protein beta by MAPK and glycogen synthase kinase 3beta is required for adipogenesis. Proc Natl Acad Sci U S A. 2005;102:9766–71.

Tang QQ, Lane MD. Adipogenesis: from stem cell to adipocyte. Annu Rev Biochem. 2012;81: 715–36.

Tang W, Zeve D, Suh JM, Bosnakovski D, Kyba M, Hammer RE, Tallquist MD, Graff JM. White fat progenitor cells reside in the adipose vasculature. Science. 2008;322:583–6.

Tran KV, Gealekman O, Frontini A, Zingaretti MC, Morroni M, Giordano A, Smorlesi A, Perugini J, De Matteis R, Sbarbati A, Corvera S, Cinti S. The vascular endothelium of the adipose tissue gives rise to both white and brown fat cells. Cell Metab. 2012;15:222–9.

Tseng YH, Kokkotou E, Schulz TJ, Huang TL, Winnay JN, Taniguchi CM, Tran TT, Suzuki R, Espinoza DO, Yamamoto Y, Ahrens MJ, Dudley AT, Norris AW, Kulkarni RN, Kahn CR. New role of bone morphogenetic protein 7 in brown adipogenesis and energy expenditure. Nature. 2008;454:1000–4.

van Marken Lichtenbelt WD, Vanhommerig JW, Smulders NM, Drossaerts JM, Kemerink GJ, Bouvy ND, Schrauwen P, Teule GJ. Cold-activated brown adipose tissue in healthy men. N Engl J Med. 2009;360:1500–8.

Vegiopoulos A, Muller-Decker K, Strzoda D, Schmitt I, Chichelnitskiy E, Ostertag A, Berriel Diaz M, Rozman J, Hrabe de Angelis M, Nusing RM, Meyer CW, Wahli W, Klingenspor M, Herzig S. Cyclooxygenase-2 controls energy homeostasis in mice by de novo recruitment of brown adipocytes. Science. 2010;328:1158–61.

Virtanen KA, Lidell ME, Orava J, Heglind M, Westergren R, Niemi T, Taittonen M, Laine J, Savisto NJ, Enerback S, Nuutila P. Functional brown adipose tissue in healthy adults. N Engl J Med. 2009;360:1518–25.

Walden TB, Hansen IR, Timmons JA, Cannon B, Nedergaard J. Recruited vs. nonrecruited molecular signatures of brown, "brite," and white adipose tissues. American journal of physiology. Endocrinol Metab. 2012;302:E19–31.

Wu J, Bostrom P, Sparks LM, Ye L, Choi JH, Giang AH, Khandekar M, Virtanen KA, Nuutila P, Schaart G, Huang K, Tu H, van Marken Lichtenbelt WD, Hoeks J, Enerback S, Schrauwen P, Spiegelman BM. Beige Adipocytes Are a Distinct Type of Thermogenic Fat Cell in Mouse and Human. Cell. 2012;150(2):366–76.

Xu Y, Zhou YL, Erickson RL, Macdougald OA, Snead ML. Physical dissection of the CCAAT/enhancer-binding protein alpha in regulating the mouse amelogenin gene. Biochem Biophys Res Commun. 2007;354:56–61.

Zhang JW, Tang QQ, Vinson C, Lane MD. Dominant-negative C/EBP disrupts mitotic clonal expansion and differentiation of 3T3-L1 preadipocytes. Proc Natl Acad Sci U S A. 2004;101: 43–7.

Zhu Y, Qi C, Korenberg JR, Chen XN, Noya D, Rao MS, Reddy JK. Structural organization of mouse peroxisome proliferator-activated receptor gamma (mPPAR gamma) gene: alternative promoter use and different splicing yield two mPPAR gamma isoforms. Proc Natl Acad Sci U S A. 1995;92:7921–5.

Zimmerlin L, Donnenberg VS, Pfeifer ME, Meyer EM, Peault B, Rubin JP, Donnenberg AD. Stromal vascular progenitors in adult human adipose tissue. Cytometry A. 2010;77:22–30.

Zingaretti MC, Crosta F, Vitali A, Guerrieri M, Frontini A, Cannon B, Nedergaard J, Cinti S. The presence of UCP1 demonstrates that metabolically active adipose tissue in the neck of adult humans truly represents brown adipose tissue. FASEB J. 2009;23:3113–20.

Zuk PA, Zhu M, Mizuno H, Huang J, Futrell JW, Katz AJ, Benhaim P, Lorenz HP, Hedrick MH. Multilineage cells from human adipose tissue: implications for cell-based therapies. Tissue Eng. 2001;7:211–28.

Chapter 3
Stromal Vascular Cells

Sahohime Matsumoto and Ichiro Manabe

Abstract In addition to being an energy reservoir, adipose tissue is now regarded as an active endocrine organ that is crucial to the maintenance of systemic energy homeostasis. However, obese adipose tissue, particularly visceral adipose tissue, is thought to be a key contributor to the development of insulin resistance and the clinical consequences of metabolic syndrome, including cardiovascular disease. In addition to the adipocytes themselves, adipose tissue also contains vascular cells and a variety of other cells within the stroma, as well as extracellular matrix (ECM), which provides mechanical support for the tissue. These stromal cells and ECM play important roles in the dynamic processes involved in adipose tissue development, expansion, and inflammation.

Keywords Inflammation • Adipose tissue remodeling • Immune cells • Macrophages • Endothelial cell dysfunction • ECM • Adipogenesis • Hypoxia

3.1 Introduction

It is well established that obesity, particularly visceral obesity, is crucially involved in increasing the clinical risk of metabolic and cardiovascular diseases. For that reason, visceral adipose tissue is thought to play a prominent role in the development of metabolic syndrome and its clinical consequences. Adipose tissue has long been viewed as a passive energy reservoir. But with the discovery of leptin and subsequent identification of other adipose tissue-derived humoral mediators (e.g., adiponectin and resistin), collectively referred to as adipokines (de Heredia et al. 2012), it has become clear that adipose tissue is also an active endocrine organ.

S. Matsumoto • I. Manabe (✉)
Department of Cardiovascular Medicine, Graduate School of Medicine,
The University of Tokyo, Tokyo, Japan
e-mail: manabe-tky@umin.ac.jp

Y. Cao (ed.), *Angiogenesis in Adipose Tissue*, DOI 10.1007/978-1-4614-8069-3_3,

By storing and releasing of lipids and secreting adipokines, healthy adipose tissue maintains the metabolic homeostasis in the body. On the other hand, obese visceral adipose tissue contributes to the development of systemic metabolic dysfunction and its complications, such as cardiovascular disease. This is in part because obesity alters the expression of adipokines (Hotamisligil et al. 1993; Steppan et al. 2001; Weisberg et al. 2006). In addition, obesity induces inflammation within adipose tissue, which not only modifies the functionality of adipose tissue but also affects its sensitivity to insulin and alters inflammatory processes in distant tissues (Hotamisligil 2006; Lumeng and Saltiel 2011).

Adipose tissue consists of adipocytes and a variety of cells residing in the stroma. To understand the physiological and pathological functions of adipose tissue, one must address adipose tissue not merely as a mass of adipocytes but as a complex organ, within which stromal vascular cells are continuously communicating with one another. For instance, to function as a lipid reservoir, blood vessels must be able to collect lipids from adipocytes and then release them into the circulation. Consistent with that scenario, it has been shown that adipocytes and vascular cells interact with one another during the development of adipose tissue (Hausman and Richardson 2004).

Expansion of adipose tissue in obesity is also a dynamic process involving complex interplay between adipocytes and stromal cells. For instance, angiogenesis is essential for the generation of new adipocytes (adipogenesis) during obesity (Cao 2007; Nishimura et al. 2007; Hagberg et al. 2010; Tran et al. 2012). Consequently, inhibiting angiogenesis suppresses adipose tissue expansion (Rupnick et al. 2002; Brakenhielm et al. 2004). In addition to vascular cells, cells residing within the stroma include adipocyte progenitor/stem cells, fibroblasts, and various immune cells. These cells are also likely required for the development and physiological function of adipose tissue. However, they may also actively contribute to the pathological alterations seen in obese adipose tissue. For example, macrophages and lymphocytes within obese adipose tissue secrete proinflammatory cytokines. In fact, the major producers of many of the proinflammatory adipokines once thought to be produced by adipocytes are actually immune cells within the stroma (Fain 2010). Here we review the functions of adipose tissue stromal cells in the development of adipose tissue and its expansion and pathology in obesity.

3.2 Adipose Tissue Development and Blood Vessels

During fetal development, adipocytes differentiate from the mesoderm germ line, and in many species subcutaneous adipose depots develop before the visceral depots (Poulos et al. 2010). First identification of adipocytes can be made based on the appearance of small lipid droplets surrounding the cell nucleus in fetuses (Ailhaud et al. 1992). In pigs, presumptive mesenchymal fat tissue develops into primitive fat cell organs or lobules that increase in number and size throughout fetal development. Notably, there is a close spatial and temporal relationship between the developing adipocytes and the associated capillaries during fetal adipose tissue

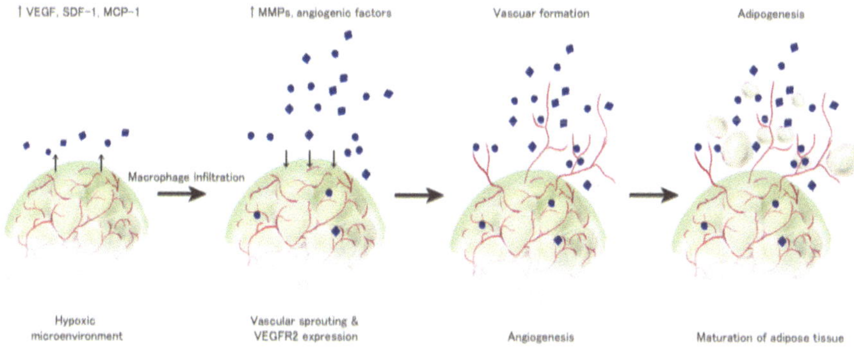

Fig. 3.1 A schematic representation of the interaction between macrophage infiltration, dense vascular network formation, and the development of adipose tissue in the tip of epididymal adipose tissue. A hypoxic microenvironment induces factors that recruit macrophages within the tip area of adipose tissue. These macrophages secrete matrix metalloproteases, angiogenic factors, and cytokines that subsequently initiate the maturation of the vascular network. The mature blood vessel networks allow triglycerides to be provided to adipocyte progenitor cells that are induced to grow into mature adipocytes

development (Crandall et al. 1997; Hausman and Richardson 2004), with morphological development of the vasculature clearly preceding overt adipocyte differentiation within internal fat depots. These observations strongly suggest that there is close interplay between adipogenesis and angiogenesis during adipose tissue development.

Similarly, adipogenesis that occurs during postnatal expansion of fat depots and obesity also requires vascular development, so that inhibiting angiogenesis suppresses fat mass expansion in adult mice (Rupnick et al. 2002; Brakenhielm et al. 2004). Cho et al. reported that the formation of a dense vascular network precedes the appearance of adipocytes in the tip portion of epididymal adipose tissue in adult mice (Cho et al. 2007). They also showed that LYVE-1^+ macrophages initially accumulate in the tip portion, where they secrete matrix metalloproteinases (MMPs) and angiogenic factors, thereby stimulating angiogenesis (Fig. 3.1). We also found that adipogenesis is closely linked to the accumulation of isolectin$^+$ cells and angiogenesis in obese epididymal fat pads (Nishimura et al. 2007). Mechanistically, there appears to be reciprocal regulation of adipogenesis and angiogenesis, at least in part via paracrine interactions between endothelial cells and preadipocytes (Fukumura et al. 2003; Nishimura et al. 2007). These reports also suggest that inhibiting VEGF signaling not only reduces angiogenesis but also inhibits adipocyte differentiation. It thus appears that the generation of adipocytes and blood vessels is tightly interlinked during the formation and development of adipose tissue and the expansion of adipose depots in obesity.

This raises the question, how is angiogenesis triggered in adipose tissue? One hypothesis is that obesity without matched expansion of the vascular supply may induce local hypoxia (Kabon et al. 2004; Hosogai et al. 2007; Trayhurn et al. 2008; Regazzetti et al. 2009), which in turn activates hypoxia-inducible factor (HIF)-1 (Nishimura et al. 2008; Wood et al. 2009; Sun et al. 2011) (Fig. 3.2). HIF1 alters

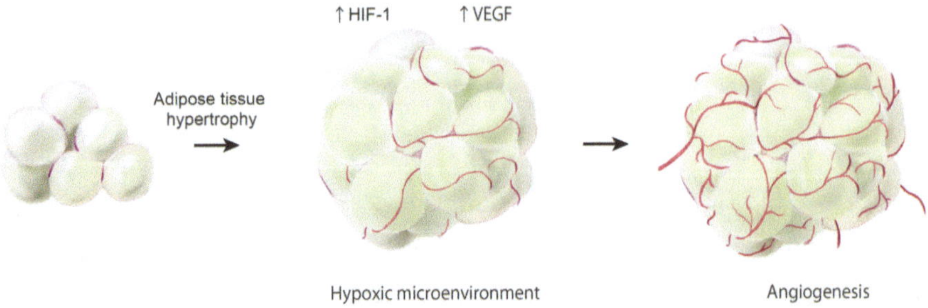

Fig. 3.2 A model of the interplay between adipose tissue hypoxia and angiogenesis. Adipose tissue expansion decreases vascular density and induces local hypoxia that activates HIF-1. HIF-1 is a key regulator of the cellular response to hypoxia, and activates angiogenesis

expression of numerous genes involved in the hypoxic response, including pro-angiogenic factors and cellular metabolic regulators. However, not all available results are consistent with that idea, and it is still unclear to what extent cells are subject to hypoxia within obese adipose tissue. In addition recent studies have demonstrated inflammatory signals can also activate HIF-1α.

3.3 Adipose Tissue Remodeling and Stromal Cells

Physiologically, adipose tissue is a major energy storage organ, which stores excess energy as lipids and then releases it as needed. Compared to glycogen storage, triacylglycerol (TG) storage in adipose tissue has the advantage of being a long-term energy buffer for individuals (Coleman et al. 2000). The amount of TG stored in adipose tissue is determined by the balance between energy intake and expenditure. Another important function of lipid storage in adipose tissue is the suppression of adverse effects of lipid accumulation in non-adipose tissues and lipotoxicity. Marked reductions in adipose tissue mass, such as that observed in lipodystrophy, lead to metabolic dysfunction because the lipids that spill out of the adipose tissue accumulate in non-adipose tissue, where they induce steatosis and lipotoxicity (Fiorenza et al. 2011).

To dynamically store and mobilize lipids in response to ever-changing demand and supply, adipose tissue has an extremely high capacity to expand its mass by increasing adipocyte size (hypertrophy) and number (hyperplasia). During these dynamic morphological transformations of adipose tissue, extracellular matrix (ECM), stromal components, and blood vessels need to be altered in concert with the hypertrophy and hyperplasia of adipocytes. In fact, adipocytes are enmeshed in a dense network of ECM (Sun et al. 2011), which not only provides mechanical support for adipose tissue but also regulates physiological and pathological processes within the tissue. During adipose tissue expansion, the ECM is actively remodeled. Khan et al. reported that genetic deletion of collagen VI resulted in the uninhibited expansion of individual adipocytes and was associated with substantial improvement in systemic metabolism in obese mice (Khan et al. 2009).

On the other hand, a high-fat diet triggered collagenolytic activity in adipose tissue (Chun et al. 2010).

MMPs are the proteolytic enzymes responsible for degradation of ECM proteins, such as collagen, proteoglycans, and elastin. Four MMPs (MMP2, MMP9, MMP14, and MT1-MMP) are expressed by adipocytes and stromal vascular cells during adipocyte expansion (Lijnen et al. 2002; Chavey et al. 2003). Chun et al. showed that the modulation of pericellular collagen rigidity is essential for maturation of adipocytes (Chun et al. 2006). Thus dynamic regulation of ECM appears to be integral to adipogenesis and to adipose tissue expansion.

3.4 Adipose Tissue Inflammation and Stromal Cells

It is now clear that obesity induces inflammation within visceral adipose tissue, and that it involves production of various proinflammatory cytokines and infiltration by immune cells (Hotamisligil et al. 1993; Steppan et al. 2001; Weisberg et al. 2003). It is also becoming increasingly clear that adipose tissue inflammation contributes to the development of systemic insulin resistance and the complications of metabolic syndrome (Cancello et al. 2004; Hotamisligil 2006; Weisberg et al. 2006). However, it remains unclear how inflammation is initiated in adipose tissue, though several models have been proposed: (1) Hypertrophic adipocytes may produce chemokines, including MCP-1, that recruit immune cells (Fig. 3.3a). (2) Adipocyte death may promote infiltration by macrophages that remove components of the dead adipocytes. Indeed, macrophages form crown-like structures surrounding dead adipocytes (Cinti et al. 2005; Pajvani et al. 2005; Nishimura et al. 2007) (Fig. 3.3b). (3) Hypoxia may upregulate expression of proinflammatory mediators, including MIF, MMPs, and VEGF in adipocytes (Chen et al. 2006; Hosogai et al. 2007; Ye et al. 2007) (Fig. 3.3c). (4) Fatty acids may activate macrophages (Orr et al. 2012). (5) T cells may be activated by lipid antigens (Lumeng et al. 2009). Although the initial events in adipose tissue inflammation remain unknown, it is likely that those mechanisms are involved in the progression of inflammatory processes. Cross talk between adipocytes and stromal cells is of particular interest in that regard. Suganami et al. proposed a paracrine loop formed through the production of free fatty acids by adipocytes and TNF-α by macrophages (Suganami et al. 2005). In addition, we showed that adipocytes, $CD8^+$ T cells, and macrophages interact with one another to activate and augment inflammatory signaling (Nishimura et al. 2009).

3.5 Functions of Adipose Stromal Cells

As we have seen, stromal cells play several roles in the physiology and pathology of adipose tissue. In the following sections, we present an overview of the function of each stromal cell type.

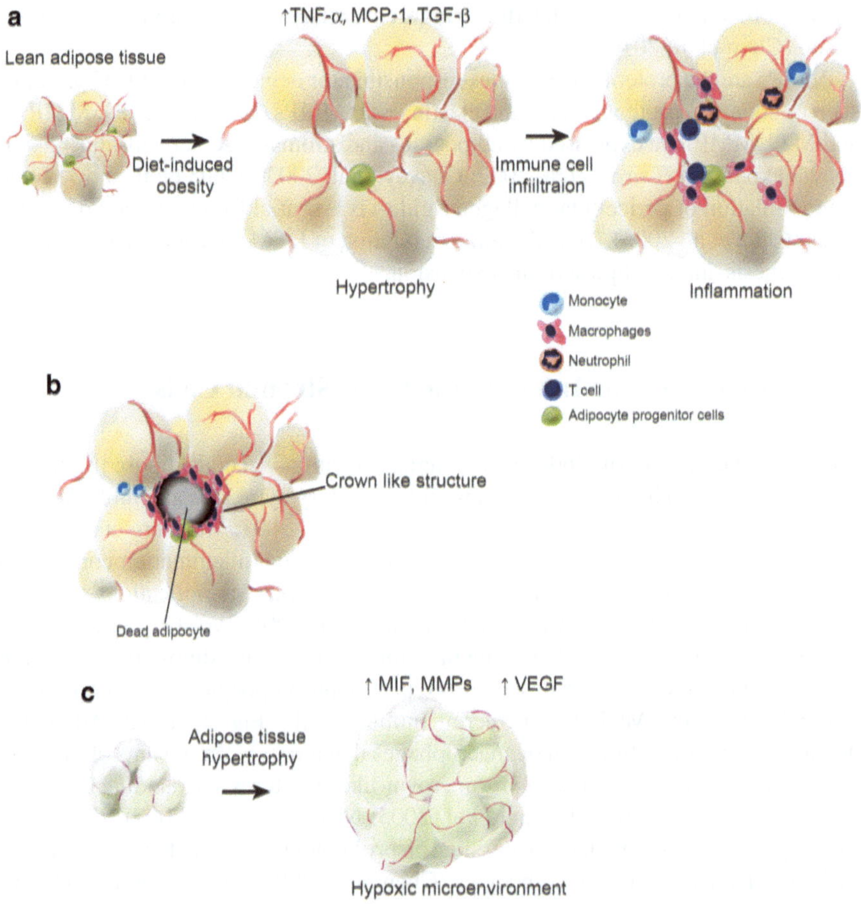

Fig. 3.3 Hypothetical models for the initiation of adipose tissue inflammation. (**a**) Hypertrophic, obese adipocytes produce proinflammatory mediators (e.g., MCP1) that induce immune cell infiltration into adipose tissue. (**b**) Dead adipocytes promote the infiltration of macrophages that remove residual components of dead adipocytes. (**c**) Local hypoxia in adipose tissue upregulates expression of proinflammatory mediators

3.5.1 Adipocyte Progenitor Cells

Adipocyte progenitor cells reside in the adipose tissue stroma and can locally differentiate into adipocytes (Cawthorn et al. 2012). Several different sets of surface markers for identification of adipocyte progenitor cells have been reported. For instance, Rodeheffer et al. identified $Lin^{-}(CD31/CD45/Ter119)^{-}CD34^{+}CD29^{+}Sca\text{-}1^{+}CD24^{+}$ cells as adipocyte progenitors in mice (Rodeheffer et al. 2008), whereas Lin et al. identified $PPAR^{+}CD34^{+}Sca\text{-}1^{+}CD45^{-}Ter119^{-}$ cells to be adipocyte progenitors able to proliferate and differentiate into adipocytes (Lin et al. 2008b). Thus a definitive set of adipocyte progenitor/stem cell markers has still not been identified.

Nonetheless, the results of recent studies point to a cell population that can differentiate into adipocytes in vitro and within obese adipose tissue in adult animals.

Recent studies have also revealed the location of adipocyte progenitor cells within the adipose tissue structure. Tang et al. showed that adipocyte progenitor cells reside in the mural cell compartment of the adipose tissue vasculature. They therefore predicted that there is cross talk between adipocyte progenitors and endothelial cells (Tang et al. 2008). Moreover, morphological evidence indicates that adipose progenitors physically surround vascular endothelial cells in a fashion similar to pericytes and function as a modulators of endothelial cell proliferation and vascular contractility (Lin et al. 2008a; Traktuev et al. 2008). Adipocyte progenitor cells also reportedly express pericyte surface markers, including CD146, NG2, and PDGF-Rβ, whereas they lack hematopoietic, endothelial, and myogenic cell markers (Crisan et al. 2008; Lin et al. 2008a; Traktuev et al. 2008). These findings indicate that adipocyte progenitor cells exist, at least in part, within the perivascular cell population.

Adipocyte progenitor cells residing in the adipose stroma have also gained considerable attention in the context of therapeutic application because they have the potential to differentiate into multiple lineages, including osteoblasts (Zheng et al. 2006), skeletal myocytes (Di Rocco et al. 2006), neural cells, and endothelial cells (Ning et al. 2006), in addition to adipocytes (Chamberlain et al. 2007).

3.5.2 *Endothelial Cells*

Endothelial cells, which play a critical role in the exchange of leukocytes, nutrients, and oxygen between blood and tissues, are one of the major populations of adipose stromal cells and play an important role in adipogenesis. Furthermore, their dysfunction likely contributes to adipose tissue dysfunction and pathology. For example, endothelial cell dysfunction inhibits angiogenesis, leading to hypoxia, as mentioned. In addition, we found that endothelial cells in obese epididymal adipose tissue express high levels of adhesion molecules (Nishimura et al. 2007), which enhances the interaction between endothelial cells and leukocytes, perturbing blood flow. Similarly, pro-angiogenic and proinflammatory gene expression is reportedly increased in obese visceral adipose tissue in humans (Villaret et al. 2010). Endothelial dysfunction also alters transendothelial transport and exclusion (Chudek and Wiecek 2006). Thus endothelial cells play major roles in both the physiology and pathology of adipose tissue.

3.5.3 *Macrophages*

Macrophages are thought to be the major effector cell type in adipose tissue inflammation and insulin resistance (Weisberg et al. 2003; Zhu et al. 2003). Weisberg et al. reported that adipose tissue macrophages (ATMs) are derived from bone marrow

(Weisberg et al. 2003). Later, Lumeng et al. noted two major types of ATMs in adipose tissue (Lumeng et al. 2007). As in other tissues, those macrophages can be classified as M1 (classically activated) and M2 (alternatively activated). Resident ATMs in lean adipose tissue primarily exhibit an M2-like phenotype and express the anti-inflammatory cytokine IL-10 and the M2 markers Ym1 and arginase 1 (Lumeng et al. 2007). By contrast, the macrophages that accumulate in obese adipose tissue express higher levels of proinflammatory cytokines, including TNF-α, IL-1β, IL-6, and iNOS, indicating that in obese adipose tissue the macrophage activation polarity is shifted toward M1. IL-10 has been shown to prevent the effects of TNF-α on adipocytes that express IL-10 receptor (Lumeng et al. 2007), suggesting that whereas M2 ATMs restrain inflammation, M1 ATMs promote and execute inflammatory processes.

Odeggard et al. and Kang et al. showed that macrophage-specific *Pparg* and *Ppard* knockout increased expression of M1 markers and reduced expression of M2 markers in the epididymal fat of diet-induced obese (DIO) mice (Odegaard et al. 2007; Kang et al. 2008). In addition, when they were cocultured with 3T3-L1 adipocytes, bone marrow-derived macrophages expressed IL-10 along with M2 markers in a partially PPARδ-dependent manner. Furthermore, the thiazolidinedione (TZD) drugs used to treat diabetes are PPARγ ligands, and macrophage PPARγ is required for the full antidiabetic actions of TZDs (Hevener et al. 2007). Collectively, these results suggest that therapeutic modulation of macrophage activation could lead to novel strategies for treating diabetes.

Finally, a recent paper presented the surprising finding that ATMs synthesize and secrete catecholamines (Nguyen et al. 2011). Exposure to the cold increases the numbers of adipose tissue M2 macrophages, which secrete catecholamines, thereby inducing expression of thermogenic genes in brown adipose tissue and lipolysis in white adipose tissue. These changes promote thermogenesis for adaptation to the cold.

3.5.4 Other Immune Cells

In addition to macrophages, obesity also leads to the activation of other types of immune cells, including $CD4^+$ Th1 cells (Winer et al. 2009), $CD8^+$ T cells (Nishimura et al. 2009), B cells (Winer et al. 2011), mast cells (Liu et al. 2009), natural killer (NK) cells (Ohmura et al. 2010), invariant NKT (iNVT) cells (Lynch et al. 2012), and neutrophils (Talukdar et al. 2012). On the other hand, as with M2 macrophages, numbers of $CD4^+$ regulatory T (Treg) cells and Th2 cells are not increased in obese adipose tissue, resulting in a reduction in the relative size of their fraction, as compared to the proinflammatory cell types (Feuerer et al. 2009; Winer et al. 2009).

$CD4^+$ Th1 cells and $CD8^+$ T cells recruit macrophages to adipose tissue by producing cytokines such as IFN-γ (Nishimura et al. 2009; Winer et al. 2009; Strissel et al. 2010). B cells in visceral fat also promote adipose tissue inflammation through production of pathogenic IgG antibodies (Winer et al. 2011). Similarly, mast cells appear to contribute to the progression of adipose tissue inflammatory processes (Liu et al. 2009), and neutrophils reportedly accumulate early in response to a high-fat diet and then recruit inflammatory macrophages through secretion of neutrophil elastase (Talukdar et al. 2012).

Conversely, Treg cells apparently suppress adipose tissue inflammation (Feuerer et al. 2009; Ilan et al. 2010), and a recent study (Cipolletta et al. 2012) showed that PPAR-γ is important for the function of adipose tissue Treg cells. Lynch et al. reported that iNKT cells are enriched in obese adipose tissue, and exert beneficial effects via anti-inflammatory cytokine production (Lynch et al. 2012). iNKT cells also recruit M2 macrophages via IL-4 (Ji et al. 2012a, b). Wu et al. showed that eosinophils are also major IL4-expressing cells and recruit M2 macrophages in an IL4/IL13-dependent manner (Wu et al. 2011). These data indicate that homeostasis and inflammation within adipose tissue are controlled through a complex network of immune cells, adipocytes, and other stromal cell types.

3.6 Concluding Remarks

The findings summarized in this review indicate that adipose stromal cells play key roles in the dynamic processes of adipose tissue development, expansion, and inflammation. Stromal cells also crucially contribute to the maintenance of homeostasis within adipose tissue. To accomplish this, they continuously communicate with adipocytes to regulate the physiological and pathological processes ongoing within adipose tissue. Therefore, to better understand the mechanisms involved in regulating adipose tissue physiology and pathology, it will be important to further elucidate the functions of stromal cells and the pathways via which they communicate with the other cell types within adipose tissue.

References

Ailhaud G, Grimaldi P, et al. Cellular and molecular aspects of adipose tissue development. Annu Rev Nutr. 1992;12(1):207–33.

Brakenhielm E, Cao R, et al. Angiogenesis inhibitor, TNP-470, prevents diet-induced and genetic obesity in mice. Circ Res. 2004;94(12):1579–88.

Cancello R, Tounian A, et al. Adiposity signals, genetic and body weight regulation in humans. Diabetes Metab. 2004;30(3):215–27.

Cao Y. Angiogenesis modulates adipogenesis and obesity. J Clin Invest. 2007;117(9):2362–8.

Cawthorn WP, Scheller EL, et al. Adipose tissue stem cells meet preadipocyte commitment: going back to the future. J Lipid Res. 2012;53(2):227–46.

Chamberlain G, Fox J, et al. Concise review: mesenchymal stem cells: their phenotype, differentiation capacity, immunological features, and potential for homing. Stem Cells. 2007; 25(11):2739–49.

Chavey C, Mari B, et al. Matrix metalloproteinases are differentially expressed in adipose tissue during obesity and modulate adipocyte differentiation. J Biol Chem. 2003;278(14):11888–96.

Chen B, Lam KS, et al. Hypoxia dysregulates the production of adiponectin and plasminogen activator inhibitor-1 independent of reactive oxygen species in adipocytes. Biochem Biophys Res Commun. 2006;341(2):549–56.
Cho CH, Koh YJ, et al. Angiogenic role of LYVE-1-positive macrophages in adipose tissue. Circ Res. 2007;100(4):e47–57.
Chudek J, Wiecek A. Adipose tissue, inflammation and endothelial dysfunction. Pharmacol Rep. 2006;58(Suppl):81–8.
Chun TH, Hotary KB, et al. A pericellular collagenase directs the 3-dimensional development of white adipose tissue. Cell. 2006;125(3):577–91.
Chun TH, Inoue M, et al. Genetic link between obesity and MMP14-dependent adipogenic collagen turnover. Diabetes. 2010;59(10):2484–94.
Cinti S, Mitchell G, et al. Adipocyte death defines macrophage localization and function in adipose tissue of obese mice and humans. J Lipid Res. 2005;46(11):2347–55.
Cipolletta D, Feuerer M, et al. PPAR-gamma is a major driver of the accumulation and phenotype of adipose tissue Treg cells. Nature. 2012;486(7404):549–53.
Coleman RA, Lewin TM, et al. Physiological and nutritional regulation of enzymes of triacylglycerol synthesis. Annu Rev Nutr. 2000;20:77–103.
Crandall DL, Hausman GJ, et al. A review of the microcirculation of adipose tissue: anatomic, metabolic, and angiogenic perspectives. Microcirculation. 1997;4(2):211–32.
Crisan M, Yap S, et al. A perivascular origin for mesenchymal stem cells in multiple human organs. Cell Stem Cell. 2008;3(3):301–13.
de Heredia FP, Gomez-Martinez S, et al. Obesity, inflammation and the immune system. Proc Nutr Soc. 2012;71(2):332–8.
Di Rocco G, Iachininoto MG, et al. Myogenic potential of adipose-tissue-derived cells. J Cell Sci. 2006;119(Pt 14):2945–52.
Fain JN. Release of inflammatory mediators by human adipose tissue is enhanced in obesity and primarily by the nonfat cells: a review. Mediators Inflamm. 2010;2010:513948.
Feuerer M, Herrero L, et al. Lean, but not obese, fat is enriched for a unique population of regulatory T cells that affect metabolic parameters. Nat Med. 2009;15(8):930–9.
Fiorenza CG, Chou SH, et al. Lipodystrophy: pathophysiology and advances in treatment. Nat Rev Endocrinol. 2011;7(3):137–50.
Fukumura D, Ushiyama A, et al. Paracrine regulation of angiogenesis and adipocyte differentiation during in vivo adipogenesis. Circ Res. 2003;93(9):e88–97.
Hagberg CE, Falkevall A, et al. Vascular endothelial growth factor B controls endothelial fatty acid uptake. Nature. 2010;464(7290):917–21.
Hausman GJ, Richardson RL. Adipose tissue angiogenesis. J Anim Sci. 2004;82(3):925–34.
Hevener AL, Olefsky JM, et al. Macrophage PPAR gamma is required for normal skeletal muscle and hepatic insulin sensitivity and full antidiabetic effects of thiazolidinediones. J Clin Invest. 2007;117(6):1658–69.
Hosogai N, Fukuhara A, et al. Adipose tissue hypoxia in obesity and its impact on adipocytokine dysregulation. Diabetes. 2007;56(4):901–11.
Hotamisligil GS. Inflammation and metabolic disorders. Nature. 2006;444(7121):860–7.
Hotamisligil GS, Shargill NS, et al. Adipose expression of tumor necrosis factor-alpha: direct role in obesity-linked insulin resistance. Science. 1993;259(5091):87–91.
Ilan Y, Maron R, et al. Induction of regulatory T cells decreases adipose inflammation and alleviates insulin resistance in ob/ob mice. Proc Natl Acad Sci U S A. 2010;107(21):9765–70.
Ji Y, Sun S, et al. Short term high fat diet challenge promotes alternative macrophage polarization in adipose tissue via natural killer T cells and interleukin-4. J Biol Chem. 2012a;287(29):24378–86.
Ji Y, Sun S, et al. Activation of natural killer T cells promotes M2 Macrophage polarization in adipose tissue and improves systemic glucose tolerance via interleukin-4 (IL-4)/STAT6 protein signaling axis in obesity. J Biol Chem. 2012b;287(17):13561–71.
Kabon B, Nagele A, et al. Obesity decreases perioperative tissue oxygenation. Anesthesiology. 2004;100(2):274–80.

Kang K, Reilly SM, et al. Adipocyte-derived Th2 cytokines and myeloid PPARdelta regulate macrophage polarization and insulin sensitivity. Cell Metab. 2008;7(6):485–95.
Khan T, Muise ES, et al. Metabolic dysregulation and adipose tissue fibrosis: role of collagen VI. Mol Cell Biol. 2009;29(6):1575–91.
Lijnen HR, Maquoi E, et al. Matrix metalloproteinase inhibition impairs adipose tissue development in mice. Arterioscler Thromb Vasc Biol. 2002;22(3):374–9.
Lin G, Garcia M, et al. Defining stem and progenitor cells within adipose tissue. Stem Cells Dev. 2008a;17(6):1053–63.
Lin YF, Jing W, et al. Identification of osteo-adipo progenitor cells in fat tissue. Cell Prolif. 2008b;41(5):803–12.
Liu J, Divoux A, et al. Genetic deficiency and pharmacological stabilization of mast cells reduce diet-induced obesity and diabetes in mice. Nat Med. 2009;15(8):940–5.
Lumeng CN, Bodzin JL, et al. Obesity induces a phenotypic switch in adipose tissue macrophage polarization. J Clin Invest. 2007;117(1):175–84.
Lumeng CN, Maillard I, et al. T-ing up inflammation in fat. Nat Med. 2009;15(8):846–7.
Lumeng CN, Saltiel AR. Inflammatory links between obesity and metabolic disease. J Clin Invest. 2011;121(6):2111–7.
Lynch L, Nowak M, et al. Adipose tissue invariant NKT cells protect against diet-induced obesity and metabolic disorder through regulatory cytokine production. Immunity. 2012;37(3): 574–87.
Nguyen KD, Qiu Y, et al. Alternatively activated macrophages produce catecholamines to sustain adaptive thermogenesis. Nature. 2011;480(7375):104–8.
Ning H, Lin G, et al. Neuron-like differentiation of adipose tissue-derived stromal cells and vascular smooth muscle cells. Differentiation. 2006;74(9–10):510–8.
Nishimura S, Manabe I, et al. Adipogenesis in obesity requires close interplay between differentiating adipocytes, stromal cells, and blood vessels. Diabetes. 2007;56(6):1517–26.
Nishimura S, Manabe I, et al. In vivo imaging in mice reveals local cell dynamics and inflammation in obese adipose tissue. J Clin Invest. 2008;118(2):710–21.
Nishimura S, Manabe I, et al. CD8+ effector T cells contribute to macrophage recruitment and adipose tissue inflammation in obesity. Nat Med. 2009;15(8):914–20.
Odegaard JI, Ricardo-Gonzalez RR, et al. Macrophage-specific PPARgamma controls alternative activation and improves insulin resistance. Nature. 2007;447(7148):1116–20.
Ohmura K, Ishimori N, et al. Natural killer T cells are involved in adipose tissues inflammation and glucose intolerance in diet-induced obese mice. Arterioscler Thromb Vasc Biol. 2010;30(2): 193–9.
Orr JS, Puglisi MJ, et al. Toll-like receptor 4 deficiency promotes the alternative activation of adipose tissue macrophages. Diabetes. 2012;61(11):2718–27.
Pajvani UB, Trujillo ME, et al. Fat apoptosis through targeted activation of caspase 8: a new mouse model of inducible and reversible lipoatrophy. Nat Med. 2005;11(7):797–803.
Poulos SP, Hausman DB, et al. The development and endocrine functions of adipose tissue. Mol Cell Endocrinol. 2010;323(1):20–34.
Regazzetti C, Peraldi P, et al. Hypoxia decreases insulin signaling pathways in adipocytes. Diabetes. 2009;58(1):95–103.
Rodeheffer MS, Birsoy K, et al. Identification of white adipocyte progenitor cells in vivo. Cell. 2008;135(2):240–9.
Rupnick MA, Panigrahy D, et al. Adipose tissue mass can be regulated through the vasculature. Proc Natl Acad Sci U S A. 2002;99(16):10730–5.
Steppan CM, Bailey ST, et al. The hormone resistin links obesity to diabetes. Nature. 2001; 409(6818):307–12.
Strissel KJ, DeFuria J, et al. T-cell recruitment and Th1 polarization in adipose tissue during diet-induced obesity in C57BL/6 mice. Obesity (Silver Spring). 2010;18(10):1918–25.
Suganami T, Nishida J, et al. A paracrine loop between adipocytes and macrophages aggravates inflammatory changes: role of free fatty acids and tumor necrosis factor alpha. Arterioscler Thromb Vasc Biol. 2005;25(10):2062–8.

Sun K, Kusminski CM, et al. Adipose tissue remodeling and obesity. J Clin Invest. 2011; 121(6):2094–101.
Talukdar S, Oh DY, et al. Neutrophils mediate insulin resistance in mice fed a high-fat diet through secreted elastase. Nat Med. 2012;18(9):1407–12.
Tang W, Zeve D, et al. White fat progenitor cells reside in the adipose vasculature. Science. 2008;322(5901):583–6.
Traktuev DO, Merfeld-Clauss S, et al. A population of multipotent CD34-positive adipose stromal cells share pericyte and mesenchymal surface markers, reside in a periendothelial location, and stabilize endothelial networks. Circ Res. 2008;102(1):77–85.
Tran KV, Gealekman O, et al. The vascular endothelium of the adipose tissue gives rise to both white and brown fat cells. Cell Metab. 2012;15(2):222–9.
Trayhurn P, Wang B, et al. Hypoxia and the endocrine and signalling role of white adipose tissue. Arch Physiol Biochem. 2008;114(4):267–76.
Villaret A, Galitzky J, et al. Adipose tissue endothelial cells from obese human subjects: differences among depots in angiogenic, metabolic, and inflammatory gene expression and cellular senescence. Diabetes. 2010;59(11):2755–63.
Weisberg SP, McCann D, et al. Obesity is associated with macrophage accumulation in adipose tissue. J Clin Invest. 2003;112(12):1796–808.
Weisberg SP, Hunter D, et al. CCR2 modulates inflammatory and metabolic effects of high-fat feeding. J Clin Invest. 2006;116(1):115–24.
Winer S, Chan Y, et al. Normalization of obesity-associated insulin resistance through immunotherapy. Nat Med. 2009;15(8):921–9.
Winer DA, Winer S, et al. B cells promote insulin resistance through modulation of T cells and production of pathogenic IgG antibodies. Nat Med. 2011;17(5):610–7.
Wood IS, de Heredia FP, et al. Cellular hypoxia and adipose tissue dysfunction in obesity. Proc Nutr Soc. 2009;68(4):370–7.
Wu D, Molofsky AB, et al. Eosinophils sustain adipose alternatively activated macrophages associated with glucose homeostasis. Science. 2011;332(6026):243–7.
Ye J, Gao Z, et al. Hypoxia is a potential risk factor for chronic inflammation and adiponectin reduction in adipose tissue of ob/ob and dietary obese mice. Am J Physiol Endocrinol Metab. 2007;293(4):E1118–28.
Zheng B, Cao B, et al. Mouse adipose-derived stem cells undergo multilineage differentiation in vitro but primarily osteogenic and chondrogenic differentiation in vivo. Tissue Eng. 2006; 12(7):1891–901.
Zhu WZ, Wang SQ, et al. Linkage of beta1-adrenergic stimulation to apoptotic heart cell death through protein kinase A-independent activation of Ca2+/calmodulin kinase II. J Clin Invest. 2003;111(5):617–25.

Chapter 4
Vascular Adipose Complex

J. Michael Sorrell

Abstract An extensive, robust microvasculature is required to maintain the metabolic integrity of adipose tissue. This means that adipocytes and vascular endothelial cells maintain a constant communication that is mediated in part through paracrine signaling, a feature that requires cellular proximity. Adipocytes are atypical mesenchymal cells in that they form a basement membrane in the absence of direct contact with epithelial cells. Adipocyte development always occurs in context with a vascular plexus. Vascular endothelial cells interact with stromal support cells to produce a basement membrane complex. Thus, adipose tissue contains two basement membrane systems in close proximity. These two basement membrane systems also develop during the construction of a tissue engineered, pre-vascularized adipose tissue and form an integrated adipocyte/vascular complex. This complex contains the heparan sulfate proteoglycan perlecan. Perlecan is critical for angiogenesis since it sequesters, concentrates, and protects multiple pro-angiogenic bioactive factors. The basement membrane/vascular complex remains stable upon implantation into a live host. Thus, the basement membrane may be a target for control of adipose tissue since it is critical for vascular function.

Keywords Adipose tissue • Vasculature • Basement membrane • Perivascular cells • Tissue engineering

J.M. Sorrell (✉)
Skeletal Research Center, Case Western Reserve University,
Cleveland, OH 44106, USA
e-mail: j.michael.sorrell@gmail.com

Y. Cao (ed.), *Angiogenesis in Adipose Tissue*, DOI 10.1007/978-1-4614-8069-3_4,

4.1 Introduction: Adipocyte/Vascular Complex

The endocrine function of adipocytes depends upon their close physical contact with the microvasculature (Niemelä et al. 2008; Cao 2010). In addition, vascular cells and adipose cells functionally interact with each other via paracrine interactions, a feature that requires cellular proximity (Cao 2010). Thus, mechanisms for establishing and maintaining close cellular contact between these two populations of cells must exist. The work outlined below proposes that a structural and physiological adipocyte/vascular complex exists and is defined by the development of a shared basement membrane structure (Sorrell et al. 2011). Basement membranes are typically found at the interface between epithelial and mesenchymal cells (Nikolova et al. 2008). However, mesenchymal adipocytes are atypical in that they surround themselves with a basement membrane without direct contact with epithelial cells (Napolitano 1963; Hausman and Richardson 2004). Mesenchymal stem cells (MSCs) that are harvested from bone marrow, adipose tissues, and other tissues can be induced to differentiate as adipocytes, and basement membrane formation is an early step in the differentiation program of these cells. Thus, it is necessary to understand the dynamics of basement membrane formation in vivo and in MSC induction.

4.2 Basement Membrane Synthesis and Adipocyte Development

The physical and physiological structure of white adipose tissue displays heterogeneity that depends upon the anatomical location of this tissue (Tavassoli 1976; Niemelä et al. 2008; Sbarbati et al. 2010). The three major types of white adipose tissue are medullary (bone marrow), subcutaneous, and visceral. All three variations follow the same developmental pattern whereby vascular development precedes adipogenesis (Hausman et al. 1991; Hausman and Richardson 2004). Figure 4.1 outlines the initial series of events for in vivo adipogenesis (Napolitano 1963; Smith and Holbrook 1986; Johnson and Holbrook 1989). Capillaries surrounded by a basement membrane reside in a bed of fibroblastic cells that acquire lipid droplets and a basement membrane before they reach maturity. Ultimately, they establish a unilocular configuration and form close physical association with the capillary network. Hausman and others (1991) studied the development of adipose tissue in fetal pigs and found that laminin immunostaining preceded that of type IV collagen. Thus, in adipocytes, laminin expression is one of the earliest markers of cellular differentiation.

The early expression of laminin also applies to the induction of adipogenic cells from mesenchymal stem cells (MSCs and ASCs—adipose stromal fractions). Figure 4.2 demonstrates the expression of laminin on surfaces of stem cells from adipose tissue at day 4 of induction for ASCs. The early induction implies that

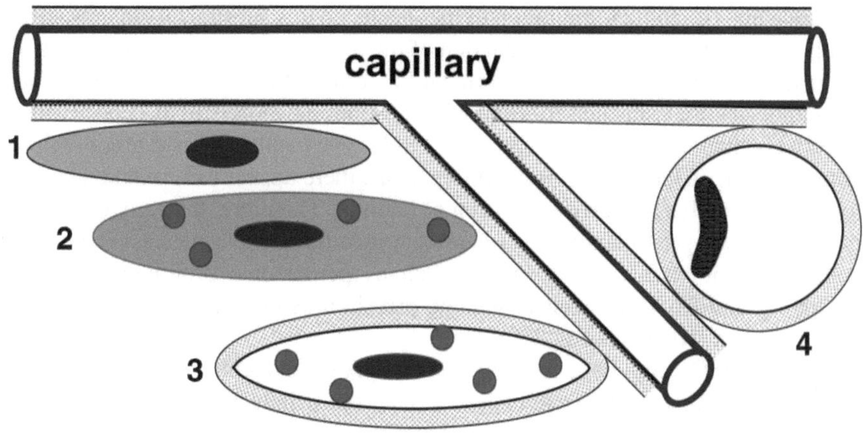

Fig. 4.1 Adipocyte development. Capillary networks appear first at sites of adipose tissue. Fibroblastic cells (**1**) acquire lipid deposits (**2**) and produce a basement membrane (**3**). These cells mature to form unilocular adipocytes (**4**). Vascular structures and adipocytes develop basement membranes early in the development of adipose tissues

Fig. 4.2 Laminin production. An early event in adipocytic differentiation is the expression of laminin at the cell surface. ASCs express surface laminin at 4 days following their induction to adipocytes. Bar = 150 μm

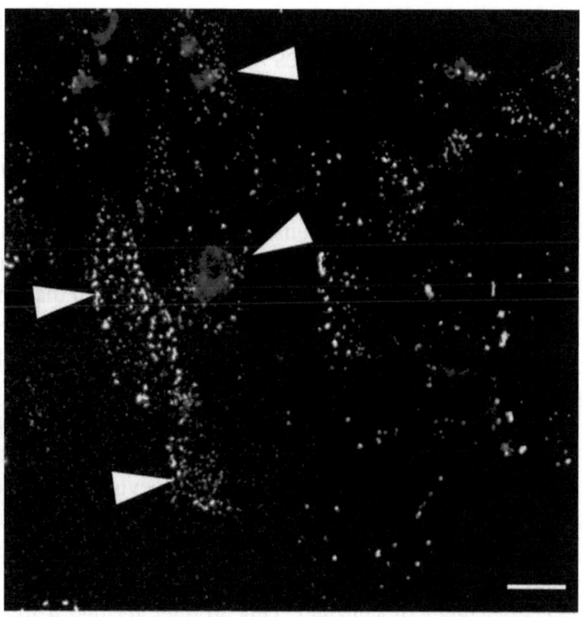

newly differentiated adipogenic cells also express cognitive receptors for laminin. Patrick and others (2003) found that preadipocytes highly expressed CD29 (β1 integrin) and CD61 (β3 integrin), both of which participate in laminin ligation at the cell surface. Further, they found that the attachment of preadipocytes to a laminin substrate and the cellular migration of these cells on this substrate were most effectively inhibited by blocking the function of the $\alpha 1\beta 1$ integrin. Thus, the adipocytic expression of this integrin may be critical for further differentiation of these cells. Preadipocytes that attached to a laminin substrate developed lipid deposits more effectively than when these cells attach to either type I collagen matrices or to tissue culture plastic (Chaubey and Burg 2008). Thus, there appear to be marked physiological reasons for the early expression of laminin. Therefore, the early expression of laminin, and subsequently other basement membrane molecules, can be used in conjuction with transcription factors such as peroxisome proliferator-activated receptor-γ and CCAAT/enhancer-binding protein-α as early markers of adipocyte differentiation (Niemelä et al. 2008).

Human MSCs from bone marrow, upon adipocyte induction, exhibit an upregulation of message for type IV collagen by 3 days post-induction (Sillat et al. 2012). Pericellular deposition of that protein occurred by day 7. Only messages for the $\alpha 1$ and $\alpha 2$ chains of type IV collagen were upregulated and only these chains were immunologically detected. This implies that type IV collagen structure surrounding MSCs would consist of the $\alpha 1\alpha 1\alpha 2$ trimer.

4.3 Expression of Non-basement Membrane Extracellular Matrix Molecules

The in vitro induction of adipogenesis results in a general upregulation in the synthesis and release of a complex set of extracellular matrix molecules that transcends basement membrane molecules (Aratani and Kitagawa 1988). These include: various collagens, proteoglycans, and hyaluronan. The principal proteoglycan in this mixture is the large proteoglycan versican (Zimmermann and Ruoslahti 1989). Versican contains an hyaluronan-binding domain that enables it to form stable complexes with hyaluronan. These complexes have also been proposed to play a role in the promotion of adipogenesis (Zizola et al. 2007). Multiple versicans may bind to a single hyaluronan molecule to form large, water-organizing complexes that increase the viscosity of the culture medium. In addition, proteomic analysis of adipogenic conditioned medium revealed the presence of multiple bioactive factors that include, but are not limited to, hepatocyte growth factor, interleukin-1RII, interleukin-20, lymphotactin, macrophage inflammatory protein-1α, progranulin, and vascular endothelial growth factor. Upon concentration, this complex mixture of matrix and bioactive factors spontaneously gels to produce a product that resembles the consistency of Matrigel. This product, termed Adipogel, has been shown to support the culture of hepatocytes (Sharma et al. 2010). Thus, upon adipogenic conversion, preadipocytes not only initiate basement membrane formation, but they also release potent bioactive factors that can potentially influence the adjacent vascular networks.

In tissue engineered applications, stem cells from various sources have been induced to differentiate towards the adipogenic lineage. The extracellular matrix to which these cells are exposed may play a role in this process. Matrix molecules extracted from adult adipose tissue were used to create a hydrogel support for adipogenesis (Uriel et al. 2008). Placement of preadipocytes onto this hydrogel did not induce adipogeneis; however, adipogenic events were hastened and improved compared with that of cells not in contact with this matrix. Matrigel, which contains basement membrane proteins extracted from a murine tumor, can promote adipogenesis in the presence of FGF-2 (Kimura et al. 2002; Vashi et al. 2006). The role of type IV collagen in adipogenic induction was studied by Mauney and others (2010). They found that native type IV collagen did not support adipogenic conversion of human MSCs. However, denatured type IV collagen was effective in this support. This implies that proteolytic activity at the cell surface may be important in regulating adipogenic conversion. Thus, extracellular matrix molecules play critical roles in the development of adipogenic cells. Decellularized adipose tissue contains multiple extracellular matrix molecules that in combination support adipogenesis (Flynn 2010).

4.4 Cellular Constituents of Adipose Tissue

4.4.1 Introduction

The cellular composition of adipose tissue is important in defining its function. Mature adipose tissue contains multiple cellular constituents: adipocytes, preadipocytes, fibroblasts, vascular endothelial cells, multipotential stem cells, neural cells, and variable numbers of macrophages (Niemelä et al. 2008). All of these cells play a role in normal adipose tissue function. Noncellular extracts from freshly isolated adipose tissue contain a large number of bioactive factors that include those that promote adipogenesis and angiogenesis (Sarkanen et al. 2012). Consequently, it would be helpful to reconstitute a tissue engineered adipose tissue which contains a complex mixture of cells. A pre-vascularized tissue contains at a minimum three cellular components: adipose cells, vascular support cells, and vascular endothelial cells (Sorrell et al. 2011). When adipocytes are induced from MSCs or ASCs, it is first necessary to expose these cells to adipocyte induction medium (AIM). This cannot be accomplished with a mixed population of cells since the vascular support cells also respond to this induction medium. Consequently, the first step is to create an "adipose tissue." Upon the removal of AIM, additional cells, included vascular endothelial cells can then be introduced. Induced adipocytes do not support vascular development. Therefore, a separate vascular support population must be introduced to the system.

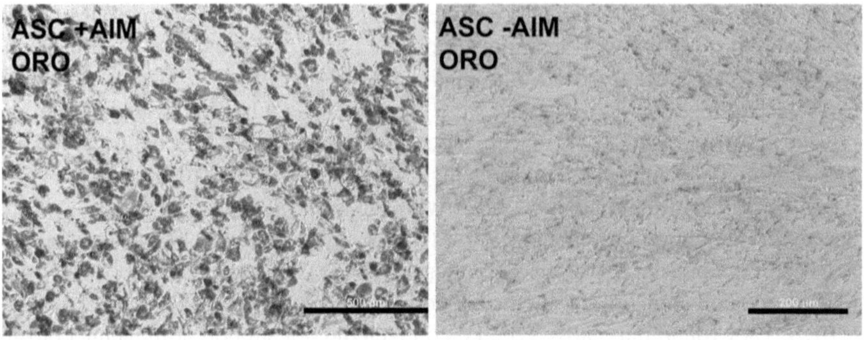

Fig. 4.3 In vitro development of adipose cells. ASCs seeded at high density and cultured in the presence of adipocyte induction medium (AIM) and elevated levels of ascorbate for 2 weeks developed into oil red O positive cells. In the absence of AIM-treatment, oil red O positive cells failed to form. Bars = 500 and 200 μm. Sorrell et al., Biomaterials 32:9667, 2011, reproduced with permission of the Publisher

4.4.2 Adipose Cells/Induction of Mesenchymal Stem Cells

Cells that can be induced towards the adipocyte phenotype reside in multiple tissues that include bone marrow, dermis and hypodermis, and ventral adipose tissue (Pittenger et al. 1999; Vermette et al. 2007; Junker et al. 2010). The removal of adipocytes from white adipose tissue leaves behind a stromal vascular fraction that contains multiple populations of cells. This population can be further purified to produce an ASC which is still not a homogeneous population of cells. These ASCs express many of the same cell surface markers as do bone marrow MSC and they can also be induced to differentiate along adipogenic, chondrogenic, and osteogenic lineages (Gomillion and Burg 2006; Zuk et al. 2001). These ASCs contain a higher proportion of cells that can be differentiated toward the adipogenic lineage than do MSCs. Furthermore, procedures have been developed to convert these cells into tissue engineered adipose tissue (Vermette et al. 2007). First, these cells were cultured at high density in the presence of elevated levels of ascorbate to enhance extracellular matrix production (Hata and Senoo 1989; Berthod et al. 2006). Then the cells were exposed to AIM to induce adipogenic cells (Fig. 4.3). The result was a thin, three-dimensional adipose tissue that served as the base for the addition of subsequent populations of cells (Sorrell et al. 2011).

4.4.3 Vascular Support Cells

Vascular support cells produce a combination of extracellular matrix and bioactive factors that are essential for vascular tubule formation (Sorrell et al. 2008; Merfeld-Clauss et al. 2010). Multiple populations of vascular support cells have been identified, and, in general, these cells possess fibroblast-like characteristics. Nonetheless, not all fibroblast-like cells exhibit vascular support characteristics (Sorrell et al. 2008). Therefore, empirical determinations for optimal vascular support populations are necessary to select cells for this purpose. The selection process was adapted from the work of Bishop and others (1999) who created dense fibroblast lawns to support vascular tubule formation by HUVECs. This same approach has also been employed for MSC and ASC populations (Sorrell et al. 2009; Merfeld-Clauss et al. 2010). Figure 4.4 provides a comparison of vascular tubule formation when different support cells were employed. Dermal fibroblasts obtained from the papillary dermis provided the best support; however, inclusion of MSCs and ASCs did not interfere with this level of support. Because of the stem cell components of MSC and ASC populations, these cells were included in the vascular support cells at a ratio of 1:1:1 of dermal fibroblasts:MSCs:ASCs. This mixture was seeded onto adipose cells and cultured in the presence of elevated ascorbate for 1 week prior to seeding HUVECs. This formulation was used in all subsequent applications.

4.4.4 Vascular Endothelial Cells

HUVECs have routinely been employed in tissue engineering of vascularized tissues. These cells will align and form authentic tubules under co-culture conditions and when properly inserted into athymic rodents (Au et al. 2008; Sanz et al. 2008). Following implantation into athymic hosts, tubules formed by HUVECs have been shown to anastomose with host vasculature and to carry host blood. Fibroblastic cells do not express type IV collagen or laminin unless they are induced to differentiate (Fig. 4.5). An exception to this occurs when fibroblastic cells physically contact epithelial cells such as keratinocytes or vascular endothelial cells. Contact with epithelial cells results in the early enhanced expression of type IV collagen by fibroblasts. However, laminin expression follows type IV collagen expression (Sorrell, unpublished). Subsequently, this translates into the formation of a basement membrane that surrounds vascular tubules. Basement membrane formation by vasculature is essential in that the basement membrane that forms contains pro-angiogenic molecules such as the proteoglycan perlecan. Perlecan is substituted with heparan sulfate chains that sequester, concentrate, and protect a variety of bioactive factors that are associated with angiogenesis (Iozzo and San Antonio 2001). In addition, basement membranes act to stabilize vascular structures. Thus, in co-cultures that do not contain adipose cells, the only basement membranes are those that surround vascular structures.

As discussed previously, induced adipose cells present basement membrane material on their surfaces, which contrasts them with vascular support cells (Fig. 4.6). This patchwork display of basement membrane molecules creates preferred regions upon which HUVECs initially attach. HUVECs contain cell surface receptors that bind to basement membrane which means that they initially attach to adipocyte regions within the co-culture, but then migrate to adjacent vascular

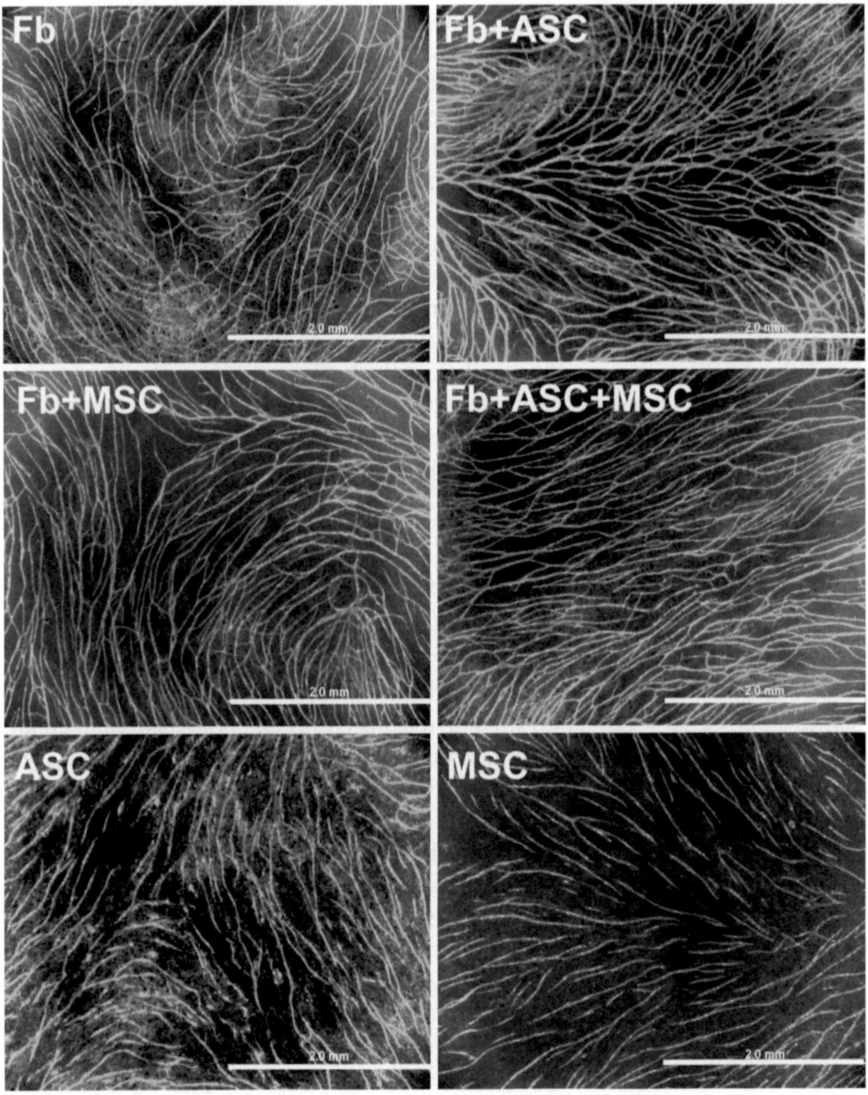

Fig. 4.4 Vascular support cells. Vascular support cells were selected for their ability to support vascular tubule formation in an in vitro angiogenic assay and for the presence of mesenchymal stem cells. A 1:1:1 mixture of dermal fibroblasts, bone marrow MSCs, and ASCs was selected for this purpose. Bars = 2 mm. Sorrell et al., Biomaterials 32:9667, 2011, reproduced with permission of the Publisher

support cells (Fig. 4.7). The initial attachment to adipose cells ensures the proximity of vascular structures to clusters of adipose cells. HUVECs migrate and align on the surfaces of vascular support cells and for stable tubules by day 5–6 following their seeding.

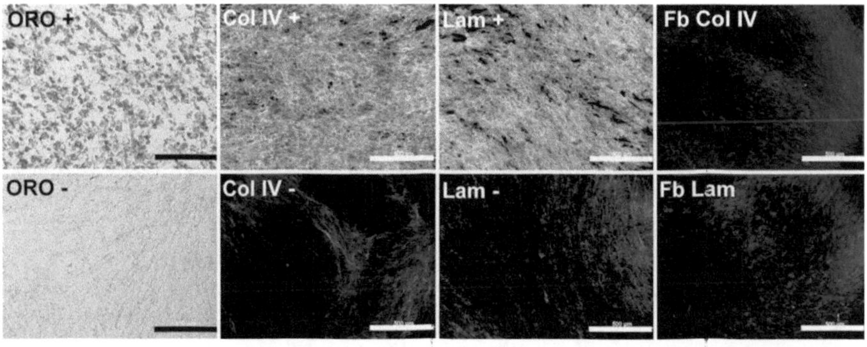

Fig. 4.5 Induction of basement membrane components. Treatment of high density ASC cultures with AIM significantly upregulated the expression of type IV collagen and laminin. In contrast, fibroblast cultures and noninduced ASC cultures did not express these matrix molecules. Bar = 500 μm. Sorrell et al., Biomaterials 32:9667, 2011, reproduced with permission of the Publisher

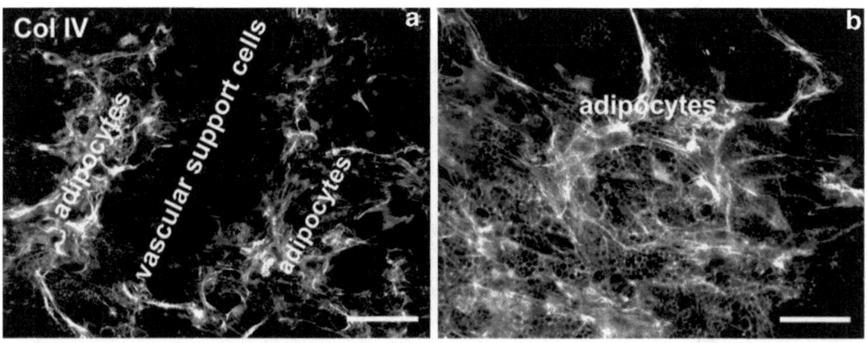

Fig. 4.6 Nonuniform distribution of basement membrane components. Vascular endothelial cells are seeded onto a co-culture with a nonuniform distribution of basement membrane molecules. Bar = 100 and 50 μm. Sorrell et al., Biomaterials 32:9667, 2011, reproduced with permission of the Publisher

4.4.5 *Perivascular Cells*

The microvasculature is stabilized by a population of perivascular cells that are commonly referred to as pericytes (Gerhardt and Betsholtz 2003). In addition, perivascular cells are now known to act as stem cells and to play roles in wound healing (Crisan et al. 2008; Corselli et al. 2010; Zimmerlin et al. 2010). Their introduction into tissue engineered vascularized construct may provide for additional vascular stability (Au et al. 2008; Sanz et al. 2008; Sorrell et al. 2009). Perivascular cells are attracted to neo-vasculature via bioactive factors released from vascular endothelial cells. Platelet-derived growth factor-BB (PDGF-BB) is one of the principal factors (Hirschi et al. 1998). The development of a basement membrane appears to augment pericyte attraction through the sequestration of the heparin-binding PDGF-BB.

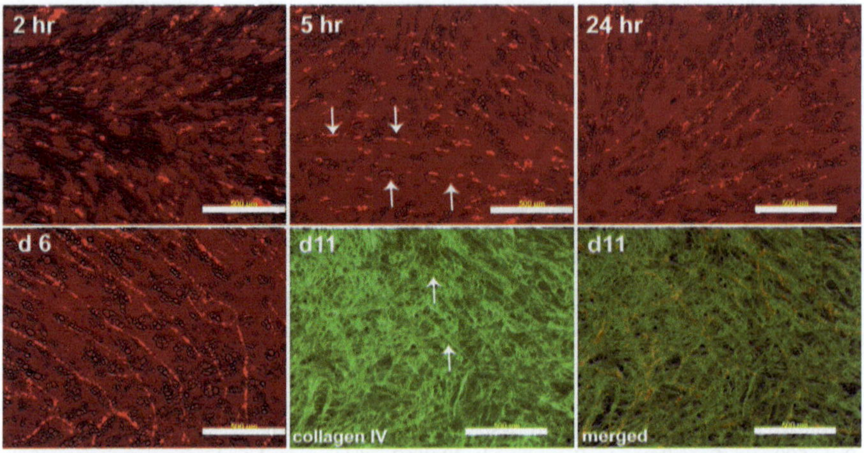

Fig. 4.7 Organization of vasculature. HUVECs seeded onto vascular support/adipocyte co-cultures were tracked in live cultures starting at 2 h following seeding. At 2 h, HUVECs primarily attach to basement membrane regions that are enriched in adipocytes. Subsequently (5 h), these cells migrate onto the adjacent vascular support layer and align (24 h). Vascular tubules have formed by day 6 and an elaborate interaction between vascular and adipocyte basement membranes has formed by day 11. The merged image indicates the location of labeled HUVECs within this matrix complex. Bars = 500 μm. Sorrell et al., Biomaterials 32:9667, 2011, reproduced with permission of the Publisher

Bone marrow-derived MSCs seeded simultaneously with HUVECs or when seeded up to 5 days later will assume a perivascular orientation. Therefore, equal numbers of HUVECs and CM-DiI labeled MSCs were seeded onto the adipocyte/vascular support co-cultures. Significant numbers of perivascular cells were attracted to co-align with vascular structures. Figure 4.8 depicts the organization of these MSCs on day 11 following their seeding. The majority of these cells are in close proximity to vascular components even though both vascular and adipose cells express basement membranes as depicted by the localization of the heparan sulfate proteoglycan perlecan.

As indicated above, it is feasible to fabricate a tissue that contains multiple populations of cells. These cells possess the inherent ability to sort and organize so as to produce a defined pattern that resembles native tissue.

4.5 Adipocyte/Vascular Complex

The physical interaction between vascular structures and adipocytes is not well appreciated when standard immunostaining for CD31 (Fig. 4.8) is employed. However, when co-cultures are immunostained for type IV collagen, perlecan (Figs. 4.8, 4.9, and 4.10), or laminin, the integration of basement membranes becomes apparent. As demonstrated above, two sets of basement membranes are formed separately; the first by adipocytes and the second by the interaction of

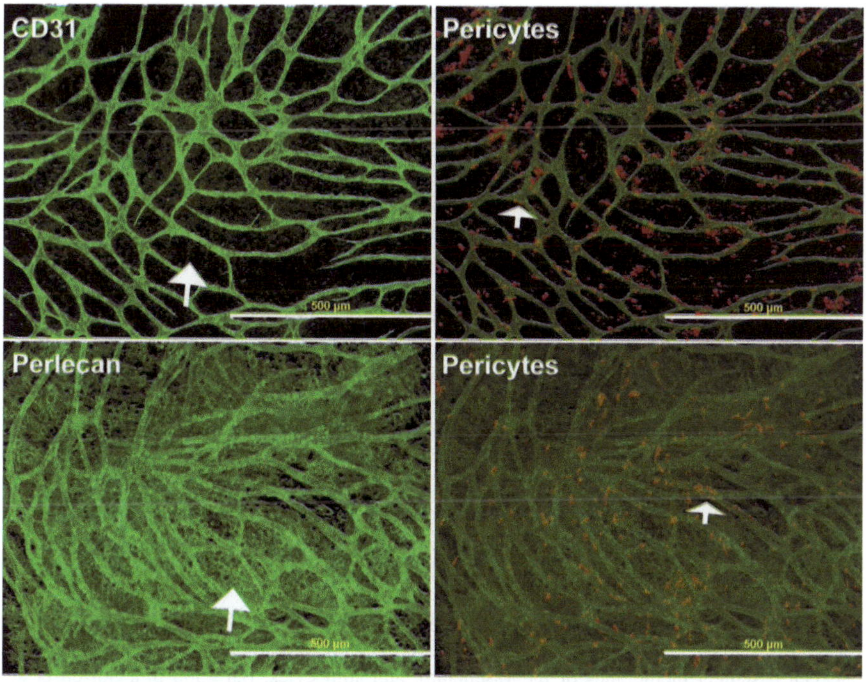

Fig. 4.8 Introduction of perivascular cells. CM-DiI-labeled MSCs were mixed with unlabeled HUVECs and the mixture was seeded onto co-cultures of adipocytes and vascular support cells. At day 12 post-seeding, live cultures were immunostained with an antibody for CD31 and an antibody for the heparan sulfate proteoglycan perlecan. *Arrows* indicate the location of adipocytes, which are surrounded by a basement membrane that contains perlecan. *Arrowheads* indicate the perivascular location of MSCs. Bars = 500 μm

HUVECs with vascular support cells. The close proximity between these two sets of cells facilitates the development of a physically integrated complex that can be visualized in whole cultures and in sections of these cultures.

The basement membrane is a complex structure, consisting of major components such as laminins, type IV collagen, entactin/nidogen, and perlecan (Iozzo 2005). Immunohistochemical staining has demonstrated the presence of type IV collagen, laminin, and perlecan in the basement membranes that surround both vascular structures and adipose cells (Sorrell et al. 2011). Perlecan contains five structural domains each of which mediate interactions with other matrix molecules and with bioactive factors (Iozzo 2005; Bix and Iozzo 2008). The N-terminal domain contains binding sites for heparan sulfate glycosaminoglycan chains and the C-terminal domain contains binding sites for chondroitin sulfate glycosaminoglycan chains. Heparan sulfate chains play an important role in angiogenesis through their interaction with heparin-binding bioactive factors. A partial list of heparin-binding angiogenic factors is presented in Table 4.1.

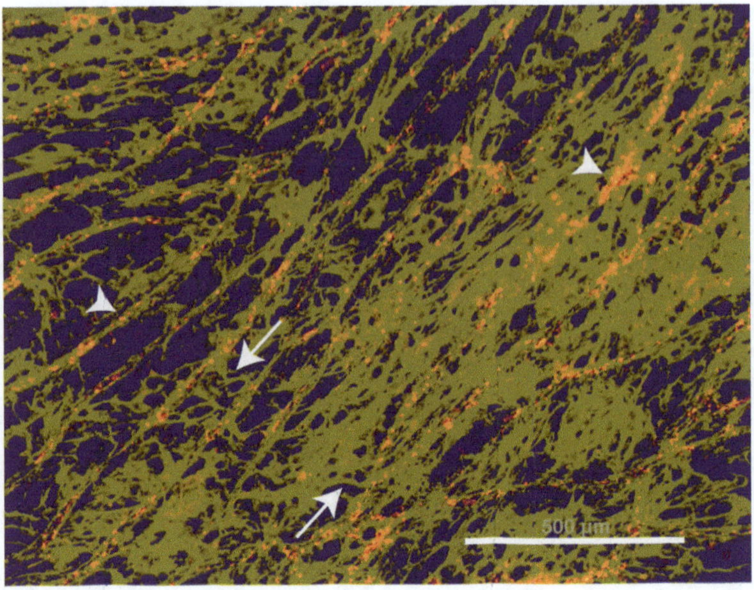

Fig. 4.9 Vascular/adipocyte complex I. A merged image of the day 11 co-culture immunostained for type IV collagen (*green*) and HUVECs (*red*) has been recolored to highlight the interactions between basement membranes of the vasculature and adipocytes. Bar = 500 µm. Sorrell et al., Biomaterials 32:9667, 2011, reproduced with permission of the Publisher

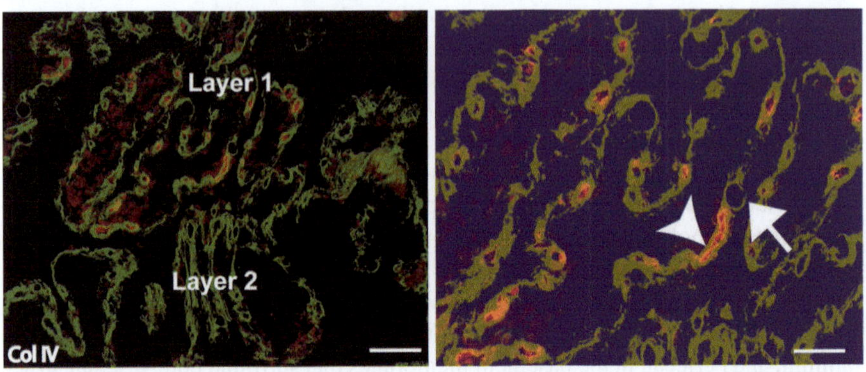

Fig. 4.10 Vascular/adipocyte complex II. Sections of vascularized adipocyte co-cultures are shown at a lower and at a higher magnification. Sections were immunostained for type IV collagen. The vasculature contains CM-DiI HUVECs (*arrowhead*) and adipocytes are indicated by an *arrow*. Bars = 200 and 100 µm. Sorrell et al., Biomaterials 32:9667, 2011, reproduced with permission of the Publisher

Table 4.1 A partial list of heparin-binding angiogenic bioactive factors

Angiogenin
Fibroblast growth factor-2 (also other FGF's bind heparin with varying affinities)
Heparin-binding epidermal growth factor-like growth factor
Hepatocyte growth factor/scatter factor
Human apo-lactoferrin
Insulin-like growth factor binding proteins
Interleukin-1α
Interleukin-1β
Interleukin-8
Placental growth factor
Platelet-derived growth factors
Stromal cell-derived factor-1
Tissue factor
Transforming growth factor- β1
Tumor necrosis factor- α
Vascular endothelial growth factor-A (145-, 165-, and 189 isoforms)

List modified from Norrby (2006)

4.6 Adipose/Vascular Complex Following Implantation

In vitro cultures are useful tools for developing an understanding of biological dynamics; however, determining the fate of such cultures in an in vivo setting is equally important. Consequently, vascularized adipose tissue co-cultures were implanted into SCID mice. Individual cultures were too thin for proper handling. Therefore, cultures were gently detached from the culture dish to allow for contraction and two contracted co-cultures were stacked so that the upper surfaces outwards (Fig. 4.11). The stacked cultures continued to float in the medium, and after 5 days they had fused into a single entity that possessed sufficient bulk for handling purposes. Table 4.2 summarizes the steps and time-sequences involved in the development of these structures. Histological assessment revealed clusters of vascularized adipocytes in a connective tissue matrix (Figs. 4.12 and 4.13).

Following 1 week of subcutaneous implantation, host vasculature invested implants (Fig. 4.14). Immunohistological analyses of implants indicate the presence of mouse vasculature within the implant (Fig. 4.15). Both mouse and human vasculatures were identified using species-specific antibodies for CD31. Both sets of vasculatures within the implant were surrounded by basement membranes (Fig. 4.15). The vascular/adipocyte complex seen in culture appears to remain intact.

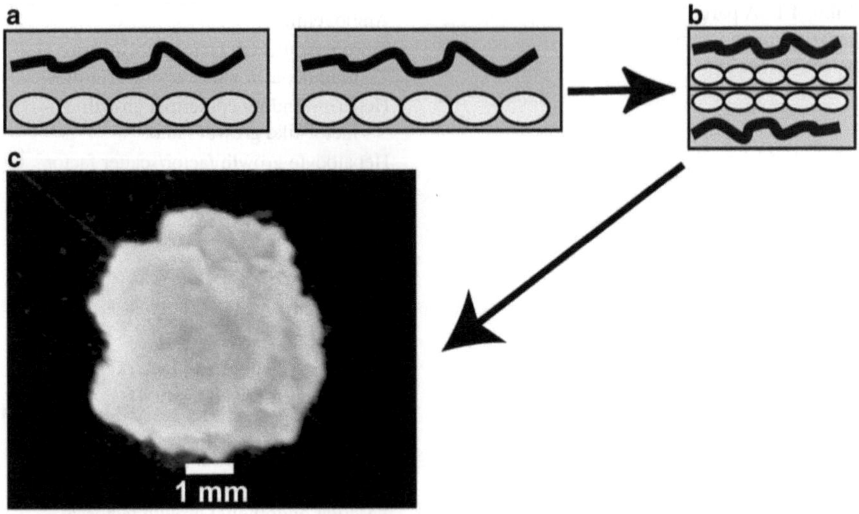

Fig. 4.11 Production of an implantable tissue. Co-cultures set up in 35-mm dishes were detached and allowed to contract. (**a**) Two cultures were stacked so that the upper surfaces faced outwards (**b**). (**c**) The stacked structures were cultured for 4–5 days to allow for fusion of the two layers. Bar = 1 mm. Sorrell et al., Biomaterials 32:9667, 2011, reproduced with permission of the Publisher

Table 4.2 Summary: preparation of a pre-vascularized adipose tissue

1. Create dense ASC fibroblasts cultured in EGM-2 MV medium + ascorbate	1 week
2. Treat with AIM to induce adipocytes	2 weeks
3. Seed with vascular support cells (dermal fibroblasts, ASCs, MSCs) culture in DMEM + ascorbate	1 week
4. Seed with HUVEC-MSC mixture, culture in Bishop's medium (Bishop et al. 1999)	10–14 days
5. Seed with dermal fibroblasts culture in Bishop's medium (Bishop et al. 1999)	1 week
6. Detach from culture dish and allow to contract	Overnight
7. Stack two constructs and allow to fuse in culture	1–2 days
8. Implants ready for histological analyses or for engraftment	

Sorrell et al., Biomaterials 32:9667, 2011, reproduced with permission of the Publisher

4.7 Summary/Conclusions/Perspectives

Adipose tissue is a complex mixture of multiple cell types; however, the two most important elements are the vasculature and adipocytes. These two components initiate adipose tissue development (Napolitano 1963; Hausman et al. 1991) and functionally interact in mature adipose tissues (Fukumura et al. 2003; Armani et al. 2010; Cao 2010; Christiaens and Lijnen 2010). Both of these cellular components independently produce basement membranes that are critical for capturing, protecting, and storing multiple bioactive factors (Bix and Iozzo 2008).

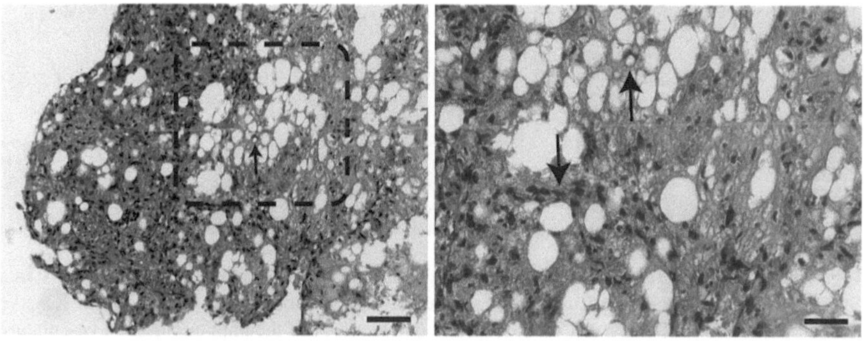

Fig. 4.12 Histological structure of the implantable tissue. The *boxed region* shows a cluster of adipocytes and the *arrow* indicates a vascular tubule within the cluster. At higher magnification, vascular structures are indicated (*arrows*). Tissue was fixed with neutral buffered formalin and sections were stained with hematoxilyn and eosin. Bars = 200 μm; 100 μm

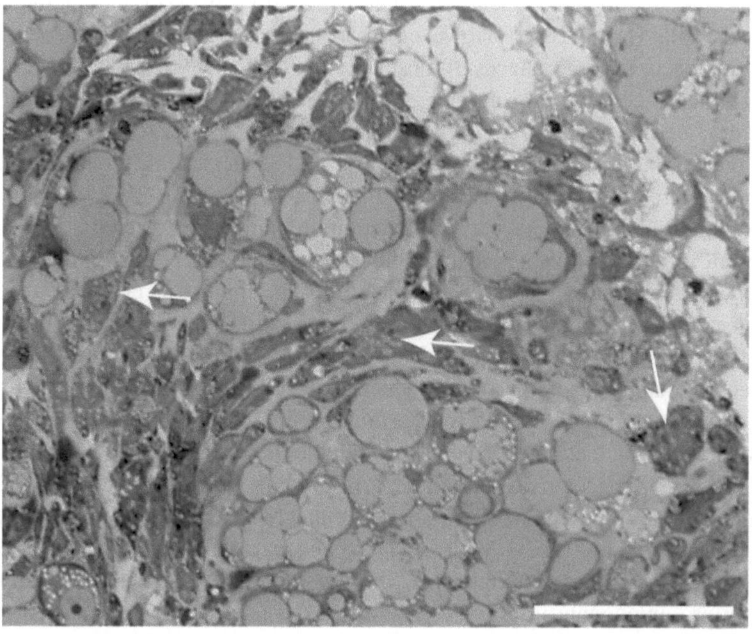

Fig. 4.13 Adipocyte precursors. Clusters of more mature adipocytes also contain adipocyte precursor cells (*arrows*) Tissue was fixed with glutaraldehyde and osmium tetroxide and sections were stained with toluidine blue. Bar = 50 μm

The creation of a co-culture that contains both adipogenic cells and vascular cells provides a platform to study the development of a basement membrane network that physically integrates adipocytes and vasculature to establish an adipocyte/vascular complex. This complex is stable in culture over a period of weeks and

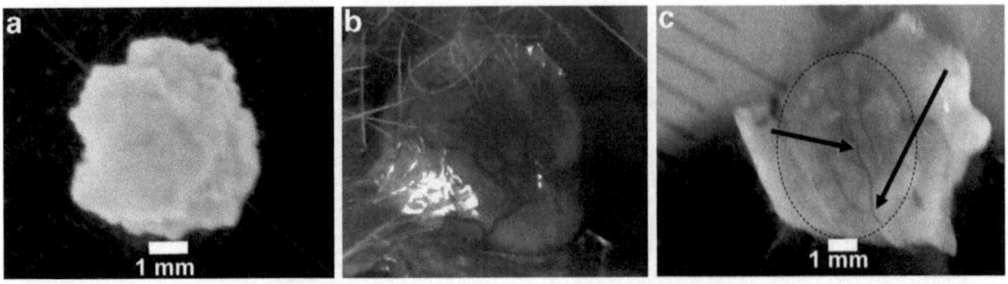

Fig. 4.14 Implant at 1 week. (**a**) Pre-vascularized adipose tissue implanted subcutaneously in a SCID mouse (work performed using an approved IACUC protocol from Case Western Reserve University) is shown in place at 1 week (**b**). (**c**) Upon recovery of the implant, the host vasculature is evident (*arrows*). Bars = 1 mm

remains stable following implantation (Sorrell et al. 2011). This has important implications for future studies of adipocyte vascular interactions.

Basement membranes typically form at intersections between mesenchymal and epithelial cells (Iozzo 2005). However, in special situations, basement membrane formation occurs at the surfaces of mesenchymal cells that are not in direct contact with epithelial cells. Two examples of this are in adipose tissue and in cartilage (Nakajima et al. 1998, 2002; Gomes et al. 2004; Bix and Iozzo 2008). In both of these tissues, basement membrane formation acts as an early differentiation event. Laminin expression initiates basement membrane formation by adipocytic cells (Hausman et al. 1991). This early reliance upon laminin expression appears to be important for interactions with extracellular matrix and other cells (Patrick and Wu 2003). This means that the early expression of laminin and its cognitive receptors provides a means to detect newly differentiated adipocytes and to separate these cells from non-differentiated cells. This ability to segregate newly differentiated cells could then be adapted for the production of tissue engineered adipose tissues.

Adipocytes and preadipocytes release multiple angiogenic factors such as vascular endothelial growth factor-A, placental growth factor, fibroblast growth factor-2, and hepatocyte growth factor that are all heparin-binding bioactive factors (Cao 2010). Heparan sulfate in basement membranes are attached to three heparan sulfate proteoglycans, perlecan, type XVIII collagen, and agrin (Iozzo 2005; Kruegel and Miosge 2010). These proteoglycans, through their heparan sulfate chains, are actively pro-angiogenic in their sequestration and protection of secreted heparin-binding angiogenic factors (Bix and Iozzo 2008; Zoeller et al. 2009). Heparin-binding facilitates the establishment of gradients that are critical in cellular signaling. The extracellular enzyme heparanase is an endo-β-glucuronidase that plays an important role in the establishment of gradients by promoting the release bound bioactive factors (McKenzie 2007). Consequently, the inhibition of this enzyme profoundly inhibits angiogenesis. Perlecan is the most prominent of the basement membrane proteoglycans. It consists of five structural domains, each of which exhibits interactions with other matrix molecules and with multiple bioactive factors (Bix and Iozzo 2008; Melrose et al. 2008; Friedl 2010). This proteoglycan is a component of the basement membrane produced by adipocytes as well as by

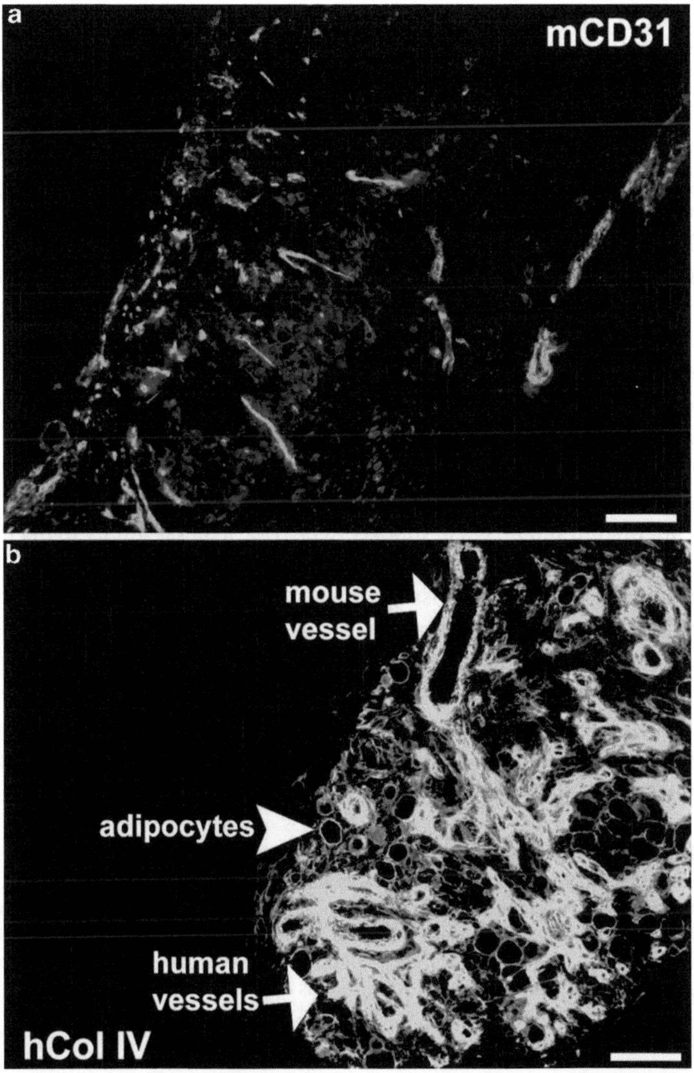

Fig. 4.15 Vasculature within the implant. (**a**) Sections of the implant were immunostained using an antibody specific for mouse CD31. The host vasculature has invaded in the implant. (**b**) Sections of the implant were immunostained using an antibody specific for human type IV collagen. Adipocyte clusters were immunostained (*arrowhead*). Both human and mouse vasculatures were surrounded by human type IV collagen (*arrows*). Bars = 200 μm

interaction of vascular endothelial cells and fibroblasts. In addition, to heparin-binding, perlecan interacts via domain II with connective tissue growth factor, an upstream regulator of transforming growth factor-β system (Grotendorst 1997). Thus, transforming growth factor β-1, while it does not bind heparan sulfate, it is still regulated by perlecan. This factor also exerts pro-angiogenic characteristics (Hirschi et al. 1998). Under special situations, basement membrane proteoglycans may also exert anti-angiogenic characteristics (Bix and Iozzo 2008). This occurs when the C-terminal domains of proteoglycans perlecan and type XVIII collagen are cleaved by proteinases. These actions produces soluble molecules endorepellin and endostatin, respectively. Thus, basement membranes are important not only for maintaining physical proximity but they are also equally important for the regulation of angiogenesis.

Adipose tissue is used as a soft-tissue filler in surgical applications (Patrick 2000). However, adipocytes are highly metabolic cells that require an extensive vascular network to provide nutrients (Verseijden et al. 2010). A failure to establish a sufficient vascular supply results in the resorption of implanted tissue. Tissue engineered adipose tissue is subject to a similar, often more severe level of resorption, due to a lack of a vascular network (Borges et al. 2003; Choi et al. 2010). Therefore, the development of a pre-vascularized adipose tissue presents a means to ameliorate tissue resorption. It is important that the tissue engineered vasculature be robust and stable at the time of implantation. Upon implantation, it is also important that host vascular ingrowth occurs quickly and, ideally, anastomosis of host and implant vasculature should occur quickly to provide a blood supply to the interior of the implant. The tissue engineered, pre-vascularized adipose tissue described here provides a model to study these issues.

The vasculature of adipose tissue has been proposed as a target for the control of obesity (Cao 2010). In the absence of a robust vascular supply, adipose tissue regresses. Therefore, developing a better understanding of the communication network between adipocytes and vasculature is important. Understanding the role of the basement membrane in creating an adipose/vascular complex and in directing intercellular communication is a critical component in this process.

Acknowledgments This work was supported in part by a grant from the Ohio Third Frontier for the Clinical Tissue Engineering Center and it was also supported in part by a grant from the David and Virginia Baldwin Research Fund. The ASCs used for this study were provided by Drs. Keith March (University of Indiana, Indianapolis IN) and Farshid Guliak (Duke University, Durham, NC).

I would also like to thank Marilyn Baber, Dimitry Traktuev, Keith March, and Arnold Caplan who were co-authors of a previous study Biomaterials 32:9667, 2011 from which this chapter was adapted. Selected figures and Table 4.2 were adapted with permission of the publisher (Elsevier, B.V.).

References

Aratani Y, Kitagawa Y. Enhanced synthesis and secretion of type IV collagen and entactin during adipose conversion of 3T3-L1 cells and production of unorthodox laminin complex. J Biol Chem. 1988;263:16163–69.

Armani A, Mammi C, Marzolla V, Calanchini M, Antelmi A, Rosano GMC, Fabbri A, Caprio M. Cellular models for understanding adipogenesis, adipose dysfunction, and obesity. J Cell Biochem. 2010;110:564–72.

Au P, Tam J, Fukamura D, Jain RK. Bone marrow-derived mesenchymal stem cells facilitate engineering of long-lasting function vasculature. Blood. 2008;111:4551–8.

Berthod F, Germain L, Tremblay N, Auger FA. Extracellular matrix deposition by fibroblasts is necessary to promote capillary-like tube formation in vitro. J Cell Physiol. 2006;207:491–8.
Bishop ET, Bell GT, Bloor S, Broom IJ, Hendry NFK, Wheatley DN. An in vitro model of angiogenesis: basic features. Angiogenesis. 1999;3:335–44.
Bix G, Iozzo RV. Novel interactions of perlecan: unraveling perlecan's role in angiogenesis. Microsc Res Tech. 2008;71:339–48.
Borges J, Mueller MC, Padron T, Tegtmeier F, Lang EM, Stark GB. Engineered adipose tissue supplied by functional microvessels. Tissue Eng. 2003;9:1263–70.
Cao Y. Adipose tissue angiogenesis as a therapeutic target for obesity and metabolic diseases. Nat Rev Drug Discov. 2010;9:107–15.
Chaubey A, Burg KJL. Extracellular matrix components as modulators of adult stem cell differentiation in an adipose system. J Bioact Compat Polym. 2008;23:20–37.
Choi JH, Gimble JM, Lee K, Marra KG, Rubin JP, Yoo JJ, Vunjak-Novakovic G, Kaplan DL. Adipose tissue engineering for soft tissue regeneration. Tissue Eng Part B. 2010;16:414–26.
Christiaens V, Lijnen HR. Angiogenesis and development of adipose tissue. Mol Cell Endocrinol. 2010;318:2–9.
Corselli M, Chen CW, Crisan M, Lazzari L, Peault B. Perivascular ancestors of adult multipotent stem cells. Arterioscler Thromb Vasc Biol. 2010;30:1104–9.
Crisan M, Yap S, Casteilla L, Chen C-W, Corselli M, Park TS, Andriolo G, Sun B, Zheng B, Zhang L, Norotte C, Teng P-N, Traas J, Schugar R, Deasy BM, Badylak S, Buhring H-J, Giacobino J-P, Lazzari L, Huard J, Peault B. A perivascular origin for mesenchymal stem cells in multiple human organs. Cell Stem Cell. 2008;3:301–13.
Flynn LE. The use of decellularized adipose tissue to provide an inductive microenvironment for the adipogenic differentiation of human adipose-derived stem cells. Biomaterials. 2010;31: 4715–24.
Friedl A. Proteoglycans: master modulators of paracrine fibroblast-carcinoma cell interactions. Semin Cell Dev Biol. 2010;21:66–71.
Fukumura D, Ushiyama A, Duda DG, Xu L, Tam J, Chatterjee KK, Garkavtsev I, Jain RK. Paracrine regulation of angiogenesis and adipocyte differentiation during in vivo adipogenesis. Circ Res. 2003;93:e88–97.
Gerhardt H, Betsholtz C. Endothelial-pericyte interactions in angiogenesis. Cell Tissue Res. 2003;314:15–23.
Gomes RR, Farach-Carson MC, Carson DD. Perlecan functions in chondrogenesis: insights from in vitro and in vivo models. Cells Tissues Organs. 2004;176:79–86.
Gomillion CT, Burg KJL. Stem cells and adipose tissue engineering. Biomaterials. 2006;27: 6052–63.
Grotendorst GR. Connective tissue growth factor: a mediator of TGF-β action on fibroblasts. Cytokine Growth Factor Rev. 1997;8:171–9.
Hata R-I, Senoo H. L-ascorbic adid 2-phosphate stimulates collagen accumulation, cell proliferation, and formation of a three-dimensional tissue-like substance by skin fibroblasts. J Cell Physiol. 1989;138:8–16.
Hausman GJ, Richardson RL. Adipose tissue angiogenesis. J Anim Sci. 2004;82:925–34.
Hausman GJ, Wright JT, Thomas GB. Vascular and cellular development in fetal adipose tissue: lectin binding studies and immunocytochemistry for laminin and type IV collagen. Microvasc Res. 1991;41:111–25.
Hirschi KK, Rohovsky SA, D'Amore PA. PDGF, TGF-β, and heterotypic cell-cell interactions mediate endothelial cell-induced recruitment of 10T1/2 cells and their differentiation to a smooth muscle fate. J Cell Biol. 1998;141:805–14.
Iozzo RV. Basement membrane proteoglycans: from cellar to ceiling. Nat Rev Mol Cell Biol. 2005;6:646–56.
Iozzo RV, San Antonio JD. Heparan sulfate proteoglycans: heavy hitters in the angiogenesis arena. J Clin Invest. 2001;108:349–55.
Johnson CL, Holbrook KA. Development of human embryonic and fetal dermal vasculature. J Invest Dermatol. 1989;93:10S–17S.

Junker JPE, Sommar P, Skog M, Johnson H. Adipogenic, chondrogenic and osteogenic differentiation of clonally derived human dermal fibroblasts. Cells Tissues Organs. 2010;191:105–18.
Kimura Y, Ozeki M, Inamoto T, Tabata Y. Time course of de novo adipogenesis in Matrigel by gelatin microspheres incorporating basic fibroblast growth factor. Tissue Eng. 2002;8:603–13.
Kruegel J, Miosge N. Basement membrane components are key players in specialized extracellular matrices. Cell Mol Life Sci. 2010;67:2879–95.
Mauney J, Volloch V. Human bone marrow-derived stromal cells show highly efficient stress-resistant adipogenesis on denatured collagen IV matrix but not on its native counterpart: implications for obesity. Matrix Biol. 2010;29:9–14.
McKenzie EA. Heparanase: a target for drug discovery in cancer and inflammation. Br J Pharmacol. 2007;151:1–14.
Melrose J, Hayes AJ, Whitelock JM, Little CB. Perlecan, the "jack of all trades" proteoglycan of cartilaginous weight-bearing connective tissues. Bioessays. 2008;30:457–69.
Merfeld-Clauss S, Collahalli N, March KL, Traktuev DO. Adipose tissue progenitor cells directly interact with endothelial cells to induce vascular network formation. Tissue Eng Part A. 2010;16:2953–66.
Nakajima I, Yamaguchi T, Ozutsumi K, Aso H. Adipose tissue extracellular matrix: newly organized by adipocytes during differentiation. Differentiation. 1998;63:193–200.
Nakajima I, Muroya S, Tanabe R-I, Chikuni K. Extracellular matrix development during differentiation into adipocytes with a unique increase in type V and VI collagen. Biol Cell. 2002;94: 197–203.
Napolitano L. The differentiation of white adipose cells. An electron microscope study. J Cell Biol. 1963;18:663–79.
Niemelä S, Miettinen S, Sarkanen JR, Ashammakhi N. Adipose tissue and adipocyte differentiation: molecular and cellular aspects and tissue engineering applications. In: Ashammakhi N, Reis R, Chiellini F, editors. Topics in tissue engineering, vol. 4. Oulu, Finland: University of Oulu; 2008. p. 1–26.
Nikolova G, Strilic B, Lammert E. The vascular niche and its basement membrane. Trends Cell Biol. 2008;17:19–25.
Norrby K. Low-molecular-weight heparins and angiogenesis. APMIS. 2006;114:79–102.
Patrick CW. Adipose tissue engineering: the future of breast and soft tissue reconstruction following tumor resection. Semin Surg Oncol. 2000;19:302–11.
Patrick CW, Wu X. Integrin-mediated preadipocyte adhesion and migration on laminin-1. Ann Biomed Eng. 2003;31:505–14.
Pittenger MF, Mackay AM, Beck SC, Jaiswal RK, Douglas R, Mosca JD, Moorman MA, Simonetti DW, Craig S, Marshak DR. Multilineage potential of adult human mesenchymal stem cells. Science. 1999;284:143–7.
Sanz L, Santos-Valle P, Alonso-Camino V, Salas C, Serrano A, Vicario JL, Cuesta AM, Compte M, Sanchez-Martin D, Alvarez-Vallina L. Long-term in vivo imaging of human angiogenesis: critical role of bone marrow-derived mesenchymal stem cells for the generation of durable blood vessels. Microvasc Res. 2008;75:308–14.
Sarkanen J-R, Kaila V, Mannerström B, Räty S, Kuokkanen H, Miettinen S, Ylikomi T. Human adipose tissue extract induces angiogenesis and adipogenesis in vitro. Tissue Eng Part A. 2012;18:17–25.
Sbarbati A, Accorsi D, Benati D, Marchetti L, Orsini G, Rigotti G, Panettiere P. Subcutaneous adipose tissue classification. Eur J Histochem. 2010;54:e48.
Sharma NS, Nagrath D, Yarmush ML. Adipocyte-derived basement membrane extract with biological activity: applications in hepatocyte functional augmentation in vitro. FASEB J. 2010;24:2364–74.
Sillat T, Saat R, Pollanen R, Hukkanen M, Takagi M, Konttinen YT. Basement membrane collagen type IV expression by human mesenchymal stem cells during adipogenic differentiation. J Cell Mol Med. 2012;16:1485–95.
Smith LT, Holbrook KA. Embryogenesis of the dermis in human skin. Pediatr Dermatol. 1986;3:271–80.

Sorrell JM, Baber MA, Caplan AI. Human dermal fibroblast subpopulations: differential interactions with vascular endothelial cells in coculture: non-soluble factors in the extracellular matrix influence interactions. Wound Repair Regen. 2008;16:300–9.

Sorrell JM, Baber MA, Caplan AI. Influence of adult mesenchymal stem cells on in vitro vascular formation. Tissue Eng Part A. 2009;15:1751–61.

Sorrell JM, Baber MA, Traktuev DO, March KL, Caplan AI. The creation of an in vitro adipose tissue that contains a vascular-adipocyte complex. Biomaterials. 2011;32:9667–76.

Tavassoli M. Ultrastructural development of bone marrow adipose cells. Acta Anat. 1976;94: 65–77.

Uriel S, Huang J-J, Moya ML, Francis ME, Wang R, Chang S-Y, Cheng M-H, Brey EM. The role of adipose protein derived hydrogels in adipogenesis. Biomaterials. 2008;29:3712–19.

Vashi AV, Abberton KM, Thomas GP, Morrison WA, O'Connor AJ, Cooper-White JJ, Thompson EW. Adipose tissue engineering based on the controlled release of fibroblast growth factor-2 in a collagen matrix. Tissue Eng. 2006;12:3035–43.

Vermette M, Trottier V, Ménard V, Saint-Pierre L, Roy A, Fradette J. Production of a new tissue-engineered adipose substitute from human adipose-derived stromal cells. Biomaterials. 2007; 28:2850–60.

Verseijden F, Posthumus-van Sluijs SJ, Pavijasevic P, Hofer SOP, van Osch GJVM, Farrell E. Adult human bone marrow- and adipose tissue-derived stromal cells support the formation of prevascular-like structures from endothelial cells in vitro. Tissue Eng Part A. 2010; 16:101–14.

Zimmerlin L, Donnenberg VS, Pfeifer ME, Meyer EM, Peault B, Rubin JP, Donnenberg AD. Stromal vascular progenitors in adult human adipose tissue. Cytometry A. 2010;77A:22–30.

Zimmermann D, Ruoslahti E. Multiple domains of the large fibroblast proteoglycan, versican. EMBO J. 1989;8:2975–81.

Zizola CF, Julianelli V, Bertolesi G, Yanagishita M, Calvo JC. Role of versican and hyaluronan in the differentiation of 3T3-L1 cells into preadipocytes and mature adipocytes. Matrix Biol. 2007;26:419–30.

Zoeller JJ, Whitelock JM, Iozzo RV. Perlecan regulates developmental angiogenesis by modulating the VEGF-VEGFR2 axis. Matrix Biol. 2009;28:284–91.

Zuk PA, Zhu M, Mizuno H, Huang J, Futrell JW, Katz AJ, Benhaim P, Lorenz HP, Hedrick MH. Multilineage cells from human adipose tissue: implications for cell-based therapies. Tissue Eng. 2001;7:211–28.

Part III
Adipose Vasculature

Chapter 5
Blood Vessels in White and Brown Adipose Tissues

Sharon Lim, Jennifer Honek, and Yihai Cao

Abstract The adipose tissue is composed of two major subtypes, white adipose tissue (WAT) and brown adipose tissue (BAT). While WAT mainly stores excess energy in, BAT plays a role in burning energy and generating heat in a process known as non-shivering thermogenesis involving uncoupling protein-1 (UCP-1). Apart from its role in energy homeostasis, adipose tissue also exerts an endocrine function secreting a number of hormones and factors. Among those are pro- and anti-angiogenic factors involved in vascular remodeling. The vasculature plays a major role in the adipose tissue and both WAT and BAT are highly vascularized.

Keywords Adipokines • Brown adipose tissue • White adipose tissue • Inflammatory cells • Anti-angiogenic factors • Pro-angiogenic factors • Vascular plasticity

5.1 Structures and Vascular Density in White Adipose Tissue and Brown Adipose Tissue

Adipose tissue is highly dynamic and has multifunctional roles in the body. Adipose tissue can be categorized into white adipose tissue (WAT) and brown adipose tissue (BAT). WAT, the most commonly known adipose tissue, provides insulation,

---

Sharon Lim and Jennifer Honek contributed equally to this work.

S. Lim • J. Honek
Department of Microbiology, Tumor and Cell Biology, Karolinska Institute, 17177 Stockholm, Sweden

Y. Cao, M.D., Ph.D. (✉)
Department of Microbiology, Tumor and Cell Biology (MTC), Karolinska Institute, 17177 Stockholm, Sweden

Department of Medicine and Health Sciences, Linköping University, 58185 Linköping, Sweden
e-mail: yihai.cao@ki.se

Y. Cao (ed.), *Angiogenesis in Adipose Tissue*, DOI 10.1007/978-1-4614-8069-3_5,

cushions as fat pad around organs providing mechanical protection, and stores excess food consumed as energy in the form of triglyceride in a unilocular lipid droplet.

5.1.1 Structures and Functions of WAT

In the WAT, the large unilocular lipid droplet occupies most of the volume of the cell which pushes the cytoplasm and nucleus to the edge of the cell against the plasma membrane. The plasma membrane has many receptors essential for mediating several cytokines and signaling molecules. The adipocyte in WAT is about 60–130 μm in diameter. The large lipid droplet does not appear to contain intracellular organelles. As storage for fuel, WAT has to respond efficiently to the body's nutritional and energy status. When other tissues in the body are in demand of energy, triglyceride reserves stored in the unilocular lipid droplet in WAT are broken down into energy-rich fatty acids and glycerol (Fig. 5.1). The storage and utilization of surplus energy are tightly regulated by insulin, catecholamines, and leptin (Lafontan and Langin 2009). Apart from adipocytes, the adipose tissue consists of non-adipocyte structures including the stromal vascular cells, fibroblastic connective tissues, infiltrating inflammatory cells (leukocytes, monocytes, macrophages), mast cells, preadipocytes, pericytes (PCs), endothelial cells (ECs), and stem cells. Preadipocytes originate from multipotent stem cells of the mesodermal origin which are capable of differentiating into mature adipocytes (Ouchi et al. 2010; Ohashi et al. 2010).

5.1.1.1 Distribution of WAT

The distribution of WAT in the adult body varies depending on sex and age. There are two major types of WAT, the visceral fat which accumulates around the abdominal cavity and mediastinum, and the subcutaneous fat which accumulates under the skin. In general, adult men tend to accumulate visceral or abdominal fat around the waist, and women tend to accumulate subcutaneous fat around the hips. Excess accumulation of visceral fat around the abdomen has strong positive correlation with obesity and obesity-related metabolic disorders including cardiovascular disease, type 2 diabetes, insulin resistance, and inflammatory diseases (Gerstein et al. 2006). WAT is not merely a depot for energy storage, it is also critical in modulating lipid and glucose metabolism. WAT secretes several hormones and cytokines—known as adipokines that have paracrine or endocrine effects in the body. Visceral fat produces more pro-inflammatory cytokines while subcutaneous fat produces cytokines such as leptin and adiponectin (Wronska and Kmiec 2012).

5.1.2 Structures and Function of BAT

The primary function of WAT is to store energy. In contrast, the main functional role of BAT is thermogenesis. BAT derives its name from the rich vascularized blood

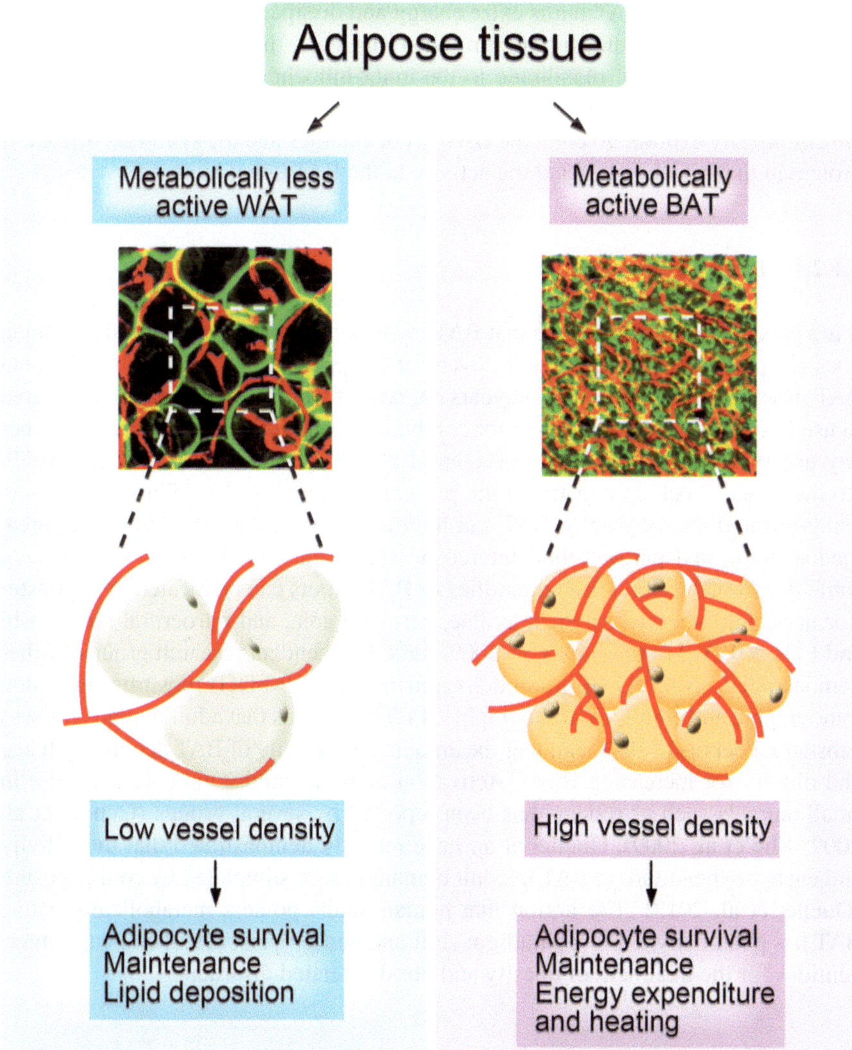

Fig. 5.1 White and brown adipose tissues. Adipose tissue can be categorized into white adipose tissue (WAT) and brown adipose tissue (BAT). WAT being metabolically less active has a low vessel density. WAT mainly stores energy in the form of triglyceride in a unilocular lipid droplet. BAT being highly metabolically active contains higher blood vessel density. Adipocytes (*green* signals), stained by anti-perilipin antibody, are encircled by blood vessels (*red* signals) stained by endothelial marker CD31 antibody

vessels and highly dense number of mitochondria, is multilocular, and has a higher metabolism than the WAT due to the high number of mitochondria. The BAT consists of a round nucleus, granular cytoplasm which contains several small vacuoles of fat droplets with mitochondria localized adjacent to these multilocular fat droplets (Fig. 5.1). The adipocyte in the BAT is approximately 25–40 μm in diameter. BAT is essential in non-shivering thermogenesis (NST) in mammals, newborn humans, and infants. BAT burns extra energy and dissipates heat through a specific mitochondrial protein, uncoupling protein-1 (UCP-1), that transfers protons from the outer mitochondrial membrane to the inner mitochondrial membrane without production of adenosine triphosphate (ATP). When small animals such as rodents are chronically exposed to cold, the BAT depot enlarges and the synthesis of UCP-1 protein in the mitochondria and the activity in the BAT depot increase.

5.1.2.1 Distribution of BAT

It is a generally accepted dogma that BAT involutes postnatally and the adult human does not possess any BAT. The discovery of the presence of active and functional BAT in adult humans in the recent years has revivified immense research in the area to use BAT as a therapeutic target for combating metabolic diseases including obesity and type 2 diabetes (Himms-Hagen et al. 1994; Howe et al. 1992). In 1979, Rothwell and Stock demonstrated the presence of functional BAT in adult human using infrared thermography. BAT can be found in the neck, supraclavicular area, mediastinum, and paravertebral intercostal spaces in adult humans. However, in some occasional cases, minor quantities of BAT depots can be located in the posterior neck, left paratracheal area, axillae, perirenal area, and retrocrural area (Joshi and Lele 2012). After 30 years, in 2009, three independent research groups further demonstrated using ^{18}F-fluorodeoxyglucose (^{18}F-FDG) positron emission tomographic and computed tomographic (PET-CT) scans that adult humans possess substantial depots of BAT, and that the amount and activity of BAT decline with age and obesity (or increasing BMI). Activation of BAT activity via cold exposure in small animals such as rodents has been reported by several groups (Cypess et al. 2009; Xue et al. 2009). Ouellet et al. have recently demonstrated that the activity and energy expenditure in BAT in adult humans can be stimulated by cold exposure (Ouellet et al. 2012). The notion that human adults possess metabolically active BAT has potentially led to a paradigm shift and hence opens new therapeutic interventions for the treatment of obesity and obesity-related disorders.

5.1.3 Vascular Density in WAT and BAT

Similar to other tissues in the body, the growth, expansion, and repair of adipose tissue are highly dependent on blood vessels to supply adipocytes with nutrients, oxygen, growth factors, cytokines, and the removal of waste products. Although WAT is not as highly vascularized as BAT, each adipocyte in the WAT is still in close proximity with at least one capillary which provides adequate blood supply to sustain adipocyte function (Silha et al. 2005). It is therefore not surprising that the highly metabolically active BAT contains a significantly higher number of capillary vessels per adipocyte than WAT, to meet the high demand of oxygen, nutrients, and the removal of metabolic waste products (Fig. 5.1).

5.1.3.1 Development of Adipose Tissue Vasculature

The development of fetal adipocytes is spatially and temporally coupled to capillary development (Crandall et al. 1977). It has been demonstrated that arteriolar differentiation occurs before adipocyte development followed by differentiation of blood vessel extracellular matrix (ECM) and eventually adipocytes' ECM differentiation. During adipogenesis, the regulation of blood vessels could potentially increase or decrease preadipocyte proliferation and migration (Hausman et al. 1996). Crandall et al. described the paracrine interaction between ECs and preadipocytes, where production and secretion of plasminogen activator inhibitor-1 (PAI-1) human preadipocytes regulate preadipocyte and EC migration (Crandall et al. 2000a, b). Expansion of WAT occurs rapidly after birth by increase in adipocyte size as well as increase in the number of adipocytes. Angiogenesis in the adipose tissue is a crucial process and it has been shown to be essential in modulating adipogenesis and obesity (Lijnen 2008; Cao 2007). The vascular density in WAT has been described to be positively correlated with weight and body mass index (BMI). WAT has been demonstrated to be able to transform into brown-like phenotype (BRITE) upon chronic exposure to cold in rodents (Xue et al. 2009). The transition of WAT to acquire BRITE phenotype is accompanied by an angiogenic switch which increased the vascular density in the BRITE.

5.2 Vascular Function in BAT and WAT

The adipose blood vasculature is vital in providing sufficient oxygen and nutrients to the adipocytes, and to drain metabolic waste products from the adipocytes to modulate growth, repair, and maintenance of the adipose tissue. The density of blood vessels in the adipose tissue is dependent on its metabolic status, for instance, the BAT, being highly metabolically active, demands greater supply of blood to provide oxygen and nutrients and to dissipate heat as its metabolic waste product (Fig. 5.2). Blood vessels also provide adipocytes with growth factors and cytokines that play important roles in the regulation of cell proliferation and intercellular functions (Cao 2007). The adipose vasculature serves as a great reservoir of circulating bone marrow-derived mesenchymal stem cells and inflammatory cells including monocytes and neutrophils to the adipocytes to maintain the homeostasis in adipose tissue. Infiltration of inflammatory cells is a common phenomenon observed in obese individuals. Circulating stem cells can be differentiated into preadipocytes and adipocytes. Adipose tissue being an endocrine organ secretes several adipokines, which in turn mediate reciprocal cross talk between adipocytes and ECs to stimulate growth and expansion of adipocytes and induce angiogenesis.

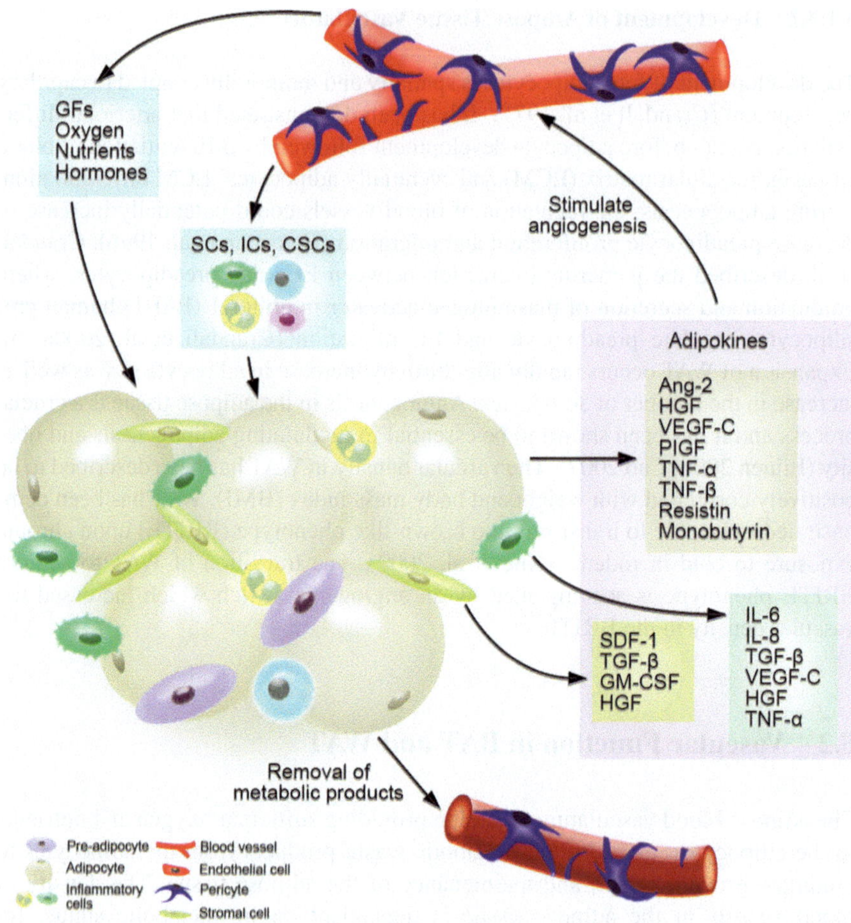

Fig. 5.2 Vascular functions in white and brown adipose tissues. Blood vessels in adipose tissue are crucial in supplying oxygen, nutrients, growth factors, cytokines, inflammatory cells, and mesenchymal stem cells to maintain the homeostasis in adipose tissue. Blood vessels serve as a niche for circulating stem cells which can be differentiated into preadipocytes and adipocytes. These adipocytes and non-adipocyte structures secrete a myriad of adipokines mediating reciprocal cross talk between adipocytes and endothelial cells to stimulate growth and expansion of adipocytes and induce angiogenesis

5.2.1 Adipose Vasculature and Adipose-Derived Stem Cells

Adipose blood vasculature serves as a niche to provide adipose tissue with progenitor cells and signals for adipocyte development. The progenitor cells have been described to reside in the adipose stromal vascular fraction (SVF). These adipose-derived stem cells (ADSCs) are biologically similar to mesenchymal stem cells derived from the bone marrow, and have the ability to differentiate into cells of several lineages including adipocytes, osteoblasts, chondrocytes, myocytes, ECs, hematopoietic cells, hepatocytes, and neuronal cells (Zuk 2001; Zuk et al. 2001; Gimble et al. 2007; Planat-Benard et al. 2004). Adult mesenchymal stem cells give rise to a myriad of cell types under regulation of different growth factors. Several groups have described the differentiation of ADSCs into adipocytes in vitro. Under the induction with insulin, methyisobutylxanthine, hydrocortisone or dexamethasone, and indomethacin and thiazolidinedione, proliferating ADSCs can be differentiated into preadipocytes which further proliferate at the site of adipogenesis. Adipogenesis is a well-characterized process and follows a temporal sequence which has been shown to be highly sensitive to insulin (Rosen and Spiegelman 2006, 2000). On the contrary, brown adipocytes could derive from different origins and it has been shown that bone morphogenetic protein 7 (BMP7) triggers commitment of murine mesenchymal progenitor cells to brown adipocyte lineage in vitro and in vivo. Both peroxisome proliferator-activated receptor gamma (PPAR-γ) coactivator 1-alpha (PGC1-α) and -beta (PGC1-β) have been demonstrated to play crucial roles in the formation of brown adipocytes. PR domain containing 16 (PRDM16), a zinc-finger protein, has been described to control brown fat determination via induction of PGC1-α, UCP-1, and type 2 iodothyronine deiodinase genes (Seale et al. 2008; Uldry et al. 2006). However, the expression of PRDM16 has recently been demonstrated to be expressed in white adipocyte progenitors and macrophages isolated from WAT of adult humans (Seale et al. 2011).

5.2.2 Infiltrating Inflammatory Cells and Obesity

Obesity and obesity-related metabolic disorders have potentially become a threat to the modern society. According to the World Health Organization, there are approximately one billion overweight adults and at least 300 million of these individuals are obese (Cao 2010). Obesity is caused by excessive accumulation of adipose tissue in the body. Obesity due to visceral fat accumulation in the abdomen triggers inflammatory cascades which lead to infiltration of immune cells including macrophages, lymphocytes, and natural killer cells. Adipose tissue is an essential endocrine organ that secretes a myriad of adipokines crucial in mediating processes important in inflammation, metabolism, and tissue remodeling. The serum level and adipose tissue of individuals have elevated circulating inflammation markers. In a lean individual, macrophages comprise up to 5–10 % of the cell population, whereas in obese individuals, the percentage of macrophages is significantly increased to 40 % (Weisberg et al. 2003). Obesity is considered a low grade chronic inflammation due to the vicious cycle between adipocytes and macrophages. The infiltrated macrophages secrete inflammatory cytokines which further propagate the inflammatory cascades (Suganami et al. 2005). Accumulation of macrophages in obese patients not only causes chronic inflammation but also causes insulin insensitivity and metabolic abnormalities (Vachharajani and Granger 2009). Macrophages can be categorized into two subpopulations, the M1 macrophages being pro-inflammatory and the M2 macrophages being anti-inflammatory. In adipose tissue of obese individuals, there is a tendency of M1 macrophage accumulation and M2 macrophage suppression. It has been described by Nishimura et al. that obese adipose tissue contains higher number of T cells. The population cytotoxic of $CD8^+$ T cells increases with progressive obesity while the $CD4^+$ regulatory T cell decreases. Infiltration of cytotoxic $CD8^+$ T cells in obese individuals could contribute to recruitment, differentiation, and activation of macrophages. The accumulation of $CD8^+$ T cells precedes the accumulation of macrophages suggesting that $CD8^+$ T cells initiate the inflammatory cascades in the adipose tissue (Nishimura et al. 2009; Feuerer et al. 2009; Winer et al. 2009). Despite the immense research on increased inflammation in obese individuals, the use of anti-inflammatory drugs in the treatment of obesity warrants further investigations.

5.3 Angiogenic Factors and Adipokines in WAT and BAT

In the adipose tissue—both in BAT and in WAT—a myriad of angiogenic factors and adipokines is expressed. These factors are involved in processes such as blood vessel formation and vascular remodeling, recruitment of inflammatory cells, metabolic processes such as lipolysis, and in insulin sensitivity.

5.3.1 Angiogenic Factors in WAT and BAT

Blood vessel formation in the adipose tissue is regulated by a balance between pro- and anti-angiogenic factors that are partly expressed in the adipose tissue. Some of the angiogenic regulators are adipokines and will be described more thoroughly in Sect. 5.3.2. The tight regulation of the angiogenic process determined by the interplay between pro- and anti-angiogenic factors is crucial in adipose tissue growth and regression and thereby plays a role in many physiological as well as pathological processes (Cao 2007, 2010).

5.3.1.1 Vascular Endothelial Growth Factor

Vascular endothelial growth factor (VEGF) is a potent angiogenic factor which induces the sprouting of new vessels from preexisting ones. It promotes EC proliferation and migration as well as survival (Hausman and Richardson 2004). Interestingly, VEGF expression is higher in BAT compared to WAT which is reflected in a higher vascular density in BAT. Expression of VEGF in BAT can further be elevated by cold exposure (Hausman and Richardson 2004; Asano et al. 1999, 1997). Xue et al. have also found an increase in VEGF expression in subcutaneous WAT of cold-exposed mice which correlated with an angiogenic switch in this tissue and a transformation of white to BRITE adipocytes (Xue et al. 2009).

5.3.1.2 Angiopoietin

Angiopoietin (Ang)-1 and -2 play a role in vascular remodeling. They regulate vessel maturation and stabilization (Xue et al. 2008; Yancopoulos et al. 2000). There is evidence that Ang-1 functions to stabilize blood vessels whereas Ang-2 has a destabilizing effect on the vasculature (Hausman and Richardson 2004). Angiopoietin-2 expression in the adipose tissue can be induced by leptin (Cohen et al. 2001).

5.3.1.3 Other Angiogenic Factors in the Adipose Tissue

Apart from VEGF and Ang-1 and -2, there are a number of other pro- and anti-angiogenic factors that are present in the adipose tissue. Basic fibroblast growth factor (bFGF) and hepatocyte growth factor (HGF) are potent angiogenesis stimulators (Bell et al. 2008; Yamashita et al. 1994). Similar to VEGF, expression levels of bFGF are elevated under cold exposure and might contribute to an increase of the amount of BAT (Yamashita et al. 1994).

On the other hand, thrombospondin (TSP) and VEGF receptor-2 exert angiogenesis inhibiting functions (Cao 2007).

5.3.2 Adipokines in WAT and BAT

As mentioned in previous sections of this chapter, there are two types of adipose tissues. WAT that primarily stores energy in the form of triglycerides and BAT which is metabolically active and dissipates energy and produces heat through a process termed NST (Cannon and Nedergaard 2004; Trayhurn and Beattie 2001). Besides these functions, however, adipose tissue also plays an important role as an endocrine organ. Around 50 years ago, it was first hypothesized that a circulating factor exists that regulates energy expenditure and appetite by acting in the central nervous system in a negative feedback loop (Galic et al. 2010; Kennedy 1953). Only in 1994, Zhang et al. could identify this circulating factor as the 16 kDa peptide hormone leptin (Galic et al. 2010; Zhang et al. 1994). Soon after this discovery, a number of other factors produced and secreted by the adipose tissue were identified. These factors—named adipokines or adipocytokines—are produced by different cell types within the adipose tissue and include metabolic regulators, angiogenic factors, as well as inflammatory cytokines and exert their functions in a paracrine, autocrine, or even systemic manner (Trayhurn and Beattie 2001; Galic et al. 2010; Waki and Tontonoz 2007; Bozaoglu et al. 2010; Mu et al. 2006; Vona-Davis and Rose 2007).

5.3.2.1 Leptin

The most studied adipokine is the peptide hormone leptin (MacDougald and Burant 2007). Leptin acts centrally via the hypothalamus to regulate food intake and energy expenditure but also on peripheral organs. The major functions of the 16-kDa hormone are mediation of satiety, stimulation of lipolysis, and suppression of lipogenesis (Ronti et al. 2006). The hormone has been discovered using positional cloning of the *ob/ob* mouse to identify the underlying mutant gene that causes the obese phenotype of this model organism (Trayhurn and Beattie 2001; Galic et al. 2010; Zhang et al. 1994). As evident in the *ob/ob* mouse, a deficiency of functional leptin leads to the development of early-onset obesity. Several studies have shown that treatment of this genetically modified mouse strain by daily injections of leptin decreases food intake and normalizes body weight (Galic et al. 2010; Campfield et al. 1995; Halaas et al. 1995). Synthesis of the leptin hormone occurs both in WAT and in BAT but also in several other organs (Trayhurn and Beattie 2001). However, leptin expression in BAT only occurs in inactive tissue (Cannon and Nedergaard 2004). Interestingly, in humans, plasma levels of circulating leptin correlate with body fat mass and weight gain. This phenomenon can be explained by the fact that in obese individuals the average fat cell size is 2–4 times enlarged compared to lean individuals and it has been shown that the amount of leptin secretion increases with adipocyte size (Galic et al. 2010; Lonnqvist et al. 1997; Fried et al. 2000; Havel et al. 1996).

Apart from its metabolic implications, leptin also affects angiogenesis (Ronti et al. 2006). Both in vitro and in vivo studies could demonstrate a proliferative effect of leptin on ECs via activation of the endothelial leptin receptor (Ob-R) (Bouloumie et al. 1998; Park et al. 2001). Upon leptin treatment, Bouloumié et al. described enhanced capillary tube formation of human umbilical vein endothelial cells (HUVEC) as well as neovascularization in an in vivo model (Bouloumie et al. 1998).

5.3.2.2 Resistin

Resistin is another adipocytokine produced in the adipose tissue. It is secreted mainly by WAT and can be considered WAT-specific even though minor amounts of resistin mRNA have been detected in BAT (Steppan et al. 2001). In diet-induced as well as in genetic mouse models of obesity and insulin resistance, resistin levels are elevated. Indeed, studies have shown that administration of resistin can induce insulin resistance whereas neutralization of the adipokine improves glucose tolerance (Steppan et al. 2001). In humans, however, the role of resistin in terms of insulin resistance remains to be elucidated as existing studies show controversial results (Ronti et al. 2006; Wang et al. 2002; Hotamisligil 2003; Patel et al. 2003).

In humans, resistin expression is not restricted to the adipose depots but has also been found in leukocytes, macrophages, bone marrow, and spleen (Jamaluddin et al. 2012). Similar to leptin, resistin is also associated with angiogenic potential. Mu et al. have shown that human recombinant resistin induces EC proliferation, migration, and tube formation in vitro and further studies are available suggesting an angiogenic effect of resistin (Mu et al. 2006; Robertson et al. 2009). The angiogenesis-promoting function of resistin is at least partly mediated by the up-regulation of pro-angiogenic vascular VEGF receptor-1 (VEGFR-1) and -2 (VEGFR-2) as well as increase in the mRNA and protein levels of matrix metalloproteases (MMP) (Mu et al. 2006).

5.3.2.3 Visfatin

Visfatin (pre-B cell colony-enhancing factor) is a 52 kDa protein that has recently been identified as a novel adipokine. It is present in visceral as well as subcutaneous WAT but also expressed in BAT and other tissues such as liver and muscle (Waki and Tontonoz 2007; Revollo et al. 2007; Fukuhara et al. 2005). The levels of circulating visfatin in the plasma are correlated to the amount of visceral fat in the body and an insulin-mimetic function of visfatin has been described in vivo (Fukuhara et al. 2005; Saddi-Rosa et al. 2010). Fukuhara et al. found that visfatin treatment of obese mice can reduce plasma glucose levels (Fukuhara et al. 2005). However, 2 years after their initial finding, they retracted their paper due to controversies about their results. Nevertheless, another group could demonstrate an insulin-like effect of visfatin on osteoblast characterized by mediation of glucose uptake involving the insulin receptor (Xie et al. 2007).

Visfatin also exhibits a regulatory function of angiogenesis. The adipokine induces VEGF as well as MMP expression while decreasing the expression of tissue inhibitor of metalloproteases (TIMP) in ECs. Furthermore, visfatin has a proliferative and migratory effect on ECs in vitro and maintains EC survival by suppressing apoptosis (Adya et al. 2008). Another group has found a positive effect of visfatin treatment on neovascularization in vivo. In further in vitro studies, they could demonstrate an activation of the angiogenesis-related extracellular signal-regulated kinase 1/2 (ERK1/2) in ECs upon visfatin treatment (Kim et al. 2007).

5.3.2.4 Chemerin

Chemerin is a relatively new member in the family of adipokines and exerts its function via autocrine and paracrine signaling. The mature protein has a size of 16 kDa and is primarily expressed in WAT but chemerin mRNA at low levels is also present in BAT (Goralski et al. 2007; Roh et al. 2007). However, in genetically obese *ob/ob* mice, chemerin expression in the brown adipose depot is significantly elevated compared to lean mice. Remarkably, the morphology of brown adipocytes in these leptin-deficient mice is altered and resembles white adipocytes in lean mice regarding lipid droplet distribution. Additionally, molecular differences between brown adipocytes derived from obese versus lean mice are evident such as reduced expression of the NST-related UCP-1. These observations might indicate a potential role of chemerin in the transformation of brown adipocytes to a white adipocyte-like phenotype as seen in obese mice (Goralski et al. 2007).

Another potential role of chemerin is of chemoattractant nature. There is evidence that the adipokine may play a role in the recruitment of macrophages expressing the G-protein-coupled receptor CMKLR1 (Goralski et al. 2007; Zabel et al. 2006). Chemerin is further involved in adipocyte differentiation and there are indications that it also plays a role in the modulation of metabolic pathways such as lipolysis (Goralski et al. 2007). Bozaoglu et al. also established a role of chemerin in angiogenesis. They observed a dose-dependent formation of microtubules in vitro when treating ECs with recombinant chemerin (Bozaoglu et al. 2010).

5.3.2.5 Adiponectin

Adiponectin is an adipocyte-specific secreted protein and 30 kDa in size (Waki and Tontonoz 2007). Expression of the protein hormone is nearly exclusively restricted to WAT; however, recently, adiponectin expression has also been detected in BAT (Cannon and Nedergaard 2004; Ronti et al. 2006; Viengchareun et al. 2002). Adiponectin is abundant as a circulating plasma protein (Hu et al. 1996). As opposed to leptin, adiponectin levels are decreased in obesity and increased upon weight loss (Yang et al. 2001; Coppola et al. 2009). The protein is involved in adipocyte differentiation, lipid accumulation in mature adipocytes, and inhibits lipolysis in vivo and in vitro (Qiao et al. 2011; Fu et al. 2005). Decreased adiponectin levels as seen in obese individuals correlate with insulin resistance (Yamauchi et al. 2001). By administering adiponectin, insulin sensitivity can be at least partly restored in rodents (Chandran et al. 2003). Tschritter et al. could show that adiponectin plasma levels can serve as a marker to predict insulin sensitivity (Tschritter et al. 2003).

Brakenhielm et al. have shown an anti-angiogenic effect of adiponectin in the chick chorioallantoic membrane assay and in the mouse corneal angiogenesis assay. They found an inhibition of new vessel growth by adiponectin. Furthermore, they could observe that EC apoptosis is induced by the adipokine via a caspase-dependent mechanism (Brakenhielm et al. 2004a).

5.3.2.6 Further Adipokines

The adipokines described above are just a small part of all adipokines produced in BAT and WAT. There are a number of other factors that are produced by the different cell types within the adipose tissue such as macrophages or stromal cells that contribute with the vast majority of adipokines (Vona-Davis and Rose 2007). Apart from factors involved primarily in lipid metabolism, insulin sensitivity, and angiogenesis, there is a broad range of adipokines that are associated with inflammation and acute phase response. Among these are transforming growth factor (TGF)-β, tumor necrosis factor alpha (TNFα), interleukin (IL)-6, -8, and -10, as well as PAI-1 and metallothionein (Trayhurn and Wood 2004).

5.3.2.7 Conclusion

The secretory and endocrine role of BAT is substantially smaller compared to WAT. Most adipokines produced in BAT act auto- or paracrine while a systemic effect is very rare. One explanation for this is the fact that BAT mass is significantly smaller compared to the amount of WAT present in the body (Cannon and Nedergaard 2004). To date, there is a remarkable amount of data available on the importance of WAT as an endocrine organ. However, the implications of BAT in this context are not equally well understood. It is also of great importance to understand potential systemic consequences that could arise from sudden, extensive weight loss as seen in cancer cachexia patients or due to excessive fasting (Trayhurn and Beattie 2001).

5.4 Plasticity of BAT and WAT Adipose Vasculatures

Adipose tissues undergo constant expansion and shrinkage throughout adulthood. The plasticity of BAT and WAT is intimately dependent on the constant growth, regression, and remodeling of blood vessels (Hausman and Richardson 2004). The intimate interplay between adipocytes and blood vessels suggests that alteration in the adipose vasculature could be actively responsible in the modulation of adipogenesis and obesity. The plasticity of adipose vasculature in WAT and BAT is highly dependent on the fine balance between pro-angiogenic factors and anti-angiogenic inhibitors.

5.4.1 Pro-angiogenic Agents on Adipose Vasculature

Adipose tissue, being an endocrine organ, secretes a myriad of angiogenic factors being potentially involved in the regulation of angiogenesis; these angiogenic factors include VEGF-A, leptin, FGF-2, HGF, TNFα, TGF-β, VEGF-C, resistin, neuropeptide Y (NYP), and Ang (Li et al. 2003; Rehman et al. 2004; Friedman and Halaas 1998; Halaas and Friedman 1997). It has been shown that anti-angiogenic agents that inhibit the growth of capillaries are able to prevent weight gain in mouse model and angiogenesis is enhanced in obesity in both animal models and in humans (Rupnick et al. 2002). In the event of excess pro-angiogenic factors, such as VEGF-A, an angiogenic switch is triggered activating a downstream signaling cascade resulting in the increased density of adipose vasculature. Leptin, an adipocyte-specific cytokine, modulates angiogenesis in adipose tissue, food consumption, thermogenesis, reproduction, hematopoiesis, and pro-inflammatory immune response in the body. The plasma concentration of leptin has been found to be correlated to BMI. Leptin promotes angiogenesis through paracrine interaction with ECs thus stimulates cell growth, proliferation, and survival of endothelial cells (Bouloumie et al. 1998).

5.4.2 Anti-angiogenic Agents on Adipose Vasculature

On the other hand, excess angiogenic inhibitors result in regression of adipose vasculature. Adipose tissue produces anti-angiogenic factors such as TSP and adiponectin. For instance, an adipose-specific adipokine, adiponectin, has negative role in modulating adipogenesis. Adiponectin inhibits angiogenesis by inhibiting EC proliferation, survival, and migration, and thus inhibits adipogenesis (Brakenhielm et al. 2004a). The level of circulating adiponectin is inversely correlated with BMI. However, overexpression or deletion of adiponectin does not seem to have any effect on the body weight of mice (Matsuda et al. 2002). Impairment of vascular functions in the adipose tissue could potentially lead to alteration and dysfunction of lipid metabolism and probable initiation and progression of insulin resistance (Jansson 2007). Dysfunction of the vasculature could be caused by alteration to the ECs due to inflammation and angiogenic profile changes induced by hypoxia. Since the expansion of adipose tissue is dependent on the growth, regression, and remodeling of blood vessels, the use of anti-angiogenic agents to inhibit vessels growth may seem to be an ideal therapeutic treatment for obesity and related metabolic disorders. Indeed, it has been demonstrated that treatment with angiogenic inhibitors including TNP-470, angiostatin, and endostatin resulted in decreased body weights in a dose-dependent and reversible manner in genetically obese and high fat diet mouse models (Rupnick et al. 2002; Brakenhielm et al. 2004b). Administration of TNP-470, angiostatin, and endostatin not only resulted in weight loss; the vasculature in the adipose tissue was also reduced significantly. However, the use of anti-angiogenic agents in treatment of obesity and other obesity-related metabolic disorders in human warrants further investigations.

5.4.3 Conclusion

In conclusion, the capacity of angiogenesis in adipose tissue may be a crucial player in fat accumulation and the plasticity of WAT in obese individuals. The plasticity of WAT and BAT adipose vasculature is tightly regulated by the balance of angiogenic factors and inhibitors. The reciprocal cross talk between ECs and adipocytes is essential in modulating vessels growth or regression.

5.5 Reciprocal Interplay Between Adipocytes and Vascular Cells

In early embryonic development, the formation of adipose tissue is closely associated with angiogenesis and studies have shown that angiogenesis precedes adipogenesis (Hausman and Richardson 2004; Nishimura et al. 2007; Neels et al. 2004).

Developmental studies clearly indicate a spatial and temporal association between angiogenesis and adipogenesis and there is increasing evidence that ECs and adipocytes communicate in a paracrine manner but also via extracellular components as well as by cell–cell interactions (Cao 2007; Varzaneh et al. 1994; Hutley et al. 2001; Bouloumie et al. 2002).

The fat tissue is highly vascularized and each adipocyte is surrounded by a vessel (Cao 2007, 2010; Crandall et al. 1997). This close proximity of adipocytes and vascular cells implies that there is an intimate interplay between those cell types that is required for proper adipose tissue growth, expansion, and function (Fig. 5.3).

Interestingly, adipose tissue in the adult is considered to be one of the most plastic tissues in the body. During lifetime, it can experience shrinkage and expansion. These processes, such as the development of obesity, are accompanied by changes in the adipose vasculature such as vascular remodeling and neovascularization (Lijnen 2008; Cao 2007, 2010; Crandall et al. 1997) (Fig. 5.4).

5.5.1 Vascular Functions in Adipose Tissue

The vasculature in the adipose tissue exerts several crucial functions. This is especially important in growing adipose tissues; however, the vasculature is also required for the maintenance of adipose tissue and its functions. The role of the adipose vasculature is described in detail in Sect. 7.2.

Here, we will focus on evaluating the effect of vascular cells on adipocytes. The vascular cells include ECs and perivascular cells. They are derived from a common vascular progenitor cell of mesodermal origin that differentiates into these cell types (Chen 2011; Descamps and Emanueli 2012; Kitagawa and Era 2010).

5.5.1.1 Endothelial Cells

ECs express among others the surface markers CD31, CD34, VEGFR-1, and VEGFR-2 as well as Tie1 and Tie2 (Descamps and Emanueli 2012). In blood vessels, the endothelium formed by a monolayer consisting of ECs functions as a semi-permeable barrier regulating homeostasis of tissue fluids (Descamps and Emanueli 2012; Komarova and Malik 2010; Sumpio et al. 2002). Apart from this barrier function, ECs are also in endocrine and paracrine signaling. Thereby, they regulate a variety of processes such as immune response and growth (Sumpio et al. 2002). Studies suggest that ECs secrete factors that induce preadipocyte differentiation and proliferation in vitro (Varzaneh et al. 1994; Hutley et al. 2001). It has been suggested that this effect is induced by ECM components. Indeed, Hausman et al. could show that ECM components increase preadipocyte spreading and hypertrophy in vitro. However, these results could only be obtained using ECM components derived from tumors but not from corneal ECs (Hausman et al. 1996). It has also been

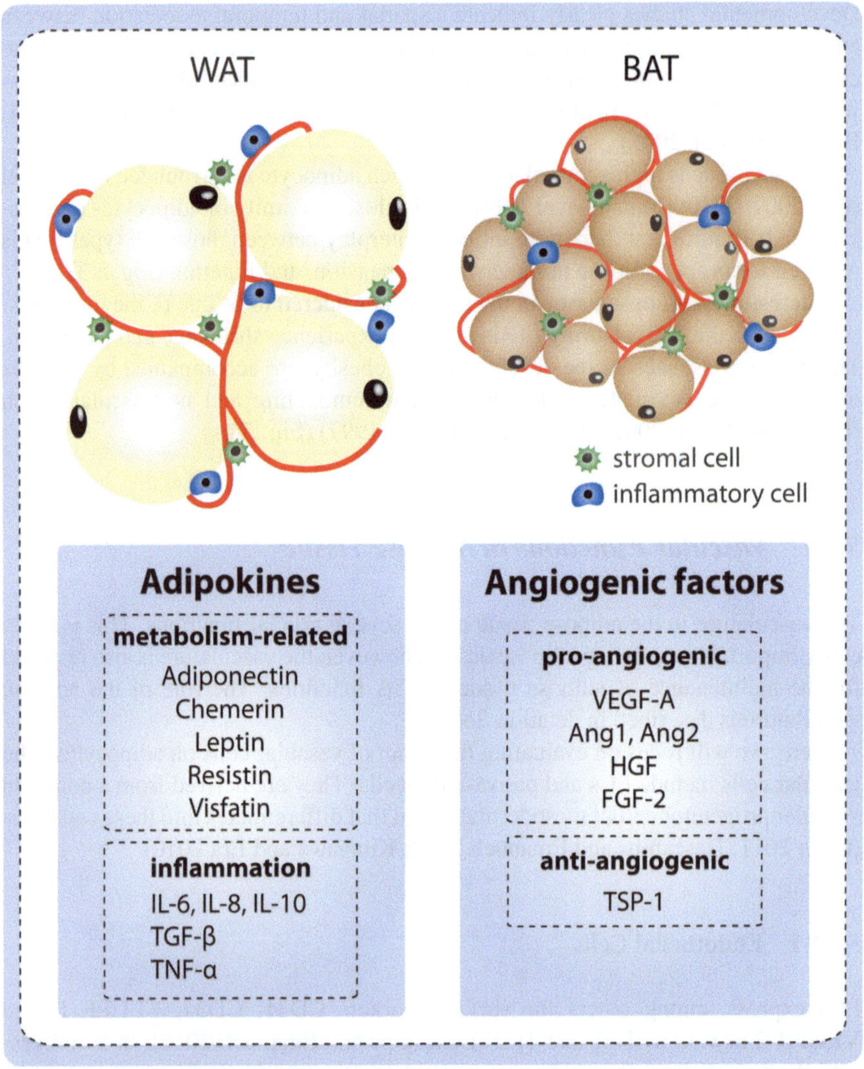

Fig. 5.3 Adipokines and angiogenic factors in WAT and BAT. The WAT plays a major role as an endocrine organ, secreting a myriad of adipokines and pro- as well as antiangiogenic factors. BAT plays only a minor role in secretion of adipose-derived factors. Due to its smaller mass, its endocrine function is mainly restricted to paracrine and autocrine signaling

hypothesized that the proliferative effect of adipose tissue ECs might be due to the secretion of mitogenic and adipogenic factors such as IGF-1 or members of the FGF family (Hutley et al. 2001; Crandall et al. 1997; Ramsay et al. 1992). The adipogenic potential of FGF-1 has been studied by Hutley et al. who could identify FGF-1 as a factor secreted by adipose-derived ECs that increases preadipocyte

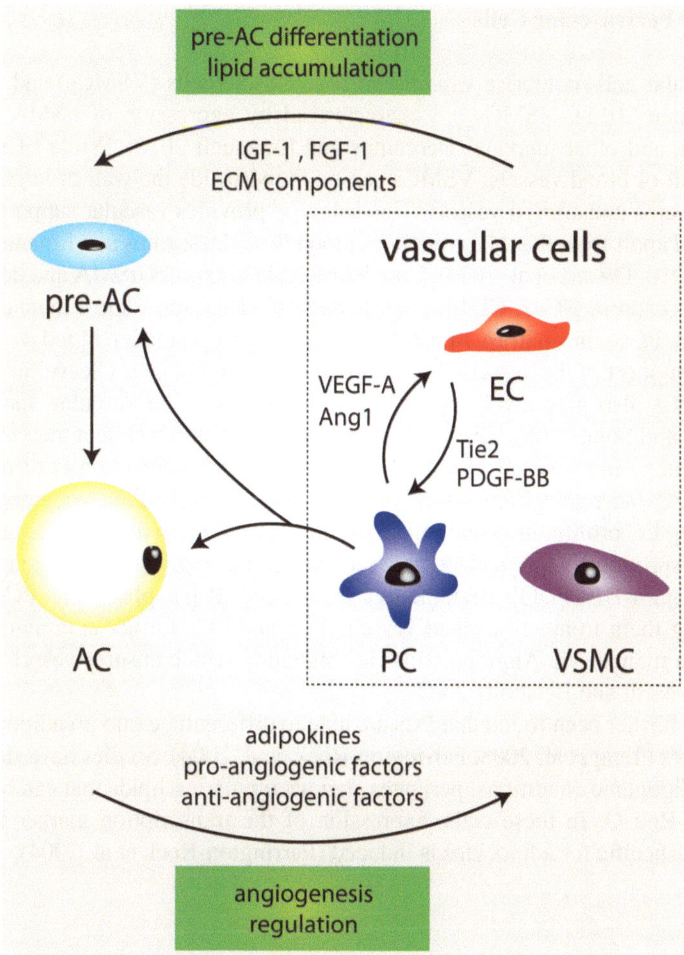

Fig. 5.4 Reciprocal interplay between vascular cells and adipocytes. Vascular cell-derived factors such as IGF-1, FGF-1, or ECM components can stimulate pre-AC differentiation and lipid accumulation. PCs can differentiate to pre-ACs or ACs and thereby further contribute to adipose tissue expansion. In turn, ACs produce factors that can stimulate or inhibit angiogenesis. Within the vascular compartment, PCs and ECs interact via VEGF, PDGF-BB, or Ang1/Tie2 signaling. *AC* adipocyte, *EC* endothelial cell, *PC* pericyte, *pre-AC* preadipocyte, *VSMC* vascular smooth muscle cell

differentiation and lipid accumulation (Hutley et al. 2004). Another in vitro study implies that coculture of ECs and adipocytes stimulates development of preadipocytes while increasing mature adipocyte growth. However, this study hypothesizes that the effect may be due to direct adhesion of ECs to mature adipocytes rather than mediated by soluble factors (Aoki et al. 2003).

5.5.1.2 Perivascular Cells

Perivascular cells comprise vascular smooth muscle cells (VSMCs) and pericytes (PC) (Chen 2011). VSMCs are characterized by expression of αSMA, desmin, vimentin, and other markers (Descamps and Emanueli 2012). While ECs line the inner wall of blood vessels, VSMCs can be found inside the wall of large but also smaller veins and arterial vessels. This cell type provides vascular support and stability and apart from that even regulates blood flow (Descamps and Emanueli 2012; Miano 2010; Owens et al. 2004). Like VSMCs, PCs express αSMA and desmin but they also express NG-2, CD146, and PDGFR-β (Descamps and Emanueli 2012). However, PCs—in contrast to VMSCs—are covering smaller blood vessels and provide support of the vessels (Descamps and Emanueli 2012; Cleaver and Melton 2003). PCs also play a role in angiogenesis as well as in vascular maintenance (Bergers and Song 2005). They interact with ECs and thereby affect their migration, proliferation, and survival (Armulik et al. 2005). This interaction is of reciprocal nature: PCs secrete VEGF which binds to the corresponding receptors on ECs mediating EC proliferation and migration during angiogenesis (Olsson et al. 2006). In turn, sprouting ECs produce and secrete during angiogenesis platelet-derived growth factor-BB (PDGF-BB), the ligand for PDGFR-β expressed on PCs thereby recruiting them to newly formed vessels. PCs and ECs further communicate in a paracrine manner via Angiopoietin–Tie2 signaling which ensures vessel stabilization (Gerhardt and Betsholtz 2003).

It has further been found that PCs are able to differentiate into preadipocytes and adipocytes (Tang et al. 2008; Farrington-Rock et al. 2004). Studies have shown that under adipogenic conditions, pericytes start accumulating lipids that can be stained with oil Red O. In these cells, expression of the transcription marker PPAR-γ2 which is specific for adipocytes is induced (Farrington-Rock et al. 2004).

5.5.2 Effects of Adipocytes on Vascular Cells

The interaction between vascular cells and adipocytes is not of unidirectional nature; on the contrary, there is a reciprocal interplay between these cell types and the adipocytes also affect the vascular compartment by secretion of several molecules (Cao 2010).

The adipose tissue consists of a heterogeneous variety of cell types including adipocytes and preadipocytes, stromal cells, inflammatory cells, and ECs. These cell types produce and secrete different growth factors and cytokines that affect vascular growth and can induce or inhibit angiogenesis in the adipose tissue (Cao 2005, 2010; Rehman et al. 2004). Here, we focus on the factors released by adipocytes themselves that exert pro- or anti-angiogenic effects, respectively.

One of these molecules is adiponectin, which is a hormone derived from adipocytes. Interestingly, it has been demonstrated that adiponectin is present at significantly lower levels in obese individuals compared to the lean population (Cao 2007;

Yamauchi et al. 2001). Circulating adiponectin levels show a negative correlation with BMI and it has been shown that plasma adiponectin is decreased in individuals that exhibit insulin resistance (Arita et al. 1999; Nedvidkova et al. 2005). Regarding its effects on the vasculature, it has been shown that adiponectin inhibits EC proliferation. An anti-proliferative and anti-migratory effect of adiponectin on VSMCs in vitro has been described (Goldstein and Scalia 2004). There are indications that inhibition of smooth muscle cell migration can at least partly be ascribed to binding of adiponectin to PDGF-BB (Goldstein and Scalia 2004; Arita et al. 2002). This effect is mediated by activating caspase-mediated apoptosis. Adiponectin also counteracts the survival of ECs. Brakenhielm et al. have described an anti-angiogenic effect of the adipocyte-derived hormone. In an in vivo model using the mouse corneal angiogenesis assay, they showed that it can act as an endogenous angiogenesis inhibitor. In this study, the effect of adiponectin on microvessels was investigated (Brakenhielm et al. 2004a). On the contrary, others have shown a pro-angiogenic effect of adiponectin in in vitro studies on ECs obtained from large vessels (Brakenhielm et al. 2004a; Ouchi et al. 2004). The discrepancy of the results might be due to different experimental systems and also due to the different origin of the ECs under investigation. ECs derived from capillaries might behave differently compared to ECs derived from larger vessels (Brakenhielm et al. 2004a).

Another hormone derived from adipocytes is resistin (Verma et al. 2003). The signaling molecule resistin has been shown to link obesity to insulin resistance and type 2 diabetes (Steppan et al. 2001; Verma et al. 2003). Rajala et al. have described that resistin inhibits skeletal muscle glucose uptake and seems to be produced during adipogenic processes (Verma et al. 2003; Rajala et al. 2003). The vasoactive effect of resistin has been first demonstrated by Verma et al. (2003). In their in vitro study, they could show that treatment of ECs with recombinant resistin up-regulates expression of endothelin-1 (ET-1), vascular cell adhesion molecule (VCAM-1), and monocyte chemoattractant chemokine (MCP-1) which play a role in endothelial dysfunction and thereby obesity and insulin resistance (Verma et al. 2003).

Apart from adiponectin and resistin, further adipocyte- and adipose tissue-derived factors affect angiogenesis. A shift in the intimate balance between pro- and anti-angiogenic factors regulates whether angiogenesis can occur or not. Typically, the angiogenic switch is determined by a down-regulation of angiogenesis inhibitors concurrently with an up-regulation of angiogenesis stimulators (Cao 2007, 2010). The variety of pro- and anti-angiogenic factors is discussed more in detail in Sect. 7.3.

5.5.3 Implications and Potential of Adipocyte–Vasculature Interplay

The intimate interplay between adipocytes and the adipose tissue vasculature opens up potential for treatment options for obesity and obesity-related metabolic disorders. Metabolic diseases such as type 2 diabetes are often related to endothelial dysfunction and adipose tissue expansion critically relies on angiogenesis (Cao 2007, 2010; Avogaro et al. 2011; Avogaro and de Kreutzenberg 2005). Thereby, angiogenesis inhibitors might serve as a powerful tool to counteract obesity and restoring proper endothelial function might be a potential target to combat metabolic disorders (Cao 2007, 2010). Indeed, a number of in vivo studies have demonstrated that the endogenous angiogenesis modulators angiostatin and endostatin, and also molecules such as the angiogenesis inhibitor TNP-470, have the potential to decrease body weight and prevent obesity in genetic models as well as diet-induced models of obesity in mice (Cao 2010; Rupnick et al. 2002; Brakenhielm et al. 2004b; O'Reilly et al. 1994, 1997).

References

Adya R, Tan BK, Punn A, Chen J, Randeva HS. Visfatin induces human endothelial VEGF and MMP-2/9 production via MAPK and PI3K/Akt signalling pathways: novel insights into visfatin-induced angiogenesis. Cardiovasc Res. 2008;78(2):356–65. Epub 2007/12/21.

Aoki S, Toda S, Sakemi T, Sugihara H. Coculture of endothelial cells and mature adipocytes actively promotes immature preadipocyte development in vitro. Cell Struct Funct. 2003; 28(1):55–60. Epub 2003/03/26.

Arita Y, Kihara S, Ouchi N, Takahashi M, Maeda K, Miyagawa J, et al. Paradoxical decrease of an adipose-specific protein, adiponectin, in obesity. Biochem Biophys Res Commun. 1999; 257(1):79–83. Epub 1999/03/27.

Arita Y, Kihara S, Ouchi N, Maeda K, Kuriyama H, Okamoto Y, et al. Adipocyte-derived plasma protein adiponectin acts as a platelet-derived growth factor-BB-binding protein and regulates growth factor-induced common postreceptor signal in vascular smooth muscle cell. Circulation. 2002;105(24):2893–8. Epub 2002/06/19.

Armulik A, Abramsson A, Betsholtz C. Endothelial/pericyte interactions. Circ Res. 2005;97(6):512–23. Epub 2005/09/17.

Asano A, Morimatsu M, Nikami H, Yoshida T, Saito M. Adrenergic activation of vascular endothelial growth factor mRNA expression in rat brown adipose tissue: implication in cold-induced angiogenesis. Biochem J. 1997;328(Pt 1):179–83. Epub 1998/01/10.

Asano A, Kimura K, Saito M. Cold-induced mRNA expression of angiogenic factors in rat brown adipose tissue. J Vet Med Sci. 1999;61(4):403–9. Epub 1999/05/26.

Avogaro A, de Kreutzenberg SV. Mechanisms of endothelial dysfunction in obesity. Clin Chim Acta. 2005;360(1–2):9–26. Epub 2005/06/29.

Avogaro A, Albiero M, Menegazzo L, de Kreutzenberg S, Fadini GP. Endothelial dysfunction in diabetes: the role of reparatory mechanisms. Diabetes Care. 2011;34 Suppl 2:S285–90. Epub 2011/05/06.

Bell LN, Cai L, Johnstone BH, Traktuev DO, March KL, Considine RV. A central role for hepatocyte growth factor in adipose tissue angiogenesis. Am J Physiol Endocrinol Metab. 2008; 294(2):E336–44. Epub 2007/12/13.

Bergers G, Song S. The role of pericytes in blood-vessel formation and maintenance. Neuro Oncol. 2005;7(4):452–64. Epub 2005/10/11.

Bouloumie A, Drexler HC, Lafontan M, Busse R. Leptin, the product of Ob gene, promotes angiogenesis. Circ Res. 1998;83(10):1059–66. Epub 1998/11/14.

Bouloumie A, Lolmede K, Sengenes C, Galitzky J, Lafontan M. Angiogenesis in adipose tissue. Ann Endocrinol (Paris). 2002;63(2 Pt 1):91–5. Epub 2002/05/08.

Bozaoglu K, Curran JE, Stocker CJ, Zaibi MS, Segal D, Konstantopoulos N, et al. Chemerin, a novel adipokine in the regulation of angiogenesis. J Clin Endocrinol Metab. 2010;95(5):2476–85. Epub 2010/03/20.

Brakenhielm E, Veitonmaki N, Cao R, Kihara S, Matsuzawa Y, Zhivotovsky B, et al. Adiponectin-induced antiangiogenesis and antitumor activity involve caspase-mediated endothelial cell apoptosis. Proc Natl Acad Sci U S A. 2004a;101(8):2476–81. Epub 2004/02/26.

Brakenhielm E, Cao R, Gao B, Angelin B, Cannon B, Parini P, et al. Angiogenesis inhibitor, TNP-470, prevents diet-induced and genetic obesity in mice. Circ Res. 2004b;94(12):1579–88. Epub 2004/05/25.

Campfield LA, Smith FJ, Guisez Y, Devos R, Burn P. Recombinant mouse OB protein: evidence for a peripheral signal linking adiposity and central neural networks. Science. 1995; 269(5223):546–9. Epub 1995/07/28.

Cannon B, Nedergaard J. Brown adipose tissue: function and physiological significance. Physiol Rev. 2004;84(1):277–359. Epub 2004/01/13.

Cao Y. Opinion: emerging mechanisms of tumour lymphangiogenesis and lymphatic metastasis. Nat Rev Cancer. 2005;5(9):735–43. Epub 2005/08/05.

Cao Y. Angiogenesis modulates adipogenesis and obesity. J Clin Invest. 2007;117(9):2362–8. Epub 2007/09/06.

Cao Y. Adipose tissue angiogenesis as a therapeutic target for obesity and metabolic diseases. Nat Rev Drug Discov. 2010;9(2):107–15. Epub 2010/02/02.

Chandran M, Phillips SA, Ciaraldi T, Henry RR. Adiponectin: more than just another fat cell hormone? Diabetes Care. 2003;26(8):2442–50. Epub 2003/07/29.

Chen YE. Vascular cell lineage determination and differentiation. Arterioscler Thromb Vasc Biol. 2011;31(7):1467–8. Epub 2011/06/17.

Cleaver O, Melton DA. Endothelial signaling during development. Nat Med. 2003;9(6):661–8. Epub 2003/06/05.

Cohen B, Barkan D, Levy Y, Goldberg I, Fridman E, Kopolovic J, et al. Leptin induces angiopoietin-2 expression in adipose tissues. J Biol Chem. 2001;276(11):7697–700. Epub 2001/01/11.

Coppola A, Marfella R, Coppola L, Tagliamonte E, Fontana D, Liguori E, et al. Effect of weight loss on coronary circulation and adiponectin levels in obese women. Int J Cardiol. 2009; 134(3):414–6. Epub 2008/04/02.

Crandall ED, Bidani A, Forster RE. Postcapillary changes in blood pH in vivo during carbonic anhydrase inhibition. J Appl Physiol. 1977;43(4):582–90. Epub 1977/10/01.

Crandall DL, Hausman GJ, Kral JG. A review of the microcirculation of adipose tissue: anatomic, metabolic, and angiogenic perspectives. Microcirculation. 1997;4(2):211–32. Epub 1997/06/01.

Crandall DL, Groeling TM, Busler DE, Antrilli TM. Release of PAI-1 by human preadipocytes and adipocytes independent of insulin and IGF-1. Biochem Biophys Res Commun. 2000a; 279(3):984–8. Epub 2001/02/13.

Crandall DL, Busler DE, McHendry-Rinde B, Groeling TM, Kral JG. Autocrine regulation of human preadipocyte migration by plasminogen activator inhibitor-1. J Clin Endocrinol Metab. 2000b;85(7):2609–14. Epub 2000/07/21.

Cypess AM, Lehman S, Williams G, Tal I, Rodman D, Goldfine AB, et al. Identification and importance of brown adipose tissue in adult humans. N Engl J Med. 2009;360(15):1509–17. Epub 2009/04/10.

Descamps B, Emanueli C. Vascular differentiation from embryonic stem cells: novel technologies and therapeutic promises. Vascul Pharmacol. 2012;56(5–6):267–79. Epub 2012/04/17.

Farrington-Rock C, Crofts NJ, Doherty MJ, Ashton BA, Griffin-Jones C, Canfield AE. Chondrogenic and adipogenic potential of microvascular pericytes. Circulation. 2004; 110(15):2226–32. Epub 2004/10/07.

Feuerer M, Herrero L, Cipolletta D, Naaz A, Wong J, Nayer A, et al. Lean, but not obese, fat is enriched for a unique population of regulatory T cells that affect metabolic parameters. Nat Med. 2009;15(8):930–9. Epub 2009/07/28.

Fried SK, Ricci MR, Russell CD, Laferrere B. Regulation of leptin production in humans. J Nutr. 2000;130(12):3127S–31. Epub 2001/01/04.

Friedman JM, Halaas JL. Leptin and the regulation of body weight in mammals. Nature. 1998; 395(6704):763–70. Epub 1998/10/31.

Fu Y, Luo N, Klein RL, Garvey WT. Adiponectin promotes adipocyte differentiation, insulin sensitivity, and lipid accumulation. J Lipid Res. 2005;46(7):1369–79. Epub 2005/04/19.

Fukuhara A, Matsuda M, Nishizawa M, Segawa K, Tanaka M, Kishimoto K, et al. Visfatin: a protein secreted by visceral fat that mimics the effects of insulin. Science. 2005;307(5708):426–30. Epub 2004/12/18.

Galic S, Oakhill JS, Steinberg GR. Adipose tissue as an endocrine organ. Mol Cell Endocrinol. 2010;316(2):129–39. Epub 2009/09/03.

Gerhardt H, Betsholtz C. Endothelial-pericyte interactions in angiogenesis. Cell Tissue Res. 2003;314(1):15–23. Epub 2003/07/29.

Gerstein HC, Yusuf S, Bosch J, Pogue J, Sheridan P, Dinccag N, et al. Effect of rosiglitazone on the frequency of diabetes in patients with impaired glucose tolerance or impaired fasting glucose: a randomised controlled trial. Lancet. 2006;368(9541):1096–105. Epub 2006/09/26.

Gimble JM, Katz AJ, Bunnell BA. Adipose-derived stem cells for regenerative medicine. Circ Res. 2007;100(9):1249–60. Epub 2007/05/15.

Goldstein BJ, Scalia R. Adiponectin: a novel adipokine linking adipocytes and vascular function. J Clin Endocrinol Metab. 2004;89(6):2563–8. Epub 2004/06/08.

Goralski KB, McCarthy TC, Hanniman EA, Zabel BA, Butcher EC, Parlee SD, et al. Chemerin, a novel adipokine that regulates adipogenesis and adipocyte metabolism. J Biol Chem. 2007;282(38):28175–88. Epub 2007/07/20.

Halaas JL, Friedman JM. Leptin and its receptor. J Endocrinol. 1997;155(2):215–6. Epub 1998/01/01.

Halaas JL, Gajiwala KS, Maffei M, Cohen SL, Chait BT, Rabinowitz D, et al. Weight-reducing effects of the plasma protein encoded by the obese gene. Science. 1995;269(5223):543–6. Epub 1995/07/28.

Hausman GJ, Richardson RL. Adipose tissue angiogenesis. J Anim Sci. 2004;82(3):925–34. Epub 2004/03/23.

Hausman GJ, Wright JT, Richardson RL. The influence of extracellular matrix substrata on preadipocyte development in serum-free cultures of stromal-vascular cells. J Anim Sci. 1996;74(9):2117–28. Epub 1996/09/01.

Havel PJ, Kasim-Karakas S, Mueller W, Johnson PR, Gingerich RL, Stern JS. Relationship of plasma leptin to plasma insulin and adiposity in normal weight and overweight women: effects of dietary fat content and sustained weight loss. J Clin Endocrinol Metab. 1996;81(12):4406–13. Epub 1996/12/01.

Himms-Hagen J, Cui J, Danforth Jr E, Taatjes DJ, Lang SS, Waters BL, et al. Effect of CL-316,243, a thermogenic beta 3-agonist, on energy balance and brown and white adipose tissues in rats. Am J Physiol. 1994;266(4 Pt 2):R1371–82. Epub 1994/04/11.

Hotamisligil GS. The irresistible biology of resistin. J Clin Invest. 2003;111(2):173–4. Epub 2003/01/18.

Howe R, Rao BS, Holloway BR, Stribling D. Selective beta 3-adrenergic agonists of brown adipose tissue and thermogenesis. 1. [4-[2-[(2-Hydroxy-3-phenoxypropyl)amino]ethoxy]phenoxy]acetates. J Med Chem. 1992;35(10):1751–9.

Hu E, Liang P, Spiegelman BM. AdipoQ is a novel adipose-specific gene dysregulated in obesity. J Biol Chem. 1996;271(18):10697–703. Epub 1996/05/03.

Hutley LJ, Herington AC, Shurety W, Cheung C, Vesey DA, Cameron DP, et al. Human adipose tissue endothelial cells promote preadipocyte proliferation. Am J Physiol Endocrinol Metab. 2001;281(5):E1037–44. Epub 2001/10/12.

Hutley L, Shurety W, Newell F, McGeary R, Pelton N, Grant J, et al. Fibroblast growth factor 1: a key regulator of human adipogenesis. Diabetes. 2004;53(12):3097–106. Epub 2004/11/25.

Jamaluddin MS, Weakley SM, Yao Q, Chen C. Resistin: functional roles and therapeutic considerations for cardiovascular disease. Br J Pharmacol. 2012;165(3):622–32. Epub 2011/05/07.

Jansson PA. Endothelial dysfunction in insulin resistance and type 2 diabetes. J Intern Med. 2007;262(2):173–83. Epub 2007/07/25.

Joshi PV, Lele VR. Unexpected visitor on FDG PET/CT–brown adipose tissue (BAT) in mesentery in a case of retroperitoneal extra-adrenal pheochromocytoma: is the BAT activation secondary

to catecholamine-secreting pheochromocytoma? Clin Nucl Med. 2012;37(5):e119–20. Epub 2012/04/06.
Kennedy GC. The role of depot fat in the hypothalamic control of food intake in the rat. Proc R Soc Lond B Biol Sci. 1953;140(901):578–96. Epub 1953/01/15.
Kim SR, Bae SK, Choi KS, Park SY, Jun HO, Lee JY, et al. Visfatin promotes angiogenesis by activation of extracellular signal-regulated kinase 1/2. Biochem Biophys Res Commun. 2007;357(1):150–6. Epub 2007/04/06.
Kitagawa M, Era T. Differentiation of mesodermal cells from pluripotent stem cells. Int J Hematol. 2010;91(3):373–83. Epub 2010/03/13.
Komarova Y, Malik AB. Regulation of endothelial permeability via paracellular and transcellular transport pathways. Annu Rev Physiol. 2010;72:463–93. Epub 2010/02/13.
Lafontan M, Langin D. Lipolysis and lipid mobilization in human adipose tissue. Prog Lipid Res. 2009;48(5):275–97. Epub 2009/05/26.
Li J, Zhang YP, Kirsner RS. Angiogenesis in wound repair: angiogenic growth factors and the extracellular matrix. Microsc Res Tech. 2003;60(1):107–14. Epub 2002/12/25.
Lijnen HR. Angiogenesis and obesity. Cardiovasc Res. 2008;78(2):286–93. Epub 2007/11/17.
Lonnqvist F, Nordfors L, Jansson M, Thorne A, Schalling M, Arner P. Leptin secretion from adipose tissue in women. Relationship to plasma levels and gene expression. J Clin Invest. 1997;99(10):2398–404. Epub 1997/05/15.
MacDougald OA, Burant CF. The rapidly expanding family of adipokines. Cell Metab. 2007;6(3):159–61. Epub 2007/09/05.
Matsuda M, Shimomura I, Sata M, Arita Y, Nishida M, Maeda N, et al. Role of adiponectin in preventing vascular stenosis. The missing link of adipo-vascular axis. J Biol Chem. 2002;277(40):37487–91. Epub 2002/07/26.
Miano JM. Vascular smooth muscle cell differentiation. J Biomed Res. 2010;24(3):169–80.
Mu H, Ohashi R, Yan S, Chai H, Yang H, Lin P, et al. Adipokine resistin promotes in vitro angiogenesis of human endothelial cells. Cardiovasc Res. 2006;70(1):146–57. Epub 2006/03/07.
Nedvidkova J, Smitka K, Kopsky V, Hainer V. Adiponectin, an adipocyte-derived protein. Physiol Res. 2005;54(2):133–40. Epub 2004/11/17.
Neels JG, Thinnes T, Loskutoff DJ. Angiogenesis in an in vivo model of adipose tissue development. FASEB J. 2004;18(9):983–5. Epub 2004/04/16.
Nishimura S, Manabe I, Nagasaki M, Hosoya Y, Yamashita H, Fujita H, et al. Adipogenesis in obesity requires close interplay between differentiating adipocytes, stromal cells, and blood vessels. Diabetes. 2007;56(6):1517–26. Epub 2007/03/29.
Nishimura S, Manabe I, Nagasaki M, Eto K, Yamashita H, Ohsugi M, et al. CD8+ effector T cells contribute to macrophage recruitment and adipose tissue inflammation in obesity. Nat Med. 2009;15(8):914–20. Epub 2009/07/28.
O'Reilly MS, Holmgren L, Shing Y, Chen C, Rosenthal RA, Moses M, et al. Angiostatin: a novel angiogenesis inhibitor that mediates the suppression of metastases by a Lewis lung carcinoma. Cell. 1994;79(2):315–28. Epub 1994/10/21.
O'Reilly MS, Boehm T, Shing Y, Fukai N, Vasios G, Lane WS, et al. Endostatin: an endogenous inhibitor of angiogenesis and tumor growth. Cell. 1997;88(2):277–85. Epub 1997/01/24.
Ohashi K, Parker JL, Ouchi N, Higuchi A, Vita JA, Gokce N, et al. Adiponectin promotes macrophage polarization toward an anti-inflammatory phenotype. J Biol Chem. 2010;285(9):6153–60. Epub 2009/12/24.
Olsson AK, Dimberg A, Kreuger J, Claesson-Welsh L. VEGF receptor signalling—in control of vascular function. Nat Rev Mol Cell Biol. 2006;7(5):359–71. Epub 2006/04/25.
Ouchi N, Kobayashi H, Kihara S, Kumada M, Sato K, Inoue T, et al. Adiponectin stimulates angiogenesis by promoting cross-talk between AMP-activated protein kinase and Akt signaling in endothelial cells. J Biol Chem. 2004;279(2):1304–9. Epub 2003/10/15.
Ouchi N, Higuchi A, Ohashi K, Oshima Y, Gokce N, Shibata R, et al. Sfrp5 is an anti-inflammatory adipokine that modulates metabolic dysfunction in obesity. Science. 2010;329(5990):454–7. Epub 2010/06/19.

Ouellet V, Labbe SM, Blondin DP, Phoenix S, Guerin B, Haman F, et al. Brown adipose tissue oxidative metabolism contributes to energy expenditure during acute cold exposure in humans. J Clin Invest. 2012;122(2):545–52. Epub 2012/01/25.

Owens GK, Kumar MS, Wamhoff BR. Molecular regulation of vascular smooth muscle cell differentiation in development and disease. Physiol Rev. 2004;84(3):767–801. Epub 2004/07/23.

Park HY, Kwon HM, Lim HJ, Hong BK, Lee JY, Park BE, et al. Potential role of leptin in angiogenesis: leptin induces endothelial cell proliferation and expression of matrix metalloproteinases in vivo and in vitro. Exp Mol Med. 2001;33(2):95–102. Epub 2001/07/20.

Patel L, Buckels AC, Kinghorn IJ, Murdock PR, Holbrook JD, Plumpton C, et al. Resistin is expressed in human macrophages and directly regulated by PPAR gamma activators. Biochem Biophys Res Commun. 2003;300(2):472–6. Epub 2002/12/31.

Planat-Benard V, Silvestre JS, Cousin B, Andre M, Nibbelink M, Tamarat R, et al. Plasticity of human adipose lineage cells toward endothelial cells: physiological and therapeutic perspectives. Circulation. 2004;109(5):656–63. Epub 2004/01/22.

Qiao L, Kinney B, Schaack J, Shao J. Adiponectin inhibits lipolysis in mouse adipocytes. Diabetes. 2011;60(5):1519–27. Epub 2011/03/25.

Rajala MW, Obici S, Scherer PE, Rossetti L. Adipose-derived resistin and gut-derived resistin-like molecule-beta selectively impair insulin action on glucose production. J Clin Invest. 2003;111(2):225–30. Epub 2003/01/18.

Ramsay TG, Rao SV, Wolverton CK. In vitro systems for the analysis of the development of adipose tissue in domestic animals. J Nutr. 1992;122(3 Suppl):806–17. Epub 1992/03/01.

Rehman J, Traktuev D, Li J, Merfeld-Clauss S, Temm-Grove CJ, Bovenkerk JE, et al. Secretion of angiogenic and antiapoptotic factors by human adipose stromal cells. Circulation. 2004;109(10):1292–8. Epub 2004/03/03.

Revollo JR, Korner A, Mills KF, Satoh A, Wang T, Garten A, et al. Nampt/PBEF/Visfatin regulates insulin secretion in beta cells as a systemic NAD biosynthetic enzyme. Cell Metab. 2007;6(5):363–75. Epub 2007/11/07.

Robertson SA, Rae CJ, Graham A. Induction of angiogenesis by murine resistin: putative role of PI3-kinase and NO-dependent pathways. Regul Pept. 2009;152(1–3):41–7. Epub 2008/08/30.

Roh SG, Song SH, Choi KC, Katoh K, Wittamer V, Parmentier M, et al. Chemerin—a new adipokine that modulates adipogenesis via its own receptor. Biochem Biophys Res Commun. 2007;362(4):1013–8. Epub 2007/09/05.

Ronti T, Lupattelli G, Mannarino E. The endocrine function of adipose tissue: an update. Clin Endocrinol (Oxf). 2006;64(4):355–65. Epub 2006/04/06.

Rosen ED, Spiegelman BM. Molecular regulation of adipogenesis. Annu Rev Cell Dev Biol. 2000;16:145–71. Epub 2000/10/14.

Rosen ED, Spiegelman BM. Adipocytes as regulators of energy balance and glucose homeostasis. Nature. 2006;444(7121):847–53. Epub 2006/12/15.

Rupnick MA, Panigrahy D, Zhang CY, Dallabrida SM, Lowell BB, Langer R, et al. Adipose tissue mass can be regulated through the vasculature. Proc Natl Acad Sci U S A. 2002;99(16):10730–5. Epub 2002/08/01.

Saddi-Rosa P, Oliveira CS, Giuffrida FM, Reis AF. Visfatin, glucose metabolism and vascular disease: a review of evidence. Diabetol Metab Syndr. 2010;2:21. Epub 2010/03/30.

Seale P, Bjork B, Yang W, Kajimura S, Chin S, Kuang S, et al. PRDM16 controls a brown fat/skeletal muscle switch. Nature. 2008;454(7207):961–7. Epub 2008/08/23.

Seale P, Conroe HM, Estall J, Kajimura S, Frontini A, Ishibashi J, et al. Prdm16 determines the thermogenic program of subcutaneous white adipose tissue in mice. J Clin Invest. 2011;121(1):96–105. Epub 2010/12/03.

Silha JV, Krsek M, Sucharda P, Murphy LJ. Angiogenic factors are elevated in overweight and obese individuals. Int J Obes (Lond). 2005;29(11):1308–14. Epub 2005/06/15.

Steppan CM, Bailey ST, Bhat S, Brown EJ, Banerjee RR, Wright CM, et al. The hormone resistin links obesity to diabetes. Nature. 2001;409(6818):307–12. Epub 2001/02/24.

Suganami T, Nishida J, Ogawa Y. A paracrine loop between adipocytes and macrophages aggravates inflammatory changes: role of free fatty acids and tumor necrosis factor alpha. Arterioscler Thromb Vasc Biol. 2005;25(10):2062–8. Epub 2005/08/27.

Sumpio BE, Riley JT, Dardik A. Cells in focus: endothelial cell. Int J Biochem Cell Biol. 2002;34(12):1508–12. Epub 2002/10/16.

Tang W, Zeve D, Suh JM, Bosnakovski D, Kyba M, Hammer RE, et al. White fat progenitor cells reside in the adipose vasculature. Science. 2008;322(5901):583–6. Epub 2008/09/20.

Trayhurn P, Beattie JH. Physiological role of adipose tissue: white adipose tissue as an endocrine and secretory organ. Proc Nutr Soc. 2001;60(3):329–39. Epub 2001/10/30.

Trayhurn P, Wood IS. Adipokines: inflammation and the pleiotropic role of white adipose tissue. Br J Nutr. 2004;92(3):347–55. Epub 2004/10/08.

Tschritter O, Fritsche A, Thamer C, Haap M, Shirkavand F, Rahe S, et al. Plasma adiponectin concentrations predict insulin sensitivity of both glucose and lipid metabolism. Diabetes. 2003;52(2):239–43. Epub 2003/01/24.

Uldry M, Yang W, St-Pierre J, Lin J, Seale P, Spiegelman BM. Complementary action of the PGC-1 coactivators in mitochondrial biogenesis and brown fat differentiation. Cell Metab. 2006;3(5):333–41. Epub 2006/05/09.

Vachharajani V, Granger DN. Adipose tissue: a motor for the inflammation associated with obesity. IUBMB Life. 2009;61(4):424–30. Epub 2009/03/26.

Varzaneh FE, Shillabeer G, Wong KL, Lau DC. Extracellular matrix components secreted by microvascular endothelial cells stimulate preadipocyte differentiation in vitro. Metabolism. 1994;43(7):906–12. Epub 1994/07/01.

Verma S, Li SH, Wang CH, Fedak PW, Li RK, Weisel RD, et al. Resistin promotes endothelial cell activation: further evidence of adipokine-endothelial interaction. Circulation. 2003;108(6):736–40. Epub 2003/07/23.

Viengchareun S, Zennaro MC, Pascual-Le Tallec L, Lombes M. Brown adipocytes are novel sites of expression and regulation of adiponectin and resistin. FEBS Lett. 2002;532(3):345–50. Epub 2002/12/17.

Vona-Davis L, Rose DP. Adipokines as endocrine, paracrine, and autocrine factors in breast cancer risk and progression. Endocr Relat Cancer. 2007;14(2):189–206. Epub 2007/07/20.

Waki H, Tontonoz P. Endocrine functions of adipose tissue. Annu Rev Pathol. 2007;2:31–56. Epub 2007/11/28.

Wang H, Chu WS, Hemphill C, Elbein SC. Human resistin gene: molecular scanning and evaluation of association with insulin sensitivity and type 2 diabetes in Caucasians. J Clin Endocrinol Metab. 2002;87(6):2520–4. Epub 2002/06/07.

Weisberg SP, McCann D, Desai M, Rosenbaum M, Leibel RL, Ferrante Jr AW. Obesity is associated with macrophage accumulation in adipose tissue. J Clin Invest. 2003;112(12):1796–808. Epub 2003/12/18.

Winer S, Chan Y, Paltser G, Truong D, Tsui H, Bahrami J, et al. Normalization of obesity-associated insulin resistance through immunotherapy. Nat Med. 2009;15(8):921–9. Epub 2009/07/28.

Wronska A, Kmiec Z. Structural and biochemical characteristics of various white adipose tissue depots. Acta Physiol (Oxf). 2012;205(2):194–208. Epub 2012/01/10.

Xie H, Tang SY, Luo XH, Huang J, Cui RR, Yuan LQ, et al. Insulin-like effects of visfatin on human osteoblasts. Calcif Tissue Int. 2007;80(3):201–10. Epub 2007/03/07.

Xue Y, Cao R, Nilsson D, Chen S, Westergren R, Hedlund EM, et al. FOXC2 controls Ang-2 expression and modulates angiogenesis, vascular patterning, remodeling, and functions in adipose tissue. Proc Natl Acad Sci U S A. 2008;105(29):10167–72. Epub 2008/07/16.

Xue Y, Petrovic N, Cao R, Larsson O, Lim S, Chen S, et al. Hypoxia-independent angiogenesis in adipose tissues during cold acclimation. Cell Metab. 2009;9(1):99–109. Epub 2009/01/02.

Yamashita H, Sato Y, Kizaki T, Oh S, Nagasawa J, Ohno H. Basic fibroblast growth factor (bFGF) contributes to the enlargement of brown adipose tissue during cold acclimation. Pflugers Arch. 1994;428(3–4):352–6. Epub 1994/10/01.

Yamauchi T, Kamon J, Waki H, Terauchi Y, Kubota N, Hara K, et al. The fat-derived hormone adiponectin reverses insulin resistance associated with both lipoatrophy and obesity. Nat Med. 2001;7(8):941–6. Epub 2001/08/02.

Yancopoulos GD, Davis S, Gale NW, Rudge JS, Wiegand SJ, Holash J. Vascular-specific growth factors and blood vessel formation. Nature. 2000;407(6801):242–8. Epub 2000/09/23.

Yang WS, Lee WJ, Funahashi T, Tanaka S, Matsuzawa Y, Chao CL, et al. Weight reduction increases plasma levels of an adipose-derived anti-inflammatory protein, adiponectin. J Clin Endocrinol Metab. 2001;86(8):3815–9. Epub 2001/08/15.

Zabel BA, Ohyama T, Zuniga L, Kim JY, Johnston B, Allen SJ, et al. Chemokine-like receptor 1 expression by macrophages in vivo: regulation by TGF-beta and TLR ligands. Exp Hematol. 2006;34(8):1106–14. Epub 2006/07/26.

Zhang Y, Proenca R, Maffei M, Barone M, Leopold L, Friedman JM. Positional cloning of the mouse obese gene and its human homologue. Nature. 1994;372(6505):425–32. Epub 1994/12/01.

Zuk PA. Stem cell research has only just begun. Science. 2001;293(5528):211–2. Epub 2001/07/17.

Zuk PA, Zhu M, Mizuno H, Huang J, Futrell JW, Katz AJ, et al. Multilineage cells from human adipose tissue: implications for cell-based therapies. Tissue Eng. 2001;7(2):211–28. Epub 2001/04/17.

Chapter 6
Lymphatic System in Adipose Tissues

Bernhard Nausch, Sonia Rehal, and Pierre-Yves von der Weid

Abstract The lymphatic system is uniquely aligned to integrate with many organ systems. Originally discovered as a vessel network separate from the circulatory system, it took centuries before the lymphatic system was given the attention it truly deserves. Within the last 2 decades, a resurgence of literature has revealed the lymphatic system to be intimately linked to tissue fluid homeostasis, immune cell trafficking, and nutrient absorption. Particularly, the lymphatic system within the digestive system is specially devoted to transport absorbed lipids, making it an important organ system in adipose tissue biology. This chapter will review the origins and structural organization of the lymphatic system, followed by the key events taking place in the development of new lymphatic vessels, or lymphangiogenesis. The biological links between the lymphatic system and the adipose tissue will be further highlighted, with emphasis on its functional relevance. Finally, the implications of dysfunctions of the lymphatic system will be considered in relation to adipose tissue disorders.

Keywords Lymphangiogenesis • Lymphatic pumping • Lymph drainage • Lipid transport • Fat deposition • Inflammation • Inflammatory mediators

Bernhard Naush and Sonia Rehal contributed equally to the redaction of this manuscript.

B. Nausch, Ph.D. • S. Rehal, M.Sc. • P.-Y. von der Weid (✉)
Department of Physiology & Pharmacology, Inflammation Research Network and Smooth Muscle Research Group, Snyder Institute for Chronic Diseases, University of Calgary, 3330 Hospital Drive N.W., Calgary, AB, Canada T2N 4N1
e-mail: vonderwe@ucalgary.ca

Y. Cao (ed.), *Angiogenesis in Adipose Tissue*, DOI 10.1007/978-1-4614-8069-3_6,

6.1 Discovery of the Lymphatic System

The very first reference to lymphatic vessels can be found in the early works of Aristotle (384–322 BC) who described vessels that contain a colorless fluid (Lord 1968). Similarly, Hippocrates (460–270 BC) observed structures containing "white blood" (Skobe and Detmar 2000), most probably describing lymphatic vessels. After being overlooked for several centuries, lymphatic vessels were rediscovered by Gaspare Aselli (1581–1626), professor of anatomy and surgery at the University of Pavia. He identified "opaque vessels," which he named "lacteis venis" or milky veins, in the postprandial dog's mesentery (as reviewed in Alexander et al. (2010)). Since lymphatic vessels were first identified in the mesentery in close proximity to the digestive system, it was assumed that their main function was the transport of nutrients from the gut to the liver, although Herophilus (335–280 BC) noted that these "peculiar veins" terminate in glandular bodies, most likely describing lymph nodes (Lord 1968; Cruikshank 1786). Subsequently, in 1651, Jean Pecquet described the cisterna chyli and thoracic duct, laying the foundation for the current knowledge that lymph is returned to the great veins in the neck (as reviewed in Skobe and Detmar (2000)). Two years later Olof Rudbeck and Thomas Bartholin implicated lymphatic vessels in absorption of interstitial fluid throughout the entire body, not just the digestive system (Mayerson 1963).

Mechanisms that underlie fluid absorption via lymphatics have been studied since the nineteenth century. Starling demonstrated in 1896 that lymph is a special filtrate from the blood (as reviewed in Skobe and Detmar (2000)), while Cecil Drinker and Joseph Yoffey described the lymphatic system as an absorbing system that returns fluid and proteins, filtered from the capillaries, to the bloodstream (Yoffey and Drinker 1938; Yoffey et al. 1938). More importantly, lymph has been shown to contain a plethora of cells, including immune cells. Early observations include those of William Hewson of Hexham (1739–1774) noting lymphocytic structures in lymph, followed by the proposition by Henri François Le Dran (1685–1770) implicating lymphatic vessels as routes for the metastasis of cancer. Thus, very early observations of the lymphatic system indicated its major functions: (1) transport of lipids from the gut to the blood, (2) maintenance of fluid homeostasis in the interstitium, and (3) transport of immune cells to allow for the initiation of specific immune responses.

6.2 Structure and Function of Lymphatic Vessels and Lymph Nodes

Fluid homeostasis is delicately balanced by fluid pressures within the blood and lymphatic vessels. Blood pressure forces fluid and proteins to leak out of capillaries. Interstitial fluid that is not reabsorbed by the venous system enters the lymphatic system driven by hydrostatic and oncotic pressures and is subsequently termed

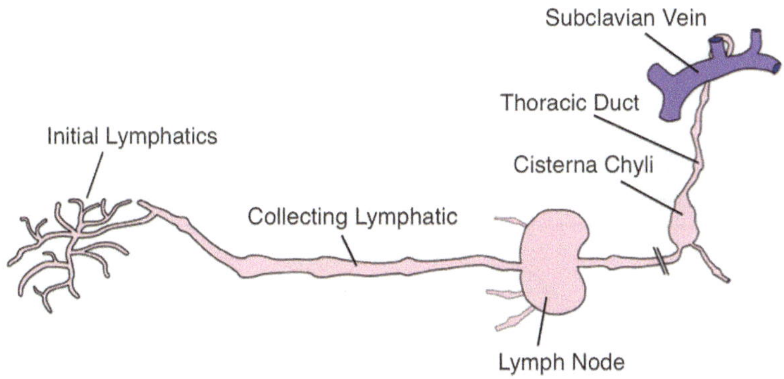

Fig. 6.1 Structural organization of the lymphatic system. Initial lymphatics absorb fluid in the interstitium and drain lymph into collecting lymphatics. These muscular vessels propel lymph by means of active contractions to the lymph node and further to the cisterna chili, which receives all lymph from the legs and abdominal organs. From here lymph is transported by the thoracic duct, the body's largest collecting lymphatic, to the subclavian vein

lymph. The lymphatic system is responsible for draining 10 % of the interstitial fluid that is left behind by the circulatory system, which amounts to about 2 L of lymph a day (Bierman et al. 1953). If these were not removed by the lymphatic system, fluid and proteins would accumulate and lymph stasis and edema would accrue.

The structure of the lymphatic system is intelligently designed to accomplish this drainage of fluid along with its other functions, such as lipid absorption and immune cell trafficking. As illustrated in Fig. 6.1, the lymphatic vessel network originates with initial lymphatic vessels, where interstitial fluid enters the lymphatic system. Lymph is then actively propelled against hydrostatic pressure gradients to make its way to local immune structures, the lymph nodes. Exiting lymph nodes the lymph travels through larger efferent vessels and lymphatic ducts to ultimately enter the blood circulation, when ducts merge with the superior vena cava at the level of the subclavian veins (Ryan 1989).

6.2.1 Initial Lymphatic Vessels

The initial lymphatic vessels (initial lymphatics, Fig. 6.2a) are found penetrating almost all tissue beds and organs of the body. They are often described as blind-ended tubes in the interstitium, much like the fingers of a glove. Initial lymphatics are larger than blood capillaries and have a diameter of ~60–110 μm in the skin (Ryan 1989). Their thin wall consists of a single layer of oak leaf-shaped endothelial cells and a discontinuous, rudimentary basement membrane (Leak 1976). Neighboring endothelial cells are connected by button-like junction (Fig. 6.2a), made of vascular endothelial (VE)-cadherin and tight junction proteins such as occluding, claudin-5,

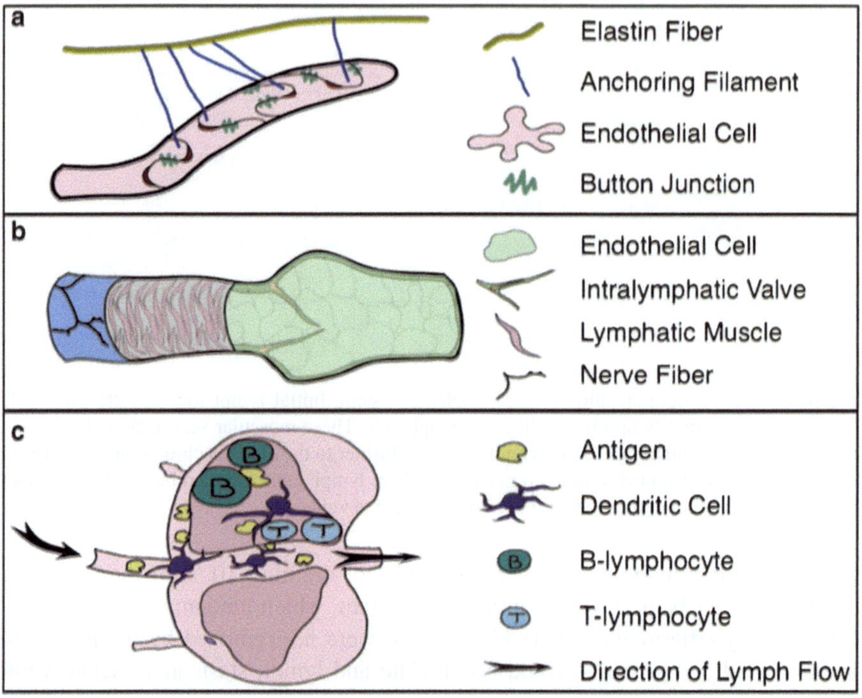

Fig. 6.2 Components of the lymphatic system. (**a**) Schematic of an initial lymphatic depicting oak leaf-shaped endothelial cells connected by button junctions, which leave free flaps that form primitive valves. Endothelial cells are tethered to elastin fibers of surrounding connecting tissue by anchoring filaments. (**b**) Schematic of collecting lymphatics depicting the three cell layers in their wall. The adventitia consists of connective tissue and contains nerve fibers. The media is formed by lymphatic muscle embedded in collagen and elastin fibers. The intima consists of tightly connected endothelial cells. Intralymphatic valves are formed by a matrix core sandwiched between endothelial cells. (**c**) Schematic of lymph node depicting the flow of lymph through the lymph node and activation of resident T- and B-lymphocytes by dendritic cells and antigens, respectively. B-cells reside in germinal centers in the peripheral cortex and T-cells deeper in the parenchyma. Lymph enters the subcapsular sinus through afferent lymphatics and leaves the lymph node at the hilum through the efferent lymphatic

zonula occludens (ZO)-1, endothelial cell-selective adhesion molecule (ESAM), and junctional adhesion molecule-A (JAM-A) (Baluk et al. 2007). Pores of about 2 μm exist between these button-like junctions, and the nonconnected flaps of the oak leaf-shaped endothelial cells form primitive valves that allow for entry of fluid from the interstitium into the lumen, elegantly preventing leakage of lymph in the opposite direction (Ryan 1989; Schulte-Merker et al. 2011; Schmid-Schonbein 2003; Mendoza and Schmid-Schonbein 2003). The "pores" formed between neighboring endothelial cells not only provide the major entry way into the lymphatic system for fluids, but also for immune cells, proteins, and chylomicrons.

The endothelial cells of the initial lymphatics are tethered to the surrounding connective tissue by anchoring filaments (Fig. 6.2a). These filaments consist of fibrillin and tie the outer valve-forming flaps of the endothelial cells to elastin fibers in the surrounding connective tissue (Leak and Burke 1968). This arrangement allows for expansion of the lymphatic vessel lumen and consequent entry of fluid when the interstitium is expanded due to accumulation of interstitial fluid. When the initial lymphatic is compressed by movement of adjacent pulsating arteries, surrounding skeletal muscle, or during manual lymphatic drainage, the lymph is pushed towards the next lymphatic vessel structure, the collecting lymphatic vessels.

6.2.2 *Collecting Lymphatic Vessels*

Collecting lymphatic vessels (collecting lymphatics, Fig. 6.2b) are considerably larger than initial lymphatics with a diameter of ~ 80–200 μm in the rat mesentery (Huxley and Scallan 2011). Collecting vessels are more complex in nature than initial lymphatics, comprising an intima, media, and adventitia. Their innermost layer, the intima, consists of endothelial cells and a basement membrane. In contrast to initial lymphatics, neighboring endothelial cells are tightly connected by zipper-like junctions that contain VE-cadherin, occluding, claudin-5, ZO-1, ESAM, and JAM-A (Baluk et al. 2007). In addition, collecting vessel endothelial cells also form intraluminal semilunar valves (Bazigou et al. 2009). These valves consist of a matrix core that is sandwiched between endothelial cells. These valves direct lymph flow in a unidirectional manner while also dividing the collecting lymphatics into segments which are called "lymphangions" (Mislin 1961).

The media layer of collecting lymphatics consists of lymphatic muscle cells embedded in a meshwork of collagen and elastin fibers (Arkill et al. 2010). Lymphatic muscle confers to collecting lymphatic vessels their ability to rhythmically contract. This function is the main driving force of the intrinsic lymph pump and allows for active transport of lymph along the lymphatic vessel network (Orlov et al. 1975; Mislin 1976; Hargens and Zweifach 1977). The contractile machinery within the lymphatic muscle is very similar to that seen in vascular smooth muscle, with the existence of additional actin isoforms typically found in cardiac muscle (Muthuchamy et al. 2003). This characteristic makes these cells a unique muscle entity. The intrinsic and rhythmic contractile activity of lymphatic muscle is another similarity to cardiac muscle. This feature and the presence of valves on each side of a lymphatic chamber bear similarity to the ventricles of the heart and lymphangions are also referred to as "lymph hearts."

The outermost layer of the collecting lymphatic's wall is the adventitia that consists of mostly connective tissue and contains adrenergic (Costa and Funess 1973; Ohhashi et al. 1982), cholinergic (Ohhashi et al. 1982), and peptidergic (Hukkanen et al. 1992) nerve fibers. In addition, adipocytes have been observed entangled in the meshwork of collagen and elastin fibers of the vessel wall (Arkill et al. 2010).

Although contractile activity of lymphatic muscle is autonomous (Mislin 1961), it is subject to regulation by physical factors such as temperature and stretch (Mislin 1961; Davis et al. 2009), as well as by chemical mediators. Endothelium-derived nitric oxide (NO), for example, inhibits contractility (Ferguson et al. 1994; Gashev et al. 2002; von der Weid and Van Helden 1996; Gasheva et al. 2006), as does prostaglandin E_2 (Mizuno et al. 1998; Rehal et al. 2009; Elias and Johnston 1988; Johnston et al. 1983; Johnston and Gordon 1981; Johnston and Feuer 1983). Prostaglandin H2/Thromboxane A2, on the other hand, increases contractile activity (Mizuno et al. 1998; Johnston et al. 1983; Johnston and Gordon 1981; Johnston and Feuer 1983; Plaku and von der Weid 2006). In addition to the endothelium, factors that could regulate contractile activity can be derived from other sources, such as neighboring tissues or the lymph itself. As will be discussed later in the chapter, one particularly relevant source of factors is the adipose tissue, which very often surrounds collecting lymphatics. Adipocytes and adipose tissue macrophages present in these depots or entangled in the adventitia (see above) actively secrete many different neurohormonal substances, such as NO (Hodis et al. 2011; Lumeng et al. 2007), potentially able to modulate lymphatic contractile activity (Wang et al. 2007) (Fig. 6.6).

6.2.3 Lymph Nodes

Lymph nodes are bean-shaped secondary lymphoid organs that can be found interspersed between afferent and efferent lymphatic vessels (Fig. 6.2c). They play similar roles to the thymus and spleen, providing an interaction platform for foreign antigens with systemically circulating immune cells and serve as local immune centers for corresponding tissues and organs. Lymph enters the subcapsular sinus of lymph nodes through afferent lymphatics, travels through lymphatic sinuses, and subsequently leaves the lymph node through the efferent lymphatic vessel (Grigorova et al. 2010; Ohtani and Ohtani 2008). Conduits within the lymph node facilitate the flow of lymph and antigens deeper into the lymph node parenchyma (as reviewed in Shields (2011)).

Lymph nodes are rich in lymphocytes of different activation phenotypes (Drayson and Ford 1984). Specifically, lymph nodes contain T-cell zones within the medullary and cortical sinuses, and B-cell zones located in the outer cortical area of the lymph node (Drayson and Ford 1984). Antigens, either presented by antigen-presenting cells or transported by convection with the lymph, meet and interact with resident lymphocytes within the lymph node, initiating an appropriate and specific immune response to foreign antigens, or immune tolerance to self-antigens. Depending on the antigen being encountered, T-cells and B-cells can differentiate and mature to eventually destroy foreign antigens while leaving self-antigens untouched. Critical to proper lymphocyte development, and therefore lymph node function is a stimulating microenvironment. Above and beyond specific colony-stimulating factors, lipids provide the proper building materials and fuel for lymphocyte development (Pond 2000). Accordingly, it is not surprising that lymph nodes are embedded in adipose tissue depots (Fig. 6.3b). The role of perinodal adipose tissue (PAT) will be discussed later in this chapter.

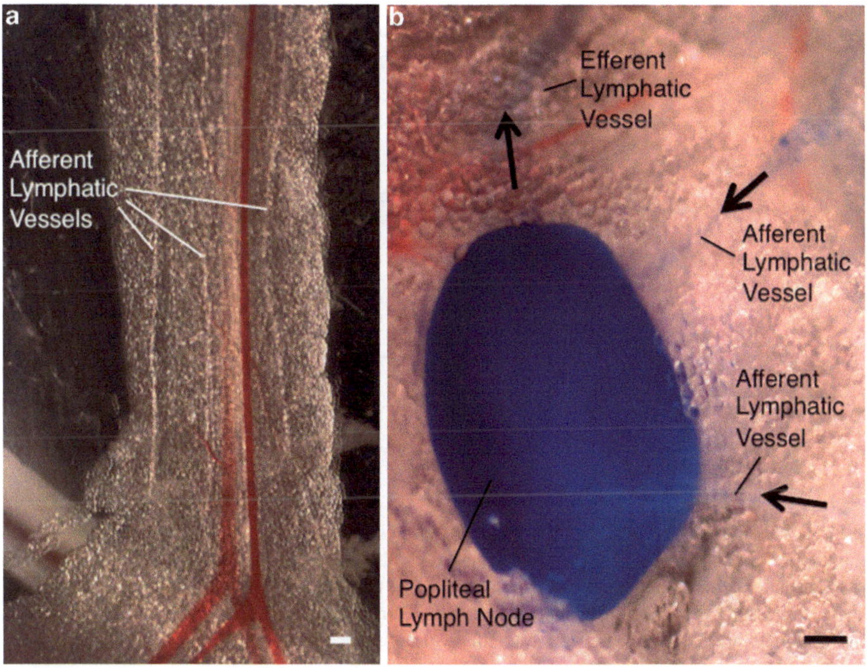

Fig. 6.3 Lymphatic vessels and lymph nodes embedded in fat. (**a**) Photograph of a rat mesenteric vascular arcade where afferent collecting lymphatics embedded in fat can be identified along a vein and an artery. (**b**) Photograph of a mouse popliteal lymph node with two afferent lymphatics and one efferent lymphatic labeled with Indian ink. The afferent lymphatics enter the lymph node from the *top* and from the side of the lymph node, while the efferent lymphatic exits from underneath the lymph node at its hilum. The lymph node and its associated lymphatic vessels are embedded in the popliteal fat pad. Part of the fat pad had been removed to allow for a better visualization of the lymph node and the lymphatics. Scale bars 200 μm

6.3 Adipolymphatic Connections

Adipose tissue is composed of mainly adipocytes that make up ~ 90 % of the tissue with the remaining ~ 10 % consisting of collagen-rich connective tissue that contains blood capillaries, fibroblasts, and nerve fibers (Ryan 1989, 1995). Adipocytes are cells that store a large droplet of lipids. Most lipids are derived from the diet and are absorbed in the jejunum. Lipids are solubilized by the action of bile acids and fatty acids are cleaved off by pancreatic lipases. Free fatty acids and monoacylglycerols are absorbed by the enterocytes in the gut epithelium (Fig. 6.4) and triglycerides are resynthesized from these substrates (Kindel et al. 2010). Together with cholesterol the triglycerides are packaged into chylomicrons, large lipoprotein particles that consist of a core of triglycerides and cholesterol esters surrounded by more polar free fatty acids and un-esterified cholesterol as well as proteins, especially Apo B, but also Apo AI, AII, CI, CII, and CIII (Cox and Garcia-Palmieri

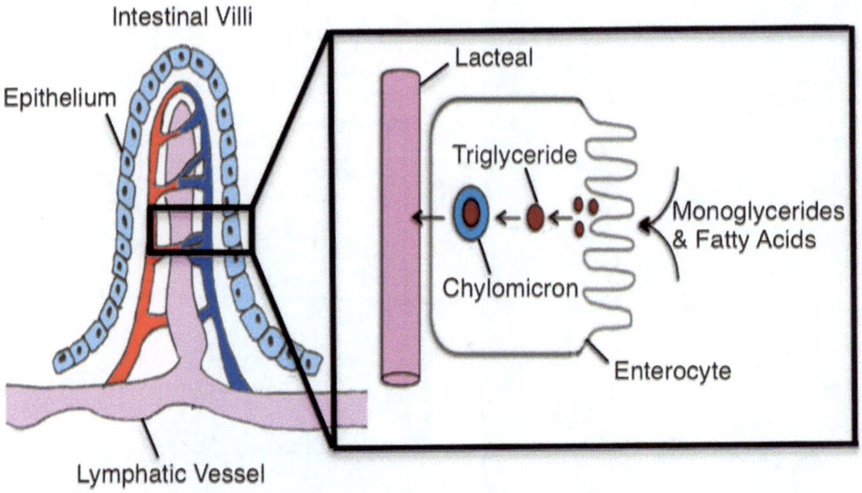

Fig. 6.4 Lipid absorption through the lymphatic system. Schematic depicting chylomicron absorption by an initial lymphatic vessel (lacteal) in the villi of the jejunum. Monoacylglycerides and free fatty acids are absorbed by jejunal enterocytes that resynthesize triacylglycerides. Together with cholesterol, phospholipids, and proteins these triglycerides are incorporated into chylomicrons, which are secreted into the extracellular space in the villi and absorbed into the lacteal

1990). Chylomicrons are exocytosed by the enterocytes into the interstitium of the villi where they enter the lacteals (Kohan et al. 2011) and are transported in the lymph via collecting lymphatics, lymph nodes, the cisterna chyli, and thoracic duct to the bloodstream (as reviewed in Rutkowski et al. (2006)). In blood capillaries, endothelial lipases cleave fatty acids off the lipoprotein particles and these fatty acids are either used by cells for energy or stored in adipocytes (Wang et al. 2009a). Thereby the lymphatic system is essential for the transport of dietary lipids from the small intestine to the blood. Any lymphatic dysfunction could lead to visceral edema, malabsorption of lipids, or protein-losing enteropathy (Rockson 2010).

The lymphatic system not only transports lipids from the gut, but also removes lipids from the interstitium of peripheral tissues (Nanjee et al. 2001; Wong et al. 1995), such as the skin. The lipids found in peripheral lymph stem from the blood circulation, whereas fatty acids released from adipocytes are usually bound by albumin and removed by blood vessels (Ranallo and Rhodes 1998). Interestingly, in early studies, initial lymphatics were not found within adipose tissue, but at its borders (Ryan 1989, 1995). This prompted researchers to hypothesize that adipose tissue develops where there is no drainage of lipids from the tissue by lymphatic vessels. Unna postulated that uptake of lipids by cells occurs where lymph is stagnant (Unna 1896) and thus where lipids are not removed from the interstitium. Clark and Clark showed that fat deposition in the wounded area of a rabbit ear occurs at sites devoid of lymphatic vessels (Clark and Clark 1940), and recently it has been shown that saphenous vein incompetence is associated with loss of lymphatic vessels and accumulation of lipids (Tanaka et al. 2012). A direct relationship between extracellular lipid concentration and the amount of lipid uptake was shown in tissue culture by Poznanski in 1973 (Poznanski et al. 1973).

These observations are reminiscent of what occurs during lymphedema, where lymphatic drainage is disrupted or malfunctioning. This situation can result from a heritable or genetic malformation of lymphatics (primary lymphedema), or more commonly, as a consequence of infection, injury, or lymphatic disruption due to surgery (secondary lymphedema) (Rockson 2001; Szuba and Rockson 1998) and leads to imbalances in lymphatic load and transport, and edema formation. Increased fluid in the interstitial space leads to architectural changes in the skin and subcutaneous tissues, which are often significant (Rockson 2001; Piller 1980; 1990). As lymph stasis and the resultant edema become chronic, there is a predisposition to increasing numbers of adipocytes, fibroblasts, and keratinocytes in the edematous tissues (Shin et al. 2003). Such changes in tissue composition lead to thickening of the skin and subcutaneous tissue fibrosis (Rockson 2001), but these events are not well understood. In addition to lymph stasis and tissue structure disruption inflammation occurs in lymphedema (Tabibiazar et al. 2006).

Hypothesizing that lymph stasis promotes inflammation and adipogenesis, Schirger et al. (1962) examined surgically removed subcutaneous tissue from patients with primary lymphedema. Perilymphatic infiltration of plasma cells and lymphocytes, fibrosis in the dermis, subcutaneous septa, and deep fascia, as well as fibro-adipose tissue in the subcutis were observed in the resected tissue. More recently, liposuction was used to treat chronic lymphedema, for example after mastectomy with axillary lymph node resection. These studies reported that the aspirate consisted mainly of an adipose fraction (90–100 %), and a much smaller fluid fraction (Brorson 2010; Peled et al. 2012), clearly demonstrating that chronic edematous limbs contain large amounts of adipose tissue in addition to lymph.

Adipocyte accumulation has also been reported in animal models, such as the Chy mouse, which exhibits lymphedema caused by hypoplastic cutaneous lymphatics as a result of heterozygous inactivating mutations in the VEGFR3 gene. These animals also display increased adiposity in the edematous subcutaneous layer close to the hypoplastic dysfunctional dermal lymphatics (Karkkainen et al. 2001). Histopathology, architectural changes, and lymphoscintigraphic characteristics similar to that observed in human acquired lymphedema are well simulated in another animal model of acute lymph stagnation, where excision of a 2 mm circumferential segment of skin from a mouse-tail that effectively disrupts lymphatic drainage but leaves blood supply intact causes an edema distal to the site of injury (Tabibiazar et al. 2006; Slavin et al. 1999). This experimental model was recently used to further study the effect of lymphatic fluid stasis on adipogenesis (Aschen et al. 2012; Zampell et al. 2012). The authors showed that lymph stasis leads to an increase in adipose tissue mass as well as fibrosis and inflammation of the subcutaneous adipose tissue (Zampell et al. 2012). In addition, the same group analyzed factors that contribute to the development of adipose tissue and found an up-regulation of CCAAT/enhancer-binding protein-α (CEBP-α) and adiponectin

(Aschen et al. 2012). CREB-α participates in the activation of peroxisome proliferator-activated receptor-γ (PPAR-γ), which is a crucial factor in adipogenesis (Cristancho and Lazar 2011). Adiponectin is an adipocytokine with anti-inflammatory properties, which normally induces fatty acid β-oxidation (Aghamohammadzadeh et al. 2012), but is also known to be expressed at high levels during lipid accumulation (as reviewed in Harford et al. (2011)). These data corroborate clinical studies indicating that chronic lymph stasis results in inflammation, fibrosis, and adipose deposition.

Other important experimental data further strongly suggest that altered lymphatic vessel development promotes obesity. Harvey et al. reported that a genetic mouse model of abnormal lymphatic development, the Prox-1 haploinsufficient mouse, develops late-onset obesity (Harvey et al. 2005). Prox-1 is a transcription factor that is crucial for the development of lymphatics, and haploinsufficiency results in malformed, dysfunctional, and leaky lymphatics. This greatly hampers lymphatic drainage, promotes lymph stasis, and the subsequent development of obesity (Harvey et al. 2005). In addition, differentiation of preadipocytes in culture was induced by the addition of chyle, i.e., lymph containing lipids. In contrast, addition of safflower oil did not induce differentiation, indicating that the availability of fat alone is not the adipogenic factor driving adipocyte differentiation (Harvey et al. 2005). Similarly, Nougues et al. presented evidence that differentiation of rabbit adipocyte precursors to adipocytes is enhanced when either lymph or chylomicrons are added to the culture medium (Nougues et al. 1988).

Alternatively, several studies investigated the effects of lipids, adipose tissue, and obesity on lymphatic function. Tso et al. reported an increase in mesenteric lymph flow following a fatty meal (Tso et al. 1985). Although the underlying mechanisms are still unknown, the presence in lymph of lipid components such as low-density lipoprotein (LDL), which has been shown to enhance lymphatic contractile activity (Wang et al. 2009b), could mediate this increase.

A direct link between fat metabolism and lymphatic function was suggested from a study showing that hypercholesterolemic apolipoprotein E knockout mice have dysfunctional lymphatic vessels with abnormal expression of lymphatic vessel endothelial hyaluronic acid receptor 1 (LYVE-1), enlarged initial lymphatics, and reduced smooth muscle content of collecting lymphatics (Lim et al. 2009). A recent clinical study reported that the normal increase in blood and lymph flow that occurs in subcutaneous adipose tissue in response to glucose ingestion is blunted in obese subjects (Arngrim et al. 2013). Altogether, these data suggest that lymphatic function can be regulated by adipose tissue, lipids, or cholesterol (see Fig. 6.6), but detailed investigations are still outstanding.

The adipogenic property of lymph could account for the development of adipose tissue along collecting lymphatic vessels and around lymph nodes (Fig. 6.3). This anatomical arrangement is similar to the intimate relation between small blood vessels and their perivascular adipose tissue, a source of adipocytokines that regulate blood vessel diameter and thereby blood flow (Aghamohammadzadeh et al. 2012). Interestingly, peripheral lymph contains several adipocytokines, namely monocyte chemoattractant protein-1 (MCP-1), leptin, tumor necrosis factor α (TNFα), as well as the interleukins (IL-) 1β, 6, and 8 (Miller et al. 2011), all of which could potentially modulate lymphatic activity. Zawieja et al. examined lymphatic contractile activity in a rat model of metabolic syndrome. In these high-fructose diet-fed rats, plasma triglycerides were elevated and TNFα over-expressed compared to control. In addition, contraction frequency of lymphatic vessels and lymph flow were significantly reduced (Zawieja et al. 2012). However, whether TNFα plays a role in the observed impairment of lymphatic pumping remains unknown. Several adipose-derived modulators have been suggested to affect lymphatic pumping, such as TNFα and LDL; however further studies need to confirm these hypotheses (Fig. 6.6). Furthermore, prostaglandins and other inflammatory mediators could be released from adipose-resident macrophages and alter pumping (Fig. 6.6).

Like collecting lymphatics, lymph nodes are also embedded in a fat pad (Fig. 6.3b). This perinodal adipose tissue (PAT) serves the needs of the lymph node rather than the body's need for energy (Mattacks and Pond 1999), and accordingly, lipolysis in PAT is stimulated by chronic inflammation as seen by the increased fragmentation of lipid droplets in perinodal adipocytes in arthritic mice (Straub et al. 2011). Further, noradrenalin and the cytokines TNFα and IL-4 are more potent inducers of lipolysis in perinodal than in perirenal or gonadal adipocytes (Mattacks and Pond 1999). PAT is further distinguished by its lipid composition. In mesenteric lymph nodes, PAT contains more polyunsaturated fatty acids (PUFAs), such as linoleic acid (18:2n-6) and α-linoleic acid (18:3n-3), but less saturated fatty acids such as margaric acid (17:0) and stearic acid (18:0) than subcutaneous adipose depots (Westcott et al. 2005). Interestingly, the fatty acid composition of triglycerides of perinodal adipocytes correlates with the fatty acid composition of phospholipids from stimulated lymph nodes, indicating that PAT supplies lipids for proliferating lymphoid cells (Pond and Mattacks 2003).

6.4 Introduction to Lymphangiogenesis

Biological functions of the lymphatic system such as lipid absorption and immune cell trafficking set it apart from the vascular system. Although lymphangiogenesis, the development of lymphatic vessels, has received far less attention than angiogenesis, interest in the genetic programming of lymphangiogenesis has been growing recently and research has linked many unique signaling molecules to lymphangiogenesis (Karkkainen et al. 2004; Fang et al. 2000). Dr. Florence Sabin, a renowned scientist in the early 1920s, made the important discovery that lymphatic vessels originate from the cardinal vein, budding off the venous vasculature almost like branches on a tree (Chikly 1997). Since then many molecules have been identified in the key steps of lymphangiogenesis, the commitment to lymphatic endothelial cells, migration of these cells into tissues, proliferation of newly formed primitive lymph vessels, separation from the cardinal vein, and maturation of the lymphatic system (Fig. 6.5) as discussed in more detail below.

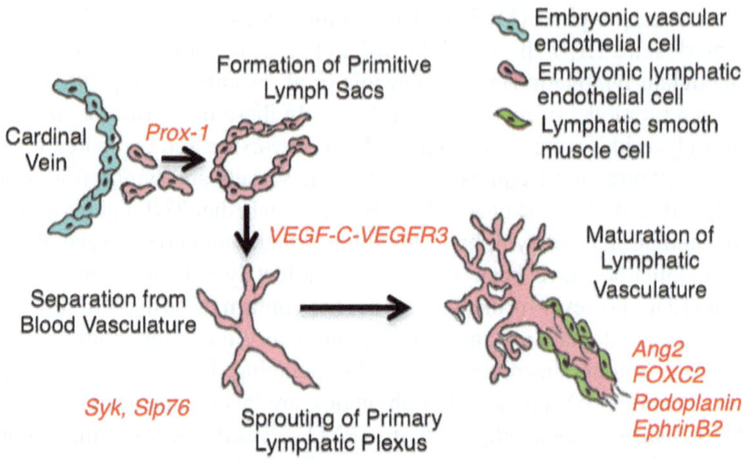

Fig. 6.5 Molecular events in lymphangiogenesis. Schematic representation of the main events occurring during lymphangiogenesis. Lymphatic endothelial cell-commitment occurs through the master regulator transcription factor, Prox-1. One of the major determinants for proper lymphatic development regulated by Prox-1, VEGF-C interacts with its receptor, VEGFR3 to induce budding, migration, and formation of primary lymph sacs. Complete separation of the lymphatic vasculature ensues through the expression of proteins Syk and Slp76. Finally maturation of lymphatic vessels such as valve formation and smooth muscle cell recruitment takes place under the transcription factor FOX2 and proteins such as angiopoietin-2 (Ang2), EphrinB2, and podoplanin

6.4.1 Commitment to Lymphatic Endothelium

Lymphatic development begins shortly after the establishment of the primary vasculature (Oliver 2004), around E 9.5–E 10.5 in mice, when endothelial cells from cardinal veins start to express cell-commitment proteins such as the lymphatic endothelial hyaluronan receptor-1, LYVE-1, and the transcription factor prospero-related homeobox 1, Prox1 (Wigle et al. 2002; Wigle and Oliver 1999). Budding of the lymphatic vasculature from embryonic veins has been well established through thorough genetic and developmental studies (Karkkainen et al. 2004; Oliver 2004; Wigle and Oliver 1999).

LYVE-1 guides the first set of endothelial cells to become committed to the lymphatic endothelial phenotype and remains expressed in mature initial lymphatics, but is mostly absent from the collecting lymphatic vessels (Jackson 2004). Prox-1 has been termed the "master regulator" of lymphatic endothelium, as specific deletion of Prox-1 during embryogenesis results in the complete absence of a lymphatic vasculature in mouse embryos (Wigle et al. 2002). Prox-1 target genes have been only loosely identified, but include proteins such as the vascular endothelial growth factor receptor 3 (VEGFR3), a receptor of great importance for inflammation-induced lymphangiogenesis, and the fibroblast growth factor receptor 3, whose function has not been described in detail in lymphatic development (Wigle et al. 2002).

6.4.2 Migration and Proliferation

Lymphatic sprouting is initiated and maintained by the interaction of VEGFR3 and its associated ligand, vascular endothelial growth factor C or D, VEGF-C/D, which is produced in large quantities by the surrounding mesenchymal cells (Karpanen and Alitalo 2008; Tammela et al. 2005a). VEGFR3 expression is very high during lymphatic commitment and remains exclusively expressed on lymphatic endothelium while being lost by most blood endothelium (Karpanen and Alitalo 2008; Tammela et al. 2005a). Lymphatic endothelial cell survival is completely dependent on this intimate connection with VEGFR3 and its ligands during embryogenesis. VEGF-C is abundantly expressed where lymphatic vessels are sprouting and VEGF-C homozygous deletions are very similar in phenotype to the Prox-1 knockout, with a complete absence of a lymphatic vascular system (Karkkainen et al. 2004). Other growth factors associated with lymphatic sprouting are neuropilin-2 (Yuan et al. 2002), insulin-like growth factors, hepatocyte growth factor, and growth hormone (Bjorndahl et al. 2005). However, most of the effects of the aforementioned factors might be synergistic as they ultimately converge to the tyrosine kinase pathway of VEGFR3. Lymphatic vessel growth under the command of VEGFR3 and its ligands has been shown to occur in adulthood during pathologies such as cancer and inflammation, which will be discussed further in this chapter. Adult lymphangiogenesis mainly occurs through the expansion and sprouting of existing lymphatic vessels rather than de novo genesis (Junghans and Collin 1989; Jeltsch et al. 1997).

6.4.3 Separation from the Blood Vasculature

To become an entity separate from the blood vasculature, the newly formed lymphatic vessels must physically disconnect from the embryonic veins they originated from. Separation of the lymphatic system occurs around E 10-E 11 in mice and relies heavily on the intracellular signaling of the tyrosine kinase Syk and its adaptor protein Slp76 (Abtahian et al. 2003). Abnormal functioning of these two proteins due to mutations can produce abnormal lymphatico-venous connections, leading to mixing of blood and lymphatic compartments, ultimately ending in hemorrhaging and perinatal death (Abtahian et al. 2003). Furthermore, separation of the lymphatic system has been shown to involve the interaction of the glycoprotein podoplanin with the C-type lectin receptor 2 (CLEC-2) in platelets to promote platelet aggregation at separation zones between the two systems (Suzuki-Inoue et al. 2006).

6.4.4 Maturation and Remodeling

At this stage a primitive lymphatic plexus exists, but it is devoid of proper architecture and secondary lymphatic structures such as smooth muscle cells, basement membrane, and unidirectional valves. These changes are the most profound and occur at the latest stage of development, as lymphatic vessels start sprouting from the main lymphatic plexus to embed various organs and mucosal surfaces (Karpanen and Alitalo 2008). Maturation of lymphatic vessels is governed by key proteins such as ephrinB2, angiopoietin-2 (Ang2), and the forkhead transcription factor FOXC2 (Fang et al. 2000; Makinen et al. 2005; Petrova et al. 2004; Tammela et al. 2005b). Ephrin ligands and their receptors are critical in the formation and function of luminal valves as well as the remodeling of the lymphatic plexus into a proper lymphatic vascular phenotype (Makinen et al. 2005). Ang2 can also signal through its receptor Tie2 to initiate many lymphatic changes, such as proper formation of initial lymphatics and smooth muscle cell recruitment (Gale et al. 2002). During adulthood, FOXC2 is abundantly expressed in the valves in collecting lymphatics (Petrova et al. 2004) and has been shown to be critical for proper valve formation and smooth muscle recruitment. It has been shown recently that FOXC2 is a downstream target of VEGFR3 signaling and is an important mediator for defining lymphatic collector phenotype (Norrmen et al. 2009).

6.5 Adipose Tissue and Lymphangiogenesis

Adipose tissue acts as an endocrine organ and releases a variety of adipocytokines that can influence local microcirculation. In return, the neighboring microcirculation provides adipocytes with the necessary fuel and nutrition to develop normally. Specifically, the lymphatic circulation provides a critical route for lipid transport to the blood and fat deposits, allowing for normal fat expansion. Adipocytokines released by adipocytes include but are not limited to leptin, adiponectin, IL-6, and TNFα (Drevon 2005; Guerre-Millo 2004). In addition to cytokines, adipocytes can also release many different growth factors that have been shown recently to stimulate lymphatic growth (Fig. 6.6). Ang-2 for instance has been shown to be expressed by adipocytes and leads to increased lymphangiogenesis in a mouse corneal model (Morisada et al. 2005). Furthermore, adipose tissue has been shown to directly secrete the lymphangiogenic factor VEGF-C (Voros et al. 2005).

Leptin, an anti-satiety signal, has also emerged recently as a lymphangiogenic growth factor, most likely signaling through the Akt/ERK1/2 proliferation pathway (Drabkin and Gemmill 2010). Adipose-derived stem cells have also been shown to directly promote lymphangiogenesis when exposed to VEGF-C (Hwang et al. 2011; Yan et al. 2011). These studies are based on the observation that adipose stem cells differentiate into lymphatic-committed cells when harvested with VEGF-C on matrigel plugs (Hwang et al. 2011; Yan et al. 2011). This effect could also be mimicked with TGF-β1 inhibition, implicating an immune component to lymphatic endothelial cell development (Yan et al. 2011; Oka et al. 2008).

Finally, adipose-associated immune cells such as macrophages can stimulate lymphangiogenesis through their production of VEGF-C (Cho et al. 2007) (Fig. 6.6). These events could possibly increase overall lymphatic transport of lipids to the bloodstream; however, further studies are needed to demonstrate this. Interestingly, evidence points to macrophages as potential lymphatic precursors, and as

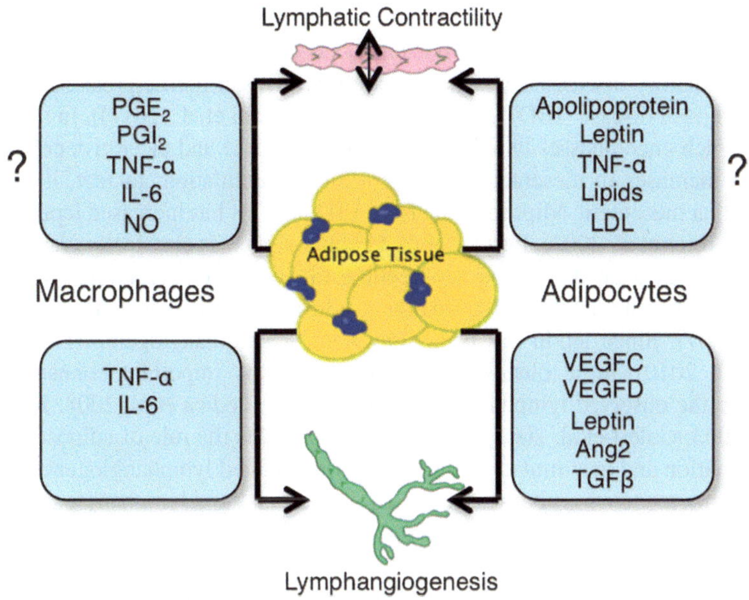

Fig. 6.6 Modulation of lymphatic contractility and lymphangiogenesis by adipocytes and associated macrophages. Adipocytes and associated macrophages can release many substances that have been shown to induce lymphatic growth. Notably, adipocytes release VEGF-C, VEGF-D, leptin, Angiopoetin-2 (Ang2), and transforming growth factor β (TGFβ) that can all promote lymphangiogenesis. Furthermore, macrophages that reside within the adipocyte milieu can release inflammatory substances such as tumor necrosis factor α (TNFα) and interleukin-6 (IL-6) that can promote lymphatic growth in inflammatory situations. Many factors that originate from adipose tissue and its resident macrophages are also known to alter lymphatic contractility. The demonstration of such adipose-mediated modulation of lymphatic pumping has yet to be documented

macrophages are intimately linked with adipose tissue, these cells could be another factor promoting lymphatic proliferation in adipose-associated tissues (Maruyama et al. 2005, 2007).

6.6 Inflammation, Tumor-Associated Lymphangiogenesis, and Adipose Tissue

Adult lymphangiogenesis is not a normal occurrence; it is rather related to pathologies such as inflammation and cancer.

6.6.1 Inflammation

Lymphangiogenesis has been implicated in inflammatory disorders, including corneal, airway, and skin inflammation (Bock et al. 2008; Baluk et al. 2005; Enholm et al. 2001; Halin et al. 2008). Furthermore, adipocytokines have been shown to have inflammatory effects in many disease models (Trayhurn and Wood 2004; Schaffler et al. 2005). Leptin has been associated with pro-inflammatory signaling in inflammatory diseases such as rheumatoid arthritis, atherosclerosis, and inflammatory bowel disease (IBD) (as reviewed in Procaccini et al. (2012)). In the case of IBD, which encompasses two diseases: Crohn's disease and ulcerative colitis, both exhibit increases in mesenteric adipose tissue accumulation. In fact, it has been shown that mesenteric adipose tissue from IBD patients has increased leptin expression (Barbier et al. 2003). Furthermore, Crohn's disease is classically characterized by an increased fat deposition, or creeping fat, which is "wrapping" around the inflamed intestine, and is often involving adipocyte hyperplasia (Peyrin-Biroulet et al. 2007). Since leptin has been shown to be lymphangiogenic (Drabkin and Gemmill 2010), these observations could make the important connecting link between the observed lymphangiogenesis and IBD (Pedica et al. 2008; Kaiserling et al. 2003; Geleff et al. 2003). This further highlights the role of adipose tissue in inflammation and in stimulated inflammation-associated lymphangiogenesis. In the near future, associations between other adipocytokines and lymphangiogenesis will be integral to our thorough understanding of adipose tissue and its role in inflammation.

Inflammatory conditions not only induce lipolysis, but affect the lipid composition in perinodal adipose tissue (PAT) as well. In contrast to healthy subjects, the n-3 PUFA content of PAT is not higher than that of other adipose depots in patients with Crohn's disease (Westcott et al. 2005). On the other hand, trinitrobenzene sulfonic acid (TNBS)-induced colitis increases n-6 PUFA content of PAT (Acedo et al. 2011). This is very interesting, because PUFAs have the ability to modulate immune responses. The n-3 PUFA docosahexaenoic acid, for example, inhibits dendritic cell-mediated activation of T-cells and thereby limits lipopolysaccharide-induced immune responses (Weatherill et al. 2005). Accordingly, the reduced content of n-3 PUFAs in Crohn's disease could exacerbate inflammation.

VEGF-C is a potent lymphangiogenic factor when bound to its receptor, VEGFR3, on lymphatic endothelial cells. Interestingly, a 2005 clinical study that assessed the levels of various vascular growth factors in lean and obese patients found that serum VEGF-C was elevated in obese subjects (Silha et al. 2005). Whether clinical obesity is associated with increased lymphangiogenesis is still unclear, and even if new lymphatic growth is stimulated in obesity, the functional physiology of these "new" lymphatic vessels remains to be studied. Obesity, which is now being recognized as a low-grade inflammatory disorder, is associated with an increase in adipocytokine production from increased fat mass, such as leptin, which has been shown to induce lymphangiogenesis (Silha et al. 2005). Adipose tissue hypoxia in obesity can also be a key source for dynamic changes in the lymphatic vasculature. Hypoxic adipose tissue has been shown to stimulate the secretion of lymphaniogenic leptin and VEGF-C (Wang et al. 2007). Furthermore, obesity-related adipose tissue hypoxia has been implicated in proliferative angiogenesis (Xue et al. 2009; Michailidou et al. 2012) and it seems plausible that obesity also promotes lymphangiogenesis. Since blood vessels require the nearby presence of lymphatic vessels for maintaining tissue fluid homeostasis, it is likely that obesity-associated hypoxia induces lymphangiogenesis. In fact target genes of hypoxia signaling through the transcription factor hypoxia-inducible factor 1 (HIF-1), such as VEGF-A, -C, -D, are common to both proliferation of new blood and lymphatic vessels (Hopfl et al. 2004; Okada et al. 2005; Schoppmann et al. 2006).

6.6.2 Cancer

Cancer metastasis, the spread of cancerous cells to other organs and tissues, commonly occurs through the tumor-draining lymph node (as reviewed in Albrecht and Christofori (2011)). To infiltrate the lymph node and reach the blood circulation, cancerous cells must access the lymphatic vasculature. Tumor-associated lymphatic vessels can aid in the spread or metastasis of cancer in several ways. Lymphatic vessels are lined with lymphatic endothelial cells that are a source of the chemokine ligand CCL21 that can attract many different immune cells that express the receptor CCR7, such as macrophages and dendritic cells. In a cancerous environment, the CCL21-CCR7 interaction could provide tumor cells with the ability to migrate to nearby tissues via lymphatic conduits. This has in fact been shown in CCR7 expressing melanoma cells (Wiley et al. 2001). Access to the lymph nodes, i.e., afferent lymphatic vessels, has also been shown to be increased in cancer patients (Chen et al. 2011; Da et al. 2008; Deng et al. 2009; Gao et al. 2008; Guo et al. 2009; Raica et al. 2011). Notably in animal models, tumor-associated lymphangiogenesis has been associated with an increase in lymphaniogenic factors such as VEGFC and VEGFD (Karpanen et al. 2001; Skobe et al. 2001a). However, most of these newly formed vessels are often collapsed, suggesting inefficient or obstructed lymph drainage (Albrecht and Christofori 2011).

Tumor progress and fate is intimately linked to inflammation as tumors are often associated with a strong immune response and inflammatory infiltrate (reviewed in Grivennikov et al. (2010)). Accordingly, lymphatic vessels not only provide a route for tumor cell movement but also provide an efficient route for tumor-specific immune responses. In fact, growth of new lymphatic vessels has been closely related to cancer metastasis (Skobe et al. 2001a, b; Stacker et al. 2001), and new therapies have been developed or are underway to interfere with this process (Burton et al. 2008; Lin et al. 2005; Roberts et al. 2006). Furthermore, newly generated blood vessels associated with tumors are functionally impaired or exhibit abnormal permeability, leading to excess interstitial leakage (Jain 2003, 2005). Lymphangiogenesis in tumors, therefore, could be compensatory for the increases in interstitial fluid and pressures.

Adipose tissue has been shown to be intimately associated with the microenvironment of many tumors (Nakamura et al. 2011; Ribeiro et al. 2012; Wang et al. 2012). Peritumoral adipose tissue, being an active endocrine system, can release many mediators that may promote tumor microenvironment-associated lymphangiogenesis. For example, peritumoral adipose tissue in a melanoma model has been shown to upregulate the expression of inflammatory and lymphangiogenic mediators, such as MCP-1, VEGF-C, and IL-6 (Wagner et al. 2012).

6.7 Lymphangiogenesis as a Therapeutic Target for Treatment of Obesity and Metabolic Disorders

Obesity is a current pandemic in industrialized countries, with over one billion estimated to be overweight worldwide (Bruemmer 2012). Current pharmacological treatments, diet, and exercise have limited efficacy (Bruemmer 2012). Obesity and associated metabolic disorders have been deemed very complex illnesses with many different etiopathologies such as genetic predisposition and immune dysfunction. Obesity can alter physiology on many levels, including type 2 diabetes and vascular dysfunction leading to hypertension and cardiovascular diseases.

Adipose tissue is a richly vascularized tissue, and angiogenesis has been shown to have important modulatory effects in obesity, specifically through angiogenesis-mediated adipocyte expansion (Pandya et al. 2006). Evidence for pathological angiogenesis has been shown in obesity-related disorders (Pandya et al. 2006; Rupnick et al. 2002; Yoon and Kim 2011), therefore anti-angiogenic drugs could have a possible therapeutic role in obesity. The basis of this hypothesis stems from the knowledge that blood vessels provide a positive growth milieu for adipose tissue expansion. Since this milieu involves different growth factors, hormones, and inflammatory cells, the lymphatic system becomes an important partner in adipose tissue growth.

Lymphatic vessels carry lymph, excess interstitial fluid that is no longer needed in the tissue microenvironment. If the adipose lymphatic vasculature is not functioning properly, one can assume a backup of lymph within adipose tissue. Lymph is the major exit for inflammatory cells from adipose tissue; therefore, lymphatic vessel function is integral to adipose tissue physiology. Unfortunately, there is very little evidence to date to support a connection between lymphatic function/lymphangiogenesis and obesity.

As discussed previously, the work of Harvey et al. (2005) on the Prox-1 haploinsufficient mouse and its associated late-onset obesity opened the door to connecting the promotion of obesity to lymphatic dysfunction. Prox-1 haploinsufficiency resulted in dysfunctional lymphatics, promoting lymph stasis and obesity. Subsequent studies, including the tail injury model of lymphatic disruption leading to lymph stasis and increases in adipose tissue deposition (Aschen et al. 2012; Zampell et al. 2012), corroborate the findings of Harvey et al.

Following earlier observations, a more recent clinical paper has shown increased levels of VEGF-C in sera from obese subjects (Wada et al. 2011). The increase in VEGF-C has been shown to be more closely correlated to obesity-related metabolic dysfunctions such as insulin and lipid parameters, than the observed increase in VEGF-A. These findings were further confirmed with the apolipoprotein E-deficient mouse model of obesity (Lim et al. 2009). These mice, when fed a high fat diet, exhibited a significant increase in VEGF-C levels associated with atherosclerosis and hypercholesterolemia, with no changes in VEGF-A. These data suggest a more important role for lymphangiogenesis than angiogenesis in promoting metabolic disorders and atherosclerosis in obesity.

Unlike the development of anti-angiogenic agents, anti-lymphangiogenic agents have not yet been tested clinically. However, there is great promise for anti-lymphangiogenic therapy in the context of many disorders including corneal graft rejection and cancer (Nagahashi et al. 2010; Niederleithner et al. 2012; Kojima et al. 2008). As research grows in the area of lymphatic pathophysiology and lymphangiogenesis, anti-lymphangiogenic agents could be a possible avenue for the treatment of obesity, given the important connections that have been laid down between the lymphatic system and obesity.

6.8 Conclusion

Adipose tissue is recognized as a dynamic organ system requiring proper vascularization. Recent literature is providing new evidence for the relationship that exists between the adipose system and the lymphatic system. Specifically, adipocyte-mediated lymphangiogenesis is beginning to receive much more attention, especially within the context of inflammation and immune cell trafficking. Moreover, research suggests that many adipocytokines could directly modulate lymphatic contractility, having implications on lymph movement throughout the lymphatic system. Being indispensable for lipid absorption, abnormalities in the lymphatic pump system could have drastic implications for proper adipocyte development.

Lymphangiogenesis and lymphatic function have now been recognized as important players in many inflammatory disorders, as well as tumorigenic environments. Gaining a better understanding of the relationship between adipocyte and lymphatic biology could possibly lead to the development of new therapeutic or diagnostic tools to tackle adipocyte-related disorders such as obesity and vascular inflammation.

References

Alexander JS, Ganta VC, Jordan PA, Witte MH. Gastrointestinal lymphatics in health and disease. Pathophysiology. 2010;17(4):315–35.

Aschen S, Zampell JC, Elhadad S, Weitman E, De Brot M, Mehrara BJ. Regulation of adipogenesis by lymphatic fluid stasis: part II. Expression of adipose differentiation genes. Plast Reconstr Surg. 2012;129(4):838–47.

Aghamohammadzadeh R, Withers S, Lynch F, Greenstein A, Malik R, Heagerty A. Perivascular adipose tissue from human systemic and coronary vessels: the emergence of a new pharmacotherapeutic target. Br J Pharmacol. 2012;165(3):670–82.

Arngrim N, Simonsen L, Holst JJ, Bulow J. Reduced adipose tissue lymphatic drainage of macromolecules in obese subjects: a possible link between obesity and local tissue inflammation? Int J Obes (Lond). 2013;37(5):748–50.

Abtahian F, Guerriero A, Sebzda E, Lu MM, Zhou R, Mocsai A, et al. Regulation of blood and lymphatic vascular separation by signaling proteins SLP-76 and Syk. Science. 2003;299(5604):247–51.

Acedo SC, Gotardo EM, Lacerda JM, de Oliveira CC, de Oliveira CP, Gambero A. Perinodal adipose tissue and mesenteric lymph node activation during reactivated TNBS-colitis in rats. Dig Dis Sci. 2011;56(9):2545–52.

Albrecht I, Christofori G. Molecular mechanisms of lymphangiogenesis in development and cancer. Int J Dev Biol. 2011;55(4–5):483–94.

Arkill KP, Moger J, Winlove CP. The structure and mechanical properties of collecting lymphatic vessels: an investigation using multimodal nonlinear microscopy. J Anat. 2010; 216(5):547–55.

Bierman HR, Byron Jr RL, Kelly KH, Gilfillan RS, White LP, Freeman NE, et al. The characteristics of thoracic duct lymph in man. J Clin Invest. 1953;32(7):637–49.

Baluk P, Fuxe J, Hashizume H, Romano T, Lashnits E, Butz S, et al. Functionally specialized junctions between endothelial cells of lymphatic vessels. J Exp Med. 2007;204(10):2349–62.

Bazigou E, Xie S, Chen C, Weston A, Miura N, Sorokin L, et al. Integrin-alpha9 is required for fibronectin matrix assembly during lymphatic valve morphogenesis. Dev Cell. 2009;17(2):175–86.

Brorson H. From lymph to fat: complete reduction of lymphoedema. Phlebology. 2010;25 Suppl 1:52–63.

Bjorndahl M, Cao R, Nissen LJ, Clasper S, Johnson LA, Xue Y, et al. Insulin-like growth factors 1 and 2 induce lymphangiogenesis in vivo. Proc Natl Acad Sci U S A. 2005;102(43): 15593–8.

Bock F, Onderka J, Dietrich T, Bachmann B, Pytowski B, Cursiefen C. Blockade of VEGFR3-signalling specifically inhibits lymphangiogenesis in inflammatory corneal neovascularisation. Graefes Arch Clin Exp Ophthalmol. 2008;246(1):115–9.

Baluk P, Tammela T, Ator E, Lyubynska N, Achen MG, Hicklin DJ, et al. Pathogenesis of persistent lymphatic vessel hyperplasia in chronic airway inflammation. J Clin Invest. 2005;115(2):247–57.

Barbier M, Vidal H, Desreumaux P, Dubuquoy L, Bourreille A, Colombel JF, et al. Overexpression of leptin mRNA in mesenteric adipose tissue in inflammatory bowel diseases. Gastroenterol Clin Biol. 2003;27(11):987–91.

Burton JB, Priceman SJ, Sung JL, Brakenhielm E, An DS, Pytowski B, et al. Suppression of prostate cancer nodal and systemic metastasis by blockade of the lymphangiogenic axis. Cancer Res. 2008;68(19):7828–37.

Bruemmer D. Targeting angiogenesis as treatment for obesity. Arterioscler Thromb Vasc Biol. 2012;32(2):161–2.

Cruikshank W. The anatomy of the absorbing vessels of the human body. London; 1786.

Costa M, Funess JB. Observations on the anatomy and amine histochemistry of the nerves and ganglia which supply the pelvic viscera and on the associated chromaffin tissue in the guinea-pig. Z Anat Entwicklungsgesch. 1973;140(1):85–108.

Clark ER, Clark EL. Microscopic studies of new formation of fat in living adult rabbits. Am J Anat. 1940;67:255–81.

Chikly B. Who discovered the lymphatic system. Lymphology. 1997;30(4):186–93.

Cho CH, Koh YJ, Han J, Sung HK, Jong Lee H, Morisada T, et al. Angiogenic role of LYVE-1-positive macrophages in adipose tissue. Circ Res. 2007;100(4):e47–57.

Chen Y, Yu G, Ding H. Lymphangiogenic and angiogentic microvessel density in gallbladder carcinoma. Hepatogastroenterology. 2011;58(105):20–5.

Cristancho AG, Lazar MA. Forming functional fat: a growing understanding of adipocyte differentiation. Nat Rev Mol Cell Biol. 2011;12(11):722–34.

Cox R, Garcia-Palmieri MR. Chapter 31: cholesterol, triglycerides, and associated lipoproteins. In: Walker H, Hall WD, Hurst JW, editors. Clinical methods: the history, physical, and laboratory examinations. 3rd ed. Boston: Butterworths; 1990.

Davis MJ, Davis AM, Lane MM, Ku CW, Gashev AA. Rate-sensitive contractile responses of lymphatic vessels to circumferential stretch. J Physiol. 2009;587(Pt 1):165–82.

Drayson MT, Ford WL. Afferent lymph and lymph borne cells: their influence on lymph node function. Immunobiology. 1984;168(3–5):362–79.

Da MX, Wu XT, Wang J, Guo TK, Zhao ZG, Luo T, et al. Expression of cyclooxygenase-2 and vascular endothelial growth factor-C correlates with lymphangiogenesis and lymphatic invasion in human gastric cancer. Arch Med Res. 2008;39(1):92–9.

Deng J, Liang H, Sun D, Pan Y, Wang B, Guo Y. Vascular endothelial growth factor-D is correlated with hepatic metastasis from gastric cancer after radical gastrectomy. Surgery. 2009;146(5):896–905.

Drevon CA. Fatty acids and expression of adipokines. Biochim Biophys Acta. 2005;1740(2): 287–92.

Drabkin HA, Gemmill RM. Obesity, cholesterol, and clear-cell renal cell carcinoma (RCC). Adv Cancer Res. 2010;107:39–56.

Elias RM, Johnston MG. Modulation of fluid pumping in isolated bovine mesenteric lymphatics by a thromboxane/endoperoxide analogue. Prostaglandins. 1988;36(1):97–106.

Enholm B, Karpanen T, Jeltsch M, Kubo H, Stenback F, Prevo R, et al. Adenoviral expression of vascular endothelial growth factor-C induces lymphangiogenesis in the skin. Circ Res. 2001; 88(6):623–9.

Ferguson MK, DeFilippi VJ, Reeder LB. Characterization of contractile properties of porcine mesenteric and tracheobronchial lymphatic smooth muscle. Lymphology. 1994;27(2):71–81.

Fang J, Dagenais SL, Erickson RP, Arlt MF, Glynn MW, Gorski JL, et al. Mutations in FOXC2 (MFH-1), a forkhead family transcription factor, are responsible for the hereditary lymphedema-distichiasis syndrome. Am J Hum Genet. 2000;67(6):1382–8.

Gashev AA, Davis MJ, Zawieja DC. Inhibition of the active lymph pump by flow in rat mesenteric lymphatics and thoracic duct. J Physiol. 2002;540(Pt 3):1023–37.

Gasheva OY, Zawieja DC, Gashev AA. Contraction-initiated NO-dependent lymphatic relaxation: a self-regulatory mechanism in rat thoracic duct. J Physiol. 2006;575(Pt 3):821–32.

Gao Y, Zhong WX, Mu DB, Yuan YP, Zhang YH, Yu JM, et al. Distributions of angiogenesis and lymphangiogenesis in gastrointestinal intramucosal tumors. Ann Surg Oncol. 2008;15(4): 1117–23.

Guo X, Chen Y, Xu Z, Qian Y, Yu X. Prognostic significance of VEGF-C expression in correlation with COX-2, lymphatic microvessel density, and clinicopathologic characteristics in human non-small cell lung cancer. Acta Biochim Biophys Sin (Shanghai). 2009;41(3):217–22.

Grigorova IL, Panteleev M, Cyster JG. Lymph node cortical sinus organization and relationship to lymphocyte egress dynamics and antigen exposure. Proc Natl Acad Sci U S A. 2010;107(47): 20447–52.

Geleff S, Schoppmann SF, Oberhuber G. Increase in podoplanin-expressing intestinal lymphatic vessels in inflammatory bowel disease. Virchows Arch. 2003;442(3):231–7.

Guerre-Millo M. Adipose tissue and adipokines: for better or worse. Diabetes Metab. 2004;30(1): 13–9.

Grivennikov SI, Greten FR, Karin M. Immunity, inflammation, and cancer. Cell. 2010;140(6): 883–99.

Huxley VH, Scallan J. Lymphatic fluid: exchange mechanisms and regulation. J Physiol. 2011; 589(Pt 12):2935–43.

Hargens AR, Zweifach BW. Contractile stimuli in collecting lymph vessels. Am J Physiol. 1977; 233(1):H57–65.

Hukkanen M, Konttinen YT, Terenghi G, Polak JM. Peptide-containing innervation of rat femoral lymphatic vessels. Microvasc Res. 1992;43(1):7–19.

Hodis J, Vaclavikova R, Farghali H. Beta-3 agonist-induced lipolysis and nitric oxide production: relationship to PPARgamma agonist/antagonist and AMP kinase modulation. Gen Physiol Biophys. 2011;30(1):90–9.

Harford KA, Reynolds CM, McGillicuddy FC, Roche HM. Fats, inflammation and insulin resistance: insights to the role of macrophage and T-cell accumulation in adipose tissue. Proc Nutr Soc. 2011;70(4):408–17.

Harvey NL, Srinivasan RS, Dillard ME, Johnson NC, Witte MH, Boyd K, et al. Lymphatic vascular defects promoted by Prox1 haploinsufficiency cause adult-onset obesity. Nat Genet. 2005;37(10):1072–81.

Hwang JH, Kim IG, Lee JY, Piao S, Lee DS, Lee TS, et al. Therapeutic lymphangiogenesis using stem cell and VEGF-C hydrogel. Biomaterials. 2011;32(19):4415–23.
Halin C, Fahrngruber H, Meingassner JG, Bold G, Littlewood-Evans A, Stuetz A, et al. Inhibition of chronic and acute skin inflammation by treatment with a vascular endothelial growth factor receptor tyrosine kinase inhibitor. Am J Pathol. 2008;173(1):265–77.
Hopfl G, Ogunshola O, Gassmann M. HIFs and tumors–causes and consequences. Am J Physiol Regul Integr Comp Physiol. 2004;286(4):R608–23.
Johnston MG, Gordon JL. Regulation of lymphatic contractility by arachidonate metabolites. Nature. 1981;293(5830):294–7.
Johnston MG, Feuer C. Suppression of lymphatic vessel contractility with inhibitors of arachidonic acid metabolism. J Pharmacol Exp Ther. 1983;226(2):603–7.
Johnston MG, Kanalec A, Gordon JL. Effects of arachidonic acid and its cyclo-oxygenase and lipoxygenase products on lymphatic vessel contractility in vitro. Prostaglandins. 1983;25(1):85–98.
Jackson DG. Biology of the lymphatic marker LYVE-1 and applications in research into lymphatic trafficking and lymphangiogenesis. APMIS. 2004;112(7–8):526–38.
Jain RK. Molecular regulation of vessel maturation. Nat Med. 2003;9(6):685–93.
Jain RK. Normalization of tumor vasculature: an emerging concept in antiangiogenic therapy. Science. 2005;307(5706):58–62.
Junghans BM, Collin HB. Limbal lymphangiogenesis after corneal injury: an autoradiographic study. Curr Eye Res. 1989;8(1):91–100.
Jeltsch M, Kaipainen A, Joukov V, Meng X, Lakso M, Rauvala H, et al. Hyperplasia of lymphatic vessels in VEGF-C transgenic mice. Science. 1997;276(5317):1423–5.
Karkkainen MJ, Saaristo A, Jussila L, Karila KA, Lawrence EC, Pajusola K, et al. A model for gene therapy of human hereditary lymphedema. Proc Natl Acad Sci U S A. 2001;98(22):12677–82.
Karkkainen MJ, Haiko P, Sainio K, Partanen J, Taipale J, Petrova TV, et al. Vascular endothelial growth factor C is required for sprouting of the first lymphatic vessels from embryonic veins. Nat Immunol. 2004;5(1):74–80.
Kaiserling E, Krober S, Geleff S. Lymphatic vessels in the colonic mucosa in ulcerative colitis. Lymphology. 2003;36(2):52–61.
Karpanen T, Egeblad M, Karkkainen MJ, Kubo H, Yla-Herttuala S, Jaattela M, et al. Vascular endothelial growth factor C promotes tumor lymphangiogenesis and intralymphatic tumor growth. Cancer Res. 2001;61(5):1786–90.
Kojima T, Azar DT, Chang JH. Neostatin-7 regulates bFGF-induced corneal lymphangiogenesis. FEBS Lett. 2008;582(17):2515–20.
Kindel T, Lee DM, Tso P. The mechanism of the formation and secretion of chylomicrons. Atheroscler Suppl. 2010;11(1):11–6.
Kohan AB, Yoder SM, Tso P. Using the lymphatics to study nutrient absorption and the secretion of gastrointestinal hormones. Physiol Behav. 2011;105(1):82–8.
Karpanen T, Alitalo K. Molecular biology and pathology of lymphangiogenesis. Annu Rev Pathol. 2008;3:367–97.
Lord RS. The white veins: conceptual difficulties in the history of the lymphatics. Med Hist. 1968;12(2):174–84.
Leak LV. The structure of lymphatic capillaries in lymph formation. Fed Proc. 1976;35(8):1863–71.
Leak LV, Burke JF. Ultrastructural studies on the lymphatic anchoring filaments. J Cell Biol. 1968;36(1):129–49.
Lim HY, Rutkowski JM, Helft J, Reddy ST, Swartz MA, Randolph GJ, et al. Hypercholesterolemic mice exhibit lymphatic vessel dysfunction and degeneration. Am J Pathol. 2009;175(3):1328–37.
Lumeng CN, Deyoung SM, Bodzin JL, Saltiel AR. Increased inflammatory properties of adipose tissue macrophages recruited during diet-induced obesity. Diabetes. 2007;56(1):16–23.

Lin J, Lalani AS, Harding TC, Gonzalez M, Wu WW, Luan B, et al. Inhibition of lymphogenous metastasis using adeno-associated virus-mediated gene transfer of a soluble VEGFR-3 decoy receptor. Cancer Res. 2005;65(15):6901–9.

Mayerson HS. On lymph and lymphatics. Circulation. 1963;28:839–42.

Maruyama K, Asai J, Ii M, Thorne T, Losordo DW, D'Amore PA. Decreased macrophage number and activation lead to reduced lymphatic vessel formation and contribute to impaired diabetic wound healing. Am J Pathol. 2007;170(4):1178–91.

Mendoza E, Schmid-Schonbein GW. A model for mechanics of primary lymphatic valves. J Biomech Eng. 2003;125(3):407–14.

Mislin H. Active contractility of the lymphangion and coordination of lymphangion chains. Experientia. 1976;32(7):820–2.

Mislin H. Experimental detection of autochthonous automatism of lymph vessels. Experientia. 1961;17:29–30.

Muthuchamy M, Gashev A, Boswell N, Dawson N, Zawieja D. Molecular and functional analyses of the contractile apparatus in lymphatic muscle. FASEB J. 2003;17(8):920–2.

Mizuno R, Koller A, Kaley G. Regulation of the vasomotor activity of lymph microvessels by nitric oxide and prostaglandins. Am J Physiol. 1998;274(3 Pt 2):R790–6.

Maruyama K, Ii M, Cursiefen C, Jackson DG, Keino H, Tomita M, et al. Inflammation-induced lymphangiogenesis in the cornea arises from CD11b-positive macrophages. J Clin Invest. 2005;115(9):2363–72.

Mattacks CA, Pond CM. Interactions of noradrenalin and tumour necrosis factor alpha, interleukin 4 and interleukin 6 in the control of lipolysis from adipocytes around lymph nodes. Cytokine. 1999;11(5):334–46.

Makinen T, Adams RH, Bailey J, Lu Q, Ziemiecki A, Alitalo K, et al. PDZ interaction site in ephrinB2 is required for the remodeling of lymphatic vasculature. Genes Dev. 2005;19(3): 397–410.

Michailidou Z, Turban S, Miller E, Zou X, Schrader J, Ratcliffe PJ, et al. Increased angiogenesis protects against adipose hypoxia and fibrosis in metabolic disease-resistant 11beta-hydroxysteroid dehydrogenase type 1 (HSD1)-deficient mice. J Biol Chem. 2012;287(6): 4188–97.

Miller NE, Michel CC, Nanjee MN, Olszewski WL, Miller IP, Hazell M, et al. Secretion of adipokines by human adipose tissue in vivo: partitioning between capillary and lymphatic transport. Am J Physiol Endocrinol Metab. 2011;301(4):E659–67.

Morisada T, Oike Y, Yamada Y, Urano T, Akao M, Kubota Y, et al. Angiopoietin-1 promotes LYVE-1-positive lymphatic vessel formation. Blood. 2005;105(12):4649–56.

Nagahashi M, Ramachandran S, Rashid OM, Takabe K. Lymphangiogenesis: a new player in cancer progression. World J Gastroenterol. 2010;16(32):4003–12.

Nanjee MN, Cooke CJ, Wong JS, Hamilton RL, Olszewski WL, Miller NE. Composition and ultrastructure of size subclasses of normal human peripheral lymph lipoproteins: quantification of cholesterol uptake by HDL in tissue fluids. J Lipid Res. 2001;42(4):639–48.

Nakamura K, Hongo A, Kodama J, Hiramatsu Y. Fat accumulation in adipose tissues as a risk factor for the development of endometrial cancer. Oncol Rep. 2011;26(1):65–71.

Niederleithner H, Heinz M, Tauber S, Bilban M, Pehamberger H, Sonderegger S, et al. Wnt1 is anti-lymphangiogenic in a melanoma mouse model. J Invest Dermatol. 2012;132(9):2235–44.

Norrmen C, Ivanov KI, Cheng J, Zangger N, Delorenzi M, Jaquet M, et al. FOXC2 controls formation and maturation of lymphatic collecting vessels through cooperation with NFATc1. J Cell Biol. 2009;185(3):439–57.

Nougues J, Reyne Y, Dulor JP. Differentiation of rabbit adipocyte precursors in primary culture. Int J Obes. 1988;12(4):321–33.

Okada K, Osaki M, Araki K, Ishiguro K, Ito H, Ohgi S. Expression of hypoxia-inducible factor (HIF-1alpha), VEGF-C and VEGF-D in non-invasive and invasive breast ductal carcinomas. Anticancer Res. 2005;25(4):3003–9.

Oliver G. Lymphatic vasculature development. Nat Rev Immunol. 2004;4(1):35–45.

Oka M, Iwata C, Suzuki HI, Kiyono K, Morishita Y, Watabe T, et al. Inhibition of endogenous TGF-beta signaling enhances lymphangiogenesis. Blood. 2008;111(9):4571–9.

Orlov RS, Borisova RP, Mandryko ES. [The contractile and electrical activity of the smooth muscles of the major lymph vessels]. Fiziol Zh SSSR Im I M Sechenova. 1975;61(7):1045–53.

Ohhashi T, Kobayashi S, Tsukahara S, Azuma T. Innervation of bovine mesenteric lymphatics: from the histochemical point of view. Microvasc Res. 1982;24(3):377–85.

Ohtani O, Ohtani Y. Structure and function of rat lymph nodes. Arch Histol Cytol. 2008;71(2): 69–76.

Plaku KJ, von der Weid PY. Mast cell degranulation alters lymphatic contractile activity through action of histamine. Microcirculation. 2006;13(3):219–27.

Pond CM. Adipose tissue: quartermaster to the lymph node garrisons. Biologist (London). 2000; 47(3):147–50.

Poznanski WJ, Waheed I, Van R. Human fat cell precursors. Morphologic and metabolic differentiation in culture. Lab Invest. 1973;29(5):570–6.

Piller NB. Lymphoedema, macrophages and benzopyrones. Lymphology. 1980;13(3):109–19.

Piller NB. Macrophage and tissue changes in the developmental phases of secondary lymphoedema and during conservative therapy with benzopyrone. Arch Histol Cytol. 1990; 53(Suppl):209–18.

Peled AW, Slavin SA, Brorson H. Long-term outcome after surgical treatment of lipedema. Ann Plast Surg. 2012;68(3):303–7.

Pond CM, Mattacks CA. The source of fatty acids incorporated into proliferating lymphoid cells in immune-stimulated lymph nodes. Br J Nutr. 2003;89(3):375–83.

Petrova TV, Karpanen T, Norrmen C, Mellor R, Tamakoshi T, Finegold D, et al. Defective valves and abnormal mural cell recruitment underlie lymphatic vascular failure in lymphedema distichiasis. Nat Med. 2004;10(9):974–81.

Procaccini C, Jirillo E, Matarese G. Leptin as an immunomodulator. Mol Aspects Med. 2012; 33(1):35–45.

Peyrin-Biroulet L, Chamaillard M, Gonzalez F, Beclin E, Decourcelle C, Antunes L, et al. Mesenteric fat in Crohn's disease: a pathogenetic hallmark or an innocent bystander? Gut. 2007;56(4):577–83.

Pedica F, Ligorio C, Tonelli P, Bartolini S, Baccarini P. Lymphangiogenesis in Crohn's disease: an immunohistochemical study using monoclonal antibody D2-40. Virchows Arch. 2008; 452(1):57–63.

Pandya NM, Dhalla NS, Santani DD. Angiogenesis–a new target for future therapy. Vascul Pharmacol. 2006;44(5):265–74.

Rehal S, Blanckaert P, Roizes S, von der Weid PY. Characterization of biosynthesis and modes of action of prostaglandin E2 and prostacyclin in guinea pig mesenteric lymphatic vessels. Br J Pharmacol. 2009;158(8):1961–70.

Rutkowski JM, Boardman KC, Swartz MA. Characterization of lymphangiogenesis in a model of adult skin regeneration. Am J Physiol Heart Circ Physiol. 2006;291(3):H1402–10.

Rockson SG. Causes and consequences of lymphatic disease. Ann N Y Acad Sci. 2010;1207 Suppl 1:E2–6.

Ranallo RF, Rhodes EC. Lipid metabolism during exercise. Sports Med. 1998;26(1):29–42.

Rockson SG. Lymphedema. Am J Med. 2001;110(4):288–95.

Raica M, Cimpean AM, Ceausu R, Ribatti D. Lymphatic microvessel density, VEGF-C, and VEGFR-3 expression in different molecular types of breast cancer. Anticancer Res. 2011;31(5): 1757–64.

Rupnick MA, Panigrahy D, Zhang CY, Dallabrida SM, Lowell BB, Langer R, et al. Adipose tissue mass can be regulated through the vasculature. Proc Natl Acad Sci U S A. 2002;99(16): 10730–5.

Roberts N, Kloos B, Cassella M, Podgrabinska S, Persaud K, Wu Y, et al. Inhibition of VEGFR-3 activation with the antagonistic antibody more potently suppresses lymph node and distant metastases than inactivation of VEGFR-2. Cancer Res. 2006;66(5):2650–7.

Ribeiro R, Monteiro C, Cunha V, Oliveira MJ, Freitas M, Fraga A, et al. Human periprostatic adipose tissue promotes prostate cancer aggressiveness in vitro. J Exp Clin Cancer Res. 2012;31(1):32.
Ryan TJ. Lymphatics and adipose tissue. Clin Dermatol. 1995;13(5):493–8.
Ryan TJ. Structure and function of lymphatics. J Invest Dermatol. 1989;93(2 Suppl):18S–24.
Schulte-Merker S, Sabine A, Petrova TV. Lymphatic vascular morphogenesis in development, physiology, and disease. J Cell Biol. 2011;193(4):607–18.
Schmid-Schonbein GW. The second valve system in lymphatics. Lymphat Res Biol. 2003;1(1):25–9. discussion 9–31.
Shields JD. Lymphatics: at the interface of immunity, tolerance, and tumor metastasis. Microcirculation. 2011;18(7):517–31.
Skobe M, Detmar M. Structure, function, and molecular control of the skin lymphatic system. J Investig Dermatol Symp Proc. 2000;5(1):14–9.
Szuba A, Rockson SG. Lymphedema: classification, diagnosis and therapy. Vasc Med. 1998;3(2):145–56.
Shin WS, Szuba A, Rockson SG. Animal models for the study of lymphatic insufficiency. Lymphat Res Biol. 2003;1(2):159–69.
Schirger A, Harrison Jr EG, Janes JM. Idiopathic lymphedema. Review of 131 cases. JAMA. 1962;182:14–22.
Slavin SA, Van den Abbeele AD, Losken A, Swartz MA, Jain RK. Return of lymphatic function after flap transfer for acute lymphedema. Ann Surg. 1999;229(3):421–7.
Straub RH, Lowin T, Klatt S, Wolff C, Rauch L. Increased density of sympathetic nerve fibers in metabolically activated fat tissue surrounding human synovium and mouse lymph nodes in arthritis. Arthritis Rheum. 2011;63(11):3234–42.
Suzuki-Inoue K, Fuller GL, Garcia A, Eble JA, Pohlmann S, Inoue O, et al. A novel Syk-dependent mechanism of platelet activation by the C-type lectin receptor CLEC-2. Blood. 2006;107(2):542–9.
Schaffler A, Scholmerich J, Buchler C. Mechanisms of disease: adipocytokines and visceral adipose tissue–emerging role in intestinal and mesenteric diseases. Nat Clin Pract Gastroenterol Hepatol. 2005;2(2):103–11.
Silha JV, Krsek M, Sucharda P, Murphy LJ. Angiogenic factors are elevated in overweight and obese individuals. Int J Obes (Lond). 2005;29(11):1308–14.
Skobe M, Hawighorst T, Jackson DG, Prevo R, Janes L, Velasco P, et al. Induction of tumor lymphangiogenesis by VEGF-C promotes breast cancer metastasis. Nat Med. 2001a;7(2):192–8.
Skobe M, Hamberg LM, Hawighorst T, Schirner M, Wolf GL, Alitalo K, et al. Concurrent induction of lymphangiogenesis, angiogenesis, and macrophage recruitment by vascular endothelial growth factor-C in melanoma. Am J Pathol. 2001b;159(3):893–903.
Schoppmann SF, Fenzl A, Schindl M, Bachleitner-Hofmann T, Nagy K, Gnant M, et al. Hypoxia inducible factor-1alpha correlates with VEGF-C expression and lymphangiogenesis in breast cancer. Breast Cancer Res Treat. 2006;99(2):135–41.
Stacker SA, Caesar C, Baldwin ME, Thornton GE, Williams RA, Prevo R, et al. VEGF-D promotes the metastatic spread of tumor cells via the lymphatics. Nat Med. 2001;7(2):186–91.
Tabibiazar R, Cheung L, Han J, Swanson J, Beilhack A, An A, et al. Inflammatory manifestations of experimental lymphatic insufficiency. PLoS Med. 2006;3(7):e254.
Tanaka H, Zaima N, Sasaki T, Yamamoto N, Sano M, Konno H, et al. Loss of lymphatic vessels and regional lipid accumulation is associated with great saphenous vein incompetence. J Vasc Surg. 2012;55(5):1440–8.
Tso P, Pitts V, Granger DN. Role of lymph flow in intestinal chylomicron transport. Am J Physiol. 1985;249(1 Pt 1):G21–8.
Tammela T, Petrova TV, Alitalo K. Molecular lymphangiogenesis: new players. Trends Cell Biol. 2005a;15(8):434–41.
Tammela T, Saaristo A, Lohela M, Morisada T, Tornberg J, Norrmen C, et al. Angiopoietin-1 promotes lymphatic sprouting and hyperplasia. Blood. 2005b;105(12):4642–8.

Trayhurn P, Wood IS. Adipokines: inflammation and the pleiotropic role of white adipose tissue. Br J Nutr. 2004;92(3):347–55.

Unna P. The histopathology of the disease of the skin. New York: Macmillan; 1896.

Voros G, Maquoi E, Demeulemeester D, Clerx N, Collen D, Lijnen HR. Modulation of angiogenesis during adipose tissue development in murine models of obesity. Endocrinology. 2005;146(10):4545–54.

von der Weid PY, Van Helden DF. Beta-adrenoceptor-mediated hyperpolarization in lymphatic smooth muscle of guinea pig mesentery. Am J Physiol. 1996;270(5 Pt 2):H1687–95.

Wang B, Wood IS, Trayhurn P. Dysregulation of the expression and secretion of inflammation-related adipokines by hypoxia in human adipocytes. Pflugers Arch. 2007;455(3):479–92.

Wang L, Gill R, Pedersen TL, Higgins LJ, Newman JW, Rutledge JC. Triglyceride-rich lipoprotein lipolysis releases neutral and oxidized FFAs that induce endothelial cell inflammation. J Lipid Res. 2009a;50(2):204–13.

Wong L, Sivok B, Kurucz E, Sloop CH, Roheim PS, Asztalos B. Lipid composition of HDL subfractions in dog plasma and lymph. Arterioscler Thromb Vasc Biol. 1995;15(11):1875–81.

Wang W, von der Weid PY, Muthuchamy M, Zawieja DC. Low density lipoprotein modulates rat mesenteric lymphatic pumping. Microcirculation. 2009b;16:766.

Westcott E, Windsor A, Mattacks C, Pond C, Knight S. Fatty acid compositions of lipids in mesenteric adipose tissue and lymphoid cells in patients with and without Crohn's disease and their therapeutic implications. Inflamm Bowel Dis. 2005;11(9):820–7.

Wigle JT, Oliver G. Prox1 function is required for the development of the murine lymphatic system. Cell. 1999;98(6):769–78.

Weatherill AR, Lee JY, Zhao L, Lemay DG, Youn HS, Hwang DH. Saturated and polyunsaturated fatty acids reciprocally modulate dendritic cell functions mediated through TLR4. J Immunol. 2005;174(9):5390–7.

Wiley HE, Gonzalez EB, Maki W, Wu MT, Hwang ST. Expression of CC chemokine receptor-7 and regional lymph node metastasis of B16 murine melanoma. J Natl Cancer Inst. 2001;93(21):1638–43.

Wang YY, Lehuede C, Laurent V, Dirat B, Dauvillier S, Bochet L, et al. Adipose tissue and breast epithelial cells: a dangerous dynamic duo in breast cancer. Cancer Lett. 2012;324(2):142–51.

Wagner M, Bjerkvig R, Wiig H, Melero-Martin JM, Lin RZ, Klagsbrun M, et al. Inflamed tumor-associated adipose tissue is a depot for macrophages that stimulate tumor growth and angiogenesis. Angiogenesis. 2012;15(3):481–95.

Wigle JT, Harvey N, Detmar M, Lagutina I, Grosveld G, Gunn MD, et al. An essential role for Prox1 in the induction of the lymphatic endothelial cell phenotype. EMBO J. 2002;21(7):1505–13.

Wada H, Ura S, Kitaoka S, Satoh-Asahara N, Horie T, Ono K, et al. Distinct characteristics of circulating vascular endothelial growth factor-a and C levels in human subjects. PLoS One. 2011;6(12):e29351.

Xue Y, Petrovic N, Cao R, Larsson O, Lim S, Chen S, et al. Hypoxia-independent angiogenesis in adipose tissues during cold acclimation. Cell Metab. 2009;9(1):99–109.

Yuan L, Moyon D, Pardanaud L, Breant C, Karkkainen MJ, Alitalo K, et al. Abnormal lymphatic vessel development in neuropilin 2 mutant mice. Development. 2002;129(20):4797–806.

Yoffey JM, Drinker CK. The lymphatic pathway from the nose and pharynx : the absorption of dyes. J Exp Med. 1938;68(4):629–40.

Yoffey JM, Sullivan ER, Drinker CK. The lymphatic pathway from the nose and pharynx : the absorption of certain proteins. J Exp Med. 1938;68(6):941–7.

Yan A, Avraham T, Zampell JC, Haviv YS, Weitman E, Mehrara BJ. Adipose-derived stem cells promote lymphangiogenesis in response to VEGF-C stimulation or TGF-beta1 inhibition. Future Oncol. 2011;7(12):1457–73.

Yoon M, Kim MY. The anti-angiogenic herbal composition Ob-X from Morus alba, Melissa officinalis, and Artemisia capillaris regulates obesity in genetically obese ob/ob mice. Pharm Biol. 2011;49(6):614–9.

Zampell JC, Aschen S, Weitman ES, Yan A, Elhadad S, De Brot M, et al. Regulation of adipogenesis by lymphatic fluid stasis: part I. Adipogenesis, fibrosis, and inflammation. Plast Reconstr Surg. 2012;129(4):825–34.

Zawieja SD, Wang W, Wu X, Nepiyushchikh ZV, Zawieja DC, Muthuchamy M. Impairments in the intrinsic contractility of mesenteric collecting lymphatics in a rat model of metabolic syndrome. Am J Physiol Heart Circ Physiol. 2012;302(3):H643–53.

Chapter 7
Origin of Adipocyte Precursors from Adipose Vascular Endothelium

Andrea Frontini, Silvia Corvera, and Saverio Cinti

Abstract Morphological and immunohistochemical data suggest that, during mouse and rat development, capillary endothelial cells of both classic white and classic brown mouse fat depots transform into pericytes, which have been shown in numerous studies to display characteristics of white and brown preadipocytes. Lineage studies also support the hypothesis that adipocytes from both mouse depots are derived from cells that express classical endothelial cell markers during development. That this developmental process may also occur in humans is supported by the demonstration that human fat explants give rise to tubular capillary-like structures, composed by cells with structural and molecular markers of endothelial cells, that can be converted into adipocytes after treatment with rosiglitazone. These findings will enable further identification of the biochemical and physiological cues that enable conversion of endothelial and preadipocyte precursors to adipocytes during normal growth and in obesity.

Keywords Adipose stem cell • Preadipocytes • Adipogenesis • White adipose tissue • Brown adipose tissue • Endothelial cells • Electron microscopy • Lineage tracing • Human tissue explants

A. Frontini • S. Cinti (✉)
Department of Experimental and Clinical Medicine, University of Ancona (Politecnica delle Marche), Ancona 60126, Italy
e-mail: cinti@univpm.it

S. Corvera
Program in Molecular Medicine, University of Massachusetts Medical School, Worcester, MA 01605, USA

Y. Cao (ed.), *Angiogenesis in Adipose Tissue*, DOI 10.1007/978-1-4614-8069-3_7,

7.1 Plasticity of the Adipose Organ: An Old View as a New Concept

7.1.1 Adipose Tissue as an Organ

Obesity is a widespread condition that shortens life expectancy due to its associated metabolic and cardiovascular disorders. Hypertrophy and hyperplasia of white adipose tissue (WAT) are the basic cellular processes characterizing this pathological condition (Friedman 2000; Korner and Aronne 2003). Adipogenesis is an extremely dynamic process, both throughout development and during the maintenance of tissue homeostasis in adulthood (Arner et al. 2011). Thus, a better understanding of the mechanisms that underlie the adipocyte life cycle could be very important for the development of future therapies. Adipose tissue is generally regarded as a form of connective tissue, distributed throughout the body with no specific anatomy. However, systematic observations in both rodents and humans support the idea that adipose tissue is an organ composed of specific depots, with discrete anatomy, specific vascular and nerve supplies, and complex cytology and high physiologic plasticity (Frontini and Cinti 2010). The depots of the adipose organ are located in two compartments of the body: below the skin (subcutaneous depots) and inside the peritoneal cavity (visceral depots) (Fig. 7.1). In both cases, the main parenchymal cells of the organ are the adipocytes, but also highly relevant are vascular components and cells of hematopoietic origin.

7.1.2 Types of Adipocytes

Different types of adipocytes can be distinguished by morphological and biochemical characteristics (Cinti 1999). Two clearly distinct types are distinguished by: (a) large spherical cells with a discoid nucleus, where 90 % of the cell volume is comprised of a cytoplasmic single lipid droplet, and containing mitochondria with sparsely packed cristae (white adipocytes), and (b) smaller, polygonal cells with a spherical nucleus, multiple smaller lipid droplets, and numerous, large mitochondria with highly packed cristae (brown adipocytes) (Cinti 2005). Biochemically, the most classical difference between white and brown adipocytes is the presence in the latter of a unique protein (uncoupling protein 1, UCP1) able to uncouple oxidative phosphorylation and thereby generate heat (Cannon and Nedergaard 2004). The morphological and biochemical features are associated with distinct functions, where white adipocytes store energy to supply the metabolic needs of the organism and brown adipocytes consume energy for the purpose of thermogenesis.

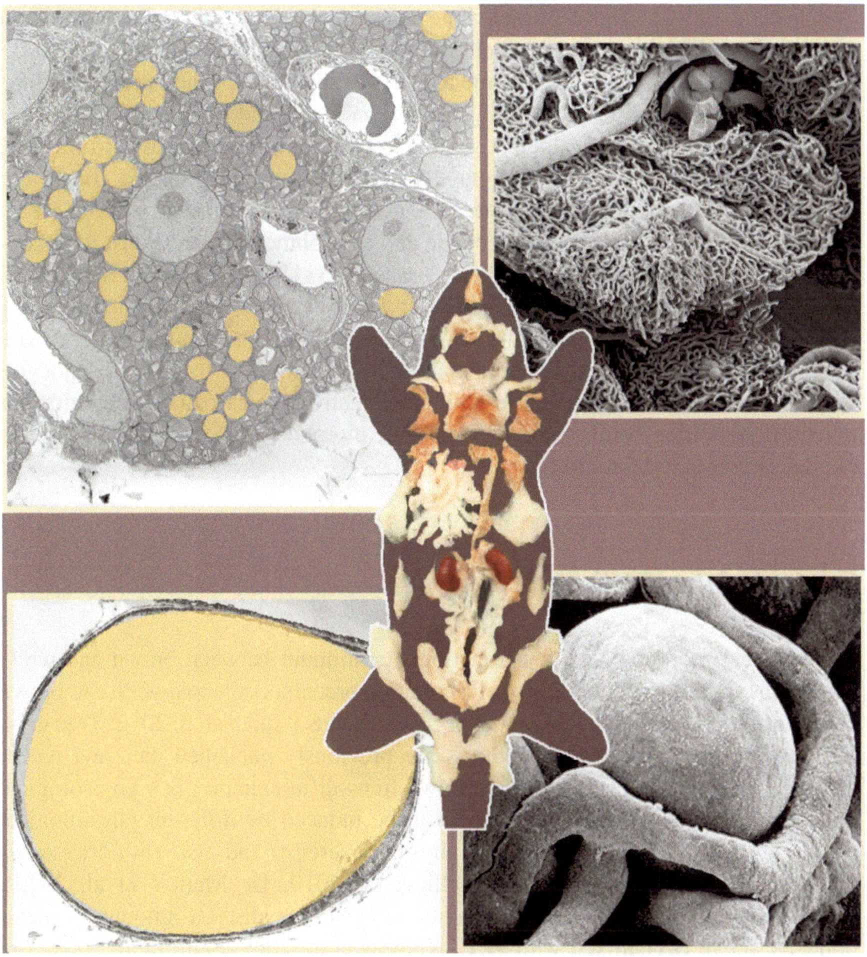

Fig. 7.1 In the *middle* of the image, it is showed the adipose organ of an adult Sv129 female mice kept at 28 °C. Individual depots are dissected out and placed on their anatomical position. The mixed composition (WAT-BAT) is evident also at gross anatomy level. In the *top left side*, a multilocular brown adipocyte with numerous and big rounded shape mitochondria is visible by Transmission Electron Microscopy (TEM). On the *top right side*, vascular casting of BAT disclosed the rich vascular network visualized by Scanning Electron Microscopy (SEM). At the *bottom* of the image on the *left* a classical white adipocyte with the unilocular lipid filling droplet (TEM). On the *right*, SEM appearance of white adipocyte surrounded by a capillary

7.1.3 Relationship Between Depots and Adipocyte Types

The several subcutaneous and visceral depots forming the adipose organ in mice contain both white and brown adipocytes, albeit at very different proportions. In addition, comparison of the adipose organ of two different strains of mice (B6 and Sv129) revealed the presence of intermediate forms of adipocytes in all depots

(Vitali et al. 2012). These intermediate adipocytes are multilocular but do not express the UCP1 protein, and their mitochondria display intermediate features between those found in classical white and brown adipocytes. Another intermediate form (paucilocular adipocyte) contains a predominant lipid droplet, and expresses detectable levels of UCP-1, albeit at significantly lower levels compared to classical brown adipocytes, suggesting they have a weak thermogenic capacity (Barbatelli et al. 2010). All these intermediate steps are present within the rodent adipose organ, and their prevalence is affected by genetic and environmental conditions; for example, they are more abundant in the obesity-resistant Sv129 mouse strain. In addition, these intermediate adipocyte forms are usually located in areas at the boundary between classic compact WAT and classic compact brown adipose tissue (BAT). This localization is suggestive of a "continuum" from white to brown adipocytes with the intermediate forms representing alternative routes of differentiation, or trans-differentiation events resulting in specific morphological features and patterns of gene expression (Cinti 2009).

7.1.4 Origin of Adipocyte Subtypes

The apparent morphological and functional continuum between brown and white adipocytes suggests the possibility of a common precursor able to give rise to brown or white adipocytes depending on the organism needs (Gupta et al. 2012; Tran et al. 2012). Consistent with this hypothesis are previously published data and recent studies, which suggest that white to brown trans-differentiation is a physiological phenomenon in adult mammals and it can be induced by different physiological conditions (such as cold exposure or physical exercise) and also with drug treatments (e.g., β3-AR agonist) (Barbatelli et al. 2010; De Matteis et al. 2012). Furthermore, all mammals, including adult humans, have discrete amounts of metabolically active BAT mixed with WAT (Cypess et al. 2009; Saito et al. 2009; van Marken Lichtenbelt et al. 2009; Virtanen et al. 2009). BAT have anti-obesity properties (Bachman et al. 2002; Lowell et al. 1993; Kopecky et al. 1995; Nedergaard et al. 2011) and brown adipocyte (BA) precursors can be found in the adipose organ of adult humans (Zingaretti et al. 2009). Thus investigation on the molecular basis of adipose tissue formation and on its stem cell niche, both for WAT and BAT, are of paramount importance for future therapies of obesity and to better understand the homeostatic mechanisms of this extremely plastic organ.

7.2 Adipocyte Precursors and Vasculature: State of the Art

Adipose tissue growth is mediated by adipocyte hypertrophy, where individual adipocytes expand to accommodate additional lipid storage, and adipocyte hyperplasia, which involves formation of new adipocytes from the preadipocyte pool.

Much is known about the morphological aspects and molecular mechanisms by which preadipocytes differentiate into mature adipocytes in vitro (Rosen et al. 2000; Farmer 2006; Tontonoz and Spiegelman 2008; Kajimura et al. 2010) and in vivo (Napolitano 1963; Soukas et al. 2001); however, the identity of the precursor cells that can give rise to committed preadipocytes is less clear. In addition, adipose tissue growth is absolutely dependent on the expansion of its capillary network, which is required to supply oxygen, deliver free fatty acids for storage, and allow the proper endocrine function of adipose tissue (Hausman and Richardson 2004; Christiaens and Lijnen 2010). How the growth and differentiation of adipocytes are coordinated with the growth of capillaries to form functional adipose tissue is another important question in this area.

7.2.1 Development of Adipose Tissue During Embryogenesis

Adipose tissue is generally thought to originate from the mesoderm. Embryonic mesoderm gives rise to mesenchymal stem cells, which in turn give rise to adipose stem cells (Gesta et al. 2007). Under appropriate conditions, adipose stem cells develop into committed white and brown preadipocytes and ultimately mature adipocytes. However, no single, functional molecular marker expressed specifically by adipocyte stem cells has been identified, and thus the exact nature of these precursors, the specific steps and factors that determine their commitment to preadipocytes and subsequent differentiation, are not clear. Of note, brown adipocytes forming the classic BAT depot have been hypothesized to originate from a cell lineage common to myocytes (Myf5) and fundamentally different from the white adipocyte lineage (Timmons et al. 2007; Seale et al. 2008). The origin of the newly formed brown adipocytes found in the adipose organ after cold exposure or treatment with β3-AR agonist (phenomena called browning) is under debate and the data are not still conclusive. However, several independent lines of research have produced contradictory results ascribing the origin of brown adipocytes to different cell lineages (Barbatelli et al. 2010; Petrovic et al. 2010; Lee et al. 2012; Wu et al. 2012).

7.2.2 Morphological Evidence That Pericytes Give Rise to White Adipocytes

Recent (Amos et al. 2008; Crisan et al. 2008; Tang et al. 2008) and old data (Iyama et al. 1979; Cinti et al. 1984; Tavassoli 1976) support the hypothesis that the vascular wall of adipose tissue capillaries is the niche where adipose stem cells reside. Indeed, there is a strong convergence of morphological evidence that suggests that pericytes of adipose tissue capillaries are adipose stem cells of both white and brown adipose depots. The epididymal fat (eWAT) is widely considered as classic

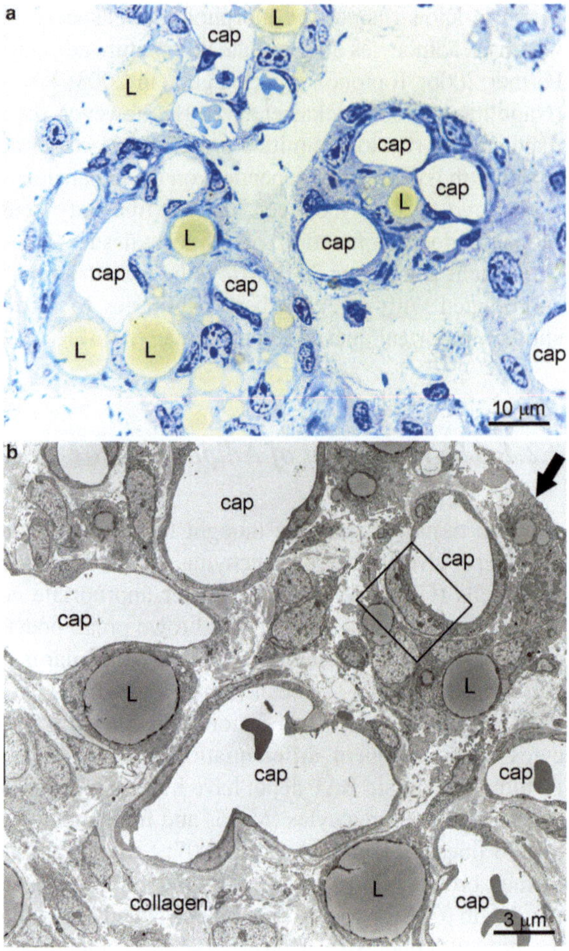

Fig. 7.2 Morphological aspects of eWAT at P7. (**a**) Light microscopy of eWAT in a postnatal mice in which are evident the large capillaries (cap) associated with preadipocytes with small lipid droplets accumulation (L). The capillaries are extremely numerous and about threefold larger than those found in adult eWAT. (**b**) TEM of vasculo-adipocytic island. The *black squared area* and the area pointed by *black arrow* are enlarged in Fig. 7.3

murine "pure" white fat depot. Its development starts in rats in the days immediately following birth (similar results were found also in mice but with a slightly different timing). Before birth its morphology shows the typical features of a poorly differentiated mesenchymal tissue with a homogeneous population of fibroblast-like cells in a loose connective matrix with small and sparse capillaries. Both endothelial cells of capillaries and mesenchymal fibroblast-like cells are often in mitosis. At the postnatal day 4–6 (P4-6) the morphology of eWAT change dramatically (Fig. 7.2a). At this time it is formed by well-circumscribed areas in which adipocytes at different stages of differentiation are visible. These areas are delimited by fibroblast-like cells and into these areas numerous unusual large capillaries are visible. Thus, adipogenesis appear to be restricted into these areas characterized by large capillaries (vasculo-adipocytic islets).

Electron microscopy revealed that the interstitial matrix into vasculo-adipocytic islets was dense of collagen fibrils that were sparse in the rest of the tissue outside

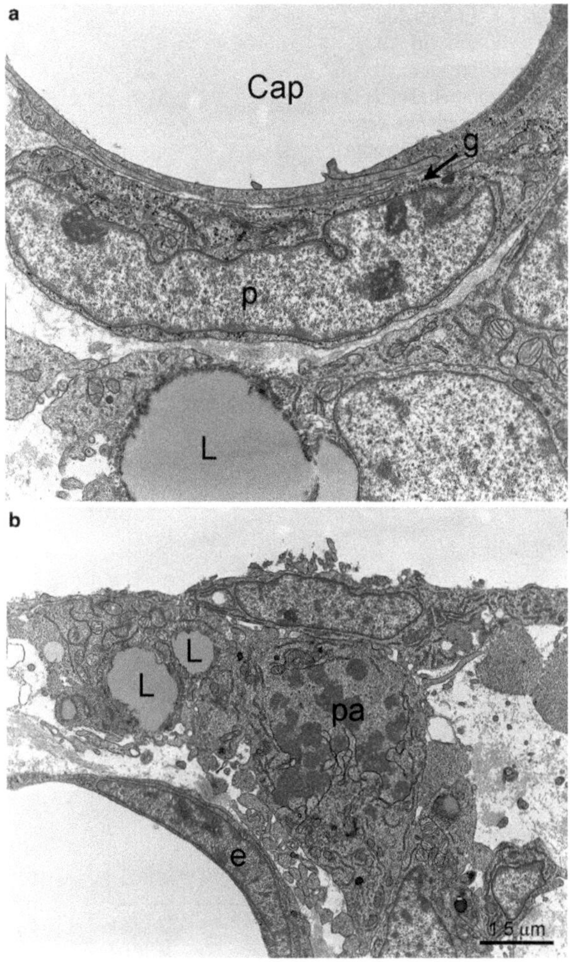

Fig. 7.3 Ultrastructure of endothelial cells, pericytes, and preadipocytes in the vasculo-adipocytic island. (**a**) Poorly differentiated cells in pericapillary position (most likely been perycites; p) with glycogen accumulation (*arrow*) and surrounded by a distinct basal membrane. (**b**) Preadipocyte (pa) with small lipids droplets (L) where often found in mitosis and in association with endothelial cells (e)

the islets (Fig. 7.2b). Of note, all adipocytes and preadipocytes as defined *bona fide* by their lipid content were present only into these vasculo-adipocytic islets, suggesting that these islets are the site of preadipocyte proliferation and development. The wall of capillaries was often thick and endowed with abundant pericytes (Fig. 7.3). Fibroblasts, macrophages, and mastocytes were also present into the islets. Lipid accumulation, small/elongated mitochondria, short strands of RER, large cytoplasmic areas containing only ribosomes and polyribosomes, together with the external lamina were the main structural characteristics of preadipocytes at the earliest stage of differentiation. These cells were always in tight contact with the wall of capillaries. Several intermediate aspects between pericytes and poorly differentiated preadipocytes were found, but pericytes appeared more often without lipid droplets and with glycogen particles in variable amount, in line with the use of glycogen by preadipocytes forming triacyl glycerols visible as lipid droplets (Fig. 7.4a).

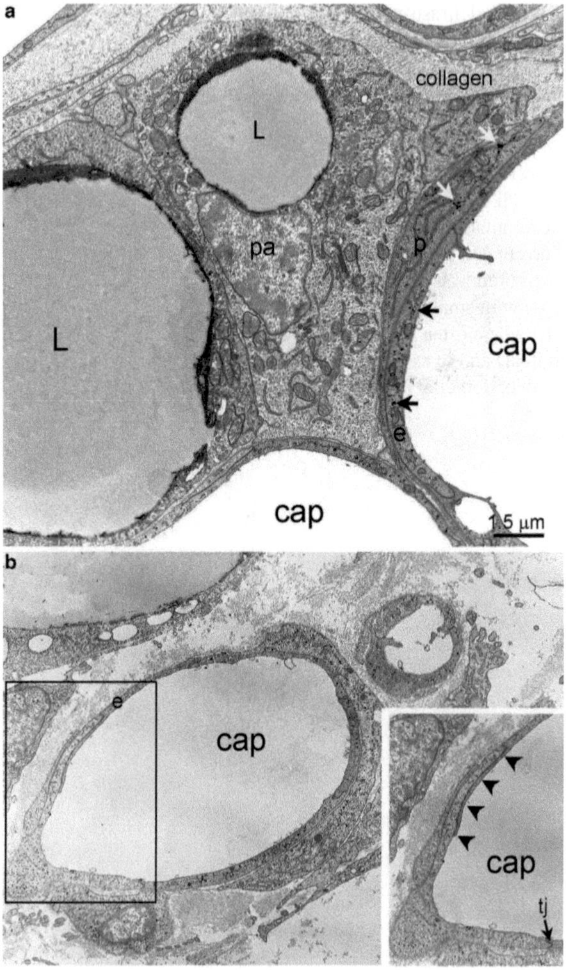

Fig. 7.4 Endothelial-pericytic cell. (**a**) All pericyte-like elements (p), some endothelial cells (e) and preadipocytes (pa) were found to contain glycogen (*white* and *black arrows*). (**b**) An endothelial-pericytic cell is enlarged in the *inset*. Endothelial cell (e) exposed to the capillary lumen, but also extended over a vicinal endothelial cell (pointed by *black arrowhead*) to adopt a pericytic position. Further details are found in the text (thigh junction; tj)

Pericyte-like elements can be distinguished from all other poorly differentiated cell types by their distinct basal lamina, in fact only pericytes and all preadipocytes at different stages of development were surrounded by a distinct basal lamina into the vasculo-adipocytic islets. Furthermore pericytes in stages of detachment from the capillary wall were also observed. Thus, structural data suggest that pericytes could be the early step of development of preadipocytes. In the subsequent days of development (up to P20), the islets disappeared progressively substituted by a more homogeneous tissue composed mainly by unilocular white adipocytes. Capillaries lost their characteristic morphology and became smaller (not shown), but pericytes-preadipocytes were often found in tight connection with capillary wall.

7.2.2.1 Endothelial Cells of Capillaries of the Vasculo-Adipocytic Islets Share Structural Features with Pericytes

The vasculo-adipocytic islets of P4-6 animals are the ideal tissue to study the details of white adipogenesis in vivo, mainly because of high pericyte-preadipocyte abundance. In these islets, endothelial cells and pericytes share some structural aspects: glycogen particles, pinocytotic vesicles, and a distinct external (basal) lamina. Of note, only some endothelial cells have all common structural aspects, in fact many endothelial cells lacked glycogen particles. In about 1–3 % of vasculo-adipocytic islet capillaries, we observed unusual, novel features, such as some endothelial cells exposed to the capillary lumen, but also extended over a vicinal endothelial cell to adopt a pericytic position. Importantly, the junction of these cells with adjacent endothelial cells was composed of typical oblique tight junctions, confirming the endothelial nature of these cells (Fig. 7.4b). Some of the endothelial-pericytic cells were found to contain glycogen, as do some endothelial cells, and almost all pericytes. Endothelial cells and pericytes were sometimes joined by tight junctions between a protrusion of the endothelial cell crossing the basal membrane, and the complementary indentation in the pericyte. These data show that there is a complex relationship between these cells in the vasculo-adipocytic islets where extensive adipogenesis is ongoing. Furthermore, cells containing glycogen particles, which have been associated with adipocyte progenitors, can be seen partially associated with the capillary wall (in pericytic position) and also abutting into the interstitial space (Tavassoli 1976). Taken together, this evidence suggests that endothelial-pericytic cells may represent an intermediate step between endothelial and preadipocyte stages. Cells in endothelial-pericytic position were found only at this step of development of the fat depot, consistent with the hypothesis that at this stage of development the number of poorly differentiated adipocytes and preadipocytes reach the maximal density. Furthermore, endothelial cells, pericytes, and early preadipocytes (with some lipid accumulation in form of small lipid droplets) of vasculo-adipocytic islets were frequently found in mitosis (Fig. 7.5a,b), supporting the active developmental role of vasculo-adipocytic islets for epididymal fat and the importance of endothelial cells and pericytes in this process.

7.2.2.2 Endothelial Cells, Pericytes, and Preadipocytes of Vasculo-Adipocytic Islets are Immunoreactive for Transcription Factors and Proteins Described as Markers of Preadipocytes In Vitro

Most of the cells located in the pericapillary position of the vasculo-adipocytic islets, most likely corresponding to the cells described above by EM, showed nuclear staining for PPARγ, C/EBPα, and C/EBPβ, which are transcription factors widely considered as markers of cells undergoing an adipogenetic program in vitro (Farmer 2006; Tontonoz and Spiegelman 2008). This evidence supports the notion that they are potential preadipocytes. Notably, some endothelial cells were also

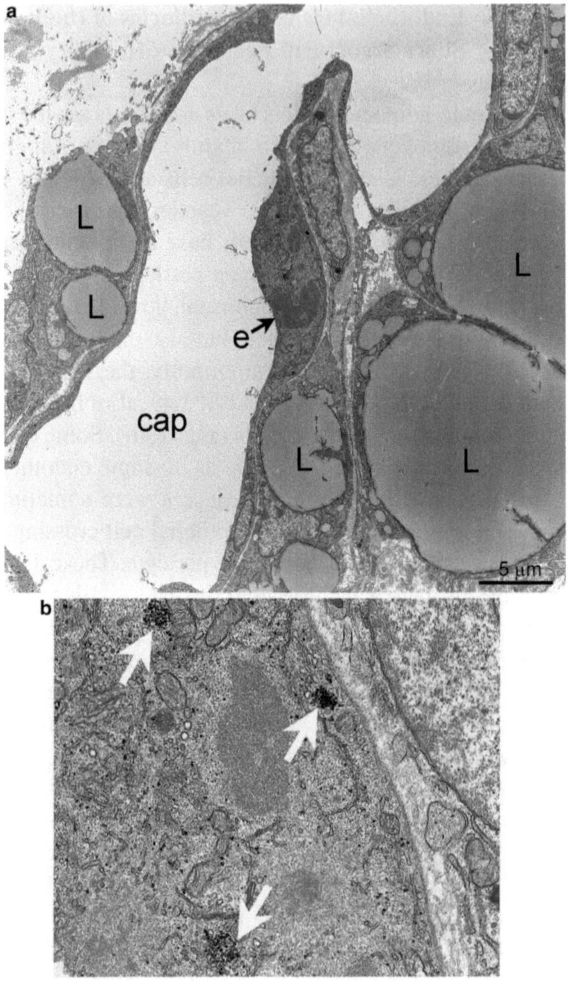

Fig. 7.5 Endothelial cells where often found in mitosis (**a**) and the enlargement in (**b**) shows the presence of glycogen which have been associated with the development of adipocytes precursors

found positive for C/EBPβ, which is considered upstream of PPARγ and C/EBPα in the sequence of transcriptional control of adipogenesis (Fig. 7.6).

Perilipin, S-100b, and leptin were found only in cells with signs of adipocyte differentiation (i.e., at least a minimum of cytoplasmic lipid accumulation). Perilipin was found only at the level of lipid droplet surface. Leptin and S100b immunoreactivity were diffuse in the cytoplasm of adipocytes (not shown). No other cell type, recognizable by their structural characteristics, in the vasculo-adipocytic islets was immunoreactive for these proteins (C/EBPα, PPARγ-tot, PPARγ2, perilipin, S-100b, and leptin).

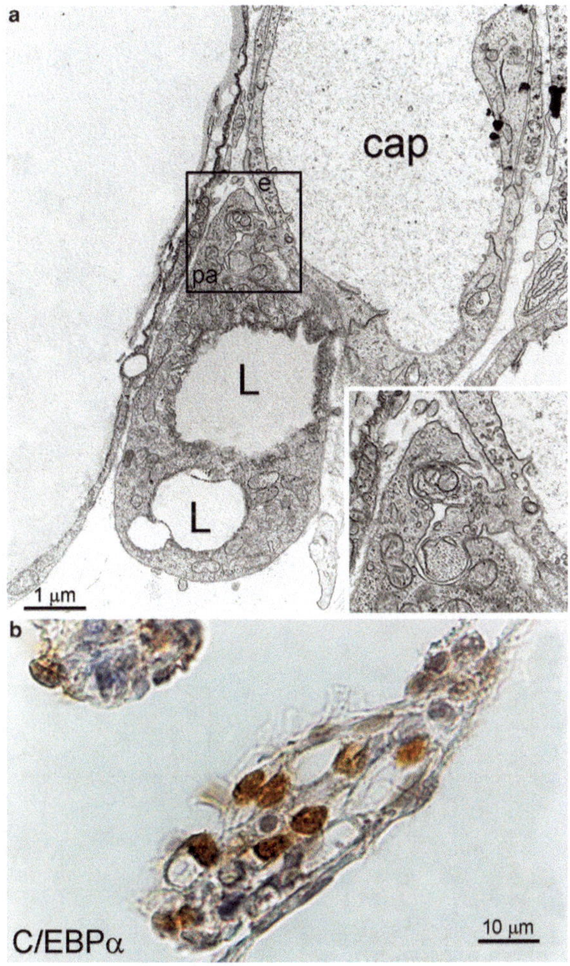

Fig. 7.6 (**a**) To note that also preadipocytes (pa) with small lipid droplets contained in the cytoplasm displayed close connection with the endothelium (e). (**b**) Immunohistochemistry performed in the eWAT anlage at P7 showing cell immune-reactive for C/EBPα

7.2.3 Morphological Evidence That Pericytes Give Rise to Brown Adipocytes

The fetal developing BAT that gives rise to the interscapular fat depot (interscapular BAT) reside in a specific anatomical location among dorsal muscles and the less differentiated anlage is visible at embryonic day (E15-16) in that site in rats. At this stage the anlage is formed mainly by large capillaries, with thick wall due to the presence of abundant pericytes (Fig. 7.7). At E17-18 the anlage is enriched with poorly differentiated cells (>90 % of cells in the anlage) occupying almost all the interstitial space among capillaries. These cells were also characterized by numerous large mitochondria (with morphology similar to that of mitochondria of brown

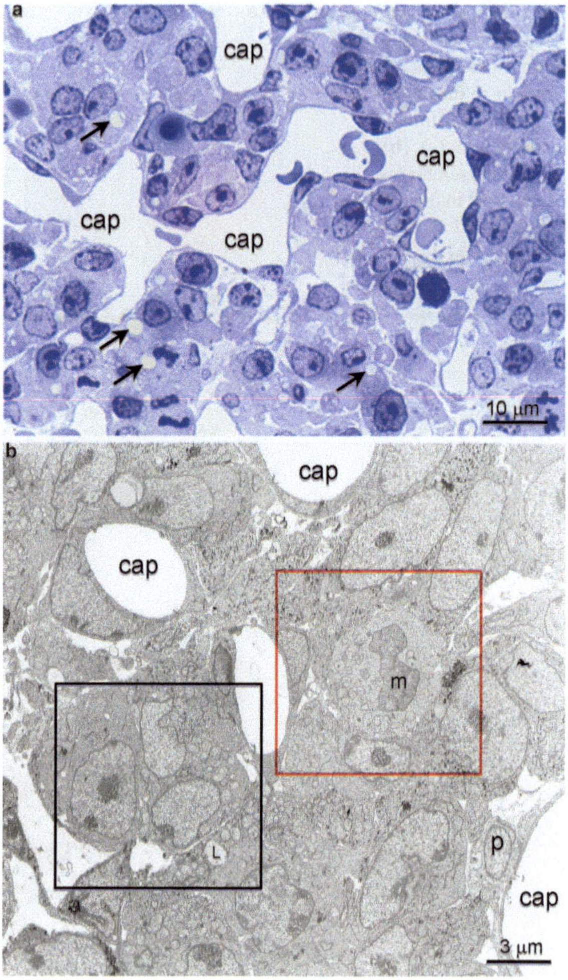

Fig. 7.7 Morphology of BAT anlage at E15-16. (**a**) Light microscopy of fetal BAT with large capillaries (cap) with the interstitial matrix endowed of poorly differentiated elements with small degree of lipid accumulation (*arrows*). (**b**) TEM appearance of BAT anlage at E15-16. The *squared areas* are enlarged in Fig. 7.8

adipocytes: pre-typical mitochondria), glycogen, and external lamina. Most (70–80 %) of these cells exhibited also small lipid droplets. Thus, most of these cells had structural aspects of brown preadipocytes. The pericytes belonging to the capillary wall, since E15-16 ahead exhibited structural features very similar to those found in preadipocytes (pre-typical mitochondria, glycogen, external lamina), thus, pericytes can be considered as an early step of preadipocyte differentiation both in white and brown adipose tissues. Endothelial cells, pericytes, and preadipocytes were often in mitosis (Fig. 7.8a). Preadipocytes in mitosis exhibited pre-typical mitochondria, glycogen, lipid droplets, and external lamina confirming their nature as preadipocytes. In the following steps of development until P20, the number of preadipocytes (including preadipocytes in mitosis) reduces in number and mature adipocytes become prevalent (Fig. 7.8b). Several intermediate aspects between early preadipocytes and mature adipocytes were easily found at all developmental ages. The less differentiated preadipocytes were always in thigh contact with the capillary wall

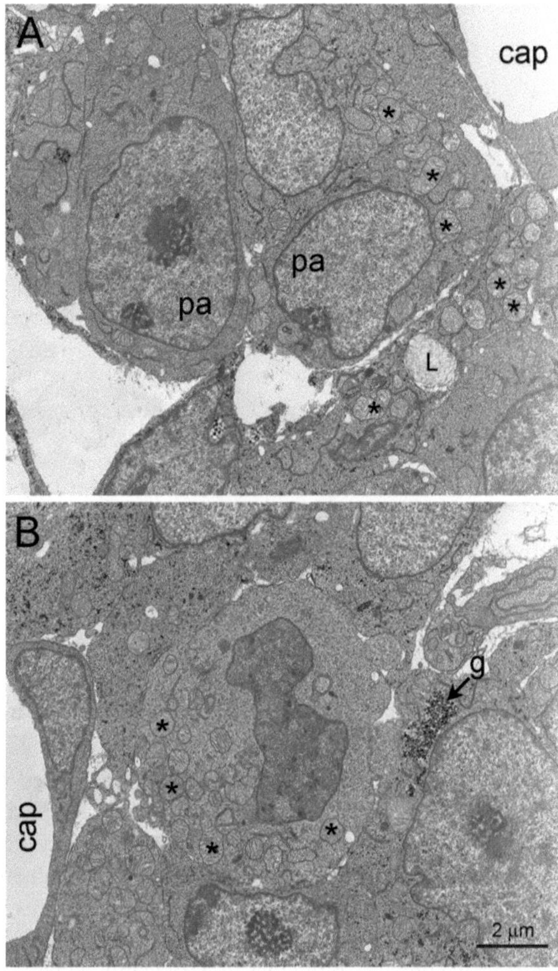

Fig. 7.8 (**a**) Preadipocytes (pa) with small lipid droplets (L) display pre-typical mitochondria (some indicated by *asterisks*) which can be considered as morphological marker of brown adipocytes. (**b**) Some cell with clear glycogen accumulation (g) and pre-typical mitochondria were often found in mitosis

(pericytic position) and in the postnatal period they were the only cell type in mitosis and with a morphology comparable to that of preadipocytes found at E17-18. Thus considering the ages between E15 and P20 the most informative period was E17-18, when most of the anlage is formed by capillaries and cells with structural futures characteristic of brown preadipocytes.

7.2.3.1 Endothelial Cells of Capillaries of Developing Interscapular BAT Anlage Share Structural Features with Pericytes and Preadipocytes

At the stage where the developing BAT is most enriched by preadipocytes (E17-18), 3–4 % of the endothelial cells the anlage contained glycogen, as well as mitochondria that displayed the very characteristic morphology of brown adipocytes (Fig. 7.9).

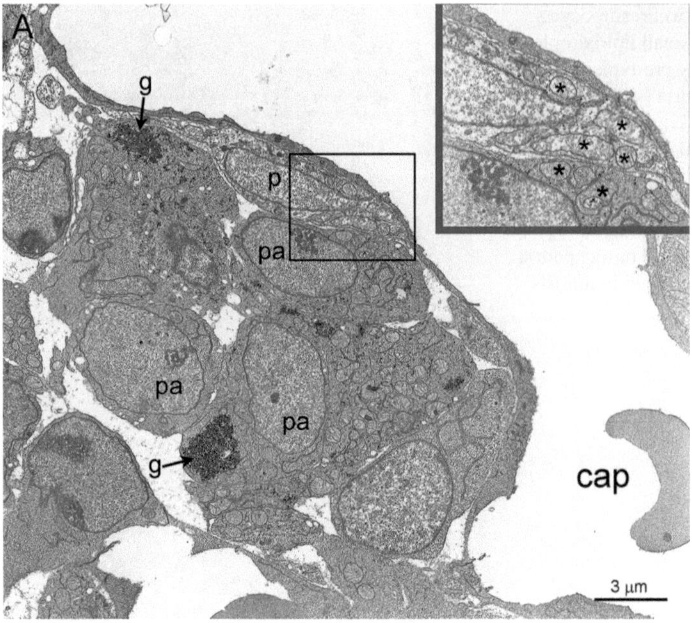

Fig. 7.9 Several stages of development are here visible. Pericyte (p) and the adjacent preadipocytes as well as other preadipocytes (pa) containing glycogen (g) disclosed in their cytoplasm the classical pre-typical mitochondria (some indicated by *asterisks*)

Because early preadipocytes are easily recognized by this highly characteristic mitochondrial morphology, the presence of pre-typical mitochondria in some endothelial cells that also contain glycogen is suggestive that these may give rise to brown preadipocytes (not represented). At a biochemical level, we found strong immunoreactivity for PPARγ in nuclei of all cells, except endothelial cells, at E17-18, when most of the cells present in the anlage were brown preadipocytes. In vitro studies have demonstrated that PPARγ is induced after C/EBPβ, and some endothelial cells, pericytes, and preadipocytes were immunoreactive for these markers at all pre- and postnatal stages. Thus endothelial cells showed structural (pre-typical mitochondria; Fig. 7.10) and molecular (C/EBPα and β) signs of brown adipogenesis preceding the preadipocyte stage. Immunogold and electron microscopy analysis showed that pre-typical mitochondria of poorly differentiated cells in pericytic position (pericytes-preadipocytes) were immunoreactive for UCP1 at E18-19 (not represented). Thus, cells with similar morphology, i.e., pericytes-preadipocytes become immunoreactive for UCP1 at late gestational age (E20-21). Of note, the density of immune-gold particles in pre-typical mitochondria was about ½ that in typical mitochondria, supporting the concept that pre-typical mitochondria are a developmental early stage of brown adipocyte mitochondria. No immunoreactivity for S-100 and leptin was ever found. Perilipin immureactivity was present only when lipid droplets were present in the cytoplasm of preadipocytes or adipocytes (not represented).

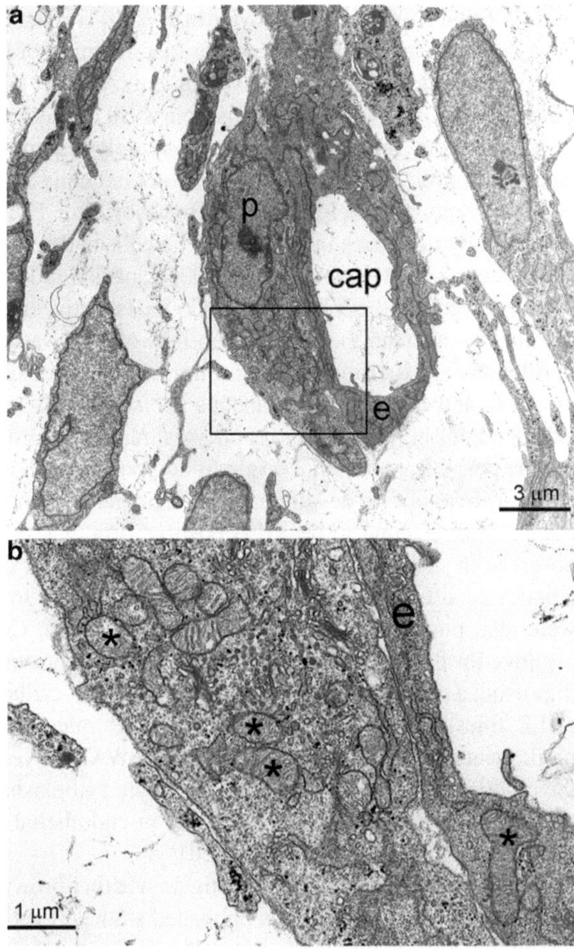

Fig. 7.10 (**a**) The endothelial cell (e) and the pericyte in close association allow the comparison of their similar mitochondria endowment. (some indicated by *asterisks*)

7.2.4 Lineage Tracing Evidence That Adipose Stem Cells Reside Within the Mural Compartment

Immunohistochemical methods, combined with GFP marking and lineage tracing analysis show that adipose stem cells are found in the wall of blood vessels that supply white adipose depots but are absent from blood vessels that supply other tissues (Tang et al. 2008). Utilizing the GFP marker, the authors characterized a microanatomic location of adipocyte stem cells and demonstrated that many coexpress three mural cell markers (SMA, PDGFRb, and NG2), arguing that adipose stem cells constitute a subset of mural cells embedded in the wall of blood vessels present within adipose tissue.

Recently, lineage-tracing experiments using Cre/loxP technology have allowed for more in-depth analysis of the developmental origin of cells. Mice expressing

LacZ (Soriano 1999) and GFP (Mao et al. 2001) driven by the VE-cadherin promoter have been created, thus allowing the identification of cells in which this promoter was active at any point during development (Alva et al. 2006; Monvoisin et al. 2006; Zovein et al. 2008). VE-cadherin is a classical cadherin involved in endothelial cell–cell interactions, and its presence is absolutely required for the formation of vasculature. The protein is expressed in endothelial cells and in some subpopulations of hematopoietic cells before E11.5. However after E11.5, VE-cadherin expression dramatically decreases in hematopoietic cells and is mainly restricted to the vasculature thereafter (Monvoisin et al. 2006; Speck and Iruela-Arispe 2009). Thus, lineage tracing with the VE-cadherin-Cre constitutive and inducible models allow us to better understand the relationship between the vasculature and adipocyte tissue development. Experiments using immunohistochemistry showed that VE-cadherin in eWAT and inguinal subcutaneous fat (scWAT) is expressed only in endothelial cells, and not in pericytes or adipocytes, making this model suitable for investigating the possible endothelial origin of adipocytes (not represented).

Prior to adipocyte development in fetal and early postnatal eWAT only endothelial cells were found to be X-gal positive (Fig. 7.11a). The same pattern was also observed in scWAT of fetal and postnatal mice. After this period, from P7 to adult, adipocytes at different stages of lipid accumulation found in eWAT and in scWAT were also positive for X-gal staining (Fig. 7.11b). Control animals were always negative for β-galactosidase activity assay (not represented). To confirm these findings using a different reporter, we crossed the VE-cadherin-Cre mice with the Rosa-eGFP lines. In P7 VE-cadherin-Cre-eGFP mice, we found co-localization of perilipin and eGFP in adipocytes found in eWAT. To verify with high resolution that the reporters were localized to the adipocyte cytoplasm, we used EM and observed a precise localization of X-gal crystals in endothelial cells, pericytes, adipocytes, and in preadipocytes (not represented).

These data prompted us to investigate whether brown adipocytes might also have an endothelial origin. Previous published studies (Cinti and Morroni 1995) showed that in the developing classic brown fat depot, some endothelial cells display characteristic structural features (glycogen and mitochondria) also present in UCP1-immunoreactive brown adipocyte precursors. Immunohistochemistry with anti-VE-cadherin-specific antibodies confirmed the absence of protein in BAT pericytes and adipocytes, while all endothelial cells lining vessels resulted positive. β-galactosidase activity in the interscapular region of E17-19 transgenic mice allowed strong positive staining in both endothelial cells and adipocytes of double transgenic mice in the region corresponding to developing BAT, while in the surrounding muscles only endothelial cells were stained (Fig. 7.12a). To verify that the X-gal stained cells correspond to brown adipocytes, sections were also immunostained with antibodies to UCP-1 (Fig. 7.12b). These experiments indicate that X-gal staining colocalized to cells containing the brown adipocyte marker. Taken together, these data strongly suggest that endothelial cells of capillaries of developing WAT and BAT are a source of adipocyte precursors.

Constitutive expression of VE-cadherin-Cre leads to labeling of cells of hematopoietic origin during embryogenesis. A hematopoietic origin for adipocytes has

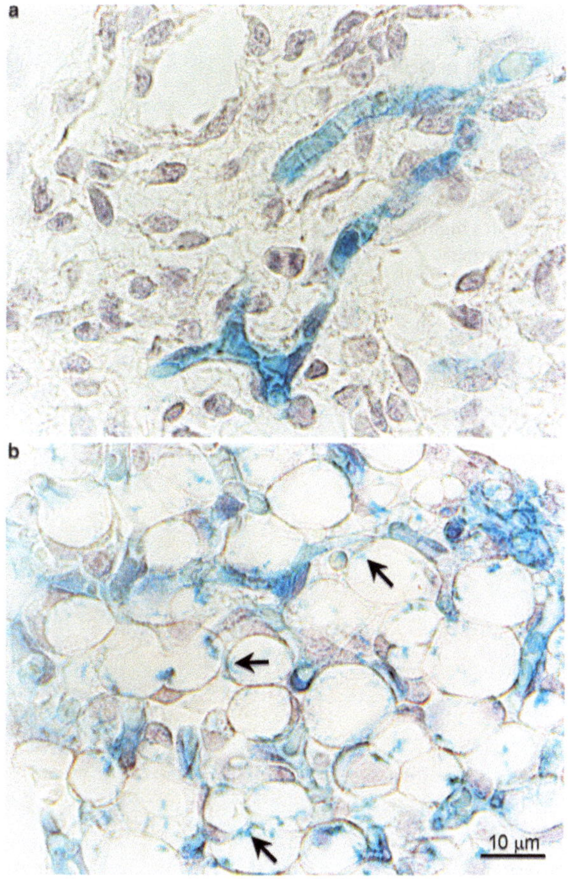

Fig. 7.11 (**a**) β-galactosidase localization on eWAT of P4 mice revealing histochemical staining only in the vasculature. (**b**) X-gal staining performed on eWAT of P7 animals disclosed the localization also in developing and mature adipocytes (*arrows*)

been previously proposed (Crossno et al. 2006), although more recent data have shown that this contribution may be negligible (Koh et al. 2007; Tomiyama et al. 2008). To rule out the possibility that the X-gal-labeled adipocytes were hematopoietic stem cell derived, we employed another mouse model in which the VE-cadherin-driven Cre recombinase is induced during adulthood. Specifically, we have used tamoxifen-inducible transgenic VE-cadherin-CreER^{T2} model described by Monvoisin et al. (2006), where expression of Cre is activated by injection of tamoxifen postnatally, resulting in its negligible excision (lower than 0.4 %) in the hematopoietic lineage. For these experiments, expression of Cre in the VE-cadherin-CreER^{T2}/R26R mice were induced by tamoxifen injection for five consecutive days at 8 weeks of age. The mice were sacrificed 3 weeks later. X-gal staining was observed in capillaries of adipose tissue, but also in numerous adipocytes distributed among the eWAT, scWAT, and BAT depots. The number of adipocytes stained was lower than that seen in the constitutive model described above, possibly due to a lower number of adipocytes being formed during this postnatal expansion period, but still clearly significantly above the background observed in

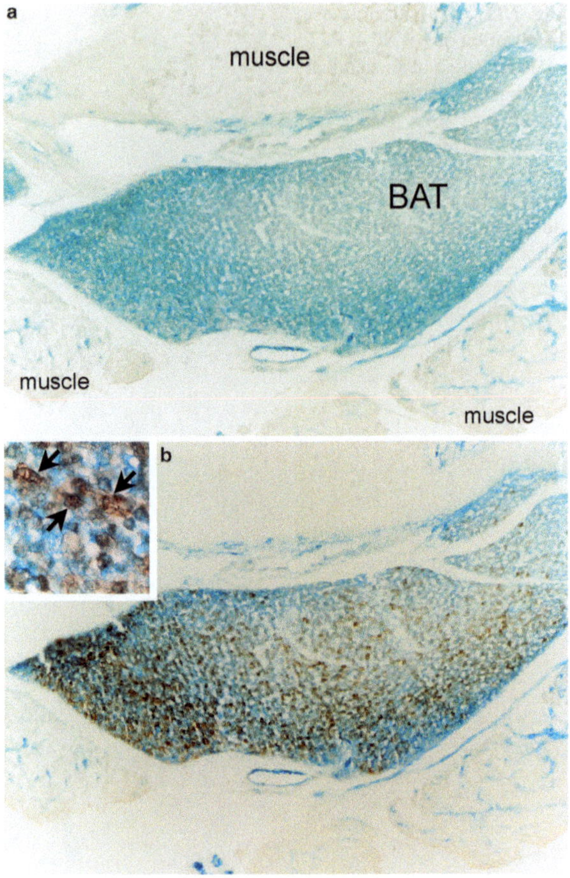

Fig. 7.12 (**a**) X-gal positive staining in brown adipocytes of BAT and in their capillaries as well as in those found in the surrounding muscles. (**b**) X-gal staining colocalized with UCP1-immunoreactive brown adipocytes highlighted in the *inset* (*arrows*)

negative controls (not represented). PPARγ agonists affect murine adipose tissue by increasing multilocularization of existing adipocytes and maybe also by stimulating preadipocyte differentiation into mature adipocytes (Koh et al. 2009; Tang et al. 2011). Mice treated with rosiglitazone for 3 weeks following induction displayed numerous multilocular adipocytes, many of which stained with X-gal (not represented). Thus under both normal and stimulated adipogenesis, adipocytes arise from VE-cadherin expressing progenitors.

7.2.5 Molecular Markers of Endothelial Cells in Adipose Stem Cells

Pioneering work by Rodeheffer and colleagues (Rodeheffer et al. 2008) and Joe et al. (2009), have prospectively identified a population of cells within the stromovascular fraction of mouse adipose tissue characterized by the presence of specific

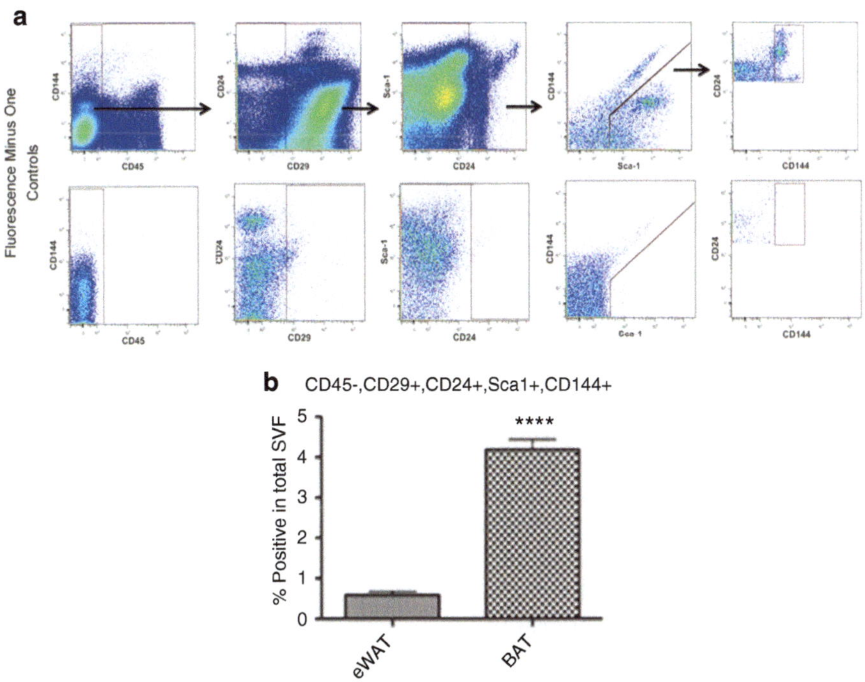

Fig. 7.13 (**a**) FACS analysis scheme of cells from SVF of mouse epididymal adipose tissue. SVF from eWAT from 8-week-old mice were isolated and stained with blue stain (Invitrogen) at 4 °C for 20 min. Cells were incubated with anti-mouse CD16/CD32 (BD Biosciences) for 15 min and then with respective FACS antibody for 2 h at room temperature. Shown in *lower panel* are fluorescence minus one controls used for gate setting. (**b**) Comparison of abundance of progenitor cells between mouse depots. Percent of cells with the immunophenotype CD45-CD29+CD24+ Sca1+CD144+ recovered from SVF from epididymal fat pads (eWAT) or interscapullar brown adipose tissue (BAT)

stem cell markers. These, expressing the cell surface phenotype CD45-CD29+ CDC24+Sca-1+, give rise to adipocytes in vitro, and can reconstitute adipose tissue in lipodystrophic mice.

Based on their studies, and findings of a developmental relationship between adipocyte and vascular cells, experiments were conducted to determine whether CD45-CD29+CDC24+Sca-1+ adipocyte precursors have some features of endothelial cells, or may have a common precursor. CD45-cells from the SVF were successively gated for the presence of CD29, CD24, and Sca1 (Fig. 7.13a). Approximately 5 % and 10 % of the cells in the SVF of epididymal fat and brown adipose tissue, respectively, were CD45-CD29+CDC24+, and this difference was statistically significant (Tran et al. 2012). Approximately 17 % and 52 % of the CD45-CD29+CDC24+Sca-1+ population in eWAT and BAT, respectively, were also positive for VE-cadherin (CD144), comprising 0.5 % and 4 % of the cells in the SVF of eWAT and BAT, respectively (Fig. 7.13b). The percentages of adipogenic stem cells expressing VE-cadherin is in line with the amount of endothelial-pericytic cells found in WAT by EM studies in vivo and the much larger proportion of VE-cadherin positive cells in BAT is consistent with its denser vascular network. These results support the hypothesis of an intimate relationship between vascular development and adipocyte formation, where adipocyte progenitors interact with the vasculature. Moreover, it is possible that VE-cadherin containing progenitors may give rise to both adipocytes and endothelial cells, and thereby play a key role in balancing the development of new adipocytes and their vasculature during adipose tissue expansion.

7.2.6 Human Adipose Stem Cells and the Vasculature

The potential high relevance of these precursor cells underscores the relevance of identifying similar populations in human adipose tissue. Numerous studies report the presence of pluripotent stem cell populations in human adipose tissue, which hold great potential for clinical applications involving repair of damaged tissues (reviewed in Baer and Geiger (2012)). These cells may also be critical determinants of adipose tissue expandability, and their numbers may influence metabolic disease risk. However, in most published studies related to adipose tissue stem cells, cells are selected on the basis of their plating properties and growth in vitro, and comprise a mixed population with no specific prospective molecular identity.

Recent results suggest that, as in rodents, human adipose stem cells are present in the adipose tissue microvasculature. In these studies, small fragments of adipose tissue are embedded in MatriGel, a complex protein mixture that mimics the tissue extracellular matrix (Greenway et al. 2007; Gealekman et al. 2010). After 5–7 days of culture in medium containing pro-angiogenic factors (EGM-2), endothelial cells emerge from the adipose tissue fragment and display sprouting behavior characteristic of developing microvessels. After 10–14 days these cells form primitive capillaries similar to those seen in developing adipose tissue (Fig. 7.14). When exposed to rosiglitazone, which through the stimulation of PPARg provides a strong pro-adipogenic stimulus, new adipocytes were seen to differentiate from cells which were tightly embedded in the walls of newly formed human adipose tissue capillaries (Tran et al. 2012) These newly emerging adipocytes contain key adipocyte-specific markers, yet remained connected to endothelial cells through tight junctions. These results suggest that human adipocytes are formed from precursors cells that are endothelial in origin or share a common developmental origin with endothelial cells (Fig. 7.14 lower panel).

7.3 Conclusions

The expansion of white fat in response to over nutrition is mediated in part by hypertrophic growth of each adipocyte, but in obese humans adipose tissue can become 70 % of the total body weight, requiring hyperplastic growth (Prins and O'Rahilly

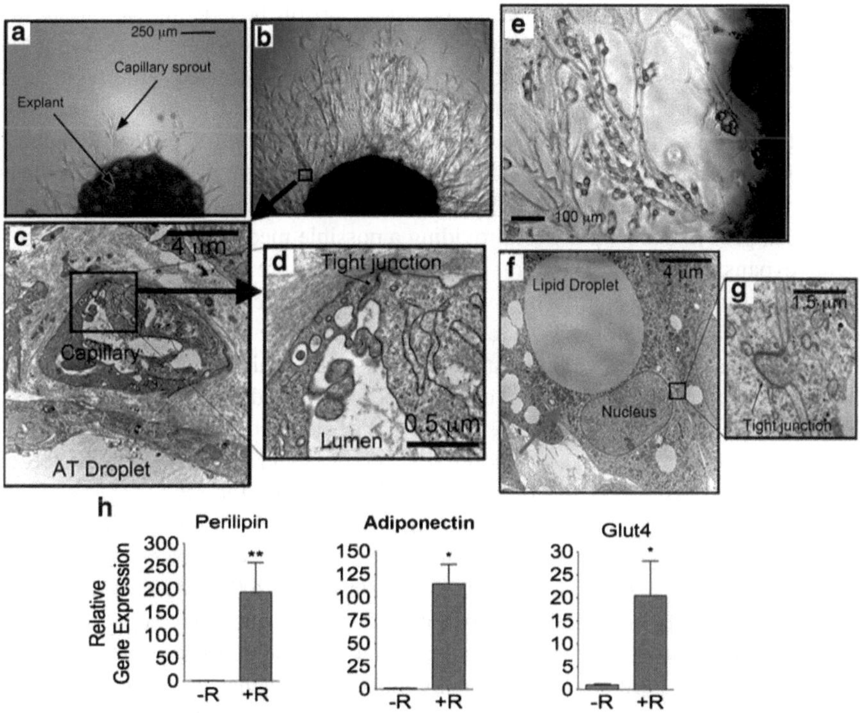

Fig. 7.14 Electron microscopy of angiogenic sprouts originating from human adipose tissue in the presence or absence of rosiglitazone. (**a**) Capillary outgrown after 15 days of culture in the absence of rosiglitazone, indicating areas distal and proximal to the embedded explants. (**b**) Capillary outgrowth after 15 days in the presence of rosiglitazone. (**c**) EM appearance of the area of capillary growth proximal to the explants revealing lumenized capillaries formed by endothelial cells joined by tight junctions. (**d**) Enlargement of the *boxed area* in (**c**). (**e**) Enlargement of area of outgrowth distal to the explant, revealing lipid droplets in cells interspersed among the capillary sprouts. (**f**) EM of lipid-laden cells revealing features of classical white adipocytes, and of endothelial cells such as the tight junction outlined. (**g**) Enlargement of tight junction outlined in (**f**). (**h**) Gene expressing profile of angiogenic sprouts originating from human adipose tissue in the presence or absence of rosiglitazone

1997; Hausman et al. 2001). In adult humans, the presence of metabolically active brown fat as well as brown preadipocytes has been recently demonstrated (Cypess et al. 2009; Saito et al. 2009; van Marken Lichtenbelt et al. 2009; Virtanen et al. 2009; Zingaretti et al. 2009). The abundance of brown adipocytes in human tissue is increased by chronic cold exposure (Saito et al. 2009; van Marken Lichtenbelt et al. 2009), thus both white (obesity) and possibly brown (cold exposure) adipose tissues can undergo hyperplasia in humans. One of the key questions in this area is the nature of the cell which gives rise to newly formed adipocytes. Adipose tissue growth also requires the concomitant expansion of its capillary network (Rupnick et al. 2002; Dallabrida et al. 2003; Fukumura et al. 2003; Brakenhielm et al. 2004). Capillary network expansion in growing tissues occurs through the process of angiogenesis, where endothelial cells that form part of the existing capillaries are stimulated to undergo mitosis, migration, and capillary tube formation and maturation. The mechanisms by which adipocyte growth is coordinated with angiogenic expansion are not clear, but adequate vascularization is critically important in ensuring normal adipose tissue function, as insufficient vascularization observed in obese individuals is associated with insulin resistance (Pasarica et al. 2009; Gealekman et al. 2010). In this study, we described that endothelial cells of adipose tissue capillaries can give rise to adipocytes, providing a possible mechanism by which angiogenic expansion and the formation of new adipocytes are coordinated. Using EM, genetic and functional approaches, we have obtained evidence that endothelial cells of capillaries in developing WAT and BAT can give rise to mature adipose cells. Based on these results, we suggest a mechanism by which capillary growth and adipocyte differentiation can be coordinated during adipose tissue expansion. Further evidence supporting the possibility that certain endothelial cell populations can give rise to adipocytes is the finding that pericytes and some endothelial cells of mouse adipose tissue express GFP driven by the promoter for Zfp423 (Gupta et al. 2012), a zinc finger transcription factor involved in early differentiation (Huang et al. 2009) that has been detected to mark cells determined to form preadipocytes (Gupta et al. 2010). Previous work (Planat-Benard et al. 2004) had shown that adipocytes can acquire an endothelial phenotype in vitro, which may suggest that the fate of the two cell types may be plastic to maintain homeostasis during adipose tissue expansion. Furthermore, recent findings indicate that vascular endothelial cells can be converted into mesenchymal stem cells, from which they can give rise to adipocytes, chondrocytes, and osteoblasts (Medici et al. 2010). Which physiological signals determine the fate of the precursors and/or regulate the switch between adipocyte and endothelial cell phenotype are unknown. One potential candidate is hypoxia, which affects the expression of multiple genes both in adipocytes and endothelial cells mainly thought activation of HIF-1α and NF-kB (Michiels et al. 2000; Lolmede et al. 2003; Geiger et al. 2011), inhibits adipogenesis and triglyceride synthesis (Ye 2009), and has pro-angiogenic effects (Liao and Johnson 2007). Hypoxia may also play role in determining the fate of precursors, thus fine-tuning the balance of adipocyte proliferation and expansion of adipose tissue capillary network. This equilibrium may become unbalanced in obesity, leading to insufficient angiogenesis relative to adipogenesis, and resulting in associated metabolic complications. The cellular and molecular mechanisms that regulate the balance between endothelial cells and adipocytes, and how these mechanisms contribute to normal and pathological adipose tissue function are important areas for continued study.

References

Alva JA, Zovein AC, Monvoisin A, Murphy T, Salazar A, Harvey NL, Carmeliet P, Iruela-Arispe ML. VE-Cadherin-Cre-recombinase transgenic mouse: a tool for lineage analysis and gene deletion in endothelial cells. Dev Dyn. 2006;235:759–67.

Amos PJ, Shang H, Bailey AM, Taylor A, Katz AJ, Peirce SM. IFATS collection: the role of human adipose-derived stromal cells in inflammatory microvascular remodeling and evidence of a perivascular phenotype. Stem Cells. 2008;26:2682–90.

Arner P, Bernard S, Salehpour M, Possnert G, Liebl J, Steier P, Buchholz BA, Eriksson M, Arner E, Hauner H, Skurk T, Rydén M, Frayn KN, Spalding KL. Dynamics of human adipose lipid turnover in health and metabolic disease. Nature. 2011;478(7367):110–3.

Bachman ES, Dhillon H, Zhang CY, Cinti S, Bianco AC, Kobilka BK, Lowell BB. BetaAR signaling required for diet-induced thermogenesis and obesity resistance. Science. 2002;297:843–5.

Baer PC, Geiger H. Adipose-derived mesenchymal stromal/stem cells: tissue localization, characterization, and heterogeneity. Stem Cells Int. 2012;2012:812693.

Barbatelli G, Murano I, Madsen L, et al. The emergence of cold-induced brown adipocytes in mouse white fat depots is determined predominantly by white to brown adipocyte transdifferentiation. Am J Physiol Endocrinol Metab. 2010;298(6):E1244–53.

Brakenhielm E, Cao R, Gao B, Angelin B, Cannon B, Parini P, Cao Y. Angiogenesis inhibitor, TNP-470, prevents diet-induced and genetic obesity in mice. Circ Res. 2004;94:1579–88.

Cannon B, Nedergaard J. Brown adipose tissue: function and physiological significance. Physiol Rev. 2004;84:277–359.

Christiaens V, Lijnen HR. Angiogenesis and development of adipose tissue. Mol Cell Endocrinol. 2010;318:2–9.

Cinti S. The adipose organ. Milan: Kurtis; 1999.

Cinti S. The adipose organ. Prostaglandins Leukot Essent Fatty Acids. 2005;73(1):9–15.

Cinti S. Reversible physiological transdifferentiation in the adipose organ. Proc Nutr Soc. 2009;68(4):340–9.

Cinti S, Morroni M. Brown adipocyte precursor cells: a morphological study. Ital J Anat Embryol. 1995;100(1):75–81.

Cinti S, Cigolini M, Bosello O, Bjorntorp P. A morphological study of the adipocyte precursor. J Submicrosc Cytol. 1984;16:243–51.

Crisan M, Deasy B, Gavina M, Zheng B, Huard J, Lazzari L, Peault B. Purification and long-term culture of multipotent progenitor cells affiliated with the walls of human blood vessels: myoendothelial cells and pericytes. Methods Cell Biol. 2008;86:295–309.

Crossno Jr JT, Majka SM, Grazia T, Gill RG, Klemm DJ. Rosiglitazone promotes development of a novel adipocyte population from bone marrow-derived circulating progenitor cells. J Clin Invest. 2006;116:3220–8.

Cypess AM, Lehman S, Williams G, Tal I, Rodman D, Goldfine AB, Kuo FC, Palmer EL, Tseng YH, Doria A, et al. Identification and importance of brown adipose tissue in adult humans. N Engl J Med. 2009;360:1509–17.

Dallabrida SM, Zurakowski D, Shih SC, Smith LE, Folkman J, Moulton KS, Rupnick MA. Adipose tissue growth and regression are regulated by angiopoietin-1. Biochem Biophys Res Commun. 2003;311:563–71.

De Matteis R, Lucertini F, Guescini M, Polidori E, Zeppa S, Stocchi V, Cinti S, Cuppini R. Exercise as a new physiological stimulus for brown adipose tissue activity. Nutr Metab Cardiovasc Dis; 2012. doi:10.1016/j.numecd.2012.01.013.

Farmer SR. Transcriptional control of adipocyte formation. Cell Metab. 2006;4:263–73.

Friedman JM. Obesity in the new millennium. Nature. 2000;404:632–4.

Frontini A, Cinti S. Distribution and development of brown adipocytes in the murine and human adipose organ. Cell Metab. 2010;11:253–6.

Fukumura D, Ushiyama A, Duda DG, Xu L, Tam J, Krishna V, Chatterjee K, Garkavtsev I, Jain RK. Paracrine regulation of angiogenesis and adipocyte differentiation during in vivo adipogenesis. Circ Res. 2003;93(9):e88–97.

Gealekman O, Guseva N, Hartigan C, Apotheker S, Gorgoglione M, Gurav K, Tran KV, Straubhaar J, Nicoloro S, Czech MP, et al. Depot-specific differences and insufficient subcutaneous adipose tissue angiogenesis in human obesity. Circulation. 2010;123:186–94.

Geiger K, Leiherer A, Muendlein A, Stark N, Geller-Rhomberg S, Saely CH, Wabitsch M, Fraunberger P, Drexel H. Identification of hypoxia-induced genes in human SGBS adipocytes by microarray analysis. PLoS One. 2011;6(10):e26465.

Gesta S, Tseng YH, Kahn CR. Developmental origin of fat: tracking obesity to its source. Cell. 2007;131(2):242–56.

Greenway FL, Liu Z, Yu Y, Caruso MK, Roberts AT, Lyons J, Schwimer JE, Gupta AK, Bellanger DE, Guillot TS, et al. An assay to measure angiogenesis in human fat tissue. Obes Surg. 2007; 17:510–5.

Gupta RK, Arany Z, Seale P, Mepani RJ, Ye L, Conroe HM, Roby YA, Kulaga H, Reed RR, Spiegelman BM. Transcriptional control of preadipocyte determination by Zfp423. Nature. 2010;464:619–23.

Gupta RK, Mepani RJ, Kleiner S, Lo JC, Khandekar MJ, Cohen P, Frontini A, Bhowmick DC, Ye L, Cinti S, Spiegelman BM. Zfp423 expression identifies committed preadipocytes and localizes to adipose endothelial and perivascular cells. Cell Metab. 2012;15(2):230–9.

Hausman GJ, Richardson RL. Adipose tissue angiogenesis. J Anim Sci. 2004;82:925–34.

Hausman DB, DiGirolamo M, Bartness TJ, Hausman GJ, Martin RJ. The biology of white adipocyte proliferation. Obes Rev. 2001;2:239–54.

Huang S, Laoukili J, Epping MT, Koster J, Holzel M, Westerman BA, Nijkamp W, Hata A, Asgharzadeh S, Seeger RC, et al. ZNF423 is critically required for retinoic acid-induced differentiation and is a marker of neuroblastoma outcome. Cancer Cell. 2009;15:328–40.

Iyama K, Ohzono K, Usuku G. Electron microscopical studies on the genesis of white adipocytes: differentiation of immature pericytes into adipocytes in transplanted preadipose tissue. Virchows Arch B Cell Pathol Incl Mol Pathol. 1979;31:143–55.

Joe AW, Yi L, Even Y, Vogl AW, Rossi FM. Depot-specific differences in adipogenic progenitor abundance and proliferative response to high-fat diet. Stem Cells. 2009;27(10):2563–70.

Kajimura S, Seale P, Spiegelman BM. Transcriptional control of brown fat development. Cell Metab. 2010;11:257–62.

Koh YJ, Kang S, Lee HJ, Choi TS, Lee HS, Cho CH, Koh GY. Bone marrow-derived circulating progenitor cells fail to transdifferentiate into adipocytes in adult adipose tissues in mice. J Clin Invest. 2007;117:3684–95.

Koh YJ, Park BH, Park JH, Han J, Lee IK, Park JW, Koh GY. Activation of PPAR gamma induces profound multilocularization of adipocytes in adult mouse white adipose tissues. Exp Mol Med. 2009;41:880–95.

Kopecky J, Clarke G, Enerback S, Spiegelman B, Kozak LP. Expression of the mitochondrial uncoupling protein gene from the aP2 gene promoter prevents genetic obesity. J Clin Invest. 1995;96:2914–23.

Korner J, Aronne LJ. The emerging science of body weight regulation and its impact on obesity treatment. J Clin Invest. 2003;111:565–70.

Lee YH, Petkova AP, Mottillo EP, Granneman JG. In vivo identification of bipotential adipocyte progenitors recruited by β3-adrenoceptor activation and high-fat feeding. Cell Metab. 2012; 15(4):480–91.

Liao D, Johnson RS. Hypoxia: a key regulator of angiogenesis in cancer. Cancer Metastasis Rev. 2007;26:281–90.

Lolmede K, Durand de Saint Front V, Galitzky J, Lafontan M, Bouloumie A. Effects of hypoxia on the expression of proangiogenic factors in differentiated 3T3-F442A adipocytes. International journal of obesity and related metabolic disorders. Int J Obes Relat Metab Disord. 2003; 27(10):1187–95.

Lowell BB, S-Susulic V, Hamann A, Lawitts JA, Himms-Hagen J, Boyer BB, Kozak LP, Flier JS. Development of obesity in transgenic mice after genetic ablation of brown adipose tissue. Nature. 1993;366:740–2.

Mao X, Fujiwara Y, Chapdelaine A, Yang H, Orkin SH. Activation of EGFP expression by Cre-mediated excision in a new ROSA26 reporter mouse strain. Blood. 2001;97:324–6.

Medici D, Shore EM, Lounev VY, Kaplan FS, Kalluri R, Olsen BR. Conversion of vascular endothelial cells into multipotent stem-like cells. Nat Med. 2010;16:1400–6.

Michiels C, Arnould T, Remacle J. Endothelial cell responses to hypoxia: initiation of a cascade of cellular interactions. Biochim Biophys Acta. 2000;1497:1–10.

Monvoisin A, Alva JA, Hofmann JJ, Zovein AC, Lane TF, Iruela-Arispe ML. VE-cadherin-CreERT2 transgenic mouse: a model for inducible recombination in the endothelium. Dev Dyn. 2006;235:3413–22.

Napolitano L. The differentiation of white adipose cells. An electron microscope study. J Cell Biol. 1963;18:663–79.

Nedergaard J, Bengtsson T, Cannon B. New powers of brown fat: fighting the metabolic syndrome. Cell Metab. 2011;13:238–40.

Pasarica M, Sereda OR, Redman LM, Albarado DC, Hymel DT, Roan LE, Rood JC, Burk DH, Smith SR. Reduced adipose tissue oxygenation in human obesity: evidence for rarefaction, macrophage chemotaxis, and inflammation without an angiogenic response. Diabetes. 2009; 58:718–25.

Petrovic N, Walden TB, Shabalina IG, Timmons JA, Cannon B, Nedergaard J. Chronic peroxisome proliferator-activated receptor gamma (PPARgamma) activation of epididymally derived white adipocyte cultures reveals a population of thermogenically competent, UCP1-containing adipocytes molecularly distinct from classic brown adipocytes. J Biol Chem. 2010; 285(10):7153–64.

Planat-Benard V, Silvestre JS, Cousin B, Andre M, Nibbelink M, Tamarat R, Clergue M, Manneville C, Saillan-Barreau C, Duriez M, et al. Plasticity of human adipose lineage cells toward endothelial cells: physiological and therapeutic perspectives. Circulation. 2004; 109:656–63.

Prins JB, O'Rahilly S. Regulation of adipose cell number in man. Clin Sci (Lond). 1997;92:3–11.

Rodeheffer MS, Birsoy K, Friedman JM. Identification of white adipocyte progenitor cells in vivo. Cell. 2008;135:240–9.

Rosen ED, Walkey CJ, Puigserver P, Spiegelman BM. Transcriptional regulation of adipogenesis. Genes Dev. 2000;14:1293–307.

Rupnick MA, Panigrahy D, Zhang CY, Dallabrida SM, Lowell BB, Langer R, Folkman MJ. Adipose tissue mass can be regulated through the vasculature. Proc Natl Acad Sci U S A. 2002;99:10730–5.

Saito M, Okamatsu-Ogura Y, Matsushita M, Watanabe K, Yoneshiro T, Nio-Kobayashi J, Iwanaga T, Miyagawa M, Kameya T, Nakada K, et al. High incidence of metabolically active brown adipose tissue in healthy adult humans: effects of cold exposure and adiposity. Diabetes. 2009;58(7):1526–31.

Seale P, Bjork B, Yang W, Kajimura S, Chin S, Kuang S, Scime A, Devarakonda S, Conroe HM, Erdjument-Bromage H, et al. PRDM16 controls a brown fat/skeletal muscle switch. Nature. 2008;454:961–7.

Soriano P. Generalized lacZ expression with the Rosa26 Cre reporter strain. Nat Genet. 1999;21:70–1.

Soukas A, Socci ND, Saatkamp BD, Novelli S, Friedman JM. Distinct transcriptional profiles of adipogenesis in vivo and in vitro. J Biol Chem. 2001;276:34167–74.

Speck NA, Iruela-Arispe ML. Conditional Cre/LoxP strategies for the study of hematopoietic stem cell formation. Blood Cells Mol Dis. 2009;43:6–11.

Tang W, Zeve D, Suh JM, Bosnakovski D, Kyba M, Hammer RE, Tallquist MD, Graff JM. White fat progenitor cells reside in the adipose vasculature. Science. 2008;322:583–6.

Tang W, Zeve D, Seo J, Jo AY, Graff JM. Thiazolidinediones regulate adipose lineage dynamics. Cell Metab. 2011;14:116–22.

Tavassoli M. Ultrastructural development of bone marrow adipose cell. Acta Anat (Basel). 1976;94:65–77.

Timmons JA, Wennmalm K, Larsson O, et al. Myogenic gene expression signature establishes that brown and white adipocytes originate from distinct cell lineages. Proc Natl Acad Sci U S A. 2007;104(11):4401–6.

Tomiyama K, Murase N, Stolz DB, Toyokawa H, O'Donnell DR, Smith DM, Dudas JR, Rubin JP, Marra KG. Characterization of transplanted green fluorescent protein+ bone marrow cells into adipose tissue. Stem Cells. 2008;26:330–8.

Tontonoz P, Spiegelman BM. Fat and beyond: the diverse biology of PPARgamma. Annu Rev Biochem. 2008;77:289–312.

Tran KV, Gealekman O, Frontini A, et al. The vascular endothelium of the adipose tissue gives rise to both white and brown fat cells. Cell Metab. 2012;15(2):222–9.

van Marken Lichtenbelt WD, Vanhommerig JW, Smulders NM, Drossaerts JM, Kemerink GJ, Bouvy ND, Schrauwen P, Teule GJ. Cold-activated brown adipose tissue in healthy men. N Engl J Med. 2009;360:1500–8.

Virtanen KA, Lidell ME, Orava J, Heglind M, Westergren R, Niemi T, Taittonen M, Laine J, Savisto NJ, Enerback S, et al. Functional brown adipose tissue in healthy adults. N Engl J Med. 2009;360:1518–25.

Vitali A, Murano I, Zingaretti MC, Frontini A, Ricquier D, Cinti S. The adipose organ of obesity-prone C57BL/6J mice is composed of mixed white and brown adipocytes. J Lipid Res. 2012;53(4):619–29.

Wu J, Boström P, Sparks LM, Ye L, Choi JH, Giang AH, Khandekar M, Virtanen KA, Nuutila P, Schaart G, Huang K, Tu H, van Marken Lichtenbelt WD, Hoeks J, Enerbäck S, Schrauwen P, Spiegelman BM. Beige adipocytes are a distinct type of thermogenic fat cell in mouse and human. Cell. 2012;150(2):366–76.

Ye J. Emerging role of adipose tissue hypoxia in obesity and insulin resistance. Int J Obes (Lond). 2009;33:54–66.

Zingaretti MC, Crosta F, Vitali A, Guerrieri M, Frontini A, Cannon B, Nedergaard J, Cinti S. The presence of UCP1 demonstrates that metabolically active adipose tissue in the neck of adult humans truly represents brown adipose tissue. FASEB J. 2009;23:3113–20.

Zovein AC, Hofmann JJ, Lynch M, French WJ, Turlo KA, Yang Y, Becker MS, Zanetta L, Dejana E, Gasson JC, et al. Fate tracing reveals the endothelial origin of hematopoietic stem cells. Cell Stem Cell. 2008;3:625–36.

Chapter 8
Vascular and Endothelial Regeneration

Louis Casteilla, Patrick Laharrague, and V. Planat-Benard

Abstract The presence of adipocyte progenitors has long been demonstrated throughout life in the stromal fraction of adipose tissue. Now, it appears that these cells are multipotent and share numerous features with mesenchymal stem cells derived from bone marrow. They also display some specificities, among them a strong pro-angiogenic potential. These progenitors, presently called adipose stromal cells, can differentiate in vitro into the three main components of the vessels, i.e., endothelial cells, smooth muscle cells, and pericytes. More importantly, they secrete numerous factors favoring angiogenesis and vasculogenesis. Since fat pads are easy to sample, numerous and promising perspectives are now opening up in regenerative medicine, particularly in ischemic situations.

Keywords Vasculogenesis • Angiogenesis • Adipose stromal cells • Ischemia

Adipose tissues and adipocytes are found in multiple locations. In mammals, two functionally different types of adipose tissue are usually distinguished: brown tissue which participates in cold- as well as in diet-induced thermogenesis, and white tissue mainly involved in energy storage (Himms-Hagen 1990). This chapter will be only focused on white adipose tissue.

In adipose tissue (AT), mature adipocytes account for only 40–60 % of the total cellularity. Other cells are endothelial cells, a complex hematopoietic cell population, and adipocyte progenitors (Bourlier and Bouloumie 2009; Caspar-Bauguil

L. Casteilla (✉)
UMR UPS/CNRS/EFS 5273, STROMALab, BP 84225, 31432 Toulouse Cedex 4, France
e-mail: louis.casteilla@inserm.fr

P. Laharrague
Université de Toulouse, UPS, Inserm U1031, BP 84225, 31432 Toulouse Cedex 4, France

V. Planat-Benard
INSERM U1031, BP 84225, 31432 Toulouse Cedex 4, France

Y. Cao (ed.), *Angiogenesis in Adipose Tissue*, DOI 10.1007/978-1-4614-8069-3_8,

et al. 2005; Planat-Benard et al. 2004). Their existence was revealed by the culture of the stroma vascular fraction obtained after fat enzymatic digestion: some cells adhered to plastic, prolifered, then differentiated towards adipocytes (Bjorntorp et al. 1978). These cells, initially called preadipocytes or adipocyte progenitors, are present throughout adult life. More recently, these same cells were named adipose-derived stem or stroma cells (ASC) because they are multipotent and share many properties with mesenchymal stem/stromal cells from the bone marrow (Guilak et al. 2006; Zuk et al. 2001). In fact, the term "stem cell" seems inappropriate here, since the self-renewal of ASC has not yet been definitively established in contrast to MSC (Sacchetti et al. 2007).

The link between adipocyte progenitors and vasculature is well established. During embryonic development, the formation of capillary convolutions is a critical step for the development of fat lobules. After birth, AT remodeling and optimal functionality require the development of the capillary network (Rupnick et al. 2002; Wassermann 1965). This point will be also discussed in another chapter of this book.

The strong link between adipogenesis and angiogenesis, and the proximity of ASC with their bone marrow counterpart, encouraged researchers to explore the angiogenic potential of these cells in a therapeutic perspective.

8.1 Angiogenic Potential of ASC

The putative capability for ASC or adipocyte progenitors to display an endothelial-like phenotype (see Table 8.1) arose from the discovery that ASC express the surface marker CD34 (Miranville et al. 2004; Planat-Benard et al. 2004). Indeed, when ASC are purified to homogeneity, most of cells express this protein, described as specifically displayed by various immature cells (hematopoietic stem cells, satellite cells…) and endothelial cells. Furthermore, when ASC are cultured in semi-solid media (Matrigel), most of them align and express another endothelial surface protein, the von Willebrand factor (Planat-Benard et al. 2004).

An angiogenic potential is demonstrated when ASC are injected in Matrigel plug or in an in vivo ischemia context. When ASC embedded in Matrigel are injected subcutaneously, they form numerous tube-like structures, and the presence of erythrocytes in the lumen demonstrates the functionality of these vascular structures. When human ASC are injected intramuscularly in tolerant mice after ligature of the femoral artery, they promote a functional neo-angiogenesis with the recovery of the subcutaneous blood flow at the extremity of the limb. Using an antibody directed to the human isoform of CD31, we revealed the integration of some of these cells in the inner surface of the vessels. Moreover, we dedifferentiated mature adipocytes after their isolation and demonstrated that these dedifferentiated adipocytes could rapidly acquire endothelial phenotype (expression of CD31 and vWF antigens) and functions (Planat-Benard et al. 2004). Similar results were observed in another ischemia situation, the heart ischemia (Mazo et al. 2008; Valina et al. 2007). On the

Table 8.1 Main phenotypic and functional characteristics of vascular cells derived from adipose tissues

	Endothelial cells	Smooth muscle cells	Pericytes
Differentiation agents	VEGF BMP 4 Hypoxia (5 % O_2)	Heparin TGF-beta 1 TGF-beta 3 BMP-4 Angiotensin II Sphingosylphos phoryl-choline Ascorbic acid PDGF BB	SDF-1 alpha/PDGF beta TGF-beta 1
Marker expression	CD31 (PECAM-1) CD34 CD105 (Endoglin) CD106 CD144	Alpha smooth muscle actin Myosin heavy chain Smoothelin Calponin SM22 alpha Calmodilin Caldesmon	CD146 Nestin Alpha smooth muscle actin Esmin NG-2 PDGFR-beta Aminopeptidase A/N Tropomyosin
	von Willebrand factor Flt-1 (VEGFR-1) Flk-1 7B4 antigen BNH9/BNF13 VE-cadh erin e NOS Tie 1 Tie 2 Neuropilin Matrix metallo-proteinase 1	Desmin Vimentin Myocardin	Non-muscle myosin
In vitro characterization	Cobblestone morphology Acetylated low-density lipoprotein uptake Cord formation on Matrigel Expression of nitric oxide synthase Response to TNF	Spindle shape morphology Contract to carbachol and KCl Contraction of collagen gels Production of extracellular matrix	Rhomboid morphology Association and coverage of endothelial tubes on Matrigel

(continued)

Table 8.1 (continued)

	Endothelial cells	Smooth muscle cells	Pericytes
In vivo characterization	Migration in wound healing assay Cell thrombogenicity Microvessel formation with blood cells after transplantation in SCID mice Blood perfusion and limb salvage Improvement of cardiac function in MI model	Co implantation with EC into the jugular vein of lambs Cell-seeded scaffolds implanted into bladder of nude rats	Matrigel plug assay in SCID mice

contrary, in cerebral ischemia, even if injected ASC display a vascular tropism, they are localized only in perivascular position and do not incorporate into the intima (Kubis et al. 2007). At this time, there is no explanation for these different behaviors according to the targeted tissues.

Our conclusion that preadipocytes or ASC can really differentiate towards endothelial cells is supported by similar results obtained by Bouloumié's group, using directly purified CD34+/CD31- cells from human AT. They also demonstrated, in a two-dimensional co-culture system, that capillary endothelial cells isolated from human AT stimulate the expression of the endothelial CD31 marker by the ASC (Sengenes et al. 2007).

Induction of mesenchymal to endothelial transformation is promoted by culturing ASC in the presence of various differentiating cues. Colazzo et al. reported that ASC cultured in a medium supplemented with fetal calf serum and VEGF uptaked acetylated LDL, but that only a small proportion of cells expressed endothelial markers (Colazzo et al. 2010). A more defined phenotype (flk1, tie2 expression, and acLDL uptake) was obtained with ASC cultured under serum-free conditions in the presence of VEGF or FGF2, the latter being necessary for the increase of vascular endothelial marker genes (Konno et al. 2010). Activation of SDF1/CXCR4 and PI3 kinase pathways was also proposed to promote endothelial differentiation of ASC (Cao et al. 2005; Sengenes et al. 2007). Shear stress could also play a role in the endothelial differentiation of ASC: after culture in the presence of endothelial cell growth supplement, human ASC form cords in Matrigel, but fail to take up acLDL or express endothelial molecular markers. Subsequent exposure to physiological shear force results in cell realignement, acLDL uptake, and expression of CD31 (Fischer et al. 2009).

In fact, a re-analysis of these results is necessary. First, the endothelial phenotype is most often only assesed by the presence of some endothelial markers and by the in vivo location of administrated ASC. Second, no study at clonal level is available at the present time. Moreover, culture conditions could promote emergence of intermediate biphenotypic cells, which could express endothelial markers without becoming fully functional. Such a situation was reported by Rose et al. for bone marrow-derived mesenchymal stroma/stem cells giving rise in vitro to cells that express cardiac-specific markers but retain the stromal phenotype and do not become functional cardiomyocytes (Rose et al. 2008). Thus, the term "endothelial-like cells" would be more appropriate to define ASC after their «differentiation», which could only mimic true endothelial phenotype without displaying all the features.

The relationship between endothelial cells and adipocyte progenitors has been recently reinforced by in vivo lineage tracing experiments using VE-cadherin promoter. Cinti et al. presented morphological and genetic evidence that adipocytes in white and brown AT originate from cells that display endothelial characteristics (Tran et al. 2012). The reverse differentiation in native tissue has not been reported to date.

Undoubtedly, the direct participation of ASC via a differentiation process is not sufficient to explain all the observed angiogenic effects. A paracrine activity of ASC is also frequently proposed, but few investigations have clearly demonstrated the true involvement of ASC-secreted factors. In the same journal issue than our seminal work, it was published that genetic knock-down of HGF expression in human ASC impaired ASC-induced neo-angiogenesis (Bell et al. 2008; Cai et al. 2007). Then MMP proteins as well as VEGF and IGF-I were identified as key factors in this process (Kilroy et al. 2007; Sadat et al. 2007). Depending on the complexity of revascularization, we can reasonably postulate that a cocktail of molecules participates to the ASC-induced neo-angiogenesis.

It is also noteworthy that hypoxia promotes both angiogenesis and the secretion of angiogenic factors by ASC (Rasmussen et al. 2011; Rehman et al. 2004; Trayhurn et al. 2008). We also provided strong evidence that mitochondrial reactive oxygen species (ROS) negatively control adipocyte differentiation, improve their pro-angiogenic potential, and increase the number of ASC-derived CD31-positive cells in the ischemic area (Traktuev et al. 2008). This suggests that mitochondrial ROS and the subsequent signaling pathway belong to the intracellular O_2-sensing mechanism of adipose lineage cells (Carriere et al. 2004). This is consistent with the fact that ROS have been shown to play major and positive roles in blood vessel growth as well as in vivo preconditioning protection (Murphy and Steenbergen 2008; Ushio-Fukai and Urao 2009). Moreover, the same stimulus increases the secretion of the pro-angiogenic and antiapoptotic factors VEGF and HGF and efficiently protects ASC against oxidative stress-induced cell death (Miranville et al. 2004). As adipocyte hypertrophia occurring during obesity induces hypoxia and then the generation of superoxide anion by respiratory chain (Hamanaka and Chandel 2009; Sun et al. 2011), we can speculate that the same signal both inhibits the adipogenic differentiation of ASC and promotes their angiogenic potential. This in turn triggers neo-angiogenesis, which provides nutrients and oxygen to the enlarging tissue, increasing oxygen tension, decreasing hypoxia, and promoting adipogenesis. In this way, adipose tissue can enlarge through successive waves of adipogenesis and angiogenesis.

8.2 Adipose-Derived Cells, Smooth Muscle Cells, and Pericytes

Stable mature vasculature requires a stabilizing layer of mural cells, smooth muscle cells, and pericytes. ASC can express many markers and functionalities (see Table 8.1) consistent with a smooth muscle phenotype (Jeon et al. 2006; Kim et al. 2008; Lee et al. 2006). As expected for putative smooth muscle cell precursors, the expression of these markers can be modulated and is sensitive to different inducers such as sphingosylphosphorylcholine, heparin, TGF-beta, or mechanical stimuli (Jeon et al. 2006; Lee et al. 2007, 2006). A complete differentiation towards the smooth muscle cell phenotype, including functional properties such as contractibility, was achieved by culturing ASC in the presence of TGF-beta plus BMP-4, or Angiotensin II alone (Kim et al. 2008; Wang et al. 2010).

Smooth muscle phenotype seems to represent a true potential since this phenotype is more defined than those of ASC-derived endothelial cells and that studies were carried out at clonal level (Rodriguez et al. 2006). Besides blood vessels, smooth muscle cells are contained in the walls of various organs and tubes in the body, where their primary role is contraction and relaxation. In this respect, Jack et al. reported the feasibility of urinary bladder tissue engineered from ASC (Jack et al. 2009).

It was recently reported that ASC cultured at high density with angiogenic stimuli were capable of self-assembling into complex, three-dimensional vascular structures. These structures resulted from complex interactions of three distinct subpopulations (CD31+, α-smooth muscle actin+ cells, and double negative cells) involving PDGF (Hutton et al. 2012). Two years ago, it was reported that similar structures could be obtained by co-culture of endothelial cells and ASC, without additional exogenous cytokines or extracellular matrix proteins: ASC strongly stimulated morphogenesis of endothelial cells into branching networks of cord structures. The emergence of this structure seemed dependent on matrix metalloproteinase activity and cell communications through VEGF, HGF, and PDGF-BB pathways and required bidirectional interactions between both types of cells (Merfeld-Clauss et al. 2010). The apparent discrepancy of these results with those reported above is probably explained by differences between culture conditions, but also by the fact that the different groups did not isolate the same ASC subset. Indeed, a crucial point with adipose tissue is the need of an enzymatic proteolysis to separate cells. This key process is not standardized, poorly taken into account, and leads to the isolation of different cell fractions with different differentiation potentials.

More recently, the hypothesis that the CD146+ pericyte population contains mesenchymal stem cells with mesodermal potential, including in adipose tissue, has been published (Crisan et al. 2008). This idea was also debated by Traktuev et al. who proposed that resident CD34+/CD31-/CD144- cells, the phenotype of which is identical to ASC, are located in the pericytic position and can function as pericytes to stabilize vasculature by interacting with endothelial cells (Traktuev et al. 2008). In our hand, the histological identification of ASC reveals two locations: rare cells are perivascular, but most of them are sparsed inside the stroma, independently to the vessel location (Maumus et al. 2011). Again, this is consistent with the view that it would exist different ASC subsets.

8.3 Conclusion

Cell-based therapies would benefit from a source of autologous pluripotential and easily accessible cells. For more than 20 years, marrow stromal cells have been extensively studied and proposed to repair dammaged tissues. However, clinical trials are hampered by the low cell number upon harvest and pain and morbidity to the donor. On the contrary, adipose tissue can be obtained in relatively large quantities with minimal discomfort. ASC can be harvested and multiplied easily, they have a large differentiation spectrum and a high proliferative potential. Importantly, ASC exhibit a strong angiogenic potential, via direct differentiation towards endothelial and mural cells and paracine secretion of pro-angiogenic factors. Therefore, cell therapies involving ASC may offer alternative treatment modalities for cardiovascular therapy. Results already obtained for the treatment of myocardial infarction and peripheral vascular disease and for the creation of tisssue-engineered vascular grafts are promising and confirm the significant potential of adipose tissue for use in autologous cell therapies.

Acknowledgement This work was supported by National Research Program on cardiovascular diseases (# PN PNC0 C0402) C), Cell and Genetic Therapy program from Région Midi-Pyrénées (# 03 03011 011999 999), Interreg 3 (# 13 13A- A-7-85-O) O), and Fondation pour la Recherche Médicale (DCV20070409252, programme Vieillissement Cardiovasculaire Normal et Pathologique).

References

Bell LN, Cai L, Johnstone BH, Traktuev DO, March KL, Considine RV. A central role for hepatocyte growth factor in adipose tissue angiogenesis. Am J Physiol Endocrinol Metab. 2008;294:E336–44.

Bjorntorp P, Karlsson M, Pertoft H, Pettersson P, Sjostrom L, Smith U. Isolation and characterization of cells from rat adipose tissue developing into adipocytes. J Lipid Res. 1978;19:316–24.

Bourlier V, Bouloumie A. Role of macrophage tissue infiltration in obesity and insulin resistance. Diabetes Metab. 2009;35(4):251–60.

Cai L, Johnstone BH, Cook TG, Liang Z, Traktuev D, Cornetta K, Ingram DA, Rosen ED, March KL. Suppression of hepatocyte growth factor production impairs the ability of adipose-derived stem cells to promote ischemic tissue revascularization. Stem Cells. 2007;25:3234–43.

Cao Y, Sun Z, Liao L, Meng Y, Han Q, Zhao RC. Human adipose tissue-derived stem cells differentiate into endothelial cells in vitro and improve postnatal neovascularization in vivo. Biochem Biophys Res Commun. 2005;332:370–9.

Carriere A, Carmona MC, Fernandez Y, Rigoulet M, Wenger RH, Penicaud L, Casteilla L. Mitochondrial reactive oxygen species control the transcription factor chop-10/gadd153 and adipocyte differentiation: a mechanism for hypoxia-dependent effect. J Biol Chem. 2004;279:40462–9.

Caspar-Bauguil S, Cousin B, Galinier A, Segafredo C, Nibbelink M, Andre M, Casteilla L, Penicaud L. Adipose tissues as an ancestral immune organ: site-specific change in obesity. FEBS Lett. 2005;579:3487–92.

Colazzo F, Chester AH, Taylor PM, Yacoub MH. Induction of mesenchymal to endothelial transformation of adipose-derived stem cells. J Heart Valve Dis. 2010;19:736–44.

Crisan M, Casteilla L, Lehr L, Carmona M, Paoloni-Giacobino A, Yap S, Sun B, Leger B, Logar A, Penicaud L, Schrauwen P, Cameron-Smith D, Paul RA, Peault B, Giacobino JP. A reservoir of brown adipocyte progenitors in human skeletal muscle. Stem Cells. 2008;26(9):2425–33.

Fischer LJ, McIlhenny S, Tulenko T, Golesorkhi N, Zhang P, Larson R, Lombardi J, Shapiro I, DiMuzio PJ. Endothelial differentiation of adipose-derived stem cells: effects of endothelial cell growth supplement and shear force. J Surg Res. 2009;152:157–66.

Guilak F, Lott KE, Awad HA, Cao Q, Hicok KC, Fermor B, Gimble JM. Clonal analysis of the differentiation potential of human adipose-derived adult stem cells. J Cell Physiol. 2006;206:229–37.

Hamanaka RB, Chandel NS. Mitochondrial reactive oxygen species regulate hypoxic signaling. Curr Opin Cell Biol. 2009;21:894–9.

Himms-Hagen J. Brown adipose tissue thermogenesis: interdisciplinary studies. FASEB J. 1990;4:2890–8.

Hutton DL, Logsdon EA, Moore EM, Gabhann FM, Gimble JM, Grayson WL. Vascular morphogenesis of adipose-derived stem cells is mediated by heterotypic cell-cell interactions. Tissue Eng Part A. 2012;18(15–16):1729–40.

Jack GS, Zhang R, Lee M, Xu Y, Wu BM, Rodriguez LV. Urinary bladder smooth muscle engineered from adipose stem cells and a three dimensional synthetic composite. Biomaterials. 2009;30:3259–70.

Jeon ES, Moon HJ, Lee MJ, Song HY, Kim YM, Bae YC, Jung JS, Kim JH. Sphingosylphosphorylcholine induces differentiation of human mesenchymal stem cells into smooth-muscle-like cells through a tgf-beta-dependent mechanism. J Cell Sci. 2006;119:4994–5005.

Kilroy GE, Foster SJ, Wu X, Ruiz J, Sherwood S, Heifetz A, Ludlow JW, Stricker DM, Potiny S, Green P, Halvorsen YD, Cheatham B, Storms RW, Gimble JM. Cytokine profile of human adipose-derived stem cells: expression of angiogenic, hematopoietic, and pro-inflammatory factors. J Cell Physiol. 2007;212:702–9.

Kim YM, Jeon ES, Kim MR, Jho SK, Ryu SW, Kim JH. Angiotensin ii-induced differentiation of adipose tissue-derived mesenchymal stem cells to smooth muscle-like cells. Int J Biochem Cell Biol. 2008;40:2482–91.

Konno M, Hamazaki TS, Fukuda S, Tokuhara M, Uchiyama H, Okazawa H, Okochi H, Asashima M. Efficiently differentiating vascular endothelial cells from adipose tissue-derived mesenchymal stem cells in serum-free culture. Biochem Biophys Res Commun. 2010;400:461–5.

Kubis N, Tomita Y, Tran-Dinh A, Planat-Benard V, Andre M, Karaszewski B, Waeckel L, Penicaud L, Silvestre JS, Casteilla L, Seylaz J, Pinard E. Vascular fate of adipose tissue-derived adult stromal cells in the ischemic murine brain: a combined imaging-histological study. NeuroImage. 2007;34:1–11.

Lee WC, Rubin JP, Marra KG. Regulation of alpha-smooth muscle actin protein expression in adipose-derived stem cells. Cells Tissues Organs. 2006;183:80–6.

Lee WC, Maul TM, Vorp DA, Rubin JP, Marra KG. Effects of uniaxial cyclic strain on adipose-derived stem cell morphology, proliferation, and differentiation. Biomech Model Mechanobiol. 2007;6:265–73.

Maumus M, Peyrafitte JA, D'Angelo R, Fournier-Wirth C, Bouloumie A, Casteilla L, Sengenes C, Bourin P. Native human adipose stromal cells: localization, morphology and phenotype. Int J Obes (Lond). 2011;35:1141–53.

Mazo M, Planat-Benard V, Abizanda G, Pelacho B, Leobon B, Gavira JJ, Penuelas I, Cemborain A, Penicaud L, Laharrague P, Joffre C, Boisson M, Ecay M, Collantes M, Barba J, Casteilla L,

Prosper F. Transplantation of adipose derived stromal cells is associated with functional improvement in a rat model of chronic myocardial infarction. Eur J Heart Fail. 2008;10:454–62.
Merfeld-Clauss S, Gollahalli N, March KL, Traktuev DO. Adipose tissue progenitor cells directly interact with endothelial cells to induce vascular network formation. Tissue Eng Part A. 2010;16:2953–66.
Miranville A, Heeschen C, Sengenes C, Curat CA, Busse R, Bouloumie A. Improvement of postnatal neovascularization by human adipose tissue-derived stem cells. Circulation. 2004;110:349–55.
Murphy E, Steenbergen C. Mechanisms underlying acute protection from cardiac ischemia-reperfusion injury. Physiol Rev. 2008;88:581–609.
Planat-Benard V, Silvestre JS, Cousin B, Andre M, Nibbelink M, Tamarat R, Clergue M, Manneville C, Saillan-Barreau C, Duriez M, Tedgui A, Levy B, Penicaud L, Casteilla L. Plasticity of human adipose lineage cells toward endothelial cells: physiological and therapeutic perspectives. Circulation. 2004;109:656–63.
Rasmussen JG, Frobert O, Pilgaard L, Kastrup J, Simonsen U, Zachar V, Fink T. Prolonged hypoxic culture and trypsinization increase the pro-angiogenic potential of human adipose tissue-derived stem cells. Cytotherapy. 2011;13:318–28.
Rehman J, Traktuev D, Li J, Merfeld-Clauss S, Temm-Grove CJ, Bovenkerk JE, Pell CL, Johnstone BH, Considine RV, March KL. Secretion of angiogenic and antiapoptotic factors by human adipose stromal cells. Circulation. 2004;109:1292–8.
Rodriguez LV, Alfonso Z, Zhang R, Leung J, Wu B, Ignarro LJ. Clonogenic multipotent stem cells in human adipose tissue differentiate into functional smooth muscle cells. Proc Natl Acad Sci U S A. 2006;103:12167–72.
Rose RA, Jiang H, Wang X, Helke S, Tsoporis JN, Gong N, Keating SC, Parker TG, Backx PH, Keating A. Bone marrow-derived mesenchymal stromal cells express cardiac-specific markers, retain the stromal phenotype, and do not become functional cardiomyocytes in vitro. Stem Cells. 2008;26:2884–92.
Rupnick MA, Panigrahy D, Zhang CY, Dallabrida SM, Lowell BB, Langer R, Folkman MJ. Adipose tissue mass can be regulated through the vasculature. Proc Natl Acad Sci U S A. 2002;99:10730–5.
Sacchetti B, Funari A, Michienzi S, Di Cesare S, Piersanti S, Saggio I, Tagliafico E, Ferrari S, Robey PG, Riminucci M, Bianco P. Self-renewing osteoprogenitors in bone marrow sinusoids can organize a hematopoietic microenvironment. Cell. 2007;131:324–36.
Sadat S, Gehmert S, Song YH, Yen Y, Bai X, Gaiser S, Klein H, Alt E. The cardioprotective effect of mesenchymal stem cells is mediated by igf-i and vegf. Biochem Biophys Res Commun. 2007;363:674–9.
Sengenes C, Miranville A, Maumus M, de Barros S, Busse R, Bouloumie A. Chemotaxis and differentiation of human adipose tissue cd34+/cd31- progenitor cells: role of stromal derived factor-1 released by adipose tissue capillary endothelial cells. Stem Cells. 2007;25:2269–76.
Sun K, Kusminski CM, Scherer PE. Adipose tissue remodeling and obesity. J Clin Investig. 2011;121:2094–101.
Traktuev DO, Merfeld-Clauss S, Li J, Kolonin M, Arap W, Pasqualini R, Johnstone BH, March KL. A population of multipotent cd34-positive adipose stromal cells share pericyte and mesenchymal surface markers, reside in a periendothelial location, and stabilize endothelial networks. Circ Res. 2008;102:77–85.
Tran KV, Gealekman O, Frontini A, Zingaretti MC, Morroni M, Giordano A, Smorlesi A, Perugini J, De Matteis R, Sbarbati A, Corvera S, Cinti S. The vascular endothelium of the adipose tissue gives rise to both white and brown fat cells. Cell Metab. 2012;15:222–9.
Trayhurn P, Wang B, Wood IS. Hypoxia in adipose tissue: a basis for the dysregulation of tissue function in obesity? Br J Nutr. 2008;100:227–35.
Ushio-Fukai M, Urao N. Novel role of nadph oxidase in angiogenesis and stem/progenitor cell function. Antioxid Redox Signal. 2009;11:2517–33.
Valina C, Pinkernell K, Song YH, Bai X, Sadat S, Campeau RJ, Le Jemtel TH, Alt E. Intracoronary administration of autologous adipose tissue-derived stem cells improves left ventricular func-

tion, perfusion, and remodelling after acute myocardial infarction. Eur Heart J. 2007;28:2667–77.

Wang C, Yin S, Cen L, Liu Q, Liu W, Cao Y, Cui L. Differentiation of adipose-derived stem cells into contractile smooth muscle cells induced by transforming growth factor-beta1 and bone morphogenetic protein-4. Tissue Eng Part A. 2010;16:1201–13.

Wassermann P. The development of adipose tissue. Handbook of physiology. Washington, DC: American Physiological Society; 1965. p. 105.

Zuk PA, Zhu M, Mizuno H, Huang J, Futrell JW, Katz AJ, Benhaim P, Lorenz HP, Hedrick MH. Multilineage cells from human adipose tissue: implications for cell-based therapies. Tissue Eng. 2001;7:211–28.

Part IV
Angiogenic Factors and Adipokines in Adipose Tissue

Chapter 9
Role of NPY in Brown Adipocytes and Obesity

Sheng Bi

Abstract Neuropeptide Y (NPY) plays an important role in maintaining energy homeostasis. Within the hypothalamus, *Npy* is primarily expressed in the arcuate nucleus (Arc) and the dorsomedial hypothalamus (DMH). ARC NPY acts as an orexigenic neuromodulator integrating energy-related systemic signals (such as the adiposity signal leptin) to modulate food intake and energy balance. In contrast, DMH NPY acts independently of leptin. In addition to its feeding effects, DMH NPY has specific actions on brown adipocyte formation and thermoregulation. Knockdown of NPY in the DMH promotes development of brown adipocytes in white adipose tissue and increases brown adipocyte activity, leading to increased energy expenditure and lowered fat contents. DMH NPY knockdown prevents high-fat diet-induced obesity. Peripheral NPY also directly acts on adipose tissue and promotes white adipogenesis. Overall, NPY plays critical roles in the control of energy balance through affecting food intake, body adiposity, and energy expenditure. These actions provide evidence for the potential target(s) for combatting obesity and diabetes.

Keywords Dorsomedial hypothalamus • Neuropeptide Y • Brown adipose tissue • Uncoupling protein 1 • Adipogenesis • Adipocyte transformation • Thermogenesis • Food intake • Energy expenditure • Obesity

S. Bi (✉)
Department of Psychiatry and Behavioral Science, Johns Hopkins University School of Medicine, Baltimore, MD 21205, USA
e-mail: sbi@jhmi.edu

Y. Cao (ed.), *Angiogenesis in Adipose Tissue*, DOI 10.1007/978-1-4614-8069-3_9,

9.1 Introduction

Two types of fat, white adipose tissue (WAT) and brown adipose tissue (BAT), exist in mammals, including in adult humans as recently reported (Cypess et al. 2009; van Marken Lichtenbelt et al. 2009; Virtanen et al. 2009). WAT consists of unilocular adipocytes that contain a large lipid droplet. WAT acts as an energy reservoir to store excess calories and supply the stored lipid once a body is starved or increases energy needs. An excessive accumulation of WAT over time due to constantly increased energy intake and/or decreased energy expenditure causes overweight and obesity, a disorder that has been linked to various life-threatening diseases such as cardiovascular diseases, type II diabetes, and some cancers. Since the discovery of leptin in 1994 (Zhang et al. 1994), a hormone produced by adipocytes, the understanding of the adipocyte has significantly increased. In addition to storing energy, the adipocyte acts as a key endocrine gland secreting various hormones and/or cytokines such as leptin (Ahima et al. 1996; Zhang et al. 1994), adiponectin (Yamauchi et al. 2001), resistin (Steppan et al. 2001), tumor necrosis factor-alpha (TNF-α) (Uysal et al. 1997), and interleukin 6 (IL-6) (Wallenius et al. 2002) and playing a fundamental role in regulating energy balance and glucose homeostasis. Dysfunctions of adipocytes have been demonstrated to cause disordered energy balance and impaired glucose homeostasis. For instance, deficits in leptin signaling derived from either leptin deficiency in *ob/ob* mice (Ingalls et al. 1950) or its receptor mutations in *db/db* mice (Hummel et al. 1966) lead to obesity and diabetes syndromes.

BAT is comprised of multilocular and mitochondrial-rich adipocytes that contain multiple small lipid droplets. BAT dissipates lipid energy to produce heat via mitochondrial uncoupling protein 1 (UCP1)-mediated nonshivering thermogenesis as a defense against cold and the potential for combatting obesity/diabetes. BAT is traditionally considered as a thermogenic organ and primarily present in small mammals and human infants for protection against cold environment. Recently, using a combination of ^{18}fluoro-labeled 2-deoxyglucose (^{18}FDG) positron emission tomography and computed tomography (PET-CT) technology, active BAT has been detected in adult humans (Cypess et al. 2009; van Marken Lichtenbelt et al. 2009; Virtanen et al. 2009). Adult humans have increased BAT activity during cold exposure, and lowered BAT activity when they are overweight or obese (van Marken Lichtenbelt et al. 2009; Virtanen et al. 2009), although the origin of these brown adipocytes remains to be determined. Nevertheless, the findings of BAT in adult humans have led to a great interest in its potential for fighting against obesity/diabetes, i.e., searching for ways to elevate BAT activity or to turn WAT into BAT that burns calories instead of storing them.

Both types of fat are innervated by the sympathetic nervous system (SNS) (Bartness and Bamshad 1998; Cannon and Nedergaard 2004). Activation of the sympathetic innervation induces lipolysis in WAT (Fredholm and Karlsson 1970; Weiss and Maickel 1968) and produces thermogenesis through mitochondrial UCP1 in BAT (Cannon and Nedergaard 2004). Intriguingly, the sympathetic activation via treatment of β3-adrenergic receptor (β3-AR) agonists or cold stress also causes development of brown adipocytes in white fat cells (Himms-Hagen et al. 1994; Jimenez et al. 2003; Nagase et al. 1996). Follow up studies demonstrate the transdifferentiation of white to brown adipocytes as multilocular fat cells (brown-like adipocytes) in WAT induced by β3-AR agonist derive from unilocular white adipocytes (Himms-Hagen et al. 2000). Consistent with this finding, the emergence of cold-induced brown adipocytes in white fat depots is determined predominantly by white to brown adipocyte transdifferentiation with evidence of transdifferentiating paucilocular adipocytes (Barbatelli et al. 2010). In contrast, Petrovic et al. (2010) argued against this idea and proposed two different progenitor cells in BAT and WAT, respectively. The classical brown adipocytes such as interscapular BAT are derived from "adipomyocytes" that share their origin with myocytes (Seale et al. 2008). The other is atypical brown adipocytes called "brite" adipocytes (brown-in-white, i.e., the brown adipocyte-like adipocytes induced in white adipocytes, or "beige" cells) developing from an origin that shares a common precursor with white adipocytes (Petrovic et al. 2010). In support of this view, Wu and colleagues (2012) have identified these beige cells as a distinct type of thermogenic fat cell in both mouse and human. Using viral-mediated RNA interference (RNAi) for specifically knocking down neuropeptide Y (NPY) in the dorsomedial hypothalamus (DMH), Chao et al. (2011) reported that DMH NPY knockdown promotes development of brown adipocytes in WAT, elevates interscapular BAT activity, and prevents high-fat diet-induced obesity and impaired glucose homeostasis. This chapter will focus on the role of NPY in brown adipocyte formation and actions on obesity.

NPY is a 36-amino acid neuropeptide that was discovered by Tatemoto and colleagues (1982) and belongs to the pancreatic polypeptide family that includes peptide tyrosine-tyrosine (PYY) and pancreatic polypeptide (PP) (Tatemoto 1982; Tatemoto et al. 1982). NPY is ubiquitously distributed in both central and peripheral nervous systems. It is prevalent in central cortical, limbic, and hypothalamic regions (Adrian et al. 1983; Allen et al. 1983) and in the peripheral SNS (Lundberg et al. 1982). NPY exhibits a variety of biological and physiological actions including feeding, thermoregulation, locomotor activity, cardiovascular function, cognition and memory, and stress-related behaviors (Bi 2007; Colmers and Wahlestedt 1993; Gray and Morley 1986). Hypothalamic NPY plays a pivotal role in the control of food intake and body weight. Central administration of NPY via intracerebroventricular (icv) (Clark et al. 1984; Levine and Morley 1984) or intrahypothalamic injection (Stanley et al. 1986; Stanley and Leibowitz 1985) causes robust increases in food intake and body weight and, with chronic administration, can eventually produce obesity (Zarjevski et al. 1993). In addition to its orexigenic effects, icv administration of NPY promotes WAT lipid storage and decreases BAT thermogenesis (Billington et al. 1991). Central administration of NPY suppresses sympathetic activity in interscapular BAT in rats (Egawa et al. 1991). Thus, hypothalamic NPY not only acts as an orexigenic peptide, but also affects adipocyte lipid mobilization and BAT functions.

9.2 Role of Arcuate NPY in Adiposity

Within the hypothalamus, NPY-expressing neurons have been identified in the arcuate nucleus (ARC) and the DMH (Bi et al. 2003; White and Kershaw 1990). As shown in Fig. 9.1, *Npy* gene expression is found in both the ARC and the DMH of rat (Fig. 9.1a) and nonhuman primate brains (Fig. 9.1b), although the expression level of DMH NPY is much lower than that within the ARC. The actions of ARC NPY in the control of food intake and energy balance have been documented. The ARC contains two distinct populations of neurons: orexigenic neuropeptide NPY/agouti-related protein (AgRP) neurons and anorexigenic proopiomelanocortin (POMC) neurons. Both types of neurons contain leptin receptors (LepRbs). Leptin acts on these neurons to down-regulate *Npy/Agrp* gene expression and upregulate *Pomc* gene expression (Schwartz et al. 2000). Collectively, these two neural systems integrate adiposity signals (such as leptin) and nutrient signals as well as other hormonal signals (such as ghrelin) to modulate food intake and energy balance (Cone 2006; Elmquist et al. 1999; Flier 2004; Friedman and Halaas 1998; Nakazato et al. 2001; Schwartz et al. 2000). Consistent with this view, while ARC *Npy* mRNA or peptide levels are elevated in genetic obesity animals with leptin signaling deficits

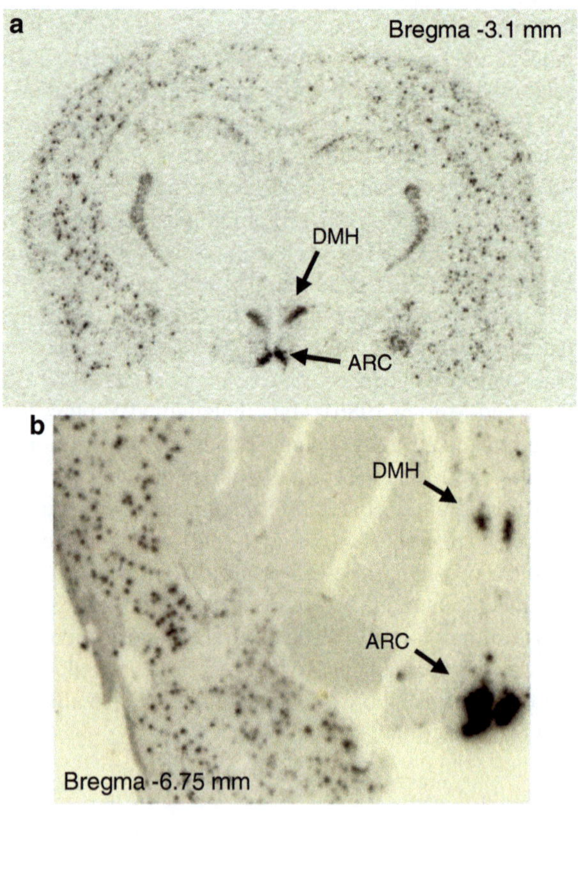

Fig. 9.1 In situ hybridization with [^{35}S]-labeled antisense riboprobe of rodent or human *Npy* shows *Npy* gene expression in the arcuate nucleus (ARC) and the dorsomedial hypothalamus (DMH) in both rats (**a**) and adult rhesus monkey brain (**b**)

including fatty Zucker (*fa/fa*) rats, obese (*cp/cp*) JCR:LA corpulent (Koletsky) rat, *ob/ob* mice, and *db/db* mice (Beck 2006; Sanacora et al. 1990; Wilding et al. 1993), such elevations are not evident in other obesity models such as those resulting from hyperphagia (Bi et al. 2001), overfeeding (McMinn et al. 1998), or access to a high-fat diet (Giraudo et al. 1994; Stricker-Krongrad et al. 1998) in which levels of ARC *Npy* expression are actually decreased, likely as a response to increased levels of leptin and body weight. Although data from mouse models with targeted disruption of NPY [either knockout (Erickson et al. 1996a) or transcriptional alterations through doxycycline-regulated system (Ste Marie et al. 2005)] have failed to demonstrate significant effects on food intake or body weight, the deletion of NPY does attenuate a hyperphagic and obese phenotype of leptin-deficient *ob/ob* mice (Erickson et al. 1996b) and genetic ablation of neurons expressing NPY/AgRP in adult mice results in a hypophagic and lean phenotype (Bewick et al. 2005; Gropp et al. 2005). Moreover, consistent with ARC NPY mediation of food deprivation-induced feeding, specific knockdown of NPY in the ARC via adeno-associated virus (AAV)-mediated RNAi attenuates the feeding response to food deprivation (Yang et al. 2009). AAV-mediated expression of antisense *Npy* cRNA in the ARC of adult rats decreases NPY expression and results in decreased food intake and body weight (Gardiner et al. 2005), whereas AAV-mediated overexpression of NPY in the ARC causes overeating, sustained body weight gain, and severe obesity in adult rats (Sousa-Ferreira et al. 2011). Together, these data demonstrate that ARC NPY acts as an orexigenic peptide to modulate food intake and energy balance to affect adiposity, but whether ARC NPY has adipocyte-specific effects, in addition to feeding effects, has yet to be reported.

9.3 Role of DMH NPY in Adiposity

Elevation or induction of *Npy* gene expression in the DMH in specific animal models with increased energy demands (Bi et al. 2003; Kawaguchi et al. 2005; Smith 1993) and several rodent models of obesity (Bi et al. 2001; Guan et al. 1998a, 1998b; Kesterson et al. 1997; Tritos et al. 1998) implicates DMH NPY in the control of energy balance. In support of this view, recent evidence has demonstrated a role for DMH NPY in modulating food intake and energy balance (Yang et al. 2009) and intriguingly in affecting brown adipocyte formation and thermoregulation (Chao et al. 2011).

9.3.1 Leptin-Independent Control of DMH NPY

While ARN NPY serves as one of the downstream mediators of leptin's actions or is under the control of the adiposity signal leptin, DMH NPY acts independently of leptin. DMH NPY neurons do not contain LepRbs, although abundant LepRbs are present in the DMH (Bi et al. 2003). Dual in situ hybridization histochemistry revealed that while *Npy* and *LepRbs* are co-expressed in ARC neurons, DMH *Npy-expressing* neurons do not co-express *LepRbs* (Bi et al. 2003). *Npy* gene expression is increased in the ARC in response to acute food deprivation, a time when circulating leptin levels are low, whereas DMH *Npy* expression is only significantly increased in rats with chronic food restriction (Bi et al. 2003). Moreover, *Npy* gene expression is elevated or induced in the DMH of specific rodent models of obesity including the lethal yellow agouti (A^y) (Kesterson et al. 1997), MC4R knockout (Kesterson et al. 1997), diet-induced obese (Guan et al. 1998a), tubby (Guan et al. 1998b), and brown adipose tissue-deficient obese mice (Tritos et al. 1998), and Otsuka Long-Evans Tokushima Fatty (OLETF) rats (Bi et al. 2001), but such elevation or induction is not evident in leptin-deficient *ob/ob* mice (Kesterson et al. 1997). In fact, while *Npy* gene expression is significantly increased in the ARC of obese animals with leptin signaling deficiency (Beck 2006; Sanacora et al. 1990; Wilding et al. 1993), obese animals with DMH NPY overexpression generally have significantly decreased expression of *Npy* in the ARC (Bi et al. 2001; Guan et al. 1998a; Kesterson et al. 1997). Thus, ARC and DMH NPY signals are differentially regulated and they each likely elicit distinct actions in the control of energy balance.

9.3.2 DMH NPY in Brown Adipocyte Development

Previous studies have shown that in addition to its orexigenic effects, icv administration of NPY increases WAT lipoprotein lipase activity (suggesting increased lipid storage) and decreases BAT GDP binding activity (indicating decreased thermogenic activity) (Billington et al. 1991) and that central administration of NPY suppresses sympathetic activity in interscapular BAT in rats (Egawa et al. 1991). These data suggest a role for NPY in lipid mobilization and thermoregulation, but the source of central NPY contributing to these actions has not been determined. Manipulation of *Npy* gene expression in the DMH using AAV-mediated RNAi revealed that knockdown of NPY in the DMH produces not only a nocturnal and meal size-specific reduction in feeding (Yang et al. 2009), but also site-specific effects on adiposity (Chao et al. 2011). DMH NPY knockdown decreases subcutaneous inguinal fat mass in rats on regular chow and ameliorates high-fat diet-induced increases in fat accumulations including inguinal and epididymal fat and interscapular BAT mass (Chao et al. 2011). Intriguingly, DMH NPY knockdown promotes development of brown adipocytes in inguinal WAT. As shown in Fig. 9.2, while inguinal adipocytes of control rats contain unilocular adipocytes, typical white adipocytes (Fig. 9.2a), inguinal adipose tissue of NPY knockdown rats contains large clusters of multilocular adipocytes (brown-like adipocytes) (Fig. 9.2b). The BAT-specific marker UCP1 is highly detected in these new formed cells and

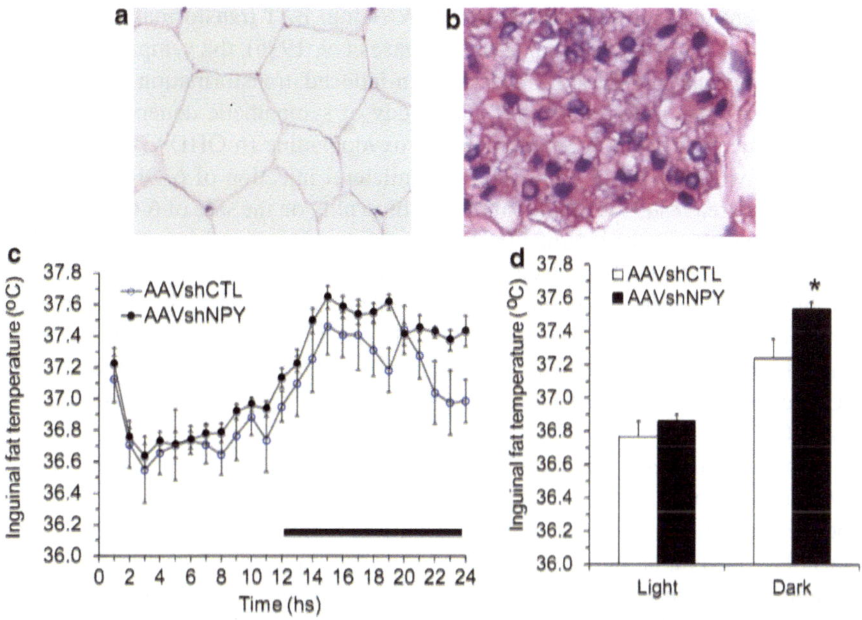

Fig. 9.2 Representative H&E (hematoxylin and eosin) stain shows unilocular adipocytes in inguinal white adipose tissue (WAT) of rats receiving bilateral DMH injections of control adeno-associated viral vector containing scrambled shRNA (AAVshCTL) (**a**) and multilocular adipocytes (brown-like adipocytes) in inguinal fat of rats receiving bilateral DMH injections of AAV-mediated RNAi vector (AAVshNPY) (**b**). Inguinal fat temperature over the 24-h period (**c**) and mean ± SEM temperature during the dark and the light (**d**) in rats at an ambient room temperature condition. *Black bar* indicates the dark cycle. *$P<0.05$ compared to AAVshCTL rats

also in a number of unilocular adipocytes around these clusters (Chao et al. 2011). UCP1 expression in this inguinal fat is further confirmed by using both real-time RT-PCR (reverse transcriptase-polymerase chain reaction) and Western blot analyses (Chao et al. 2011). But, whether these browning adipocytes in WAT are directly derived from specific precursor cells such as beige cells, or transdifferentiated from unilocular white adipocytes, or both, remains to be determined.

Elevation of UCP1 contributes to increased thermogenesis of BAT (Cannon and Nedergaard 2004). Consistent with this view, we have demonstrated increased thermogenesis in inguinal browning fat of NPY knockdown rats. In the study, the inguinal fat temperature was directly examined using a radio transmitter device (E-mitter, Mini-Mitter) that was buried under the inguinal fat. At an ambient room temperature condition (23 ± 1 °C), the inguinal fat temperature is significantly increased during the dark period in rats with DMH NPY knockdown compared to control animals (Fig. 9.2c, d).

Consistent with the SNS mediation of WAT into BAT transformation (Himms-Hagen et al. 1994; Jimenez et al. 2003; Nagase et al. 1996), the sympathetic innervation contributes DMH NPY knockdown-induced transformation of white to brown adipocytes in inguinal WAT. The study of sympathetic denervation with a regional injection of the neurotoxin 6-hydroxydopamine (6-OHDA) revealed that in the rats with DMH NPY knockdown, unilateral injection of 6-OHDA into the inguinal fat area results in significantly less browning on the side of 6-OHDA injection, whereas the site of inguinal fat receiving saline injection becomes brown-like fat (Chao et al. 2011). This observation was further confirmed by the findings that (1) numerous clusters of brown-like adipocytes (multilocular adipocytes) remain on the side of saline-treated inguinal fat, whereas brown-like adipocytes are dramatically reduced by 6-OHDA treatment and (2) 6-OHDA treatment also prevents UCPl expression in inguinal adipocytes at both the protein and mRNA levels (Chao et al. 2011).

Although the specific neural signaling pathway underlying the effects of DMH NPY on inguinal adiposity remains to be determined, evidence from viral transsynaptic retrograde tracing studies provides support for the idea that DMH NPY modulates SNS innervation on inguinal WAT. Less viral tracing is detected in the DMH in animals receiving epididymal viral injection than those receiving inguinal injection (Bamshad et al. 1998), implying that the central nervous control of inguinal WAT is more DMH related than that of epididymal WAT, i.e., DMH NPY may serve as a central modulator influencing SNS outflow to inguinal WAT.

On the molecular level, a number of transcription factors regulate development of brown adipocytes. Peroxisome proliferator-activated receptor-γ (PPAR-γ) is an essential factor for the development of both white and brown fat cells (Rosen et al. 1999). Chronic treatment with the PPAR-γ agonist rosiglitazone promotes brown adipogenesis in WAT characterized with distinct adipocytes, namely, "brite" adipocytes (Petrovic et al. 2010). The rosiglitazone treatment causes not only the expression of the PPAR-γ coactivator-1α (PGC-1α) and mitochondriogenesis, but also norepinephrine (NE)-augmentable expression of UCP1 in these brite cells (Petrovic et al. 2010). PGC-1a is another key factor involved in brown adipocyte development and thermogenesis (Handschin and Spiegelman 2006). Ectopic expression of PGC-1α in white fat cells induces a number of mitochondrial genes and thermogenic genes, such as Ucp1, whereas genetic ablation of PGC-1α results in reduced capacity for cold-induced thermogenesis in vivo (Puigserver et al. 1998). DMH NPY knockdown promotes both *Ppar-γ* and *Pgc-1α* gene expression in inguinal fat (Chao et al. 2011), suggesting that the browning effects of DMH NPY knockdown on inguinal WAT are likely mediated through affecting these key transcription factors, but the detailed molecular control of this white to brown adipocyte transformation merits further investigation.

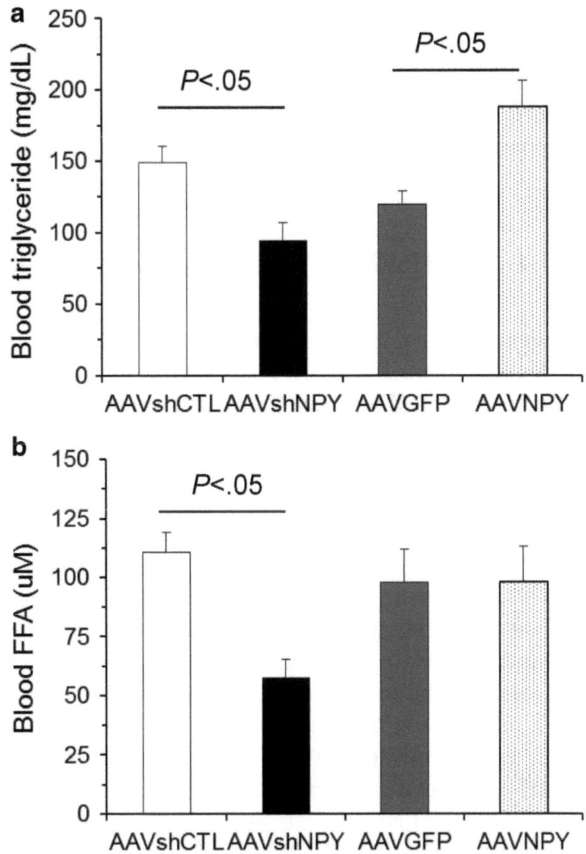

Fig. 9.3 Effects of manipulation of NPY expression in the DMH on blood triglyceride (**a**) and free fatty acid levels (**b**). AAVGFP, rats receiving bilateral DMH injections of control AAV vector (AAVGFP); AAVNPY, rats receiving bilateral DMH injections of AAV-mediated NPY expression vector (AAVNPY)

9.3.3 DMH NPY Affects Inguinal Lipid Mobilization and Blood Fat Levels

Fatty acid synthase (FAS) and carnitine palmitoyltransferase 1 (CPT1) are two important enzymes involved in fat metabolism. FAS plays a key role in fatty acid synthesis, whereas CPT1 is the rate-limiting enzyme controlling fatty acid oxidation. Compared to control rats, *Cpt1* gene expression is significantly increased in the inguinal fat of NPY knockdown rats with a trend toward a decrease in *Fas* gene expression (Chao et al. 2011), indicating a shift from lipogenesis to fatty acid oxidation in this tissue. Consistent with this view, the weight of inguinal fat mass is significantly decreased in NPY knockdown rats as compared with their control counterparts (Chao et al. 2011). Importantly, DMH NPY knockdown results in decreased levels of both blood triglyceride (Fig. 9.3a) and free fatty acid (Fig. 9.3b), whereas NPY overexpression in the

DMH causes increased levels of blood triglyceride (Fig. 9.3a). These data indicate that DMH NPY is an important neuromodulator in modulation of fat utilization. Specifically, the browning effects of NPY knockdown on fat may contribute to these reductions of blood triglyceride and free fatty acid levels since BAT activity has been demonstrated to control triglyceride clearance (Bartelt et al. 2011).

9.3.4 DMH NPY Affects Interscapular BAT Activity

Evidence has indicated the importance of the DMH in regulating sympathetic nerve activity to interscapular BAT and thermogenesis. Stimulation or disinhibition of neurons in the DMH by parenchymal microinjection of glutamate or γ-aminobutyric acid (GABA)-A receptor antagonist results in great increases in sympathetic nerve activity to interscapular BAT and increases in BAT and core body temperature (Dimicco and Zaretsky 2007; Morrison and Nakamura 2011). Collectively, the DMH has been proposed as an intermediate relay receiving the inputs from the hypothalamic preoptic area (POA), a center of integrating centrally and peripherally thermal signals, and sending the outputs to the rostral raphe pallidus (rRP) in the medulla, the area containing premotor neurons that innervate SNS to interscapular BAT (Dimicco and Zaretsky 2007; Morrison and Nakamura 2011). Although the nature of DMH neurons mediating these effects has not yet been fully characterized, DMH NPY may serve as one of the mediators. DMH NPY knockdown results in increased *Ucp1* gene expression in the interscapular BAT, whereas NPY overexpression in the DMH causes decreased expression of *Ucp1* in this fat tissue (Fig. 9.4a). This suggests that DMH NPY affects interscapular BAT activity or thermogenesis. In support of this view, DMH NPY knockdown results in increased interscapular BAT thermogenesis as directly determined by increased BAT temperature. At an ambient room temperature condition (23 ± 1 °C), interscapular BAT temperature is significantly increased during the dark with a trend for increases during the light compared to control animals (Fig. 9.4b, c). Thus, DMH NPY also modulates interscapular BAT thermogenesis or activity to affect energy expenditure.

9.3.5 Role of DMH NPY in Energy Expenditure

Consistent with the actions of DMH NPY on activity of both inguinal fat and interscapular BAT, manipulation of DMH NPY expression affects energy expenditure and thermogenesis. Using an oxymax equal flow indirect calorimetric system, we determined that DMH NPY knockdown results in increased oxygen consumption during both dark and light phases of the circadian cycle at an ambient room temperature condition (23 ± 1 °C) (Fig. 9.5a), indicating that rats with DMH NPY knockdown have increased daily energy metabolism. NPY knockdown rats also have decreased respiratory exchange rate (RER) during the dark and across 24 hours

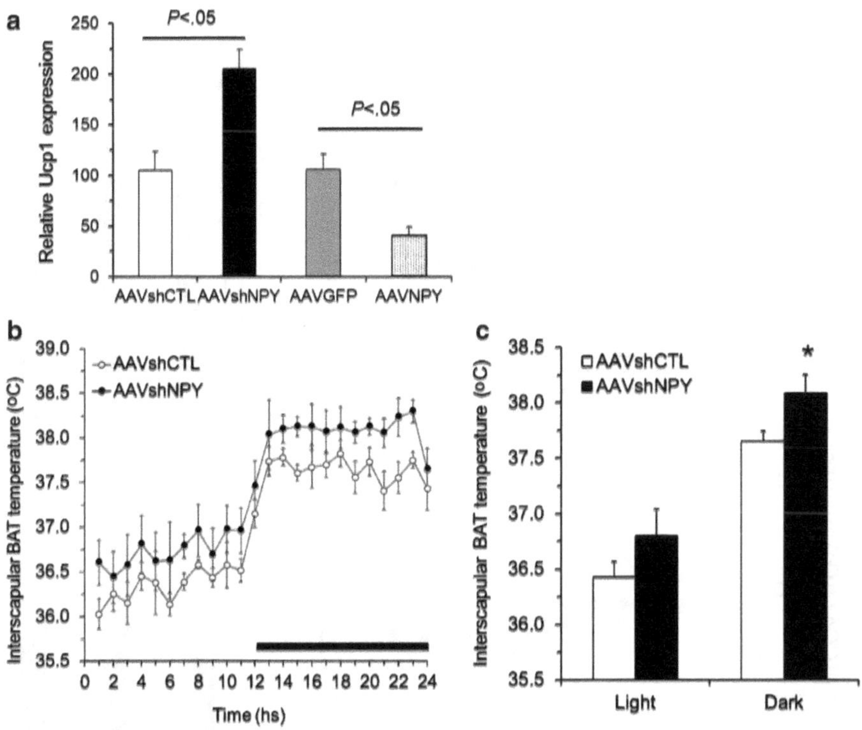

Fig. 9.4 Effects of manipulation of DMH NPY expression on the expression of uncoupling protein 1 (UCP1) in interscapular brown adipose tissue (BAT) (**a**), and effects of NPY knockdown in the DMH on interscapular BAT temperature over the 24-h period (**b**) and mean ± SEM BAT temperature during the dark and light period (**c**) in rats at an ambient room temperature condition. *$P < 0.05$ compared to AAVshCTL rats

(Fig. 9.5b), indicating increased lipid oxidation. As a result, energy expenditure (EE) is significantly increased over the 24 h period (Fig. 9.5c). Furthermore, although core body temperature does not differ between NPY knockdown and control rats at room temperature (23 ± 1 °C), NPY knockdown rats have a greater increase in thermogenic response to 6 h of cold exposure (6 °C) compared to their control counterparts (Chao et al. 2011). In response to 6 h of cold exposure, *Ucp1* gene expression is significantly elevated in the interscapular BAT of control rats (Fig. 9.6a) and DMH NPY knockdown enhances this elevation (Fig. 9.6a). Six hours of cold exposure does not induce *Ucp1* gene expression in inguinal WAT of control rats, but *Ucp1* gene expression is largely induced in inguinal WAT of rats with DMH NPY knockdown (Fig. 9.6b). Overall, DMH NPY plays a functional role in the regulation of thermogenesis and energy expenditure likely through affecting the activity of both inguinal fat and interscapular BAT.

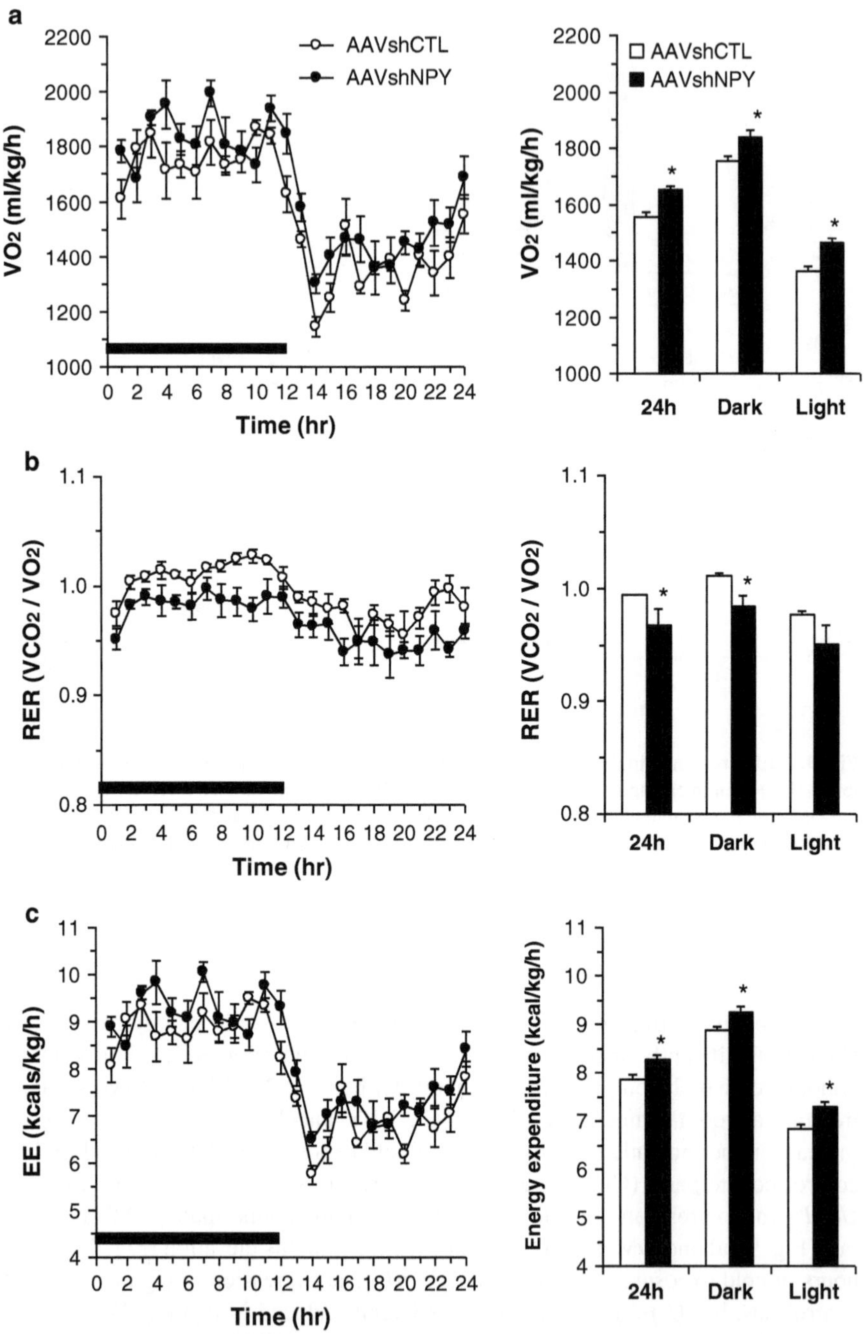

Fig. 9.5 Effects of DMH NPY knockdown on oxygen consumption (**a**), respiratory exchange rate (RER) (**b**), and energy expenditure (EE) (**c**) in rats at room temperature. *Black bar* indicates the dark cycle. $*P<0.05$ compared to AAVshCTL rats

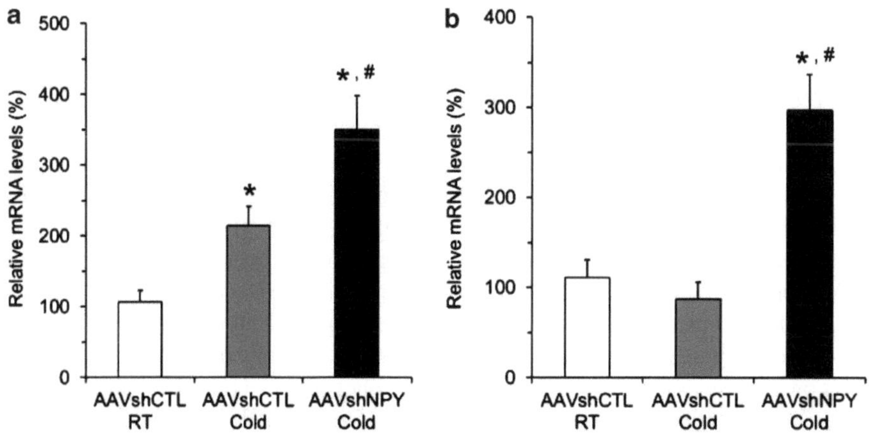

Fig 9.6 Effects of DMH NPY knockdown on *Ucp1* gene expression in interscapular BAT (**a**) and inguinal fat (**b**) in response to 6 h of cold exposure (6 °C). AAVshCTL RT, AAVshCTL rats at room temperature; AAVshCTL Cold, AAVshCTL rats at 6 °C; and AAVshNPY Cold, AAVshNPY rats at 6 °C. *$P<0.05$ compared to AAVshCTL rats. #$P<0.05$ compared to AAVshCTL Cold

9.4 Role of Peripheral NPY in Adiposity

Peripheral NPY is primarily produced in the SNS and co-operates with NE to affect sympathetic control of various tissues (Lundberg et al. 1982) including adipose tissue (Bartness and Bamshad 1998). Adipose tissue also contains NPY and its Y1 and Y2 receptors (Kuo et al. 2007; Yang et al. 2008). Peripheral NPY acts directly on adipose tissue and mediates stress-induced obesity and metabolic syndrome through its interacting with Y2 receptors (Kuo et al. 2007). While activation of the sympathetic innervation and the resultant release of NE induces lipolysis in WAT (Fredholm and Karlsson 1970; Weiss and Maickel 1968), the release of NPY from sympathetic nerves upregulates NPY and Y2 receptors in adipose tissue, leading to adipogenesis directly by stimulating proliferation and differentiation of preadipocytes and indirectly by angiogenesis (Kuo et al. 2007). Consistent with this view, pharmacological antagonism or genetic knockdown of peripheral Y2 receptors prevents high-fat diet-induced obesity (Kuo et al. 2007; Shi et al. 2011). Adipose Y1 receptors have also been shown to contribute to the effects of peripheral NPY on adipogenesis. Adipose NPY promotes proliferation of adipocyte precursor cells via the Y1 receptor (Yang et al. 2008) and adipose Y1 receptors regulate lipid oxidation and fat accretion (Zhang et al. 2010). Thus, while central NPY-Y1/2 systems have specific effects on feeding to modulate energy balance (Lin et al. 2004), adipose Y1/2 receptors mediate the actions of peripheral NPY on adipogenesis and fat accumulation, eventually leading to development of obesity.

9.5 Conclusions

NPY plays an important role in maintaining energy homeostasis. In response to energy needs or a change in energy status, ARC NPY acts as an orexigenic peptide to modulate food intake and energy balance. Particularly, ARC NPY serves as one of the downstream mediators of leptin's actions. In contrast, DMH NPY acts independently of leptin. In addition to its feeding effect, DMH NPY has specific actions on brown adipocyte formation and thermoregulation. Knockdown of NPY in the DMH promotes development of brown adipocytes in WAT and elevates BAT activity, leading to increased energy expenditure and lowered fat contents. DMH NPY knockdown prevents high-fat diet-induced obesity. Peripheral NPY also directly acts on adipose tissue and promotes white adipogenesis. Overall, NPY plays critical roles in the control of energy balance through affecting food intake, body adiposity, and energy expenditure. These actions provide potential target(s) for combatting obesity and diabetes.

Acknowledgments This work was supported by the US National Institute of Diabetes and Digestive and Kidney Diseases Grants DK074269 and DK087888.

References

Adrian TE, Allen JM, Bloom SR, Ghatei MA, Rossor MN, Roberts GW, Crow TJ, Tatemoto K, Polak JM. Neuropeptide Y distribution in human brain. Nature. 1983;306:584–6.

Ahima RS, Prabakaran D, Mantzoros C, Qu D, Lowell B, Maratos-Flier E, Flier JS. Role of leptin in the neuroendocrine response to fasting. Nature. 1996;382:250–2.

Allen YS, Adrian TE, Allen JM, Tatemoto K, Crow TJ, Bloom SR, Polak JM. Neuropeptide Y distribution in the rat brain. Science. 1983;221:877–9.

Bamshad M, Aoki VT, Adkison MG, Warren WS, Bartness TJ. Central nervous system origins of the sympathetic nervous system outflow to white adipose tissue. Am J Physiol. 1998; 275:R291–9.

Barbatelli G, Murano I, Madsen L, Hao Q, Jimenez M, Kristiansen K, Giacobino JP, De Matteis R, Cinti S. The emergence of cold-induced brown adipocytes in mouse white fat depots is determined predominantly by white to brown adipocyte transdifferentiation. Am J Physiol Endocrinol Metab. 2010;298:E1244–53.

Bartelt A, Bruns OT, Reimer R, Hohenberg H, Ittrich H, Peldschus K, Kaul MG, Tromsdorf UI, Weller H, Waurisch C, et al. Brown adipose tissue activity controls triglyceride clearance. Nat Med. 2011;17:200–5.

Bartness TJ, Bamshad M. Innervation of mammalian white adipose tissue: implications for the regulation of total body fat. Am J Physiol. 1998;275:R1399–411.

Beck B. Neuropeptide Y in normal eating and in genetic and dietary-induced obesity. Phil Trans Roy Soc London B Biol Sci. 2006;361:1159–85.

Bewick GA, Gardiner JV, Dhillo WS, Kent AS, White NE, Webster Z, Ghatei MA, Bloom SR. Post-embryonic ablation of AgRP neurons in mice leads to a lean, hypophagic phenotype. FASEB J. 2005;19:1680–2.

Bi S. Role of dorsomedial hypothalamic neuropeptide Y in energy homeostasis. Peptides. 2007;28:352–6.

Bi S, Ladenheim EE, Schwartz GJ, Moran TH. A role for NPY overexpression in the dorsomedial hypothalamus in hyperphagia and obesity of OLETF rats. Am J Physiol Regul Integr Comp Physiol. 2001;281:R254–60.
Bi S, Robinson BM, Moran TH. Acute food deprivation and chronic food restriction differentially affect hypothalamic NPY mRNA expression. Am J Physiol Regul Integr Comp Physiol. 2003;285:R1030–6.
Billington CJ, Briggs JE, Grace M, Levine AS. Effects of intracerebroventricular injection of neuropeptide Y on energy metabolism. Am J Physiol. 1991;260:R321–7.
Cannon B, Nedergaard J. Brown adipose tissue: function and physiological significance. Physiol Rev. 2004;84:277–359.
Chao PT, Yang L, Aja S, Moran TH, Bi S. Knockdown of NPY expression in the dorsomedial hypothalamus promotes development of brown adipocytes and prevents diet-induced obesity. Cell Metab. 2011;13:573–83.
Clark JT, Kalra PS, Crowley WR, Kalra SP. Neuropeptide Y and human pancreatic polypeptide stimulate feeding behavior in rats. Endocrinology. 1984;115:427–9.
Colmers WF, Wahlestedt C. The biology of neuropeptide Y and related peptides. New Jersey: Humana; 1993.
Cone RD. Studies on the physiological functions of the melanocortin system. Endocr Rev. 2006;27:736–49.
Cypess AM, Lehman S, Williams G, Tal I, Rodman D, Goldfine AB, Kuo FC, Palmer EL, Tseng YH, Doria A, et al. Identification and importance of brown adipose tissue in adult humans. N Engl J Med. 2009;360:1509–17.
Dimicco JA, Zaretsky DV. The dorsomedial hypothalamus: a new player in thermoregulation. Am J Physiol Regul Integr Comp Physiol. 2007;292:R47–63.
Egawa M, Yoshimatsu H, Bray GA. Neuropeptide Y suppresses sympathetic activity to interscapular brown adipose tissue in rats. Am J Physiol. 1991;260:R328–34.
Elmquist JK, Elias CF, Saper CB. From lesions to leptin: hypothalamic control of food intake and body weight. Neuron. 1999;22:221–32.
Erickson JC, Clegg KE, Palmiter RD. Sensitivity to leptin and susceptibility to seizures of mice lacking neuropeptide Y. Nature. 1996a;381:415–21.
Erickson JC, Hollopeter G, Palmiter RD. Attenuation of the obesity syndrome of ob/ob mice by the loss of neuropeptide Y. Science. 1996b;274:1704–7.
Flier JS. Obesity wars: molecular progress confronts an expanding epidemic. Cell. 2004; 116:337–50.
Fredholm BB, Karlsson J. Metabolic effects of prolonged sympathetic nerve stimulation in canine subcutaneous adipose tissue. Acta physiologica Scandinavica. 1970;80:567–76.
Friedman JM, Halaas JL. Leptin and the regulation of body weight in mammals. Nature. 1998; 395:763–70.
Gardiner JV, Kong WM, Ward H, Murphy KG, Dhillo WS, Bloom SR. AAV mediated expression of anti-sense neuropeptide Y cRNA in the arcuate nucleus of rats results in decreased weight gain and food intake. Biochem Biophys Res Commun. 2005;327:1088–93.
Giraudo SQ, Kotz CM, Grace MK, Levine AS, Billington CJ. Rat hypothalamic NPY mRNA and brown fat uncoupling protein mRNA after high-carbohydrate or high-fat diets. Am J Physiol. 1994;266:R1578–83.
Gray TS, Morley JE. Neuropeptide Y: anatomical distribution and possible function in mammalian nervous system. Life Sci. 1986;38:389–401.
Gropp E, Shanabrough M, Borok E, Xu AW, Janoschek R, Buch T, Plum L, Balthasar N, Hampel B, Waisman A, et al. Agouti-related peptide-expressing neurons are mandatory for feeding. Nat Neurosci. 2005;8:1289–91.
Guan XM, Yu H, Trumbauer M, Frazier E, Van der Ploeg LH, Chen H. Induction of neuropeptide Y expression in dorsomedial hypothalamus of diet-induced obese mice. Neuroreport. 1998a;9:3415–9.
Guan XM, Yu H, Van der Ploeg LH. Evidence of altered hypothalamic pro-opiomelanocortin/neuropeptide Y mRNA expression in tubby mice. Brain Res Mol Brain Res. 1998b;59:273–9.

Handschin C, Spiegelman BM. Peroxisome proliferator-activated receptor gamma coactivator 1 coactivators, energy homeostasis, and metabolism. Endocr Rev. 2006;27:728–35.

Himms-Hagen J, Cui J, Danforth Jr E, Taatjes DJ, Lang SS, Waters BL, Claus TH. Effect of CL-316,243, a thermogenic beta 3-agonist, on energy balance and brown and white adipose tissues in rats. Am J Physiol. 1994;266:R1371–82.

Himms-Hagen J, Melnyk A, Zingaretti MC, Ceresi E, Barbatelli G, Cinti S. Multilocular fat cells in WAT of CL-316243-treated rats derive directly from white adipocytes. Am J Physiol Cell Physiol. 2000;279:C670–81.

Hummel KP, Dickie MM, Coleman DL. Diabetes, a new mutation in the mouse. Science. 1966;153:1127–8.

Ingalls AM, Dickie MM, Snell GD. Obese, a new mutation in the house mouse. J Hered. 1950;41:317–8.

Jimenez M, Barbatelli G, Allevi R, Cinti S, Seydoux J, Giacobino JP, Muzzin P, Preitner F. Beta 3-adrenoceptor knockout in C57BL/6J mice depresses the occurrence of brown adipocytes in white fat. Eur J Biochem. 2003;270:699–705.

Kawaguchi M, Scott KA, Moran TH, Bi S. Dorsomedial hypothalamic corticotropin-releasing factor mediation of exercise-induced anorexia. American journal of physiology Regulatory, integrative and comparative physiology. 2005;288:R1800–1805.

Kesterson RA, Huszar D, Lynch CA, Simerly RB, Cone RD. Induction of neuropeptide Y gene expression in the dorsal medial hypothalamic nucleus in two models of the agouti obesity syndrome. Mol Endocrinol. 1997;11:630–7.

Kuo LE, Kitlinska JB, Tilan JU, Li L, Baker SB, Johnson MD, Lee EW, Burnett MS, Fricke ST, Kvetnansky R, et al. Neuropeptide Y acts directly in the periphery on fat tissue and mediates stress-induced obesity and metabolic syndrome. Nat Med. 2007;13:803–11.

Levine AS, Morley JE. Neuropeptide Y: a potent inducer of consummatory behavior in rats. Peptides. 1984;5:1025–9.

Lin S, Boey D, Herzog H. NPY and Y receptors: lessons from transgenic and knockout models. Neuropeptides. 2004;38:189–200.

Lundberg JM, Terenius L, Hokfelt T, Martling CR, Tatemoto K, Mutt V, Polak J, Bloom S, Goldstein M. Neuropeptide Y (NPY)-like immunoreactivity in peripheral noradrenergic neurons and effects of NPY on sympathetic function. Acta physiologica Scandinavica. 1982;116:477–80.

McMinn JE, Seeley RJ, Wilkinson CW, Havel PJ, Woods SC, Schwartz MW. NPY-induced overfeeding suppresses hypothalamic NPY mRNA expression: potential roles of plasma insulin and leptin. Regul Pept. 1998;75–76:425–31.

Morrison SF, Nakamura K. Central neural pathways for thermoregulation. Front Biosci. 2011;16:74–104.

Nagase I, Yoshida T, Kumamoto K, Umekawa T, Sakane N, Nikami H, Kawada T, Saito M. Expression of uncoupling protein in skeletal muscle and white fat of obese mice treated with thermogenic beta 3-adrenergic agonist. J Clin Investig. 1996;97:2898–904.

Nakazato M, Murakami N, Date Y, Kojima M, Matsuo H, Kangawa K, Matsukura S. A role for ghrelin in the central regulation of feeding. Nature. 2001;409:194–8.

Petrovic N, Walden TB, Shabalina IG, Timmons JA, Cannon B, Nedergaard J. Chronic peroxisome proliferator-activated receptor gamma (PPARgamma) activation of epididymally derived white adipocyte cultures reveals a population of thermogenically competent, UCP1-containing adipocytes molecularly distinct from classic brown adipocytes. J Biol Chem. 2010;285:7153–64.

Puigserver P, Wu Z, Park CW, Graves R, Wright M, Spiegelman BM. A cold-inducible coactivator of nuclear receptors linked to adaptive thermogenesis. Cell. 1998;92:829–39.

Rosen ED, Sarraf P, Troy AE, Bradwin G, Moore K, Milstone DS, Spiegelman BM, Mortensen RM. PPAR gamma is required for the differentiation of adipose tissue in vivo and in vitro. Mol cell. 1999;4:611–7.

Sanacora G, Kershaw M, Finkelstein JA, White JD. Increased hypothalamic content of preproneuropeptide Y messenger ribonucleic acid in genetically obese Zucker rats and its regulation by food deprivation. Endocrinology. 1990;127:730–7.

Schwartz MW, Woods SC, Porte Jr D, Seeley RJ, Baskin DG. Central nervous system control of food intake. Nature. 2000;404:661–71.
Seale P, Bjork B, Yang W, Kajimura S, Chin S, Kuang S, Scime A, Devarakonda S, Conroe HM, Erdjument-Bromage H, et al. PRDM16 controls a brown fat/skeletal muscle switch. Nature. 2008;454:961–7.
Shi YC, Lin S, Castillo L, Aljanova A, Enriquez RF, Nguyen AD, Baldock PA, Zhang L, Bijker MS, Macia L, et al. Peripheral-specific y2 receptor knockdown protects mice from high-fat diet-induced obesity. Obesity. 2011;19:2137–48.
Smith MS. Lactation alters neuropeptide-Y and proopiomelanocortin gene expression in the arcuate nucleus of the rat. Endocrinology. 1993;133:1258–1265.
Sousa-Ferreira L, Garrido M, Nascimento-Ferreira I, Nobrega C, Santos-Carvalho A, Alvaro AR, Rosmaninho-Salgado J, Kaster M, Kugler S, de Almeida LP, et al. Moderate long-term modulation of neuropeptide Y in hypothalamic arcuate nucleus induces energy balance alterations in adult rats. PLoS One. 2011;6:e22333.
Stanley BG, Leibowitz SF. Neuropeptide Y injected in the paraventricular hypothalamus: a powerful stimulant of feeding behavior. Proc Natl Acad Sci U S A. 1985;82:3940–3.
Stanley BG, Kyrkouli SE, Lampert S, Leibowitz SF. Neuropeptide Y chronically injected into the hypothalamus: a powerful neurochemical inducer of hyperphagia and obesity. Peptides. 1986;7:1189–92.
Ste Marie L, Luquet S, Cole TB, Palmiter RD. Modulation of neuropeptide Y expression in adult mice does not affect feeding. Proc Natl Acad Sci U S A. 2005;102:18632–7.
Steppan CM, Bailey ST, Bhat S, Brown EJ, Banerjee RR, Wright CM, Patel HR, Ahima RS, Lazar MA. The hormone resistin links obesity to diabetes. Nature. 2001;409:307–12.
Stricker-Krongrad A, Cumin F, Burlet C, Beck B. Hypothalamic neuropeptide Y and plasma leptin after long-term high-fat feeding in the rat. Neurosci Lett. 1998;254:157–60.
Tatemoto K. Neuropeptide Y: complete amino acid sequence of the brain peptide. Proc Natl Acad Sci U S A. 1982;79:5485–9.
Tatemoto K, Carlquist M, Mutt V. Neuropeptide Y–a novel brain peptide with structural similarities to peptide YY and pancreatic polypeptide. Nature. 1982;296:659–60.
Tritos NA, Elmquist JK, Mastaitis JW, Flier JS, Maratos-Flier E. Characterization of expression of hypothalamic appetite-regulating peptides in obese hyperleptinemic brown adipose tissue-deficient (uncoupling protein-promoter-driven diphtheria toxin A) mice. Endocrinology. 1998;139:4634–41.
Uysal KT, Wiesbrock SM, Marino MW, Hotamisligil GS. Protection from obesity-induced insulin resistance in mice lacking TNF-alpha function. Nature. 1997;389:610–4.
van Marken Lichtenbelt WD, Vanhommerig JW, Smulders NM, Drossaerts JM, Kemerink GJ, Bouvy ND, Schrauwen P, Teule GJ. Cold-activated brown adipose tissue in healthy men. N Engl J Med. 2009;360:1500–8.
Virtanen KA, Lidell ME, Orava J, Heglind M, Westergren R, Niemi T, Taittonen M, Laine J, Savisto NJ, Enerback S, et al. Functional brown adipose tissue in healthy adults. N Engl J Med. 2009;360:1518–25.
Wallenius V, Wallenius K, Ahren B, Rudling M, Carlsten H, Dickson SL, Ohlsson C, Jansson JO. Interleukin-6-deficient mice develop mature-onset obesity. Nat Med. 2002;8:75–9.
Weiss B, Maickel RP. Sympathetic nervous control of adipose tissue lipolysis. Int J Neuropharmacol. 1968;7:395–403.
White JD, Kershaw M. Increased hypothalamic neuropeptide Y expression following food deprivation. Mol Cell Neurosci. 1990;1:41–8.
Wilding JP, Gilbey SG, Bailey CJ, Batt RA, Williams G, Ghatei MA, Bloom SR. Increased neuropeptide-Y messenger ribonucleic acid (mRNA) and decreased neurotensin mRNA in the hypothalamus of the obese (ob/ob) mouse. Endocrinology. 1993;132:1939–44.
Wu J, Bostrom P, Sparks LM, Ye L, Choi JH, Giang AH, Khandekar M, Virtanen KA, Nuutila P, Schaart G, et al. Beige adipocytes are a distinct type of thermogenic fat cell in mouse and human. Cell. 2012;150:366–76.

Yamauchi T, Kamon J, Waki H, Terauchi Y, Kubota N, Hara K, Mori Y, Ide T, Murakami K, Tsuboyama-Kasaoka N, et al. The fat-derived hormone adiponectin reverses insulin resistance associated with both lipoatrophy and obesity. Nat Med. 2001;7:941–6.

Yang K, Guan H, Arany E, Hill DJ, Cao X. Neuropeptide Y is produced in visceral adipose tissue and promotes proliferation of adipocyte precursor cells via the Y1 receptor. FASEB J. 2008;22:2452–64.

Yang L, Scott KA, Hyun J, Tamashiro KL, Tray N, Moran TH, Bi S. Role of dorsomedial hypothalamic neuropeptide Y in modulating food intake and energy balance. J Neurosci. 2009;29:179–90.

Zarjevski N, Cusin I, Vettor R, Rohner-Jeanrenaud F, Jeanrenaud B. Chronic intracerebroventricular neuropeptide-Y administration to normal rats mimics hormonal and metabolic changes of obesity. Endocrinology. 1993;133:1753–8.

Zhang Y, Proenca R, Maffei M, Barone M, Leopold L, Friedman JM. Positional cloning of the mouse obese gene and its human homologue. Nature. 1994;372:425–32.

Zhang L, Macia L, Turner N, Enriquez RF, Riepler SJ, Nguyen AD, Lin S, Lee NJ, Shi YC, Yulyaningsih E, et al. Peripheral neuropeptide Y Y1 receptors regulate lipid oxidation and fat accretion. Int J Obes. 2010;34:357–73.

Chapter 10
Leptin, Adiponectin, and Other Adipokines in Regulation of Adipose Tissue Angiogenesis

Ebba Brakenhielm and Yihai Cao

Abstract Adipose tissue plays an important endocrine role through the production of secreted factors, termed adipokines. These include *hormones* such as leptin, adiponectin, resistin, omentin, apelin, and vaspin; classical *cytokines* such as tumor necrosis factor-α, interleukin-6, and transforming growth factor-β; *enzymes* such as lipoprotein lipase, visfatin, and angiogenin; and finally traditional *growth factors* such as vascular endothelial growth factor (VEGF)-A, hepatocyte growth factor, and angiopoietin-like protein-4. Recently, it has been discovered that apart from their role in regulation of physiological processes, such as energy metabolism, food intake, and immunity, some adipokines are important regulators of angiogenesis. For example, leptin is a potent stimulator of angiogenesis that acts in synergy with VEGF-A. In contrast, adiponectin has been suggested as an adipose-derived angiogenic inhibitor. Given that adipose tissue expansion is angiogenesis-dependent, the local production in white and brown adipose tissues of adipokine angiogenic regulators may play a crucial role in adipogenesis. This raises the question as to what extent the different adipokines influence adipose tissue vascular growth and remodeling during normal adipose tissue growth and in development of obesity. In this chapter, we summarize the evidence in support of adipokines as important endogenous angiogenic regulators. Further, we hint at what might be a deregulated adipokine-mediated angiogenic process occurring during pathological expansion of adipose tissue as seen in obesity.

E. Brakenhielm, Ph.D. (✉)
Inserm U1096, Institute for Research and Innovation in Biomedicine,
University of Rouen, 22 Boulevard Gambetta, 76183 Rouen, France
e-mail: Ebba.Brakenhielm@inserm.fr

Y. Cao, M.D., Ph.D.
Department of Microbiology, Tumor and Cell Biology (MTC),
Karolinska Institute, 171 77 Stockholm, Sweden

Department of Medicine and Health Sciences, Linköping University,
58185 Linköping, Sweden

Y. Cao (ed.), *Angiogenesis in Adipose Tissue*, DOI 10.1007/978-1-4614-8069-3_10,

Keywords Endothelial • Vascular growth • Obesity • Adipogenesis • VEGF • HGF • IL-6 • Angptl-4

10.1 Introduction

Adipose tissue not only serves as a site of energy storage, but also plays a dynamic endocrine role, as recognized almost 20 years ago, through the production and release of adipose-derived factors, termed *adipokines* (Kershaw and Flier 2004). These include *hormones* such as leptin, adiponectin, resistin, omentin, apelin, and vaspin; classical *cytokines* such as tumor necrosis factor (TNF)-α, interleukin (IL)-6, and transforming growth factor (TGF)-β; *enzymes* such as lipoprotein lipase (LPL), 17-β-hydroxysteroid dehydrogenase (HSD), visfatin, and angiogenin; and finally traditional *growth factors* such as vascular endothelial growth factor (VEGF)-A, hepatocyte growth factor (HGF), and angiopoietin-like protein (Angptl)-4. These systemically secreted factors are critically involved in diverse body processes (Table 10.1), such as energy metabolism and food intake, inflammation and immunity, insulin sensitivity, reproduction, as well as vascular tone (Waki and Tontonoz 2007; Gualillo et al. 2007; Lago et al. 2007; Maenhaut and Van de Voorde 2011). Moreover, these adipose tissue-derived signaling molecules play key roles in modulation of obesity and related metabolic disorders, including insulin resistance (Matsuzawa 2006; Deng and Scherer 2011; Northcott et al. 2012).

Apart from the well-recognized role of many growth factors in stimulation of angiogenesis, there is accumulating evidence that several other adipokines may directly or indirectly impact angiogenesis. Some of these adipose tissue-derived factors, such as TNF-α and vaspin, may preferentially operate in a paracrine manner influencing local adipose tissue responses, whereas others, such as leptin and adiponectin, may potentially also act in an endocrine fashion to regulate angiogenesis at distant sites. Given the recent findings that adipose tissue expansion is angiogenesis-dependent (Rupnick et al. 2002; Brakenhielm et al. 2004a), the question arises whether adipokines differentially may regulate adipose tissue vascular growth and remodeling during normal adipose tissue expansion vs. in obesity. In this chapter, we will go through the evidence of adipokines as important angiogenic regulators with both local and potential systemic impact. Further, we will summarize the data concerning the altered adipokine balance found in obesity, hinting at what might be a deregulated angiogenic process during pathological expansion of adipose tissue.

10.2 VEGF-A, HGF, and Other Known Angiogenic Regulators

It has long been realized by surgeons that adipose tissue, especially omental fat depots, can improve wound healing by stimulating vascular regrowth (Silverman et al. 1988; Cao 2007). This pro-angiogenic effect has been partly attributed to an

Table 10.1 Physiological roles of adipokines[a]

Adipokines	Physiological processes
Adiponectin	Glucose metabolism, pancreatic β cell function, insulin sensitivity, lipolysis, vasorelaxation, anti-inflammatory, anti-angiogenic
Apelin	Feeding behavior, diuresis, energy expenditure, anti-lipolytic, vasorelaxation, cardiac contractility, anti-inflammatory, pro-angiogenic
Chemerin	Glucose metabolism, pro-inflammatory, pro-angiogenic?
Leptin	Feeding behavior, thermogenesis, energy expenditure, reproduction, insulin sensitivity, lipolysis, hematopoiesis, regulation of vascular tone[b], pro-inflammatory, pro-angiogenic
Omentin	Microbe recognition, insulin sensitivity, vasorelaxation, anti-inflammatory, anti-angiogenic?
Resistin	Glucose metabolism, insulin sensitivity, pro-inflammatory, pro-angiogenic?
Vaspin	Insulin sensitivity, pro-endothelial survival
IL-6	Energy expenditure, lipolysis, inhibits LPL, insulin sensitivity, pancreatic β cell function, regulation of vascular tone[b], pro-inflammatory, pro-angiogenic
TGF-β	Anti-inflammatory, vascular permeability, glucose uptake, energy expenditure, pro-angiogenic
TNF-α	Feeding behavior, glucose metabolism, insulin sensitivity, lipolysis, regulation of vascular tone[b], pro-inflammatory, pro- or anti-angiogenic
Angiogenin	Ribonuclease activity, pro-angiogenic, neuroprotection
Visfatin	Biosynthesis of NAM and NAD, pancreatic β cell function, vasorelaxation, pro-inflammatory, pro-angiogenic?
PAI-1	Coagulation, energy expenditure?, insulin sensitivity?, pro- or anti-angiogenic
Angptl-4	Feeding behavior, energy expenditure, glucose metabolism, insulin sensitivity, inhibits LPL, vascular permeability, anti-angiogenic
HGF	Glucose metabolism, hematopoiesis, anti-inflammatory, pro-angiogenic
VEGF-A	Hematopoiesis, vasorelaxation, vascular permeability, pro-inflammatory, pro-angiogenic

[a]*Angptl-4* angiopoietin-like protein 4, *HGF* hepatocyte growth factor, *IL-6*, interleukin-6, *LPL* lipoprotein lipase, *PAI-1* plasminogen activator inhibitor-1, *TGF-β* transforming growth factor, *TNF-α* tumor necrosis factor, *VEGF-A* vascular endothelial growth factor-A
[b]Vasorelaxation or vasoconstriction depending on context

elevated VEGF-A production in adipose tissue. Among the different fat depots in the body, the large adipocytes from omental white adipose tissue (WAT) secrete the highest levels of VEGF-A (Zhang et al. 1997). It is also produced in brown adipose tissue (BAT) adipocytes, where further norepinephrine stimulation and cold-stress both increase VEGF-A levels (Xue et al. 2009). Notably, both WAT and BAT adipocyte differentiation is associated with upregulation of VEGF-A production. In addition, hypoxia is a potent inducer of VEGF-A production, through both transcriptional and post-translational mechanisms. It has been found that adipose tissue expansion leads to hypoxia (Trayhurn and Wood 2004; Hosogai et al. 2007) (see Chaps. 3 and 13). This may be due to the fact that both vessel density and diameters initially are reduced during fat expansion. Hypoxia-driven VEGF-A production in adipocytes and stromal cells may thus contribute to the local stimulation of angiogenesis in expanding adipose tissue that creates a permissive environment for further adipose tissue growth (Fukumura et al. 2003). In agreement, the use of small molecule VEGF receptor (VEGFR) tyrosine kinase inhibitors or soluble VEGFRs reduced adipose tissue vascular density in mice (Kamba et al. 2006). Furthermore, therapeutic blockage of VEGFR-2, the main signaling receptor for VEGF-A in endothelial cells, but not blockage of VEGFR-1, its more ubiquitously expressed receptor, was reported to limit diet-induced obesity in mice by inhibiting adipose tissue angiogenesis leading to a reduction of adipocyte hypertrophy (Fukumura et al. 2003; Tam et al. 2009). This argues for an important role of the canonical angiogenic growth factor VEGF-A in adipose tissue angiogenesis.

Another potent mitogenic and angiogenic growth factor, HGF, is also produced in adipose tissue, where it is mainly secreted by preadipocytes, but also by adipocytes. Notably, HGF production levels in human WAT are almost 10-fold greater as compared with VEGF-A (Fain et al. 2004). Furthermore, circulating HGF levels were reportedly increased more than threefold in obesity (Rehman et al. 2003). This suggests that HGF, in addition to VEGF-A, may be a major pro-angiogenic factor in adipose tissue with a role in fat expansion. In agreement, it was reported that the endothelial-stimulatory effect of conditioned media from adipocyte cultures was reduced to 40 % of maximal following silencing of HGF (Saiki et al. 2006). Additionally, blockage of HGF reduced angiogenesis in developing fat pads in mice, whereas HGF over-expression in preadipocytes enhanced it, leading to accelerated adipose tissue formation (Bell et al. 2008).

Beyond the proposed roles for VEGF-A and HGF in stimulation of adipose tissue angiogenesis, other known angiogenic and arteriogenic factors may also play a part. For example, the expression levels of VEGF-B, placental-derived growth factors (PlGF), and angiopoietins (Ang)-1 and Ang-2 increase during adipocyte differentiation (Stacker et al. 2000; Guo and Liao 2000; Soukas et al. 2001; Voros et al. 2005). Moreover, upregulation of Ang-2 expression in adipose tissue was found to be essential in mediating the vascular maturation and patterning effects of the transcription factor FOXC2 (Xue et al. 2008). Additionally, several members of the fibroblast growth factor (FGF) family, including FGF-1 and FGF-2, known potent inducers of angiogenesis, are expressed in adipose tissue (Gabrielsson et al. 2000), and their levels increase in obesity (Teichert-Kuliszewska et al. 1992). On the other hand, several endogenous angiogenic inhibitors are expressed in adipose tissues, including thrompospondins (TSP), endostatin, pigment epithelium-derived factor (PEDF), as well as soluble VEGFR-1/fms-like tyrosine kinase 1 (sFlt-1) (Silha et al. 2005; Varma et al. 2008; Famulla et al. 2011; Herse et al. 2011). Moreover, as we will discuss in this chapter, many of the new adipokines discovered over the last 15 years have been found to regulate angiogenesis. The local spatiotemporal interplay of adipose-derived factors, including a redundancy of both stimulators and inhibitors of angiogenesis, may be requisite to ensure that the adipose tissue vascular growth and remodeling occurring during fat mass expansion is finely tuned to meet the active metabolic needs of this dynamic organ.

10.3 Leptin

Leptin is a 167 amino acid preprotein (16 kDa, 146 amino acid mature form) four helix bundle monomer, belonging to the long-chain helical cytokine family of proteins. It was discovered as the product of the *ob/ob* gene, mutated in a genetically obese strain of mouse (Campfield et al. 1995; Halaas et al. 1995; Pelleymounter et al. 1995). Leptin regulates food intake and energy metabolism, serving as a key feedback signal between the fat organ and the brain responsible for body weight homeostasis (Table 10.1). The peripheral actions of leptin further include regulation of immune responses, hematopoiesis, sexual maturation, and organ growth partly through its effects on growth hormone production (Conde et al. 2010).

Leptin is mainly produced by subcutaneous WAT adipocytes, in a cell size-dependent manner. Leptin is also expressed at lower levels in BAT, brain, skeletal muscle, and placenta (Mora and Pessin 2002). Indeed, during pregnancy, leptin levels may increase 1.6–1.8 fold, arguing for the placenta as a substantial source of leptin (Masuzaki et al. 1997). Circulating leptin levels are positively correlated with body mass index (BMI). However, in obese subjects there seems to be a central leptin resistance, potentially involving reduced hypothalamic leptin signaling by protein tyrosine phosphatases (St-Pierre and Tremblay 2012), which prevents the anorexic effects of the hormone (Myers et al. 2012). Leptin production, similar to VEGF-A, is increased in adipocytes by hypoxia (Grosfeld et al. 2002). Further, pro-inflammatory cytokines, such as TNF-α, induce leptin production in adipocytes, and conversely leptin stimulates the production of pro-inflammatory cytokines, including IL-6 and TNF-α, leading to increased local tissue inflammation.

Leptin binds to its transmembrane-spanning receptor (Ob-R), inducing receptor homodimerization and activation of intracellular signaling pathways (Friedman and Halaas 1998). There exists several splice isoforms of the leptin receptor. The longest form, Ob-Rb, expressed at high levels in the hypothalamus where leptin exerts its anorexic effects, serves as the main signaling receptor for leptin. It contains three box motifs for cytosolic Jak/STAT binding. The Ob-Ra, a shorter form of the receptor lacking two of the box motifs, still retains some signaling activity involving the Fos and Jun pathway. Leptin receptors, both short and long forms, are expressed in peripheral organs including adipose tissue, kidneys, ovaries, and vascular endothelium.

10.3.1 Effects on Angiogenesis

The stimulatory impact of leptin on endothelial cell migration, proliferation, and tube formation in vitro and on angiogenesis in vivo, in the rat corneal model and the chicken chorioallantoic membrane (CAM) angiogenesis model, was first described in 1998 (Sierra-Honigmann et al. 1998; Bouloumie et al. 1998). Similarly, we found that leptin stimulated new blood vessel growth in the mouse corneal model

(Cao et al. 2001). Further, co-implantation of leptin with either VEGF-A or FGF-2 resulted in synergistic stimulation of angiogenesis, suggesting that leptin may cooperate with these angiogenic growth factors in tissues co-expressing both factors, such as adipose tissue, placenta, and fetal organs. In addition to its direct effects on endothelial cells, leptin may induce the expression of pro-angiogenic growth factors, including VEGF-A and FGF-2. Indeed, leptin treatment stimulated VEGF-A production in the adipose tissue of *ob/ob* mice (Stallmeyer et al. 2001). However, the fact that *ob/ob* mice, deficient for leptin, as well as *db/db* or *fa/fa* animals, lacking functional Ob-Rb leptin receptors, rapidly gain body weight and display extensive WAT vascularity indicates that leptin is not necessary for adipose tissue angiogenesis to occur (Cao et al. 2001; Cao 2007).

The role of leptin in adipose tissue vascular remodeling during adipose tissue regression was investigated in obese *ob/ob* mice, where systemic supraphysiological leptin administration induces body weight loss with selective depletion of adipose tissue. Notably, weight loss was found to be associated with vascular regression and apoptosis in subcutaneous adipose tissues. This was linked to induction of Ang-2 production in WAT adipocytes by leptin (Cohen et al. 2001). Ang-2 favors vascular regression, in the absence of VEGF-A, by stimulating the detachment of vascular pericytes, necessary for endothelial survival. In contrast, in the presence of angiogenic stimuli such as VEGF-A, Ang-2 potentiates angiogenic responses. Leptin seems to be an important physiological regulator of Ang-2 expression in subcutaneous adipose tissues. In accordance, reduced WAT Ang-2 levels were observed in the absence of leptin signaling, i.e., in obese *ob/ob* or *db/db* mice. Leptin may thus play an essential role, partly via regulation of Ang-2, as an adipogenesis "gatekeeper" synergistically stimulating angiogenesis and adipose tissue expansion in the presence of VEGF-A during weight gain, or conversely favoring blood vessel and adipose tissue regression, in the absence of VEGF-A, during weight loss. Additionally, it may be speculated that deficient leptin signaling or leptin resistance, as seen in obesity, paradoxically could lead to prevention of adipose tissue regression in fat depots with low VEGF-A expression, such as subcutaneous WAT, by limiting Ang-2-mediated vascular remodeling.

Apart from its local role in adipose tissue, circulating leptin may also influence angiogenesis in non-adipose tissues. For instance, leptin was found to stimulate the activity of circulating angiogenic cells, leading to improved revascularization in a hind limb ischemia model (Heida et al. 2010). Conversely, hyperleptinemia may contribute to the development of certain types of cancer, such as breast, prostate, and colon cancers.

10.3.2 Effects on Vascular Function

Leptin, similar to VEGF-A, seems to play a role in regulation of endothelial ultrastructure and vascular permeability. We found that leptin-induced blood vessels in the mouse cornea assay, similar to VEGF-A-induced neovessels, displayed endothelial fenestrations (Cao et al. 2001). Moreover, whereas subcutaneous WAT blood vessels in *wt* mice displayed endothelial fenestrations, the adipose tissue vasculature in *ob/ob* mice lacked fenestrations. This suggests a physiological role of leptin in induction and maintenance of specialized endothelial features characteristic of endocrine organ blood vessels. The presence of endothelial fenestrations is associated with increased vascular permeability. In agreement, we found that leptin, similar to VEGF-A, potently stimulated vascular leakage in a modified Miles assay in mice (Cao et al. 2001).

Leptin has been found to play a role in regulation of vascular tone and may directly stimulate adipose tissue blood flow. Indeed, leptin increases endothelial nitric oxide ($NO^{\cdot}$) production, via Akt-dependent activation of endothelial nitric oxide synthase (eNOS), leading to vasodilation (Vecchione et al. 2002; Mehebik et al. 2005). However, leptin also displays pro-oxidant effects, including stimulation of endothelial reactive oxygen species (ROS) production, and may further upregulate endothelin-1 expression levels, leading to vasoconstriction. Taken together with leptin's pro-inflammatory activities, hyperleptinemia may cause vascular endothelial dysfunction and contribute to the pathophysiology of cardiovascular disease, including atherosclerosis (Patel et al. 2008).

10.4 Adiponectin

Adiponectin, also known as AdipoQ, adipocyte complement-related protein (ACRP)-30, apM1, and gelatin-binding protein (GBP)-28, is a 244 amino acid preprotein (230 amino acid mature form), WAT and BAT-derived hormone. Adiponectin is the most abundant adipokine in the circulation (5–30 μg/mL in healthy humans, Table 10.2) (Arita et al. 1999), where it represents 0.01 % of total plasma protein. Paradoxically, although exclusively produced by adipose tissue, adiponectin levels decrease with increasing BMI. Adiponectin belongs to the soluble defense collagen superfamily, displaying an N-terminal signal sequence followed by a collagenous stalk region (similar to collagen VIII and X) and a globular C-terminal domain, structurally homologous to complement factor Cq1. It forms disulphide-linked covalent trimers, hexamers, and high-molecular weight (HMW) oligomeric species (Pajvani et al. 2003). Notably, the trimer folding topology is strikingly similar to members of the TNF family (Shapiro and Scherer 1998). Recently, gluthation transferase-κ, also known as DsbA-like protein, was shown to play an important role in adiponectin multimer assembly (Liu et al. 2008). Furthermore, a globular short form of adiponectin is produced through enzymatic digestion by leukocyte elastase (Fruebis et al. 2001; Waki et al. 2005). Adiponectin is present in human serum mainly in its hexameric full-length form, with a smaller contribution of HMW species and a lower level of trimers. In mouse serum, hexameric and larger complexes prevail (Pajvani et al. 2003). Females generally display higher circulating levels of adiponectin and more HMW multimers.

In contrast to other adipokines such as VEGF-A, leptin, and TNF-α, the production and secretion of adiponectin decrease with increasing adipocyte sizes. In addition, adipose tissue hypoxia has been associated with decreased

Table 10.2 Circulating serum levels of adipokines in humans

Adipokines	Physiological level	Levels in obesity
Adiponectin	5–30 μg/mL	Decreased
Omentin	200–400 ng/mL	Decreased
Angptl-4	5–25 ng/mL	Decreased/unaltered
Chemerin	70–150 ng/mL	Unaltered
Vaspin	0.01–4 ng/mL	Unaltered
Apelin-13	20–60 ng/mL	Increased/unaltered
Resistin	1–15 ng/mL	Increased/unaltered
VEGF-A	100–300 pg/mL	Increased/unaltered
TGF-β	2–9 pg/mL	Increased/unaltered
TNF-α	1–3 pg/mL	Increased/unaltered
Angiogenin	300–400 ng/mL	Increased
Leptin	1–50 ng/mL	Increased
PAI-1	10–15 ng/mL	Increased
HGF	500–700 pg/mL	Increased
IL-6	1–5 pg/mL	Increased
Visfatin	1–3 ng/mL	Increased

Serum levels reported in healthy men and women vs. overweight or obese subjects

production of adiponectin (Wree et al. 2012). Furthermore, inflammatory stimuli, such as TNF-α, reduces adiponectin production and secretion in adipocytes (Kappes and Loffler 2000).

Adiponectin is a multifunctional protein known for its stimulatory effects on energy metabolism, including regulation of hydrocarbon and lipid turnover (leading to reduction of circulating triglyceride and free fatty acid levels), as well for its protective role against insulin resistance. Indeed, adiponectin exerts pleiotropic insulin-sensitizing effects, including lowering of hepatic glucose production and increasing skeletal muscle glucose uptake and fatty acid oxidation (Table 10.1). Furthermore, adiponectin seems to play an important role as a protector of cardiovascular function in mice, and it displays anti-inflammatory and anti-atherogenic properties (see beneath). As a consequence, adiponectin has been proposed to constitute the link between obesity and metabolic syndrome (Funahashi and Matsuzawa 2007), including development of type 2 diabetes and coronary artery disease.

The molecular mechanisms of adiponectin are mainly mediated via its two specific receptors (ADIPOR1 and ADIPOR2), homologous to seven-transmembrane spanning G-protein-coupled receptors, and their adaptor protein APPL1 (Yamauchi et al. 2003a; Mao et al. 2006). Whereas ADIPOR1 is ubiquitously expressed, with elevated levels present in skeletal muscle, ADIPOR2 is predominantly found in the liver. While ADIPOR1 displays high affinity for the globular short form of adiponectin and a low affinity for the full-length form, ADIPOR2 shows intermediate affinity for both (Yamauchi et al. 2003a). The anti-atherogenic effects of adiponectin may be mediated through ADIPOR1, whereas the anti-diabetic effects may be mediated via ADIPOR2 (Kadowaki and Yamauchi 2005). Further, T-cadherin, present in endothelial and smooth muscle cells, has been proposed as an additional vascular receptor for adiponectin. It displays selectivity for the hexameric and HMW species of adiponectin and does not bind either full-length or globular trimeric forms (Hug et al. 2004). T-cadherin is suggested to play a role in vascular protection against oxidative and glucose stress-induced apoptosis.

Due to these different binding affinities for its receptors, not all forms of adiponectin display the same cellular effects. For instance, it has been reported that adiponectin trimers are more bioactive in reduction of serum glucose in vivo than the higher order forms (Pajvani et al. 2003). On the other hand, the insulin-sensitizing effects of adiponectin seem mainly attributed to HMW species. Notably, there is a selective reduction in the circulating levels of the HMW form of adiponectin in humans with hyperinsulinemia, type II diabetes, or coronary artery disease, whereas adiponectin hexamer levels remain unchanged and trimer levels may be increased (Kobayashi et al. 2004; El-Menyar et al. 2009).

The downstream signaling induced by adiponectin includes activation of adenosine monophosphate-activated kinase (AMPK), activation of ceramidase, activation of the peroxisome proliferator-activated receptor (PPAR)-α pathway, as well as inhibition of TNF-α-induced NF-κB activation in endothelial cells, via stimulation of cAMP/PKA signaling (Ouchi et al. 2000), associated with reduced platelet aggregation and macrophage activation. Whereas the globular domain of adiponectin, acting mainly via ADIPOR1, is most potent in inducing AMPK activation and fatty acid β-oxidation in skeletal muscles, the HMW complex, through ADIPOR2, is most effective in suppressing hepatic glucose production involving PPAR-α. Moreover, while adiponectin HMW forms stimulate AMPK activity in hepatocytes (Waki and Tontonoz 2007), only trimers activate AMPK in skeletal muscle (Tsao et al. 2003).

10.4.1 Effects on Angiogenesis

The vascular effects of adiponectin may be mediated via either ADIPOR1 or ADIPOR2, as both are expressed in endothelial cells. We found that full-length human, as well as murine, adiponectin potently inhibited FGF-2-induced cell proliferation in microvascular endothelial cells (Brakenhielm et al. 2004b). Notably, adiponectin reached a maximal inhibitory effect at 1 μg/mL, well within range of physiological circulating levels. Similarly, adiponectin dose-dependently decreased VEGF-A-induced endothelial cell migration. Others have confirmed the angiostatic effects of adiponectin in endothelial cells, including suppression of migration, proliferation, and tube formation (Motoshima et al. 2004; Mahadev et al. 2008; Man et al. 2010). Similarly, adiponectin was found to inhibit smooth muscle cell migration and proliferation (Arita et al. 2002; Matsuda et al. 2002; Brakenhielm et al. 2004b), albeit with weaker potency as compared with in endothelial cells. In contrast, adiponectin did not block fibroblast proliferation. This indicates that adiponectin may be a selective inhibitor of endothelial cells at low concentrations. In order to investigate the effects of adiponectin on angiogenesis in vivo, we used the chicken CAM assay and the mouse corneal assay. We found that full-length adiponectin administration induced the formation of avascular zones in the angiogenic chorioallantois membrane (Brakenhielm et al. 2004b). Similarly, co-implantation of adiponectin with FGF-2 in mouse corneas drastically reduced FGF-2-induced angiogenic responses. Moreover, administration of a peptide fragment of globular adiponectin was found to reduce choroidal neovascularisation in mice (Bora et al. 2007). These findings suggest that both globular and full-length adiponectin are potent angiogenic inhibitors. The mechanisms may include direct as well as indirect effects of adiponectin in endothelial cells. For instance, adiponectin-suppression of VEGF-A-induced endothelial cell migration, either by the globular or full-length forms, was dependent on activation of the cAMP/PKA pathway (Mahadev et al. 2008). Further, adiponectin, in particular HMW forms, has been suggested to bind and sequester several angiogenic growth factors, including platelet-derived growth factor (PDGF)-BB, and FGF-2 (Arita et al. 2002; Wang et al. 2005). Adiponectin may thus modulate the biological actions of growth factors by controlling their bioavailability at the pre-receptor level. Similarly, adiponectin was recently found to bind TSP-1 (Wang et al. 2006), a known angiogenesis inhibitor. The indirect anti-angiogenic mechanisms of adiponectin may further involve suppression of VEGF-A expression (Bora et al. 2007). Finally, induction of programmed cell death in endothelial cells seems to be involved in the anti-angiogenic effects of adiponectin. We found that adiponectin in a dose- and time-dependent manner induced endothelial cell apoptosis (Brakenhielm et al. 2004b), involving activation of pro-caspases 3, 8, and 9. In agreement, intratumoral administration of adiponectin resulted in induction of vascular endothelial cell death and inhibition of tumor angiogenesis, leading to suppression of tumor growth (Brakenhielm et al. 2004b; Man et al. 2010). However, the effects of adiponectin on endothelial apoptosis are controversial. Although some studies have confirmed the pro-apoptotic impact of adiponectin in endothelial cells, other authors have in contrast found that adiponectin may protect against high glucose- and oxidative stress-induced endothelial cell death (Sun and Chen 2010). Notably, the HMW adiponectin forms, but not trimers or globular fragments, showed pro-survival effects, linked to cellular activation of AMPK, in serum-starved macrovascular endothelial cells (Kobayashi et al. 2004). Although sustained AMPK activation induces apoptosis in several cell types, it seems that AMPK under metabolic-stress conditions may positively regulate Akt signaling in endothelial cells (Nagata et al. 2003), leading to promotion of cell survival. Further, although AMPK activation in most cell types, including endothelial cells, has been linked to inhibition of protein synthesis and decreased cellular proliferation, as a consequence of mTOR inhibition (Motoshima et al. 2006; Guo et al. 2007), AMPK also may enhance cellular antioxidant defenses and increase NO· production in endothelial cells, leading to protection against cell death during ischemia. In addition to the beneficial impact of AMPK on energy metabolism and antioxidant defenses (Fisslthaler and Fleming 2009), this stress-induced kinase may further play a role in hypoxic stress-mediated angiogenesis, notably via stimulation of hypoxia-inducible factor (HIF)-1α transcriptional activity leading to increased VEGF-A expression (Nagata et al. 2003; Lee et al. 2003b). Adiponectin-induced AMPK activation in endothelial cells may thus, under metabolic-stress conditions, paradoxically lead to indirect stimulation of angiogenesis through Akt-dependent suppression of hypoxia-induced cell death or by Hif-1-dependent, hypoxia-induced VEGF-A production. Indeed, some authors have found that adiponectin, through AMPK-modulated Akt activation, may increase cell migration and tube formation in serum-starved macrovascular-derived endothelial cells (Ouchi et al. 2004). Additionally, adiponectin was reported to stimulate angiogenesis in the mouse matrigel plug model as well as in rabbit corneas (Ouchi et al. 2004). Moreover, adiponectin knockout mice display reduced angiogenic responses to ischemic injury, both in a chronic limb ischemia model (Shibata et al. 2004), and in a concentric cardiac hypertrophy model (Shimano et al. 2010). Notably, in mice lacking adiponectin, ischemia-induced VEGF-A production was reduced. On the other hand, over-expression of adiponectin protected against ischemic stroke in a middle cerebral artery occlusion model in mice (Shen et al. 2013). This was linked to activation of AMPK and locally increased VEGF-A expression, leading to stimulation of angiogenesis. The observed stimulatory effect of adiponectin on ischemia-induced angiogenic responses may further relate to the protective effects of adiponectin against oxidative stress and inflammation, leading to prevention of endothelial dysfunction (see beneath). Indeed, vascular function plays a crucial role in assuring metabolically adapted tissue blood flow but also in regulating angiogenic responses to growth factors and other stimuli. It thus seems that whereas adiponectin may exert inhibitory, angiostatic effects in endothelial cells, in part via binding and neutralization of angiogenic growth factors and inducing endothelial cell death, it may also, particularly under metabolic-stress conditions, favor endothelial cell migration and survival. The impact of adiponectin on angiogenesis might therefore be both highly context-dependent and potentially further influenced by the protein's oligomerization state, which modulates its biological actions.

The impact of adiponectin on adipose tissue angiogenesis is currently unclear. Adiponectin-deficient mice display normal body weight, suggesting that adiponectin might be redundant under physiological conditions (Maeda et al. 2002). Similarly, over-expression of globular adiponectin did not result in any significant differences in body weight or adiposity in mice (Yamauchi et al. 2003b). However, in lean or obese mice fed a high-fat diet, systemic treatment with globular adiponectin was found to result in body weight loss (Fruebis et al. 2001; Yamauchi et al. 2001). The effect was most pronounced in obese mice and did not involve any suppression of food intake. Rather, adiponectin was found to acutely stimulate energy expenditure by increasing free fatty acid uptake and oxidation in skeletal muscle. However, it cannot be excluded that adiponectin therapy may have contributed to fat mass regression in part through its anti-angiogenic effects, including induction of endothelial cell death, leading to vascular regression in adipose tissues. Opposing results were reported in a mouse model of genetic obesity, where forced expression in adipocytes of full-length adiponectin resulted in extreme obesity in *ob/ob* leptin-deficient mice, with the excess weight accounted for by greater subcutaneous fat mass (Kim et al. 2007a). In contrast, visceral fat mass was found to be decreased to lean *wt* levels. Further, the increased adiposity in subcutaneous depots induced by adiponectin in *ob/ob* mice was due to adipocyte hyperplasia, rather than hypertrophy. Indeed, average adipocyte sizes, in subcutaneous WAT from adiponectin-transgenic *ob/ob* mice, were reduced to less than 50 % of that seen in *ob/ob* mice. Interestingly, adiponectin over-expression led to prevention of obesity-induced insulin resistance in *ob/ob* mice despite their morbid obesity. This was linked to reduced local production of pro-inflammatory cytokines, including TNF-α and IL-6, as well as reduced macrophage accumulation in subcutaneous WAT (Kim et al. 2007a). These discrepant findings raise interesting questions regarding the role of adiponectin in adipose tissues: Why is visceral, but not subcutaneous, fat mass inhibited by adiponectin over-expression? In particular, does adiponectin exert fat depot-selective anti-angiogenic effects? Further, why was adipocyte hypertrophy, but not hyperplasia, inhibited by adiponectin? Finally, what vascular changes occurred in WAT of mice treated with globular vs. full-length adiponectin?

10.4.2 Effects on Vascular Function

Adiponectin plays an important physiological role in maintenance of endothelial function in mice and men (Ouchi et al. 2003). The mechanisms are multiple and include upregulation and activation of eNOS (Schmid et al. 2011), leading to increased NO· levels, essential for both endothelial-mediated vasodilation as well as for prevention of platelet aggregation. Further, both globular and full-length forms of adiponectin display antioxidant properties by scavenging ROS induced by oxidized LDL or hyperglycemia in endothelial cells. These effects seem to involve a cAMP/PKA-dependent, but AMPK-independent, pathway (Ouedraogo et al. 2006), and may in part be conveyed through the interaction of adiponectin with T-cadherin. Notably, acute systemic administration of globular adiponectin was reported to decreased cardiac ROS formation during ischemia-reperfusion injury in mice (Tao et al. 2007), leading to cardioprotection. Furthermore, adiponectin displays anti-inflammatory and anti-atherogenic properties, including induction of anti-inflammatory cytokines and down-regulation of adhesion molecules in vascular endothelium. Moreover, adiponectin may directly modulate macrophage recruitment and activation in adipose tissue: whereas fat depots from adiponectin knockout mice displayed increased numbers of pro-inflammatory macrophages, adenoviral-delivery of adiponectin in both *wt* and adiponectin-deficient mice resulted in increased recruitment of anti-inflammatory, "M2-like," macrophages (Ohashi et al. 2010). Finally, adiponectin may prevent atherosclerosis through its suppressive effects on smooth muscle cell proliferation (Matsuda et al. 2002). In agreement, hypoadiponectinemia, with plasma levels lower than 4 μg/mL, is associated with an increased risk of cardiovascular diseases, including coronary artery disease (Kumada et al. 2003).

10.5 Resistin

Resistin, also called adipose tissue-specific secretory factor and FIZZ3, is a 108 amino acid pre-peptide (90 amino acid mature form) hormone that belongs to a new gene family of small cysteine-rich secretory proteins, called resistin-like molecules (RELMs). Resistin forms complex multimeric structures, strikingly similar to adiponectin, through the formation of coiled-coil trimers that unite tail-to-tail to form hexamers linked via N-terminal disulfide bonds (Patel et al. 2004). In the circulation, resistin is present in the form of highly active trimers and less active, but more abundant, hexamers. Among the cellular receptors proposed for resistin are: Decorin, and tyrosine kinase-like Orphan Receptor (ROR)-1 in preadipocytes, and Toll-like Receptor-4 in macrophages (Tarkowski et al. 2010; Daquinag et al. 2011; Sanchez-Solana et al. 2012).

Resistin is expressed in adipocytes and in adipose tissue-resident macrophages, but also by circulating monocytes, as well as by cells in the spleen, lung, kidney, bone marrow, and placenta. In addition, hepatic Kupffer cells seem to be an important source of resistin (Szalowska et al. 2009). Resistin expression increases in human adipose tissues in obesity (Savage et al. 2001). In contrast, while circulating resistin levels are increased in patients with atherosclerosis and coronary artery disease, the links between plasma resistin levels and obesity or type 2 diabetes are more controversial in humans (Lee et al. 2003a; Reilly et al. 2005; Chen et al. 2009). Resistin was initially proposed as the molecular link between obesity and insulin resistance, as treatment of mice with recombinant resistin induced insulin resistance and reciprocally administration of resistin-blocking antibodies restored insulin sensitivity in obese and diabetic mice (Steppan et al. 2001). However, recent studies have indicated that resistin may be more closely related to chronic inflammation than to insulin resistance in humans, where in addition myeloid cells seem to be the main source (Reilly et al. 2005). Indeed, resistin displays pro-inflammatory actions, including upregulation of inflammatory cytokines, such as IL-6 and monocyte chemoattractant protein (MCP)-1, and induction of endothelial cell adhesion molecules. It also induces expression of endothelin-1, a powerful vasoconstrictor (Verma et al. 2003) (Table 10.1). Resistin may play a deleterious role in atherosclerosis via stimulation of intimal proliferation and vascular inflammation leading to endothelial dysfunction (Jamaluddin et al. 2012). In order to investigate the role of the human expression pattern of resistin during adipose tissue expansion, transgenic mice were generated that were null for mouse resistin but expressed human resistin, notably in adipose tissue-resident macrophages. When fed a high-fat diet, the humanized resistin mice displayed similar body weight gain and adiposity as *wt* mice (Qatanani et al. 2009). However, during the development of obesity their WAT showed increased expression of pro-inflammatory cytokines, including TNF-α, IL-6, and MCP-1, leading to accelerated adipose tissue recruitment of inflammatory macrophages. Notably, this led to increased lipolysis and increased levels of circulating free fatty acids, resulting in muscle lipid deposition and the development of insulin resistance. This suggests that macrophage-derived human resistin, through its effects on WAT inflammation, may indirectly contribute to metabolic syndrome development in obese humans.

Apart from its impact on inflammation and vascular function, resistin may be a pro-angiogenic factor. Resistin stimulated, in a dose- and time-dependent manner, endothelial cell migration, proliferation, and tube formation in vitro (Mu et al. 2006). Similarly, resistin was found to stimulate smooth muscle cell proliferation, involving both extracellular signal-regulated kinase (ERK)-1/2 and PI3K pathways (Calabro et al. 2004). Whereas the results from an ex vivo model of mouse aortic arch endothelial sprouting seem to confirm a pro-angiogenic role of resistin (Robertson et al. 2009a, b), little data exists to date on the in vivo angiogenic effects of resistin. Further research is clearly needed in order to determine whether resistin may contribute to angiogenesis sustaining adipose tissue expansion in obesity.

10.5.1 Angiopoietin-Like Protein 4

Angptl-4, also known as fasting-induced adipose factor (FIAF), is as the name indicates a member of the angiopoietin-family of growth factors, synthesized and secreted in adipose tissue in response to fasting (Yoon et al. 2000; Kim et al. 2000). The 406 amino acid full-length protein contains an N-terminal coiled-coil domain, similar to adiponectin, resistin, and Tsp-1 (Carlson et al. 2008), and a C-terminal fibrinogen-like domain. The coiled-coil domain mediates the formation of disulphide-linked dimers that may further assemble into tetramers (Ge et al. 2004; Yin et al. 2009). Most of the secreted Angptl-4 is bound to the extracellular matrix, where the coiled-coil domain mediates binding to extracellular matrix components, including vitronectin and fibronectin (Goh et al. 2010b). The protein also exists in soluble, full-length dimeric or tetrameric forms. In addition, Angptl-4 fragments are generated through proteolytic cleavage, mediated by preprotein convertases, yielding oligomers containing the coiled-coil domain, and 35 kDa monomers of the fibrinogen-like domain (Ge et al. 2004; Chomel et al. 2009). Notably, this processing of full-length, oligomerized Angptl-4 is reminiscent of that of adiponectin. In the circulation, Angptl-4 exists in different forms: glycosylated, full-length oligomers or cleaved oligomers, and monomers (Mandard et al. 2004; Grootaert et al. 2012). The cellular receptors for Angptl-4 includes heparin sulfate proteoglycans, present in the cell membrane and in the extracellular matrix (Chomel et al. 2009). Further, Angptl-4 may interact with β1 and β5 integrins to mediate its cellular effects (Goh et al. 2010a).

Angptl-4 is produced in many different tissues, but the main sources are adipose tissue, liver, and placenta. Additionally, the lungs, kidneys, as well as skeletal and cardiac muscle produce Angptl-4. Levels are usually low in the intestine, where Angptl-4 expression is selectively suppressed by the gut microbiota (Backhed et al. 2004). Similar to leptin, VEGF-A, and IL-6, the levels of Angptl-4 have been found to increase in response to hypoxia (see Chap. 14) (Mazzatti et al. 2012). Indeed, increased Angptl-4 expression has been detected in ischemic skeletal muscle or heart in both mice and humans (Le Jan et al. 2003; Galaup et al. 2012). Furthermore, Angptl-4 levels increase during adipogenesis and in response to PPAR agonists

(Kersten et al. 2000). However, both adipose tissue expression and circulating Angptl-4 levels decrease with increasing BMI (Robciuc et al. 2011). Angptl-4 has been proposed as a metabolism-sensing factor mediating the cross-talk between liver, skeletal muscle, adipose tissues, and the brain, where it stimulates anorexia and energy expenditure (Kim et al. 2010). Systemically, Angptl-4 has been found to modulate glucose metabolism as well as lipid metabolism and storage in part via inhibition of LPL activity, essential for free fatty acid uptake across the endothelium and for lipogenesis.

Angptl-4 has been suggested to function as an angiostatic factor. It inhibits growth factor-induced endothelial cell migration, proliferation and tube formation in vitro and angiogenesis in vivo in the cornea angiogenesis assay as well as the matrigel plug assay in mice (Ito et al. 2003; Cazes et al. 2006; Yang et al. 2008; Chomel et al. 2009). Both the extracellular matrix-bound full-length form of Angptl-4 and the soluble, cleaved monomeric form containing the fibrinogen-like domain were found to inhibit angiogenic responses. The mechanisms seem to involve reduced cell-matrix contact formation, as well as direct suppression by Angptl-4 of the growth factor-activated ERK1/2 MAPK signaling pathway in endothelial cells (Cazes et al. 2006; Yang et al. 2008). However, context-dependent effects of Angptl-4 are reported, as under certain conditions, notably inflammation, this adipokine may instead stimulate angiogenic processes including endothelial cell migration, proliferation, and tube formation (Kim et al. 2000; Gealekman et al. 2008; Le Jan et al. 2003). It may further protect against metabolic stress-induced endothelial cell death (Kim et al. 2000). Perhaps the biological activity of Angptl-4, similar to adiponectin, depends on the oligomeric structure of the protein. Indeed, the LPL lowering activity has been shown to depend on the soluble, oligomeric, N-terminal Angptl-4 fragment species. Further, it is possible that the extracellular matrix composition surrounding the target cells as well as the cellular integrin expression profile, but also the different commercial sources of recombinant Angptl-4 protein, may influence its biological effects.

In addition to its role in angiogenesis, Angptl-4 seems to regulate vascular barrier function (Ito et al. 2003). Again, there is discrepancy in the literature as to whether Angptl-4 prevents or induces vascular permeability. However, in the context of cardiac ischemia, Angptl-4 null mice were recently found to display increased vascular leakage in a model of myocardial infarction, leading to increased cardiac edema, cardiac inflammatory response, and larger infarct sizes (Galaup et al. 2012). On the other hand, Angptl-4 therapy led to reduced vascular leakage and cardiac edema in a rabbit model of ischemia-reperfusion injury. Mechanistically, it was proposed that Angptl-4 may prevent vascular permeability by reducing VEGF-A-induced dissociation of the VEGFR-2/VE-cadherin complex involved in disruption of adherens junctions in the vascular endothelium (Galaup et al. 2012).

The role of Angptl-4 in adipose tissue angiogenesis is still unclear. Interestingly, increased intestinal expression of Angptl-4 was suggested to mediate the resistance to diet-induced obesity found in germ-free mice (Backhed et al. 2004). Similarly, over-expression of Angptl-4 in adipose tissue in mice fed a high-fat diet led to a 50 % reduction in WAT mass without altering food intake (Mandard et al. 2006).

However, although there was increased free fatty acid oxidation, Angptl-4 overexpression was found to result in hypertriglyceridemia, due to impaired plasma triglyceride clearance and impaired glucose tolerance. Conversely, Angptl-4-null mice fed a standard diet show increased body weight gain as compared to *wt* mice (Kim et al. 2010). However, Angptl-4-deficient mice were surprisingly resistant to diet-induced obesity. Again, these discrepant findings necessitate further research on the role of Angptl-4 in adipose tissue expansion.

10.6 Tumor Necrosis Factor-α

TNF-α (also called cachectin) is produced as a 233 amino acid transmembrane protein cleaved by TACE/ADAM17 into a 157 amino acid residues (17 kDa) soluble, secreted factor. Soluble TNF-α is a homotrimer that exerts its effects via type 1 (p55) and type 2 (p75) TNF-α receptors. It is a major immune response-modifying cytokine, primarily produced by activated macrophages. Further, and in addition to its central effects on food intake and its peripheral effects on insulin sensitivity (Hotamisligil et al. 1993; Romanatto et al. 2007), TNF-α plays an important local role during adipose tissue expansion. TNF-α is not only produced by adipose tissue macrophages but also by adipocytes, stromal, and vascular cells (Xu et al. 2002). Its expression is greater in subcutaneous depots as compared with visceral depots, but this may vary depending on total and regional fat mass distribution. TNF-α is known to interfere with the energy metabolism in adipocytes, leading to altered glucose and lipid metabolisms and hormone receptor signaling (Sethi and Hotamisligil 1999). For example, TNF-α exerts direct lipolytic effects on adipocytes, via activation of hormone sensitive lipase (HSL), essential for triglyceride hydrolysis. Whereas TNF-α expression in WAT increases in obesity (Li et al. 2002), the circulating levels of TNF-α are generally not concomitantly increased, arguing for a local, paracrine effect of this cytokine. In obesity, TNF-α plays an important role in the impairment of insulin signaling, and it may further contribute to adipose tissue endothelial dysfunction via stimulation of local inflammation and reduction of adiponectin expression (Donato et al. 2012).

TNF-α may either stimulate or inhibit angiogenesis depending on the context (Frater-Schroder et al. 1987; Leibovich et al. 1987): whereas TNF-α cooperates with VEGF-A and FGF-2 in stimulation of endothelial migration, it displays suppressive effects on endothelial proliferation induced by the same growth factors. This may in part be due to the activation by TNF-α of SHP-1 protein tyrosine phosphatase (PTP1C), found to negatively regulate VEGFR-2 (Guo et al. 2000). Further, chronic exposure of endothelial cells to TNF-α reduces VEGFR-2 expression levels (Patterson et al. 1996). On the other hand, TNF-α stimulate the production of angiogenic factors including VEGF-A, FGF-2, IL-8, MCP-1, and leptin (Yoshida et al. 1997; Gerhardt et al. 2001). Finally, TNF-α may induce Ang-2 and Tie2 receptor expression in endothelial cells, leading to stimulation of vascular remodeling (Chen et al. 2004).

The angiogenic or angiostatic impact of TNF-α in expanding adipose tissues may be regulated both by its own level of expression and by the local levels of production of other angiogenic regulators. Further, the effects of TNF-α are likely mediated in an indirect manner: involving either upregulation of angiogenic stimulators, leading to increased angiogenic responses, or through the pro-inflammatory and insulin resistance effects of TNF-α, linked to endothelial dysfunction and reduced angiogenic responses to growth factor stimulation.

Mice lacking functional TNF-α display improved insulin sensitivity on a high-fat diet as compared with *wt* mice, despite the fact that they show similar body weight gain, with only a tendency for a reduction in epididymal fat depot sizes (Uysal et al. 1997). Similarly, decreased production of soluble TNF-α in mice heterozygous for ADAM17 was mainly associated with protection against diet-induced insulin resistance (Serino et al. 2007). This suggests that TNF-α, in addition to its pro-inflammatory role, may contribute to adipose tissue expansion, but absence of TNF-α is not sufficient to prevent adipogenesis.

10.7 Interleukin-6

IL-6 is a 183 amino acid residue cytokine, produced in multiple glycosylated forms, ranging in size from 22 to 27 kDa. The IL-6 receptor (IL-6Rα), homologous to the leptin receptor, exists both in a membrane-bound and a soluble form. Intracellular signaling is triggered by IL-6-induced receptor dimerization followed by activation of the transmembrane signal-transducing component gp130. This leads to activation of the JAK/STAT pathway, culminating in phosphorylation of the STAT3 transcription factor and induction of target genes including acute phase proteins. Within adipose tissue, IL-6 and IL-6Rα are expressed by adipocytes and adipose tissue-resident cells, including macrophages, fibroblasts, and endothelial cells (Kershaw and Flier 2004). Expression and secretion of IL-6 are 2–3 times greater in visceral than in subcutaneous fat depots (Fried et al. 1998). In contrast to TNF-α, IL-6 levels are high in the blood (Table 10.2), and it is estimated that about 30 % of circulating IL-6 levels may originate from adipose tissue (Mohamed-Ali et al. 1997). In obesity, IL-6 production and secretion increase within adipose tissues, resulting in further elevated plasma levels. Circulating IL-6 levels are positively correlated with impaired glucose tolerance and insulin resistance in men, even after controlling for BMI, and with systolic hypertension in women (Fernandez-Real et al. 2001), and may predict the development of type 2 diabetes and cardiovascular disease. In addition to its effects on inflammation, and its deleterious impact on in insulin signaling and endothelial function, IL-6 may directly regulate adipogenesis. It reduces LPL activity leading to decreased free fatty acid uptake and reduced lipogenesis (Greenberg et al. 1992). Acute IL-6 infusion leads to increased circulating free fatty acid levels as well as altered lipoprotein profile in humans (Stouthard et al. 1995). Chronic over-expression of IL-6 in transgenic mice leads to growth retardation, including smaller fat depots, and involves decreased circulating levels of insulin-like growth factor-1 (De Benedetti et al. 1997). On the other hand mice with a targeted deletion of IL-6 develop mature-onset obesity that can be partly reversed by IL-6 replacement (Wallenius et al. 2002). The anti-obesity effect of IL-6 seems to depend on its central actions in the brain, where IL-6 stimulates energy expenditure.

IL-6 stimulates endothelial cell migration, proliferation, and tube formation and induces angiogenesis in vivo, partly via stimulation of VEGF-A expression (Cohen et al. 1996; Hernandez-Rodriguez et al. 2003). In agreement, repeated intraperitoneal IL-6 administration in mice led to increased vascular density in retroperitoneal fat depots (Rega et al. 2007). The biological actions of IL-6 may thus include stimulation of angiogenesis in adipose tissues enabling fat expansion, while centrally it exerts anti-obesity effects. Notably, this set of opposing actions by IL-6 is reminiscent of the impact of its structurally related cousin, leptin.

10.8 Transforming Growth Factor-β

Transforming Growth Factor-β (TGF-β) exists in mammals as three isoforms; TGF-β1, TGF-β2, and TGF-β3, which show partly overlapping functions. Almost all cells express receptors for TGF-β, and at least one of the isoforms is produced in all tissues. In the immune system, most cells express TGF-β1, whereas TGF-β3 may be the predominant form in adipose tissue (Miller et al. 1989). TGF-β is secreted as a preprotein of 390 or 412 amino acids, which is subsequently cleaved by the convertase family of endoproteases to generate the active form. The mature and active TGF-β protein is a 25 kDa dimer with a characteristic cystine knot structure, similar to nerve growth factor and PDGF-BB (Hinck 2012). TGF-β binds to two serine/threonine kinase receptors: TGF-βR1 and TGF-βR2. The former receptor exists as several variants, including the ubiquitously expressed Alk-5 and the endothelial-restricted Alk-1 (Lebrin et al. 2005). Endoglin also participates as a type 3 co-receptor for Alk-1. Receptor heterodimerization leads to activation of Smad transcription factors, with Smad3 being the principal signaling mediator. TGF-β regulate many aspects of cellular function including cellular adhesion, migration, proliferation, differentiation, and apoptosis (Santibanez et al. 2011). Further, TGF-βs play an important role in fibrosis. Indeed, TGF-β has been linked to several human diseases characterized by excessive fibrosis including kidney and liver and lung diseases, as well as pathological cardiac hypertrophy (Prud'homme 2007).

TGF-β is expressed by both adipocytes and stromal cells in adipose tissue, and its production as well as secretion is increased in adipose tissue of obese rodents and humans (Alessi et al. 2000). Further, TNF-α has been suggested to mediate the link between insulin resistance and the elevated levels of TGF-β seen in obesity (Samad et al. 1999).

Following initial discrepancies regarding the impact of TGF-β on angiogenesis (Muller et al. 1987; Yang and Moses 1990), it is now proposed that TGF-β may act as a regulator of endothelial cell activation state via a fine balance between Alk-5 and Alk-1 signaling. Indeed, it seems that TGF-β may either stimulate or inhibit angiogenesis depending on both its own expression levels and the expression pattern of its receptors in endothelial cells: whereas Alk-1-induced Smad1/5 activation leads to stimulation of endothelial cell migration and proliferation, Alk-5-mediated Smad2/3 activation leads to inhibition of these processes (Pepper 1997; Lebrin et al. 2005). Recently, activation of Smad2/3 was found to be essential for vessel maturation and for maintenance of vascular integrity during development (Itoh et al. 2012). Similarly, TGF-β may also in adult life play an important role in maintenance of vascular integrity in microvascular beds (Walshe et al. 2009).

The impact of TGF-β in regulation of adipose tissue angiogenesis or vascular integrity is still unclear and further efforts are warranted. Additionally, TGF-β may play a role in hypoxia-induced, HIF-1α-mediated WAT fibrosis as seen in obesity (Halberg et al. 2009; Michailidou et al. 2011 #312; Spencer et al. 2011). Finally, a role for TGF-β in local regulation of energy expenditure in adipose tissue has recently been proposed. Deficient TGF-β signaling in Smad3 mutant mice led to resistance against diet-induced obesity without altering food intake (Tan et al. 2011; Yadav et al. 2011). Interestingly, suppression of TGF-β signaling resulted in selective expansion and activation of "brown-type" adipocytes residing in WAT, associated with increased mitochondrial biogenesis and elevated basal respiration rates. Similarly, systemic blockade of TGF-β signaling by neutralizing antibodies limited adipogenesis in both diet- and genetically induced obesity models in mice (Yadav et al. 2011). Notably, it was associated with reduced WAT inflammation and with prevention of obesity-induced insulin resistance. TGF-β thus seems to play a deleterious role in obesity by potentially stimulating adipose tissue inflammation and fibrosis and suppressing local energy metabolism.

10.9 Other Adipokines

10.9.1 Angiogenin

Angiogenin, a 123 amino acid single-chain protein that belongs to the ribonuclease superfamily, is one of the first discovered angiogenic factors secreted by tumor cells (Fett et al. 1985). It displays weak ribonucleolytic activity, as compared with pancreatic ribonuclease A. However, this enzymatic activity seems to be essential for its effects on angiogenesis as well as for its other functions (Strydom et al. 1989). Angiogenin is produced and secreted by many cell types, including adipocytes and cells from the stromal vascular fraction in adipose tissues. Further, adipose tissue angiogenin expression is increased in obese mice (Kurki et al. 2012). Similarly, elevated angiogenin plasma levels have been found in overweight and obese humans (Silha et al. 2005), suggesting that this angiogenic factor may play a role in adipose tissue blood vessel growth during fat mass expansion.

It seems that angiogenin may act on several of the key steps involved in angiogenesis to stimulate blood vessel growth. Angiogenin has been found to bind to smooth muscle type a-actin on the cell surface, leading to accelerated tissue-type plasminogen activator-catalyzed cleavage of plasminogen and degradation of the vascular basement membrane. This extracellular matrix proteolysis is a crucial step in initiation of angiogenesis. Other identified extracellular binding partners of angiogenin include heparan sulfate proteoglycans, plasminogen, elastase, and angiostatin (Strydom 1998). Further, through a yet unidentified cell receptor expressed on subconfluent endothelial cells, angiogenin induces Akt signaling, leading to stimulation of angiogenesis. However, it has been suggested that angiogenin may also directly mediate some of its effects in the cell nucleus by binding to the promoter region of ribosomal DNA (rDNA) leading to stimulation of rRNA transcription (Li and Hu 2011). Although angiogenin may be an important regulator of angiogenesis, there is currently no data on the impact of angiogenin on adipose tissue angiogenesis.

10.9.2 Apelin

Apelin is a prepropeptide hormone of 77 amino acids produced by many cell types, including adipocytes, vascular stromal cells, and cardiomyocytes. During its post-translational processing several different active fragments are generated including a 36 amino acid peptide ("apelin 36"), a 17 amino acid peptide ("apelin 17"), and a 13 amino acid peptide ("apelin 13"), which seems to be the most active form of apelin. The different apelin fragments are secreted and signal through a G-protein-coupled receptor called APJ, expressed in vascular cells and cardiomyocytes but also in hypothalamic neurons. Apelins are implicated in numerous physiological processes, including central appetite control, diuresis, blood pressure control, as well as regulation of cardiac contractility in mice. Indeed, apelins are potent endothelium-dependent vasodilators and additionally, in human heart, they are among the most potent, endogenous positive inotropic agents found to date. Apelin levels are increased in obesity and diabetes, where insulin but also TNF-α stimulate apelin production (Castan-Laurell et al. 2008). Additionally, apelin is induced by hypoxia, as occurs in ischemic cardiomyopathy and in pathologically expanded adipose tissues. Apelins display anti-inflammatory properties and may further regulate angiogenesis. Apelins have been found to stimulate endothelial cell migration, proliferation, and tube formation and to stimulate angiogenesis in vivo (Kidoya and Takakura 2012). Indeed, apelins were shown to be essential for hypoxia-induced retinal angiogenesis (Kasai et al. 2010). Furthermore, siRNA-mediated suppression of apelin expression reduced angiogenic responses in a mouse model of adipose tissue transplantation (Kunduzova et al. 2008). This indicates that apelin may play a physiopathological role during vascular growth in expanding adipose tissues, especially under hypoxic conditions. However, whereas centrally injected apelin displays anorexic effects, peripheral administration of apelin was shown to reduce body weight gain and WAT expansion in mice, without altering food intake. The mechanism includes stimulation of energy expenditure, in part through upregulation of BAT uncoupling proteins (Higuchi et al. 2007).

10.9.3 Chemerin

Chemerin is a novel adipokine hormone, whose expression is increased in the adipose tissue of obese and diabetic humans (Bozaoglu et al. 2007). It signals through the G protein-coupled receptor chemokine-like receptor (CMKLR)-1, and receptor defects have been found to lead to reduced adipocyte differentiation and glucose uptake. Further, chemerin stimulates immune cell chemotaxis toward site of inflammation. On the other hand, chemerin inhibits synthesis of pro-inflammatory mediators and increases adiponectin production. Recently, chemerin was found to stimulate endothelial migration, proliferation, and tube formation in vitro (Kaur et al. 2010; Bozaoglu et al. 2010), which may suggest a role for this new adipokine in regulation of adipose tissue angiogenesis.

10.9.4 Omentin

Omentin is a 313 amino acid preprotein hormone (280 amino acid mature form) discovered in a screen for genes highly expressed in omental fat depots (Yang et al. 2006). It was subsequently found to be identical to a previously identified calcium-dependent galactose-binding lectin called intelectin-1, or soluble intestinal lactoferrin receptor. Omentin exists as two isoforms and omentin-1 appears to be the major form in human plasma. The mature protein contains an N-terminal fibrinogen-like domain and a C-terminal carbohydrate recognition domain. In humans, omentin forms glycosylated disulfide-linked homotrimers, whereas in mice it exists as a non-glycosylated 30 kDa monomer (Tsuji et al. 2007). Omentin is expressed in multiple cell types, including stromal vascular cells in visceral adipose tissues. Similar to adiponectin, omentin levels, both in adipose tissue and in the circulation, decrease with increasing BMI (de Souza Batista et al. 2007; Auguet et al. 2011). Apart from its potential role in microbe recognition and intestinal binding of lactoferrin, the effects of omentin include stimulation of insulin sensitivity in adipocytes and induction of $NO^{\cdot}$ production in endothelial cells, leading to vasodilation. While omentin has been suggested to contribute to allergic airway inflammation, in vascular endothelial cells it was found to display anti-inflammatory properties, including suppression of TNF-α-induced cyclooxygenase (COX)-2 expression (Yamawaki et al. 2011).

Recently, omentin-1 has been suggested to act as an anti-angiogenic factor, as it reduced VEGF-A-stimulated endothelial cell migration and tube formation in vitro (Tan et al. 2010). The molecular mechanisms behind the angiostatic effects of omentin may include suppression of receptor activator for NF-κB ligand (RANKL) expression (Xie et al. 2011), known for its impact on vascular permeability and role in stimulation of angiogenesis. Although in vivo data on the impact of omentin in angiogenesis models are still lacking, recent clinical data in patients with gastric cancer, showing that elevated omentin-1 levels correlate with improved prognosis (Zheng et al. 2012), are suggestive of an anti-angiogenic role of omentin.

10.9.5 Plasminogen Activator Inhibitor-1

PAI-1, a member of the serine protease inhibitor (SERPIN) family, is the primary inhibitor of fibrinolysis that inactivates urokinase-type as well as tissue-type plasminogen activator. PAI-1 is expressed by many cell types, including adipocytes. PAI-1 expression and secretion has been reported to be greater in visceral than in subcutaneous adipose tissues. In accordance, plasma PAI-1 levels are strongly associated with visceral adiposity. PAI-1 expression is induced by adipocyte differentiation (Soukas et al. 2001), as well as by TNF-α and TGF-β stimulation (Samad et al. 1999; Alessi et al. 2000). Further, the expression levels of PAI-1 increase under hypoxia. Obesity leads to increased PAI-1 expression in adipose tissues and to a correlated increase in circulating levels (Alessi et al. 2000). Furthermore, the elevated plasma levels of PAI found in obesity and insulin resistance seem to predict the risk for type 2 diabetes and cardiovascular disease (Juhan-Vague and Alessi 1997).

PAI-1 has been shown to both positively and negatively regulate angiogenesis. PAI-1 administration was found to reduce FGF-2-induced angiogenic responses in the chicken CAM in vivo assay, in part by blocking extracellular matrix vitronectin binding to $\alpha_v\beta_3$ integrins (Stefansson et al. 2001). Similarly, PAI-1 may reduce smooth muscle cell migration via competitive binding to vitronectin (Stefansson and Lawrence 1996). However, PAI-1-deficient mice were conversely found resistant to angiogenic stimulation with FGF-2 in the mouse corneal assay (Gutierrez et al. 2000). It seems that PAI-1 plays an important role in angiogenesis by indirectly regulating fibrin proteolysis in the provisional endothelial extracellular matrix: whereas excessive degradation of fibrin in the absence of PAI-1 may be deleterious for angiogenic responses, insufficient proteolysis induced by PAI-1 over-expression may likewise reduce angiogenesis by preventing endothelial cell migration and new vessel sprouting (Pepper et al. 1996). The role of PAI-1 in regulation of adipose tissue angiogenesis seems similarly variable and may depend on local expression levels as well as on mouse strain inherent differences (Lijnen 2005). Although fat-targeted over-expression of PAI-1 in implanted fat pads led to stimulation of adipogenesis, systemic over-expression resulted in reduced adipose tissue angiogenesis and smaller adipocytes (Scroyen et al. 2009). Similarly, transgenic mice over-expressing PAI-1 in adipose tissue were found to be resistant to diet-induced obesity (Lijnen et al. 2003). This was associated with reduced stromal components in the fat pad of PAI-1 over-expressing mice. On the other hand, mice with targeted deletion in PAI-1 also display resistance to both diet-induced and genetic obesity (Ma et al. 2004; Schafer et al. 2001). This was suggested to involve increased energy expenditure and improved metabolic parameters including glucose tolerance and insulin sensitivity. It seems that whereas too high PAI-1 activity prevents obesity potentially by blocking angiogenesis, too little PAI-1 also inhibits adipose tissue angiogenesis, but more importantly seem to increase energy metabolism leading to direct reduction of the drive for adipogenesis. Although PAI-1 may contribute to the development of obesity, it seems that the links are stronger between increased PAI-1 levels and development of insulin resistance or cardiovascular disease (Kershaw and Flier 2004).

10.9.6 Vaspin

Vaspin is a visceral adipose tissue-derived serpin class protein, similar to PAI-1 and PEDF. Whereas adipose tissue expression levels are increased in obesity, circulating plasma levels of vaspin remain unaltered (Auguet et al. 2011). Administration of vaspin improves glucose tolerance and insulin sensitivity in diet-induced obese mice (Hida et al. 2005). In adipose tissues, vaspin suppresses leptin, resistin, and TNF-α expression, while it stimulates adiponectin expression. Although the potential angiogenic role of this new adipokine hormone vaspin is still unknown, it was recently shown to prevent endothelial cell apoptosis through its insulin-sensitizing effects (Jung et al. 2011).

10.9.7 Visfatin

Visfatin, also known as pre-B cell colony-enhancing factor (PBEF), or nicotinamide phosphoribosyltransferase (Nampt), is a 491 amino acid enzyme highly expressed in visceral adipose tissue, but also produced in other tissues including liver. Visfatin production is increased by hypoxia and inflammation, and elevated adipose tissue expression, as well as circulating level of visfatin, is found in type 2 diabetes and obesity (Terra et al. 2012). Visfatin has, in addition to its role in the biosynthesis of nicotinamide mono- and dinucleotide, been proposed to play a role in glucose metabolism and to display pro-inflammatory effects (Jacques et al. 2012). Further, visfatin has been found to stimulate angiogenesis in the CAM assay and the mouse matrigel plug assay (Kim et al. 2007b; Adya et al. 2008). The mechanism seems to involve activation of multiple pathways in endothelial cells, including the ERK-1/2 pathway but also the Notch-1 pathway, leading to upregulation of FGF-2 expression (Kim et al. 2007b; Bae et al. 2011). Furthermore, visfatin may increase VEGF-A production, via the mammalian target of the rapamycin (mTOR) pathway (Park et al. 2011). In addition to induction of these two angiogenic growth factors, part of the stimulatory effects of visfatin on angiogenesis may be mediated through upregulation of the pro-inflammatory, pro-angiogenic cytokines IL-6 and IL-8 (Kim et al. 2009).

10.9.8 Lipid Derivatives

In addition to the adipokines, the adipose tissue produces and secretes many non-protein factors involved in paracrine or endocrine signaling, such as prostaglandins, steroid hormones, free fatty acids, and other lipid derivatives. In particular, monobutyrin (1-butyryl-glycerol), a lipid-derivative secreted by adipocytes, was found to stimulate endothelial cell migration in vitro and angiogenesis in vivo in the CAM assay (Dobson et al. 1990). It has been suggested to represent a major fraction of the total angiogenic activity of adipose tissue.

10.10 Fat Depot-Specific Differences

Fat depot-specific characteristics are apparent in the body's different fat depots, including *subcutaneous* (e.g., truncal, subscapular, inguinal), *visceral* (e.g., omental, mesenteric), *extraperitoneal* (e.g., perirenal, peripancreatic, pericardial), and *perivascular* fat depots. For example, whereas subcutaneous WAT expands via both adipocyte hypertrophy and hyperplasia, omental fat expansion seems to mainly involve adipocyte hypertrophy. Additionally, subcutaneous adipose tissue may have both larger adipocytes and higher capillary density than visceral depots (Gealekman et al. 2011). In agreement, subcutaneous adipose tissue was found to be less hypoxic as compared with visceral depots, both in lean and in obese mice (Michailidou et al. 2011). Further, the various fat depots display different adipokine production and secretion profiles: whereas HGF and FGF-2 expression levels are similar in subcutaneous and omental adipose tissues, the levels of leptin, adiponectin, and TNF-α, and Angptl-4 are highest in subcutaneous WAT adipocytes (Gealekman et al. 2011). In contrast, visceral, and especially omental, depots seem to produce larger amounts of VEGF-A, FGF-1, FGF-9, IL-6, PAI-1, visfatin, and TSP-1 (Fried et al. 1998; Gabrielsson et al. 2000; Fain et al. 2004). Further, mesenteric, intermuscular, and renal fat depots express significantly higher levels of VEGF-A and FGF-2 as compared with subcutaneous WAT (Yamada et al. 2010), and these levels correlated with adipocyte sizes. In morbidly obese humans, PAI-1 and TGF-β expression levels were reportedly similar in different fat depots, while TNF-α production was found to be increased in visceral, as compared with subcutaneous, fat depots (Alessi et al. 2000). Interestingly, whereas plasma levels of adiponectin exhibit strong negative correlation with visceral fat mass, plasma levels of another adipokine, retinol-binding protein (RBP)-4, are positively associated with visceral obesity, rather than with BMI per se (Graham et al. 2006).

Apart from adipocyte-inherent differences between fat depots, and given that the large majority of the adipokine release by adipose tissue, except for adiponectin and leptin, may actually be derived from nonfat cells, i.e., adipose stromal cells (see Chap. 5), macrophages (see Chap. 13), and vascular cells (Fain et al. 2004), the composition of the adipose tissue matrix and residual cell populations may give rise to further site-specific differences.

In addition to secretion profile differences, the anatomical location of the adipose tissue influences its endocrine function. For example, whereas adipokines produced in subcutaneous depots are secreted into the blood stream, giving rise to potentially widespread systemic effects, visceral adipose tissue-derived adipokines are secreted directly into the portal system facilitating their impact on hepatic metabolic function.

Taken together, the functional heterogeneity observed among the body's different fat depots has led to the view that adipose tissue may not simply be an endocrine organ but rather a group of similar but unique endocrine organs (Kershaw and Flier 2004). Indeed, it is known that the body fat distribution profile influences health (Matsuzawa 2008). However, although visceral obesity is clearly associated with increased risk for metabolic syndrome (Bergman et al. 2007; Matsuzawa et al. 2011), the causal molecular mechanisms are still unknown. It has been proposed that the specialized endocrine function of visceral adipose tissue, including elevated production of pro-inflammatory factors, such as IL-6, together with low expression of insulin-sensitizing factors, such as adiponectin, may play an important role in mediating the link between visceral obesity and metabolic syndrome. Furthermore, visceral fat depots may influence systemic angiogenic responses in obesity via increased release of pro-angiogenic factors, such as VEGF-A, and reduced levels of anti-angiogenic factors, such as adiponectin. It may be speculated that visceral fat-derived circulating pro-angiogenic adipokines might contribute to angiogenesis-dependant diseases including cancer. In support, visceral obesity was recently confirmed to be a significant risk factor for liver and biliary tract cancers (Schlesinger et al. 2013).

10.11 Alterations in Obesity

Pathological expansion of adipose tissue, as seen in obesity, is associated with several local cellular and molecular alterations, including a shift in adipokine expression profile with elevated adipose tissue production of VEGF-A, leptin, resistin, IL-6, TNF-α, MCP-1, PAI-1, chemerin, and vaspin, and reduced levels of adiponectin, omentin, and Angptl-4 (Li et al. 2002; Hajer et al. 2008; Inadera 2008; Robciuc et al. 2011; Northcott et al. 2012) (Fig. 10.1). Further, in obesity, the fat depot-specific differences may be reversed, as increased subcutaneous fat depot expression of IL-6, FGF-2, and visfatin and increased visceral expression of TNF-α and VEGF-C have been reported (Gealekman et al. 2011). Moreover, this altered adipokine production and secretion occurring during adipose tissue growth not only influences local tissue levels, but also modifies circulating levels. Notably, in overweight and obese subjects, multiple changes in plasma adipokines levels are observed (Hotamisligil et al. 1995; Rehman et al. 2003; Fain et al. 2004; Silha et al. 2005; Inadera 2008; Northcott et al. 2012) (Table 10.2).

As mentioned previously, many of the adipokines over-expressed in obesity display pro-inflammatory functions. Notably, during adipocyte hypertrophy there is upregulation of pro-inflammatory adipokines, including Il-6, TNF-α, MCP-1, and resistin, and down-regulation of anti-inflammatory adipokines, such as adiponectin and omentin, leading to local low level inflammation in the WAT associated with recruitment of bone marrow-derived monocytes and expansion of resident

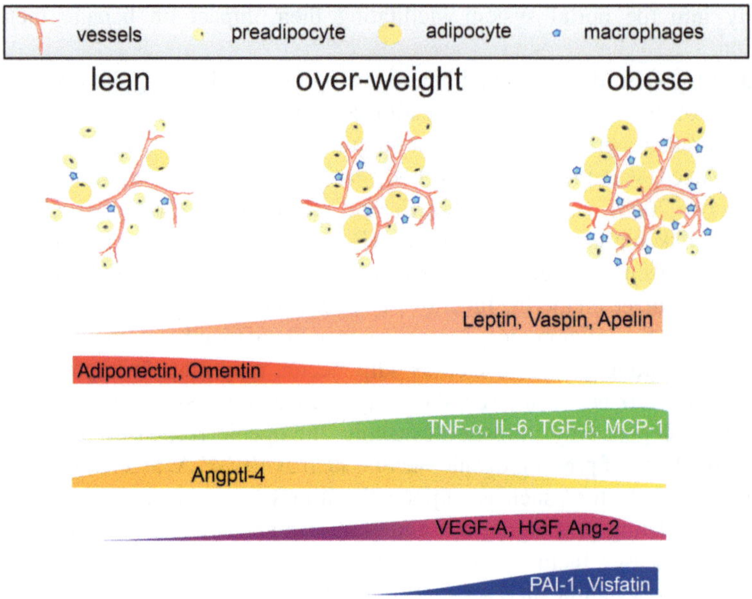

Fig. 10.1 Alterations of adipokine expression profiles in obesity. Adipokine expression changes during adipose tissue expansion and in obesity. The hypertrophic and hyperplastic adipocytes, as well as stromal cells and activated macrophages in adipose tissues, contribute to the altered production and secretion of adipose tissue-derived hormones, enzymes, cytokines as well as growth factors in overweight and obese subjects

macrophages (Weisberg et al. 2003; Suganami and Ogawa 2010). Indeed, whereas adipose tissue in the lean contains an estimated 5–10 % of macrophages, in obesity this population may increase to up to 60 % of total cell counts in WAT. These activated macrophages further secrete pro-inflammatory cytokines, e.g., TNF-α and IL-6, but also many pro-angiogenic factors (see Chap. 13), including VEGF-A, and hormones such as resistin and visfatin. Thus, the dysregulation of adipokine secretions observed in obesity is not only due to altered expression profiles in hypertrophic and hyperplastic adipocytes, but it is also influenced by the accumulation in adipose tissues of activated adipose tissue macrophages that may play a significant role (Xu et al. 2003).

Obesity is a risk factor for insulin resistance and type 2 diabetes, dyslipidemia, hypertension, and cardiovascular disease. All of these conditions are associated with microvascular alterations, including endothelial dysfunction and aberrant angiogenic responses. Whereas obesity further is a risk factor for the development of angiogenesis-dependent diseases such as cancer, type 2 diabetes has been linked to angiogenic resistance (Waltenberger 2009; Tchaikovski et al. 2009). Presumably, the deleterious vascular alterations seen in obesity and related pathologies are mediated by the changed circulating levels of adipokines, including angiogenic regulatory factors. Indeed, among the adipokines increased during obesity, both in WAT and the blood stream, are many pro-angiogenic factors, e.g., leptin, VEGF-A, HGF,

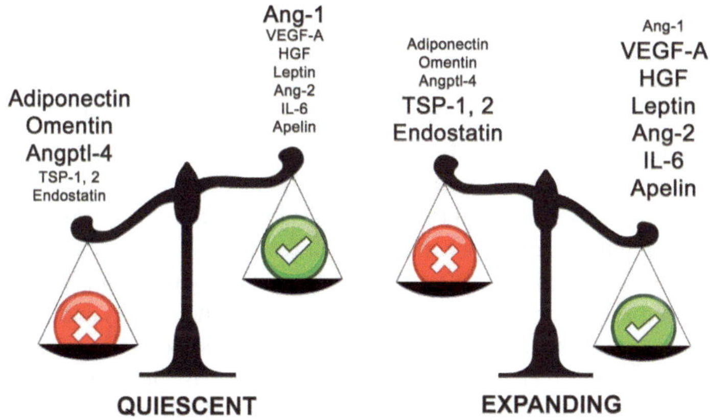

Fig. 10.2 Adipokines in regulation of angiogenesis during fat expansion. Adipokines may play important roles in regulation of adipose tissue angiogenesis. During adipose tissue expansion, the expression levels of adipokines with angiogenic stimulatory effects as well as factors with anti-angiogenic effects are altered as compared with quiescent fat depots. This may result in a local pro-angiogenic balance in expanding adipose tissues that could drive vascular growth permitting further adipogenesis

and IL-6. On the other hand, the levels of several well-known negative angiogenic regulators, including TSP-1, TSP-2, endostatin, and PEDF, may also increase during adipose tissue expansion (Silha et al. 2005; Voros et al. 2005; Varma et al. 2008; Famulla et al. 2011). In contrast, both local and circulating levels of adiponectin, Angptl-4, and omentin, suggested to function as anti-angiogenic factors, are reduced in obesity. Thus, it may be that the local balance of angiogenic regulators in obese adipose tissue is switched towards stimulation of angiogenesis, which would then favor further adipose tissue expansion (Fig. 10.2). Similarly, the balance of circulating angiogenic regulators in obese humans may also be tilted towards a more pro-angiogenic state, leading to stimulation of angiogenesis in distant organs. However, it is likely that the endothelial dysfunction occurring in obesity (Barton et al. 2012) may lead to local or generalized reductions in the angiogenic responsiveness of the endothelium to growth factors and other stimulators. This angiogenic "resistance" would result in impaired wound healing, increased vulnerability to ischemic injury, as well as cardiac dysfunction via reductions in ischemia-induced capillary growth and/or arterial remodeling in obese subjects. Furthermore, it may lead to reduced angiogenic responses in obese adipose tissues. In agreement, subcutaneous WAT capillary density and angiogenic capacity were recently found to correlate negatively with insulin sensitivity and to decrease with morbid obesity (Gealekman et al. 2011; Spencer et al. 2011). It seems thus that although adipose tissue expansion requires stimulation of angiogenesis and accordingly high levels of pro-angiogenic factors are produced during adipogenesis, there may be insufficient angiogenesis in obese WAT. Indeed, pathologically expanded adipose tissue, as seen in obesity, is hypoperfused and hypoxic (Hosogai et al. 2007; Michailidou et al. 2011). The adipose tissue hypoxia may not only result in altered secretion of adipokines and increased local inflammation, but it may also limit further adipose tissue expansion. This leads to the hypothesis that therapeutic stimulation of angiogenesis in obese adipose tissues may result in a reduction of local inflammation *and* in an increase in adipogenesis. In agreement, it was recently reported that targeted VEGF-A overexpression in WAT during high-fat diet-induced obesity led to further stimulation of adipose tissue angiogenesis. Notably, this was associated with prevention of local hypoxia-driven inflammation in the fat depot, including reduced pro-inflammatory and increased anti-inflammatory macrophage recruitment (Elias et al. 2012). This normalization of adipose tissue was found to protect against diet-induced insulin resistance in mice, likely due in part to prevention of the obesity-induced deleterious adipokine secretion profile. However, increasing adipose tissue vascularity in the fat depots did not increase their size, but rather was found to prevent fat expansion despite similar food intake. It seems that the VEGF-A-induced increase in adipose tissue angiogenesis resulted in increased thermogenic capacity and energy expenditure. Similarly, increased adipose tissue vascularity has been shown to be essential for cold-induced non-shivering thermogenesis (see Chap. 7) (Xue et al. 2009).

It is by now acknowledged that macrophages and other immune cells play an important role, not limited to VEGF-A production, in angiogenesis. It is likely that the adipose tissue-resident macrophage population influences adipose tissue angiogenesis via production of pro-angiogenic adipokines. In support of this notion, it was recently described that macrophage depletion led to a reduction in adipose tissue vascular density in lean *wt* mice (Xu et al. 2012). In addition to its role as a direct angiogenic inhibitor, TSP-1 also modulates macrophage infiltration. In agreement, mice lacking TSP-1 showed reduced macrophage accumulation in adipose tissue during diet-induced obesity (Li et al. 2011). However, this did not result in altered adiposity as compared with *wt* mice fed high-fat diet. This observation suggests that while macrophages play an important role in adipose tissue angiogenesis (Xu et al. 2012), they are not strictly necessary, for adipose tissue expansion during development of obesity. Nevertheless, reduction of macrophage infiltration in expanding adipose tissues in TSP-1-deficient mice importantly led to protection against obesity-related systemic inflammation and insulin resistance (Li et al. 2011), likely by modulating circulating adipokine levels.

In summary, during initial adipose tissue growth, the adipokines secreted locally may tip the balance in favor of adipose tissue angiogenesis. This leads to vascular growth coordinated with the expansion of fat tissue mass, thus enabling physiological adipogenesis (Fig. 10.3). In contrast, during extensive adipose tissue growth as seen in overweight and obese subjects, although the local angiogenic balance may be further tilted towards a pro-angiogenic state, there seems to be an insufficient angiogenic response. Together with the potential endothelial dysfunction of adipose tissue vasculatures occurring in obesity, this leads to a mismatch between the adipose tissue perfusion and the metabolic needs of hypertrophic adipocytes and hypoxia ensues. This microenvironmental change in WAT found during development of obesity, associated with local low level tissue inflammation, contributes to the dysregulation of adipokine expression profiles with further increase in the

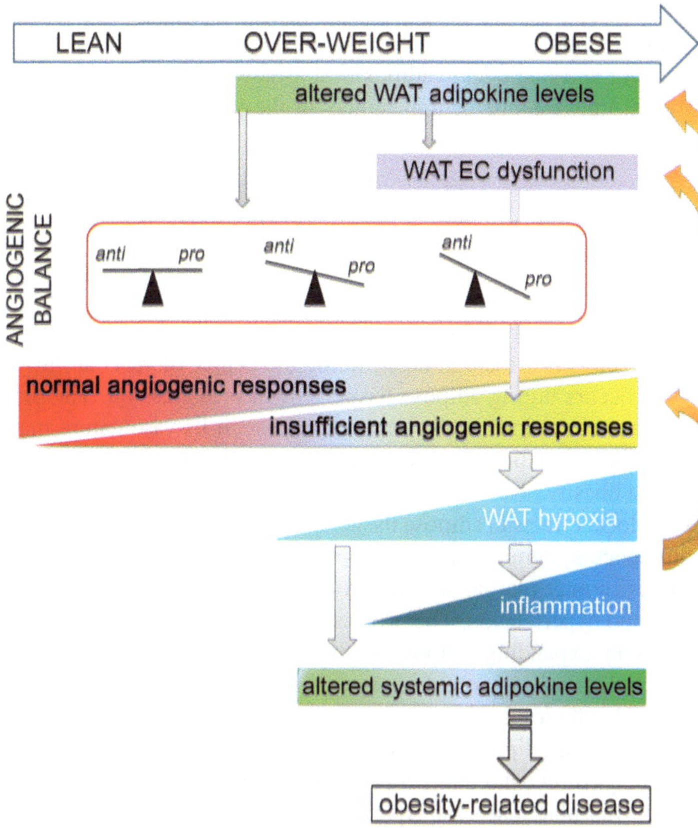

Fig. 10.3 Angiogenic responses in white adipose tissue during fat expansion. During physiological adipose tissue expansion, pro-angiogenic adipokine ("pro") levels increase out of proportion of anti-angiogenic adipokines ("anti"), tipping the local balance towards stimulation of angiogenesis, enabling expansion of healthy, well-vascularized, and sufficiently perfused fat depots. However, during excessive adipose tissue growth, as seen in severe overweight and obesity, the adipose tissue vasculature may become dysfunctional and the local angiogenic responses insufficient to prevent adipose tissue hypoxia. Together with the hypoxia-induced inflammatory response in the fat depot, this leads to pathological modifications in adipokine expression profiles. While this altered adipokine production locally exacerbate endothelial dysfunction and deficient angiogenic responses in WAT, in the blood circulation these obese-profile adipokines exert detrimental effects on insulin sensitivity and cardiovascular function, leading to the development of obesity-related diseases

expression of pro-inflammatory adipokines. This leads to a local vicious cycle, and to an increasingly dysfunctional adipose tissue vasculature. Additionally, when the circulatory levels of these adipokines are altered, as observed in overweight and obese subjects (Trayhurn and Wood 2004), the development of obesity-related diseases ensues, including insulin resistance, hypertension, and atherosclerosis. From this model, it is clear that regulation of angiogenesis not merely plays an essential role in adipose tissue growth, but that it is intimately involved in establishing the pathological phenotype of massively expanded adipose tissue, as seen in obesity, associated with a deleterious systemic impact through the secretion of pathogenic adipokines. Pro- or anti-angiogenic adipokines, as key orchestrators of adipose tissue angiogenesis, may thus be central targets in the pursuit of novel anti-obesity treatments.

10.12 Conclusion

The spatiotemporal interplay between adipose tissue-derived pro-angiogenic and anti-angiogenic factors is critical in determining the degree of vascular growth and remodeling that occurs during adipose tissue growth or regression. The discovery that adipose tissue expansion can be controlled through its vasculature makes the subject of adipokine regulation of angiogenesis a central research field in the search for a cure against obesity.

Beyond their local effects in adipose tissue, it is well appreciated that the secreted adipokines influence systemic functions. Moreover, alterations in adipokine levels observed in obesity may play a causative role in obesity-related diseases such as type 2 diabetes and hypertension. The prospect of therapeutic modulation of adipokine production in expanding adipose tissues, to favor the expression of anti-inflammatory cytokines, is a particularly appealing one as it may enable us to limit obesity-associated pathologies.

Among the different adipokines, adiponectin is of particular interest. Notably, adiponectin displays direct and indirect protective effects against both insulin resistance and cardiovascular diseases. Apart from its beneficial effects on glucose metabolism and antioxidant defenses, adiponectin is anti-inflammatory. As mentioned previously, adiponectin treatment reduces adipose tissue inflammation in obesity. As inflammatory macrophages play a key role in the altered secretion, found in obesity, of deleterious adipokines, adiponectin may also indirectly limit obesity-linked adipokine-mediated diseases including metabolic syndrome. Indeed, adiponectin administration has been suggested as a therapy for prevention of obesity-related diseases. Furthermore, in view of the anti-angiogenic effects of adiponectin, including sequestration of angiogenic factors, it may be speculated that the elevated levels of adiponectin found in lean adipose tissue may limit initial adipose tissue expansion by preventing vascular growth and remodeling. Subsequently, the decreased levels of adiponectin found in obesity may contribute to the switch in the adipose tissue from an anti-angiogenic to a pro-angiogenic phenotype during fat expansion. Interestingly, analogous to adiponectin, the expression of two other potential adipose tissue-derived angiogenic inhibitors, Angptl-4 and omentin, are also decreased in obesity. These adipokines may thus also be considered for therapies aimed at anti-angiogenic prevention of obesity.

On the other hand, although adipose tissue growth is associated with increased production of many pro-angiogenic factors, and decreased levels of several anti-angiogenic factors, it seems that adipose tissue is hypoxic, both in stages of overweight and obesity. Therefore, insufficient angiogenic tissue responses seem to be at the heart of the pathological phenotype of the adipose tissues observed in obesity and may constitute the link between excessive adipose tissue growth and adipokine dysregulation. The development of tools that correct the seeming angiogenic "resistance" of obese adipose tissues could represent a new parallel venture in the pursuit of normalization of adipokine expression patterns in the obese in order to prevent obesity-associated pathologies.

References

Adya R, Tan BK, Chen J, et al. Visfatin and endothelial angiogenesis. Cardiovasc Res. 2008;78(2):356–65.

Alessi MC, Bastelica D, Morange P, et al. Plasminogen activator inhibitor 1, transforming growth factor-beta1, and BMI are closely associated in human adipose tissue during morbid obesity. Diabetes. 2000;49(8):1374–80.

Arita Y, Kihara S, Ouchi N, et al. Adipocyte-derived plasma protein adiponectin acts as a platelet-derived growth factor-BB-binding protein and regulates growth factor-induced common post-receptor signal in vascular smooth muscle cell. Circulation. 2002;105(24):2893–8.

Arita Y, Kihara S, Ouchi N, et al. Paradoxical decrease of an adipose-specific protein, adiponectin, in obesity. Biochem Biophys Res Commun. 1999;257(1):79–83.

Auguet T, Quintero Y, Riesco D, et al. New adipokines vaspin and omentin. Circulating levels and gene expression in adipose tissue from morbidly obese women. BMC Med Genet. 2011;12:60.

Backhed F, Ding H, Wang T, et al. The gut microbiota as an environmental factor that regulates fat storage. Proc Natl Acad Sci U S A. 2004;101(44):15718–23.

Bae YH, Park HJ, Kim SR, et al. Notch1 mediates visfatin-induced FGF-2 up-regulation and endothelial angiogenesis. Cardiovasc Res. 2011;89(2):436–45.

Barton M, Baretella O, Meyer MR. Obesity and risk of vascular disease: importance of endothelium-dependent vasoconstriction. Br J Pharmacol. 2012;165(3):591–602.

Bell LN, Cai L, Johnstone BH, et al. A central role for hepatocyte growth factor in adipose tissue angiogenesis. Am J Physiol Endocrinol Metab. 2008;294(2):E336–44.

Bergman RN, Kim SP, Hsu IR, et al. Abdominal obesity: role in the pathophysiology of metabolic disease and cardiovascular risk. Am J Med. 2007;120(2 Suppl 1):S3–8; discussion S29-32.

Bora PS, Kaliappan S, Lyzogubov VV, et al. Expression of adiponectin in choroidal tissue and inhibition of laser induced choroidal neovascularization by adiponectin. FEBS Lett. 2007;581(10):1977–82.

Bouloumie A, Drexler HC, Lafontan M, et al. Leptin, the product of Ob gene, promotes angiogenesis. Circ Res. 1998;83(10):1059–66.

Bozaoglu K, Bolton K, McMillan J, et al. Chemerin is a novel adipokine associated with obesity and metabolic syndrome. Endocrinology. 2007;148(10):4687–94.

Bozaoglu K, Curran JE, Stocker CJ, et al. Chemerin, a novel adipokine in the regulation of angiogenesis. J Clin Endocrinol Metab. 2010;95(5):2476–85.

Brakenhielm E, Cao R, Gao B, et al. Angiogenesis inhibitor, TNP-470, prevents diet-induced and genetic obesity in mice. Circ Res. 2004a;94(12):1579–88.

Brakenhielm E, Veitonmaki N, Cao R, et al. Adiponectin-induced antiangiogenesis and antitumor activity involve caspase-mediated endothelial cell apoptosis. Proc Natl Acad Sci U S A. 2004b;101(8):2476–81.

Calabro P, Samudio I, Willerson JT, et al. Resistin promotes smooth muscle cell proliferation through activation of extracellular signal-regulated kinase 1/2 and phosphatidylinositol 3-kinase pathways. Circulation. 2004;110(21):3335–40.

Campfield LA, Smith FJ, Guisez Y, et al. Recombinant mouse OB protein: evidence for a peripheral signal linking adiposity and central neural networks. Science. 1995;269(5223):546–9.

Cao R, Brakenhielm E, Wahlestedt C, et al. Leptin induces vascular permeability and synergistically stimulates angiogenesis with FGF-2 and VEGF. Proc Natl Acad Sci U S A. 2001;98(11):6390–5.

Cao Y. Angiogenesis modulates adipogenesis and obesity. J Clin Invest. 2007;117(9):2362–8.

Carlson CB, Lawler J, Mosher DF. Structures of thrombospondins. Cell Mol Life Sci. 2008; 65(5):672–86.

Castan-Laurell I, Vitkova M, Daviaud D, et al. Effect of hypocaloric diet-induced weight loss in obese women on plasma apelin and adipose tissue expression of apelin and APJ. Eur J Endocrinol. 2008;158(6):905–10.

Cazes A, Galaup A, Chomel C, et al. Extracellular matrix-bound angiopoietin-like 4 inhibits endothelial cell adhesion, migration, and sprouting and alters actin cytoskeleton. Circ Res. 2006;99(11):1207–15.

Chen BH, Song Y, Ding EL, et al. Circulating levels of resistin and risk of type 2 diabetes in men and women: results from two prospective cohorts. Diabetes Care. 2009;32(2):329–34.

Chen JX, Chen Y, DeBusk L, et al. Dual functional roles of Tie-2/angiopoietin in TNF-alpha-mediated angiogenesis. Am J Physiol Heart Circ Physiol. 2004;287(1):H187–95.

Chomel C, Cazes A, Faye C, et al. Interaction of the coiled-coil domain with glycosaminoglycans protects angiopoietin-like 4 from proteolysis and regulates its antiangiogenic activity. FASEB J. 2009;23(3):940–9.

Cohen B, Barkan D, Levy Y, et al. Leptin induces angiopoietin-2 expression in adipose tissues. J Biol Chem. 2001;276(11):7697–700.

Cohen T, Nahari D, Cerem LW, et al. Interleukin 6 induces the expression of vascular endothelial growth factor. J Biol Chem. 1996;271(2):736–41.

Conde J, Scotece M, Gomez R, et al. At the crossroad between immunity and metabolism: focus on leptin. Expert Rev Clin Immunol. 2010;6(5):801–8.

Daquinag AC, Zhang Y, Amaya-Manzanares F, et al. An isoform of decorin is a resistin receptor on the surface of adipose progenitor cells. Cell Stem Cell. 2011;9(1):74–86.

De Benedetti F, Alonzi T, Moretta A, et al. Interleukin 6 causes growth impairment in transgenic mice through a decrease in insulin-like growth factor-I. A model for stunted growth in children with chronic inflammation J Clin Invest. 1997;99(4):643–50.

de Souza Batista CM, Yang RZ, Lee MJ, et al. Omentin plasma levels and gene expression are decreased in obesity. Diabetes. 2007;56(6):1655–61.

Deng Y, Scherer PE. Adipokines as novel biomarkers and regulators of the metabolic syndrome. Ann N Y Acad Sci. 2011;1212:E1–19.

Dobson DE, Kambe A, Block E, et al. 1-Butyryl-glycerol: a novel angiogenesis factor secreted by differentiating adipocytes. Cell. 1990;61(2):223–30.

Donato AJ, Henson GD, Morgan RG, et al. TNF-alpha impairs endothelial function in adipose tissue resistance arteries of mice with diet induced obesity. Am J Physiol Heart Circ Physiol. 2012;303(6):H672–9.

El-Menyar A, Rizk N, Al Nabti AD, et al. Total and high molecular weight adiponectin in patients with coronary artery disease. J Cardiovasc Med (Hagerstown). 2009;10(4):310–5.

Elias I, Franckhauser S, Ferre T, et al. Adipose tissue overexpression of vascular endothelial growth factor protects against diet-induced obesity and insulin resistance. Diabetes. 2012; 61(7):1801–13.

Fain JN, Madan AK, Hiler ML, et al. Comparison of the release of adipokines by adipose tissue, adipose tissue matrix, and adipocytes from visceral and subcutaneous abdominal adipose tissues of obese humans. Endocrinology. 2004;145(5):2273–82.

Famulla S, Horrighs A, Cramer A, et al. Hypoxia reduces the response of human adipocytes towards TNFalpha resulting in reduced NF-kappaB signaling and MCP-1 secretion. Int J Obes (Lond). 2011;36(7):986–92.

Fernandez-Real JM, Vayreda M, Richart C, et al. Circulating interleukin 6 levels, blood pressure, and insulin sensitivity in apparently healthy men and women. J Clin Endocrinol Metab. 2001;86(3):1154–9.
Fett JW, Strydom DJ, Lobb RR, et al. Isolation and characterization of angiogenin, an angiogenic protein from human carcinoma cells. Biochemistry. 1985;24(20):5480–6.
Fisslthaler B, Fleming I. Activation and signaling by the AMP-activated protein kinase in endothelial cells. Circ Res. 2009;105(2):114–27.
Frater-Schroder M, Risau W, Hallmann R, et al. Tumor necrosis factor type alpha, a potent inhibitor of endothelial cell growth in vitro, is angiogenic in vivo. Proc Natl Acad Sci U S A. 1987;84(15):5277–81.
Fried SK, Bunkin DA, Greenberg AS. Omental and subcutaneous adipose tissues of obese subjects release interleukin-6: depot difference and regulation by glucocorticoid. J Clin Endocrinol Metab. 1998;83(3):847–50.
Friedman JM, Halaas JL. Leptin and the regulation of body weight in mammals. Nature. 1998; 395(6704):763–70.
Fruebis J, Tsao TS, Javorschi S, et al. Proteolytic cleavage product of 30-kDa adipocyte complement-related protein increases fatty acid oxidation in muscle and causes weight loss in mice. Proc Natl Acad Sci U S A. 2001;98(4):2005–10.
Fukumura D, Ushiyama A, Duda DG, et al. Paracrine regulation of angiogenesis and adipocyte differentiation during in vivo adipogenesis. Circ Res. 2003;93(9):e88–97.
Funahashi T, Matsuzawa Y. Metabolic syndrome: clinical concept and molecular basis. Ann Med. 2007;39(7):482–94.
Gabrielsson BL, Carlsson B, Carlsson LM. Partial genome scale analysis of gene expression in human adipose tissue using DNA array. Obes Res. 2000;8(5):374–84.
Galaup A, Gomez E, Souktani R, et al. Protection against myocardial infarction and no-reflow through preservation of vascular integrity by angiopoietin-like 4. Circulation. 2012; 125(1):140–9.
Ge H, Yang G, Huang L, et al. Oligomerization and regulated proteolytic processing of angiopoietin-like protein 4. J Biol Chem. 2004;279(3):2038–45.
Gealekman O, Burkart A, Chouinard M, et al. Enhanced angiogenesis in obesity and in response to PPARgamma activators through adipocyte VEGF and ANGPTL4 production. Am J Physiol Endocrinol Metab. 2008;295(5):E1056–64.
Gealekman O, Guseva N, Hartigan C, et al. Depot-specific differences and insufficient subcutaneous adipose tissue angiogenesis in human obesity. Circulation. 2011;123(2):186–94.
Gerhardt CC, Romero IA, Cancello R, et al. Chemokines control fat accumulation and leptin secretion by cultured human adipocytes. Mol Cell Endocrinol. 2001;175(1–2):81–92.
Goh YY, Pal M, Chong HC, et al. Angiopoietin-like 4 interacts with integrins beta1 and beta5 to modulate keratinocyte migration. Am J Pathol. 2010a;177(6):2791–803.
Goh YY, Pal M, Chong HC, et al. Angiopoietin-like 4 interacts with matrix proteins to modulate wound healing. J Biol Chem. 2010b;285(43):32999–3009.
Graham TE, Yang Q, Bluher M, et al. Retinol-binding protein 4 and insulin resistance in lean, obese, and diabetic subjects. N Engl J Med. 2006;354(24):2552–63.
Greenberg AS, Nordan RP, McIntosh J, et al. Interleukin 6 reduces lipoprotein lipase activity in adipose tissue of mice in vivo and in 3T3-L1 adipocytes: a possible role for interleukin 6 in cancer cachexia. Cancer Res. 1992;52(15):4113–6.
Grootaert C, Van de Wiele T, Verstraete W, et al. Angiopoietin-like protein 4: health effects, modulating agents and structure-function relationships. Expert Rev Proteomics. 2012;9(2):181–99.
Grosfeld A, Zilberfarb V, Turban S, et al. Hypoxia increases leptin expression in human PAZ6 adipose cells. Diabetologia. 2002;45(4):527–30.
Gualillo O, Gonzalez-Juanatey JR, Lago F. The emerging role of adipokines as mediators of cardiovascular function: physiologic and clinical perspectives. Trends Cardiovasc Med. 2007;17(8):275–83.

Guo D, Chien S, Shyy JY. Regulation of endothelial cell cycle by laminar versus oscillatory flow: distinct modes of interactions of AMP-activated protein kinase and Akt pathways. Circ Res. 2007;100(4):564–71.
Guo DQ, Wu LW, Dunbar JD, et al. Tumor necrosis factor employs a protein-tyrosine phosphatase to inhibit activation of KDR and vascular endothelial cell growth factor-induced endothelial cell proliferation. J Biol Chem. 2000;275(15):11216–21.
Guo X, Liao K. Analysis of gene expression profile during 3T3-L1 preadipocyte differentiation. Gene. 2000;251(1):45–53.
Gutierrez LS, Schulman A, Brito-Robinson T, et al. Tumor development is retarded in mice lacking the gene for urokinase-type plasminogen activator or its inhibitor, plasminogen activator inhibitor-1. Cancer Res. 2000;60(20):5839–47.
Hajer GR, van Haeften TW, Visseren FL. Adipose tissue dysfunction in obesity, diabetes, and vascular diseases. Eur Heart J. 2008;29(24):2959–71.
Halaas JL, Gajiwala KS, Maffei M, et al. Weight-reducing effects of the plasma protein encoded by the obese gene. Science. 1995;269(5223):543–6.
Halberg N, Khan T, Trujillo ME, et al. Hypoxia-inducible factor 1alpha induces fibrosis and insulin resistance in white adipose tissue. Mol Cell Biol. 2009;29(16):4467–83.
Heida NM, Leifheit-Nestler M, Schroeter MR, et al. Leptin enhances the potency of circulating angiogenic cells via src kinase and integrin (alpha)vbeta5: implications for angiogenesis in human obesity. Arterioscler Thromb Vasc Biol. 2010;30(2):200–6.
Hernandez-Rodriguez J, Segarra M, Vilardell C, et al. Elevated production of interleukin-6 is associated with a lower incidence of disease-related ischemic events in patients with giant-cell arteritis: angiogenic activity of interleukin-6 as a potential protective mechanism. Circulation. 2003;107(19):2428–34.
Herse F, Fain JN, Janke J, et al. Adipose tissue-derived soluble fms-like tyrosine kinase 1 is an obesity-relevant endogenous paracrine adipokine. Hypertension. 2011;58(1):37–42.
Hida K, Wada J, Eguchi J, et al. Visceral adipose tissue-derived serine protease inhibitor: a unique insulin-sensitizing adipocytokine in obesity. Proc Natl Acad Sci U S A. 2005;102(30): 10610–5.
Higuchi K, Masaki T, Gotoh K, et al. Apelin, an APJ receptor ligand, regulates body adiposity and favors the messenger ribonucleic acid expression of uncoupling proteins in mice. Endocrinology. 2007;148(6):2690–7.
Hinck AP. Structural studies of the TGF-betas and their receptors - insights into evolution of the TGF-beta superfamily. FEBS Lett. 2012;586(14):1860–70.
Hosogai N, Fukuhara A, Oshima K, et al. Adipose tissue hypoxia in obesity and its impact on adipocytokine dysregulation. Diabetes. 2007;56(4):901–11.
Hotamisligil GS, Arner P, Caro JF, et al. Increased adipose tissue expression of tumor necrosis factor-alpha in human obesity and insulin resistance. J Clin Invest. 1995;95(5):2409–15.
Hotamisligil GS, Shargill NS, Spiegelman BM. Adipose expression of tumor necrosis factor-alpha: direct role in obesity-linked insulin resistance. Science. 1993;259(5091):87–91.
Hug C, Wang J, Ahmad NS, et al. T-cadherin is a receptor for hexameric and high-molecular-weight forms of Acrp30/adiponectin. Proc Natl Acad Sci U S A. 2004;101(28):10308–13.
Inadera H. The usefulness of circulating adipokine levels for the assessment of obesity-related health problems. Int J Med Sci. 2008;5(5):248–62.
Ito Y, Oike Y, Yasunaga K, et al. Inhibition of angiogenesis and vascular leakiness by angiopoietin-related protein 4. Cancer Res. 2003;63(20):6651–7.
Itoh F, Itoh S, Adachi T, et al. Smad2/Smad3 in endothelium is indispensable for vascular stability via S1PR1 and N-cadherin expressions. Blood. 2012;119(22):5320–8.
Jacques C, Holzenberger M, Mladenovic Z, et al. Proinflammatory actions of visfatin/nicotinamide phosphoribosyltransferase (Nampt) involve regulation of insulin signaling pathway and Nampt enzymatic activity. J Biol Chem. 2012;287(18):15100–8.
Jamaluddin MS, Weakley SM, Yao Q, et al. Resistin: functional roles and therapeutic considerations for cardiovascular disease. Br J Pharmacol. 2012;165(3):622–32.

Juhan-Vague I, Alessi MC. PAI-1, obesity, insulin resistance and risk of cardiovascular events. Thromb Haemost. 1997;78(1):656–60.

Jung CH, Lee WJ, Hwang JY, et al. Vaspin protects vascular endothelial cells against free fatty acid-induced apoptosis through a phosphatidylinositol 3-kinase/Akt pathway. Biochem Biophys Res Commun. 2011;413(2):264–9.

Kadowaki T, Yamauchi T. Adiponectin and adiponectin receptors. Endocr Rev. 2005;26(3): 439–51.

Kamba T, Tam BY, Hashizume H, et al. VEGF-dependent plasticity of fenestrated capillaries in the normal adult microvasculature. Am J Physiol Heart Circ Physiol. 2006;290(2):H560–76.

Kappes A, Loffler G. Influences of ionomycin, dibutyryl-cycloAMP and tumour necrosis factor-alpha on intracellular amount and secretion of apM1 in differentiating primary human preadipocytes. Horm Metab Res. 2000;32(11–12):548–54.

Kasai A, Ishimaru Y, Kinjo T, et al. Apelin is a crucial factor for hypoxia-induced retinal angiogenesis. Arterioscler Thromb Vasc Biol. 2010;30(11):2182–7.

Kaur J, Adya R, Tan BK, et al. Identification of chemerin receptor (ChemR23) in human endothelial cells: chemerin-induced endothelial angiogenesis. Biochem Biophys Res Commun. 2010;391(4):1762–8.

Kershaw EE, Flier JS. Adipose tissue as an endocrine organ. J Clin Endocrinol Metab. 2004; 89(6):2548–56.

Kersten S, Mandard S, Tan NS, et al. Characterization of the fasting-induced adipose factor FIAF, a novel peroxisome proliferator-activated receptor target gene. J Biol Chem. 2000; 275(37):28488–93.

Kidoya H, Takakura N. Biology of the apelin-APJ axis in vascular formation. J Biochem. 2012;152(2):125–31.

Kim HK, Youn BS, Shin MS, et al. Hypothalamic Angptl4/Fiaf is a novel regulator of food intake and body weight. Diabetes. 2010;59(11):2772–80.

Kim I, Kim HG, Kim H, et al. Hepatic expression, synthesis and secretion of a novel fibrinogen/angiopoietin-related protein that prevents endothelial-cell apoptosis. Biochem J. 2000;346(Pt 3):603–10.

Kim JY, Bae YH, Bae MK, et al. Visfatin through STAT3 activation enhances IL-6 expression that promotes endothelial angiogenesis. Biochim Biophys Acta. 2009;1793(11):1759–67.

Kim JY, van de Wall E, Laplante M, et al. Obesity-associated improvements in metabolic profile through expansion of adipose tissue. J Clin Invest. 2007a;117(9):2621–37.

Kim SR, Bae SK, Choi KS, et al. Visfatin promotes angiogenesis by activation of extracellular signal-regulated kinase 1/2. Biochem Biophys Res Commun. 2007b;357(1):150–6.

Kobayashi H, Ouchi N, Kihara S, et al. Selective suppression of endothelial cell apoptosis by the high molecular weight form of adiponectin. Circ Res. 2004;94(4):e27–31.

Kumada M, Kihara S, Sumitsuji S, et al. Association of hypoadiponectinemia with coronary artery disease in men. Arterioscler Thromb Vasc Biol. 2003;23(1):85–9.

Kunduzova O, Alet N, Delesque-Touchard N, et al. Apelin/APJ signaling system: a potential link between adipose tissue and endothelial angiogenic processes. FASEB J. 2008;22(12): 4146–53.

Kurki EM, Shi J, Martonen E, et al. Distinct effects of calorie restriction on adipose tissue cytokine and angiogenesis profiles in obese and lean mice. Nutr Metab (Lond). 2012;9(1):64.

Lago F, Dieguez C, Gomez-Reino J, et al. Adipokines as emerging mediators of immune response and inflammation. Nat Clin Pract Rheumatol. 2007;3(12):716–24.

Le Jan S, Amy C, Cazes A, et al. Angiopoietin-like 4 is a proangiogenic factor produced during ischemia and in conventional renal cell carcinoma. Am J Pathol. 2003;162(5):1521–8.

Lebrin F, Deckers M, Bertolino P, et al. TGF-beta receptor function in the endothelium. Cardiovasc Res. 2005;65(3):599–608.

Lee JH, Chan JL, Yiannakouris N, et al. Circulating resistin levels are not associated with obesity or insulin resistance in humans and are not regulated by fasting or leptin administration: cross-sectional and interventional studies in normal, insulin-resistant, and diabetic subjects. J Clin Endocrinol Metab. 2003a;88(10):4848–56.

Lee M, Hwang JT, Lee HJ, et al. AMP-activated protein kinase activity is critical for hypoxia-inducible factor-1 transcriptional activity and its target gene expression under hypoxic conditions in DU145 cells. J Biol Chem. 2003b;278(41):39653–61.

Leibovich SJ, Polverini PJ, Shepard HM, et al. Macrophage-induced angiogenesis is mediated by tumour necrosis factor-alpha. Nature. 1987;329(6140):630–2.

Li J, Yu X, Pan W, et al. Gene expression profile of rat adipose tissue at the onset of high-fat-diet obesity. Am J Physiol Endocrinol Metab. 2002;282(6):E1334–41.

Li S, Hu GF. Angiogenin-mediated rRNA transcription in cancer and neurodegeneration. Int J Biochem Mol Biol. 2011;1(1):26–35.

Li Y, Tong X, Rumala C, et al. Thrombospondin1 deficiency reduces obesity-associated inflammation and improves insulin sensitivity in a diet-induced obese mouse model. PLoS One. 2011;6(10):e26656.

Lijnen HR. Effect of plasminogen activator inhibitor-1 deficiency on nutritionally-induced obesity in mice. Thromb Haemost. 2005;93(5):816–9.

Lijnen HR, Maquoi E, Morange P, et al. Nutritionally induced obesity is attenuated in transgenic mice overexpressing plasminogen activator inhibitor-1. Arterioscler Thromb Vasc Biol. 2003;23(1):78–84.

Liu M, Zhou L, Xu A, et al. A disulfide-bond A oxidoreductase-like protein (DsbA-L) regulates adiponectin multimerization. Proc Natl Acad Sci U S A. 2008;105(47):18302–7.

Ma LJ, Mao SL, Taylor KL, et al. Prevention of obesity and insulin resistance in mice lacking plasminogen activator inhibitor 1. Diabetes. 2004;53(2):336–46.

Maeda N, Shimomura I, Kishida K, et al. Diet-induced insulin resistance in mice lacking adiponectin/ACRP30. Nat Med. 2002;8(7):731–7.

Maenhaut N, Van de Voorde J. Regulation of vascular tone by adipocytes. BMC Med. 2011;9:25.

Mahadev K, Wu X, Donnelly S, et al. Adiponectin inhibits vascular endothelial growth factor-induced migration of human coronary artery endothelial cells. Cardiovasc Res. 2008; 78(2):376–84.

Man K, Ng KT, Xu A, et al. Suppression of liver tumor growth and metastasis by adiponectin in nude mice through inhibition of tumor angiogenesis and downregulation of Rho kinase/IFN-inducible protein 10/matrix metalloproteinase 9 signaling. Clin Cancer Res. 2010;16(3): 967–77.

Mandard S, Zandbergen F, Tan NS, et al. The direct peroxisome proliferator-activated receptor target fasting-induced adipose factor (FIAF/PGAR/ANGPTL4) is present in blood plasma as a truncated protein that is increased by fenofibrate treatment. J Biol Chem. 2004;279(33): 34411–20.

Mandard S, Zandbergen F, van Straten E, et al. The fasting-induced adipose factor/angiopoietin-like protein 4 is physically associated with lipoproteins and governs plasma lipid levels and adiposity. J Biol Chem. 2006;281(2):934–44.

Mao X, Kikani CK, Riojas RA, et al. APPL1 binds to adiponectin receptors and mediates adiponectin signalling and function. Nat Cell Biol. 2006;8(5):516–23.

Masuzaki H, Ogawa Y, Sagawa N, et al. Nonadipose tissue production of leptin: leptin as a novel placenta-derived hormone in humans. Nat Med. 1997;3(9):1029–33.

Matsuda M, Shimomura I, Sata M, et al. Role of adiponectin in preventing vascular stenosis. The missing link of adipo-vascular axis J Biol Chem. 2002;277(40):37487–91.

Matsuzawa Y. Therapy Insight: adipocytokines in metabolic syndrome and related cardiovascular disease. Nat Clin Pract Cardiovasc Med. 2006;3(1):35–42.

Matsuzawa Y. The role of fat topology in the risk of disease. Int J Obes (Lond). 2008;32 Suppl 7:S83–92.

Matsuzawa Y, Funahashi T, Nakamura T. The concept of metabolic syndrome: contribution of visceral fat accumulation and its molecular mechanism. J Atheroscler Thromb. 2011;18(8):629–39.

Mazzatti D, Lim FL, O'Hara A, et al. A microarray analysis of the hypoxia-induced modulation of gene expression in human adipocytes. Arch Physiol Biochem. 2012;118(3):112–20.
Mehebik N, Jaubert AM, Sabourault D, et al. Leptin-induced nitric oxide production in white adipocytes is mediated through PKA and MAP kinase activation. Am J Physiol Cell Physiol. 2005;289(2):C379–87.
Michailidou Z, Turban S, Miller E, et al. Increased angiogenesis protects against adipose hypoxia and fibrosis in metabolic disease-resistant 11beta-hydroxysteroid dehydrogenase type 1 (HSD1)-deficient mice. J Biol Chem. 2011;287(6):4188–97.
Miller DA, Lee A, Matsui Y, et al. Complementary DNA cloning of the murine transforming growth factor-beta 3 (TGF beta 3) precursor and the comparative expression of TGF beta 3 and TGF beta 1 messenger RNA in murine embryos and adult tissues. Mol Endocrinol. 1989;3(12):1926–34.
Mohamed-Ali V, Goodrick S, Rawesh A, et al. Subcutaneous adipose tissue releases interleukin-6, but not tumor necrosis factor-alpha, in vivo. J Clin Endocrinol Metab. 1997;82(12):4196–200.
Mora S, Pessin JE. An adipocentric view of signaling and intracellular trafficking. Diabetes Metab Res Rev. 2002;18(5):345–56.
Motoshima H, Goldstein BJ, Igata M, et al. AMPK and cell proliferation–AMPK as a therapeutic target for atherosclerosis and cancer. J Physiol. 2006;574(Pt 1):63–71.
Motoshima H, Wu X, Mahadev K, et al. Adiponectin suppresses proliferation and superoxide generation and enhances eNOS activity in endothelial cells treated with oxidized LDL. Biochem Biophys Res Commun. 2004;315(2):264–71.
Mu H, Ohashi R, Yan S, et al. Adipokine resistin promotes in vitro angiogenesis of human endothelial cells. Cardiovasc Res. 2006;70(1):146–57.
Muller G, Behrens J, Nussbaumer U, et al. Inhibitory action of transforming growth factor beta on endothelial cells. Proc Natl Acad Sci U S A. 1987;84(16):5600–4.
Myers Jr MG, Heymsfield SB, Haft C, et al. Challenges and opportunities of defining clinical leptin resistance. Cell Metab. 2012;15(2):150–6.
Nagata D, Mogi M, Walsh K. AMP-activated protein kinase (AMPK) signaling in endothelial cells is essential for angiogenesis in response to hypoxic stress. J Biol Chem. 2003;278(33): 31000–6.
Northcott JM, Yeganeh A, Taylor CG, et al. Adipokines and the cardiovascular system: mechanisms mediating health and disease. Can J Physiol Pharmacol. 2012;90(8):1029–59.
Ohashi K, Parker JL, Ouchi N, et al. Adiponectin promotes macrophage polarization toward an anti-inflammatory phenotype. J Biol Chem. 2010;285(9):6153–60.
Ouchi N, Kihara S, Arita Y, et al. Adiponectin, an adipocyte-derived plasma protein, inhibits endothelial NF-kappaB signaling through a cAMP-dependent pathway. Circulation. 2000;102(11): 1296–301.
Ouchi N, Kobayashi H, Kihara S, et al. Adiponectin stimulates angiogenesis by promoting cross-talk between AMP-activated protein kinase and Akt signaling in endothelial cells. J Biol Chem. 2004;279(2):1304–9.
Ouchi N, Ohishi M, Kihara S, et al. Association of hypoadiponectinemia with impaired vasoreactivity. Hypertension. 2003;42(3):231–4.
Ouedraogo R, Wu X, Xu SQ, et al. Adiponectin suppression of high-glucose-induced reactive oxygen species in vascular endothelial cells: evidence for involvement of a cAMP signaling pathway. Diabetes. 2006;55(6):1840–6.
Pajvani UB, Du X, Combs TP, et al. Structure-function studies of the adipocyte-secreted hormone Acrp30/adiponectin. Implications fpr metabolic regulation and bioactivity J Biol Chem. 2003;278(11):9073–85.
Park JW, Kim WH, Shin SH, et al. Visfatin exerts angiogenic effects on human umbilical vein endothelial cells through the mTOR signaling pathway. Biochim Biophys Acta. 2011; 1813(5):763–71.
Patel SB, Reams GP, Spear RM, et al. Leptin: linking obesity, the metabolic syndrome, and cardiovascular disease. Curr Hypertens Rep. 2008;10(2):131–7.

Patel SD, Rajala MW, Rossetti L, et al. Disulfide-dependent multimeric assembly of resistin family hormones. Science. 2004;304(5674):1154–8.

Patterson C, Perrella MA, Endege WO, et al. Downregulation of vascular endothelial growth factor receptors by tumor necrosis factor-alpha in cultured human vascular endothelial cells. J Clin Invest. 1996;98(2):490–6.

Pelleymounter MA, Cullen MJ, Baker MB, et al. Effects of the obese gene product on body weight regulation in ob/ob mice. Science. 1995;269(5223):540–3.

Pepper MS. Transforming growth factor-beta: vasculogenesis, angiogenesis, and vessel wall integrity. Cytokine Growth Factor Rev. 1997;8(1):21–43.

Pepper MS, Montesano R, Mandriota SJ, et al. Angiogenesis: a paradigm for balanced extracellular proteolysis during cell migration and morphogenesis. Enzyme Protein. 1996;49(1–3): 138–62.

Prud'homme GJ. Pathobiology of transforming growth factor beta in cancer, fibrosis and immunologic disease, and therapeutic considerations. Lab Invest. 2007;87(11):1077–91.

Qatanani M, Szwergold NR, Greaves DR, et al. Macrophage-derived human resistin exacerbates adipose tissue inflammation and insulin resistance in mice. J Clin Invest. 2009;119(3):531–9.

Rega G, Kaun C, Demyanets S, et al. Vascular endothelial growth factor is induced by the inflammatory cytokines interleukin-6 and oncostatin m in human adipose tissue in vitro and in murine adipose tissue in vivo. Arterioscler Thromb Vasc Biol. 2007;27(7):1587–95.

Rehman J, Considine RV, Bovenkerk JE, et al. Obesity is associated with increased levels of circulating hepatocyte growth factor. J Am Coll Cardiol. 2003;41(8):1408–13.

Reilly MP, Lehrke M, Wolfe ML, et al. Resistin is an inflammatory marker of atherosclerosis in humans. Circulation. 2005;111(7):932–9.

Robciuc MR, Naukkarinen J, Ortega-Alonso A, et al. Serum angiopoietin-like 4 protein levels and expression in adipose tissue are inversely correlated with obesity in monozygotic twins. J Lipid Res. 2011;52(8):1575–82.

Robertson SA, Rae CJ, Graham A. Induction of angiogenesis by murine resistin: putative role of PI3-kinase and NO-dependent pathways. Regul Pept. 2009a;152(1–3):41–7.

Robertson SA, Rae CJ, Graham A. Resistin: TWEAKing angiogenesis. Atherosclerosis. 2009b;203(1):34–7.

Romanatto T, Cesquini M, Amaral ME, et al. TNF-alpha acts in the hypothalamus inhibiting food intake and increasing the respiratory quotient–effects on leptin and insulin signaling pathways. Peptides. 2007;28(5):1050–8.

Rupnick MA, Panigrahy D, Zhang CY, et al. Adipose tissue mass can be regulated through the vasculature. Proc Natl Acad Sci U S A. 2002;99(16):10730–5.

Saiki A, Watanabe F, Murano T, et al. Hepatocyte growth factor secreted by cultured adipocytes promotes tube formation of vascular endothelial cells in vitro. Int J Obes (Lond). 2006;30(11): 1676–84.

Samad F, Uysal KT, Wiesbrock SM, et al. Tumor necrosis factor alpha is a key component in the obesity-linked elevation of plasminogen activator inhibitor 1. Proc Natl Acad Sci U S A. 1999;96(12):6902–7.

Sanchez-Solana B, Laborda J, Baladron V. Mouse resistin modulates adipogenesis and glucose uptake in 3T3-L1 preadipocytes through the ROR1 receptor. Mol Endocrinol. 2012;26(1): 110–27.

Santibanez JF, Quintanilla M, Bernabeu C. TGF-beta/TGF-beta receptor system and its role in physiological and pathological conditions. Clin Sci (Lond). 2011;121(6):233–51.

Savage DB, Sewter CP, Klenk ES, et al. Resistin/Fizz3 expression in relation to obesity and peroxisome proliferator-activated receptor-gamma action in humans. Diabetes. 2001;50(10): 2199–202.

Schafer K, Fujisawa K, Konstantinides S, et al. Disruption of the plasminogen activator inhibitor 1 gene reduces the adiposity and improves the metabolic profile of genetically obese and diabetic ob/ob mice. FASEB J. 2001;15(10):1840–2.

Schlesinger S, Aleksandrova K, Pischon T, et al. Abdominal obesity, weight gain during adulthood and risk of liver and biliary tract cancer in a European cohort. Int J Cancer. 2013;132(3):645–57.

Schmid PM, Resch M, Steege A, et al. Globular and full-length adiponectin induce NO-dependent vasodilation in resistance arteries of Zucker lean but not Zucker diabetic fatty rats. Am J Hypertens. 2011;24(3):270–7.

Scroyen I, Jacobs F, Cosemans L, et al. Effect of plasminogen activator inhibitor-1 on adipogenesis in vivo. Thromb Haemost. 2009;101(2):388–93.

Serino M, Menghini R, Fiorentino L, et al. Mice heterozygous for tumor necrosis factor-alpha converting enzyme are protected from obesity-induced insulin resistance and diabetes. Diabetes. 2007;56(10):2541–6.

Sethi JK, Hotamisligil GS. The role of TNF alpha in adipocyte metabolism. Semin Cell Dev Biol. 1999;10(1):19–29.

Shapiro L, Scherer PE. The crystal structure of a complement-1q family protein suggests an evolutionary link to tumor necrosis factor. Curr Biol. 1998;8(6):335–8.

Shen L, Miao J, Yuan F, et al. Overexpression of adiponectin promotes focal angiogenesis in the mouse brain following middle cerebral artery occlusion. Gene Ther. 2013;20(1):93–101.

Shibata R, Ouchi N, Kihara S, et al. Adiponectin stimulates angiogenesis in response to tissue ischemia through stimulation of amp-activated protein kinase signaling. J Biol Chem. 2004;279(27):28670–4.

Shimano M, Ouchi N, Shibata R, et al. Adiponectin deficiency exacerbates cardiac dysfunction following pressure overload through disruption of an AMPK-dependent angiogenic response. J Mol Cell Cardiol. 2010;49(2):210–20.

Sierra-Honigmann MR, Nath AK, Murakami C, et al. Biological action of leptin as an angiogenic factor. Science. 1998;281(5383):1683–6.

Silha JV, Krsek M, Sucharda P, et al. Angiogenic factors are elevated in overweight and obese individuals. Int J Obes (Lond). 2005;29(11):1308–14.

Silverman KJ, Lund DP, Zetter BR, et al. Angiogenic activity of adipose tissue. Biochem Biophys Res Commun. 1988;153(1):347–52.

Soukas A, Socci ND, Saatkamp BD, et al. Distinct transcriptional profiles of adipogenesis in vivo and in vitro. J Biol Chem. 2001;276(36):34167–74.

Spencer M, Unal R, Zhu B, et al. Adipose tissue extracellular matrix and vascular abnormalities in obesity and insulin resistance. J Clin Endocrinol Metab. 2011;96(12):E1990–8.

St-Pierre J, Tremblay ML. Modulation of leptin resistance by protein tyrosine phosphatases. Cell Metab. 2012;15(3):292–7.

Stacker SA, Runting AS, Caesar C, et al. The 3T3-L1 fibroblast to adipocyte conversion is accompanied by increased expression of angiopoietin-1, a ligand for tie2. Growth Factors. 2000;18(3):177–91.

Stallmeyer B, Pfeilschifter J, Frank S. Systemically and topically supplemented leptin fails to reconstitute a normal angiogenic response during skin repair in diabetic ob/ob mice. Diabetologia. 2001;44(4):471–9.

Stefansson S, Lawrence DA. The serpin PAI-1 inhibits cell migration by blocking integrin alpha V beta 3 binding to vitronectin. Nature. 1996;383(6599):441–3.

Stefansson S, Petitclerc E, Wong MK, et al. Inhibition of angiogenesis in vivo by plasminogen activator inhibitor-1. J Biol Chem. 2001;276(11):8135–41.

Steppan CM, Bailey ST, Bhat S, et al. The hormone resistin links obesity to diabetes. Nature. 2001;409(6818):307–12.

Stouthard JM, Romijn JA, Van der Poll T, et al. Endocrinologic and metabolic effects of interleukin-6 in humans. Am J Physiol. 1995;268(5 Pt 1):E813–9.

Strydom DJ. The angiogenins. Cell Mol Life Sci. 1998;54(8):811–24.

Strydom DJ, Fett JW, Riordan JF. The odyssey of angiogenin: a protein that induces blood vessel growth. Anal Chem. 1989;61(20):1173A–9.

Suganami T, Ogawa Y. Adipose tissue macrophages: their role in adipose tissue remodeling. J Leukoc Biol. 2010;88(1):33–9.

Sun Y, Chen X. Effect of adiponectin on apoptosis: proapoptosis or antiapoptosis? Biofactors. 2010;36(3):179–86.

Szalowska E, Elferink MG, Hoek A, et al. Resistin is more abundant in liver than adipose tissue and is not up-regulated by lipopolysaccharide. J Clin Endocrinol Metab. 2009;94(8):3051–7.

Tam J, Duda DG, Perentes JY, et al. Blockade of VEGFR2 and not VEGFR1 can limit diet-induced fat tissue expansion: role of local versus bone marrow-derived endothelial cells. PLoS One. 2009;4(3):e4974.

Tan BK, Adya R, Farhatullah S, et al. Metformin treatment may increase omentin-1 levels in women with polycystic ovary syndrome. Diabetes. 2010;59(12):3023–31.

Tan CK, Leuenberger N, Tan MJ, et al. Smad3 deficiency in mice protects against insulin resistance and obesity induced by a high-fat diet. Diabetes. 2011;60(2):464–76.

Tao L, Gao E, Jiao X, et al. Adiponectin cardioprotection after myocardial ischemia/reperfusion involves the reduction of oxidative/nitrative stress. Circulation. 2007;115(11):1408–16.

Tarkowski A, Bjersing J, Shestakov A, et al. Resistin competes with lipopolysaccharide for binding to toll-like receptor 4. J Cell Mol Med. 2010;14(6B):1419–31.

Tchaikovski V, Olieslagers S, Bohmer FD, et al. Diabetes mellitus activates signal transduction pathways resulting in vascular endothelial growth factor resistance of human monocytes. Circulation. 2009;120(2):150–9.

Teichert-Kuliszewska K, Hamilton BS, Deitel M, et al. Augmented production of heparin-binding mitogenic proteins by preadipocytes from massively obese persons. J Clin Invest. 1992;90(4): 1226–31.

Terra X, Auguet T, Quesada I, et al. Increased levels and adipose tissue expression of visfatin in morbidly obese women. The relationship with pro-inflammatory cytokines. Clin Endocrinol (Oxf). 2012;77(5):691–8.

Trayhurn P, Wood IS. Adipokines: inflammation and the pleiotropic role of white adipose tissue. Br J Nutr. 2004;92(3):347–55.

Tsao TS, Tomas E, Murrey HE, et al. Role of disulfide bonds in Acrp30/adiponectin structure and signaling specificity. Different oligomers activate different signal transduction pathways J Biol Chem. 2003;278(50):50810–7.

Tsuji S, Yamashita M, Nishiyama A, et al. Differential structure and activity between human and mouse intelectin-1: human intelectin-1 is a disulfide-linked trimer, whereas mouse homologue is a monomer. Glycobiology. 2007;17(10):1045–51.

Uysal KT, Wiesbrock SM, Marino MW, et al. Protection from obesity-induced insulin resistance in mice lacking TNF-alpha function. Nature. 1997;389(6651):610–4.

Varma V, Yao-Borengasser A, Bodles AM, et al. Thrombospondin-1 is an adipokine associated with obesity, adipose inflammation, and insulin resistance. Diabetes. 2008;57(2):432–9.

Vecchione C, Maffei A, Colella S, et al. Leptin effect on endothelial nitric oxide is mediated through Akt-endothelial nitric oxide synthase phosphorylation pathway. Diabetes. 2002;51(1): 168–73.

Verma S, Li SH, Wang CH, et al. Resistin promotes endothelial cell activation: further evidence of adipokine-endothelial interaction. Circulation. 2003;108(6):736–40.

Voros G, Maquoi E, Demeulemeester D, et al. Modulation of angiogenesis during adipose tissue development in murine models of obesity. Endocrinology. 2005;146(10):4545–54.

Waki H, Tontonoz P. Endocrine functions of adipose tissue. Annu Rev Pathol. 2007;2:31–56.

Waki H, Yamauchi T, Kamon J, et al. Generation of globular fragment of adiponectin by leukocyte elastase secreted by monocytic cell line THP-1. Endocrinology. 2005;146(2):790–6.

Wallenius V, Wallenius K, Ahren B, et al. Interleukin-6-deficient mice develop mature-onset obesity. Nat Med. 2002;8(1):75–9.

Walshe TE, Saint-Geniez M, Maharaj AS, et al. TGF-beta is required for vascular barrier function, endothelial survival and homeostasis of the adult microvasculature. PLoS One. 2009;4(4):e5149.

Waltenberger J. VEGF resistance as a molecular basis to explain the angiogenesis paradox in diabetes mellitus. Biochem Soc Trans. 2009;37(Pt 6):1167–70.

Wang Y, Lam KS, Xu JY, et al. Adiponectin inhibits cell proliferation by interacting with several growth factors in an oligomerization-dependent manner. J Biol Chem. 2005;280(18):18341–7.

Wang Y, Xu LY, Lam KS, et al. Proteomic characterization of human serum proteins associated with the fat-derived hormone adiponectin. Proteomics. 2006;6(13):3862–70.

Weisberg SP, McCann D, Desai M, et al. Obesity is associated with macrophage accumulation in adipose tissue. J Clin Invest. 2003;112(12):1796–808.

Wree A, Mayer A, Westphal S, et al. Adipokine expression in brown and white adipocytes in response to hypoxia. J Endocrinol Invest. 2012;35(5):522–7.

Xie H, Xie PL, Wu XP, et al. Omentin-1 attenuates arterial calcification and bone loss in osteoprotegerin-deficient mice by inhibition of RANKL expression. Cardiovasc Res. 2011;92(2):296–306.

Xu F, Burk D, Gao Z, et al. Angiogenic deficiency and adipose tissue dysfunction are associated with macrophage malfunction in SIRT1$^{-/-}$ mice. Endocrinology. 2012;153(4):1706–16.

Xu H, Barnes GT, Yang Q, et al. Chronic inflammation in fat plays a crucial role in the development of obesity-related insulin resistance. J Clin Invest. 2003;112(12):1821–30.

Xu H, Uysal KT, Becherer JD, et al. Altered tumor necrosis factor-alpha (TNF-alpha) processing in adipocytes and increased expression of transmembrane TNF-alpha in obesity. Diabetes. 2002;51(6):1876–83.

Xue Y, Cao R, Nilsson D, et al. FOXC2 controls Ang-2 expression and modulates angiogenesis, vascular patterning, remodeling, and functions in adipose tissue. Proc Natl Acad Sci U S A. 2008;105(29):10167–72.

Xue Y, Petrovic N, Cao R, et al. Hypoxia-independent angiogenesis in adipose tissues during cold acclimation. Cell Metab. 2009;9(1):99–109.

Yadav H, Quijano C, Kamaraju AK, et al. Protection from obesity and diabetes by blockade of TGF-beta/Smad3 signaling. Cell Metab. 2011;14(1):67–79.

Yamada T, Kawakami S, Nakanishi N. Fat depot-specific differences in angiogenic growth factor gene expression and its relation to adipocyte size in cattle. J Vet Med Sci. 2010;72(8):991–7.

Yamauchi T, Kamon J, Ito Y, et al. Cloning of adiponectin receptors that mediate antidiabetic metabolic effects. Nature. 2003a;423(6941):762–9.

Yamauchi T, Kamon J, Waki H, et al. Globular adiponectin protected ob/ob mice from diabetes and ApoE-deficient mice from atherosclerosis. J Biol Chem. 2003b;278(4):2461–8.

Yamauchi T, Kamon J, Waki H, et al. The fat-derived hormone adiponectin reverses insulin resistance associated with both lipoatrophy and obesity. Nat Med. 2001;7(8):941–6.

Yamawaki H, Kuramoto J, Kameshima S, et al. Omentin, a novel adipocytokine inhibits TNF-induced vascular inflammation in human endothelial cells. Biochem Biophys Res Commun. 2011;408(2):339–43.

Yang EY, Moses HL. Transforming growth factor beta 1-induced changes in cell migration, proliferation, and angiogenesis in the chicken chorioallantoic membrane. J Cell Biol. 1990;111(2):731–41.

Yang RZ, Lee MJ, Hu H, et al. Identification of omentin as a novel depot-specific adipokine in human adipose tissue: possible role in modulating insulin action. Am J Physiol Endocrinol Metab. 2006;290(6):E1253–61.

Yang YH, Wang Y, Lam KS, et al. Suppression of the Raf/MEK/ERK signaling cascade and inhibition of angiogenesis by the carboxyl terminus of angiopoietin-like protein 4. Arterioscler Thromb Vasc Biol. 2008;28(5):835–40.

Yin W, Romeo S, Chang S, et al. Genetic variation in ANGPTL4 provides insights into protein processing and function. J Biol Chem. 2009;284(19):13213–22.

Yoon JC, Chickering TW, Rosen ED, et al. Peroxisome proliferator-activated receptor gamma target gene encoding a novel angiopoietin-related protein associated with adipose differentiation. Mol Cell Biol. 2000;20(14):5343–9.

Yoshida S, Ono M, Shono T, et al. Involvement of interleukin-8, vascular endothelial growth factor, and basic fibroblast growth factor in tumor necrosis factor alpha-dependent angiogenesis. Mol Cell Biol. 1997;17(7):4015–23.

Zhang QX, Magovern CJ, Mack CA, et al. Vascular endothelial growth factor is the major angiogenic factor in omentum: mechanism of the omentum-mediated angiogenesis. J Surg Res. 1997;67(2):147–54.

Zheng L, Weng M, Qi M, et al. Aberrant expression of intelectin-1 in gastric cancer: its relationship with clinicopathological features and prognosis. J Cancer Res Clin Oncol. 2012;138(1):163–72.

Part V
Regulation of Adipose Angiogenesis by Inflammatory Cells

Chapter 11
Immune Cells in Adipose Tissue: Key Players in Metabolic Disorders

Fanny Volat and Anne Bouloumié

Abstract Obesity, defined as an excessive development of adipose tissue, is associated with high risk to develop metabolic disorders including insulin resistance, type 2 diabetes, and cardiovascular diseases. Many investigations performed in mice and humans provided evidences that obesity is associated with a low grade inflammatory state with enhanced both circulating and intra-tissular pro-inflammatory cytokine and chemokine levels as well as accumulation of immune cells, including macrophages and lymphocytes. The inflammatory response that takes place within the adipose tissue during fat mass development remains chronic with obesity and has been identified as the major contributor to its pathogenesis. The present chapter describes the link between immune cells, adipose tissue and obesity-associated pathologies and highlights the role of adipose tissue macrophages in the control of adipose tissue angiogenesis.

Keywords Adipose tissue • Inflammation • Obesity • Immune cells • Endothelial cells • Angiogenesis

F. Volat • A. Bouloumié (✉)
Institut National de la Santé et de la Recherche Médicale (INSERM), Toulouse, France

Unité Mixte de Recherche (UMR) 1048, Université Paul Sabatier (UPS), Institut de Maladies Métaboliques et Cardiovasculaires (I2MC), 31432 Toulouse Cedex 4, France
e-mail: anne.bouloumie@inserm.fr

Y. Cao (ed.), *Angiogenesis in Adipose Tissue*, DOI 10.1007/978-1-4614-8069-3_11,

11.1 Obesity, Inflammation, and Metabolic Disorders

11.1.1 Contribution of Adipose Tissue in Chronic Inflammation

Obesity is characterized by a low grade inflammatory state associated with metabolic and cardiovascular complications (Hotamisligil 2006). Although major sites of inflammatory factor production are the liver and lymphoid organs, adipose tissue is also able to produce a large panel of inflammatory mediators. Thus, involvement of fat mass itself in the variation of systemic inflammatory profile during obesity has been suspected. Initial reports focused on the impact of diet-induced weight loss and demonstrated an improvement of the inflammatory state, in particular a decrease of TNFα (tumor necrosis factor alpha), IL-6 (interleukin 6), and C-reactive protein systemic levels (Bastard et al. 2000a, b; Heilbronn et al. 2001). Even after 4 weeks of very low calorie diet, the moderate weight loss is sufficient to improve the inflammatory profile in subcutaneous adipose tissue of obese women, reducing the expression of pro-inflammatory factors and increasing that of anti-inflammatory molecules, such as IL-10 (Clement et al. 2004). These data strongly suggested that the production of pro-inflammatory factors by adipose tissue itself could contribute to the promotion and maintenance of the systemic inflammation observed during obesity and consequently emphasized its influence on obesity-associated metabolic complications.

11.1.2 Adipose Tissue Inflammation

During the settlement of obesity, adipose tissue dynamically reacts to the excessive caloric input by a first stage of hypertrophy of existing adipocytes followed by the recruitment of progenitor cells to form new adipocytes, leading to adipocyte hyperplasia. Whereas adipocyte hypertrophy leads to a shift of their secretions toward a higher inflammatory profile (Skurk et al. 2007), promotion of the inflammatory state was mainly attributed to the accumulation of macrophages within the fat mass. Indeed, in mice model of obesity (high fat diet- or genetically induced), one major remodeling process during adipose tissue expansion is the accumulation of macrophages that mainly localized in crown-like structures around necrotic adipocytes and, at the same time, undergo a phenotypic change toward a pro-inflammatory or M1 state (Weisberg et al. 2003; Xu et al. 2003; Cinti et al. 2005). The development of insulin resistance that occurs during obesity was therefore believed to be largely attributable to the pro-inflammatory cytokines produced by infiltrated macrophages and hypertrophic adipocytes (Hotamisligil 2006). However, several recent data mainly obtained with approaches of deletion or adoptive transfer of immune cells in obese mice models underscore the primary role of lymphocytes in obesity-associated inflammatory state and metabolic disorders.

11.2 Lymphoid Cells

All lymphoid cells originate from a common lymphoid progenitor and participate in both innate (natural killer (NK), natural killer T (NKT), γδT cells) and adaptive immunity (B and T lymphocytes). Resident lymphocytes have been detected in the stroma-vascular fraction of both rodents and humans fat pads. Interestingly, as macrophages, their relative proportion depends on the degree of obesity, insulin resistance, and adipose tissue location. Lymphocytes from innate immunity are mainly found in intra-abdominal fat mass whereas those of adaptive immunity are present in subcutaneous fat depots (Caspar-Bauguil et al. 2009), suggesting distinct contributions in the inflammatory responses and metabolic disorders.

11.2.1 Natural Killer T Cells

NKT cells are activated by lipid antigens presented by the major histocompatibility complex (MHC) class I like CD1d-molecules on antigen presenting cells (APC) such as macrophages. Lipid-activated NKT cells produce T helper 1 and 2 cytokines (T_H1 and T_H2) including interferon (IFN) γ and interleukin (IL) 4, respectively, and appear to act as a bridge between innate and adaptive immunity (Bendelac et al. 2007). Several reports on the involvement of NKT cells in obesity and associated metabolic complications have recently emerged. While it seems clear that NKT cells could influence macrophage polarization toward a reparative or M2 phenotype, especially via IL-4 secretion (Ji et al. 2012), their involvement in adipose tissue inflammation and insulin resistance remains to be clearly defined. Indeed, in both mice and humans, adipose tissue abundance of NKT type 1 cells is inversely correlated to body mass index (BMI) and insulin resistance (Ji et al. 2012; Lynch et al. 2009). Moreover, studies performed on CD1d−/− mice demonstrated that lack of functional NKT cells does not have metabolic effect after long-term high fat feeding (Ji et al. 2012; Kotas et al. 2011; Mantell et al. 2011). However, since non-specific activation of NKT cells by α-galactosylceramide injection was associated with an improvement of glucose tolerance in obese mice (Ji et al. 2012), further studies are needed to clearly delineate the protective and/or deleterious potential roles of the distinct NKT cell subpopulations.

11.2.2 B Lymphocytes

B lymphocytes are key players of humoral immune system, producing and secreting immunoglobulins in response to antigen. First studies performed on mice reported an early infiltration of B lymphocytes after high fat feeding, occurring before the enlargement of fat mass and the onset of insulin resistance (Duffaut et al. 2009a).

The involvement of B cells in the development of insulin resistance has been recently established by Winer et al., showing that in diet-induced obesity (DIO) mice lacking B cells are protected from obesity-associated metabolic disorders despite weight gain (Winer et al. 2011). The production of pathogenic immunoglobulins by B lymphocytes and subsequent activation of pro-inflammatory macrophages and T lymphocytes contribute to the development of insulin resistance. In humans, few studies on B lymphocytes in adipose tissue have been performed. Using flow cytometry analyses on subcutaneous adipose tissue, our group observed minor B cells population (CD19+) which is not modulated by adiposity (Duffaut et al. 2009b). A recent report identified B cells within crown-like structures and perivascular space in subcutaneous adipose tissue of obese subjects (McDonnell et al. 2012). Interestingly, the predominance of B cells positively correlates with an effective treatment for obesity-related insulin resistance (McDonnell et al. 2012). So, additional studies are needed to also precise the protective and/or pathogenic role of B lymphocytes in obesity and associated complications.

11.2.3 T Lymphocytes (CD3+)

Accumulation of T lymphocytes (CD3+) occurs in adipose tissue of both obese mice (Kintscher et al. 2008; Wu et al. 2007) and humans (Duffaut et al. 2009b). The key role of T cells in the pathogenesis of insulin resistance was demonstrated by the study of Winer et al. demonstrating that the use of a CD3-depleting antibody protects mice from the development of insulin resistance (Winer et al. 2009). However, recent studies provided new advances concerning contribution of the different subpopulations of T lymphocytes in the obesity-associated inflammatory process.

11.2.3.1 CD8+ T Lymphocytes

CD8+ lymphocytes, also named "cytotoxic" T lymphocytes, are involved in the recognition and elimination of damaged dysfunctional cells. In adipose tissue of both obese mice and humans, accumulation of CD8+ T cells has been reported, notably within crown-like structure, where they are suspected to be involved in the inflammatory cascade initiation (Winer et al. 2009; Nishimura et al. 2009; Rausch et al. 2008). Indeed, in response to high fat diet (HFD), CD8+ T cells infiltration in adipose tissue precedes that of macrophages (Nishimura et al. 2009). Moreover, using neutralizing antibody treatment or CD8-deficient mice, Nishimura et al. showed that depletion of the CD8+ subset of T lymphocytes leads to a reduction of inflammatory M1 macrophage accumulation and inflammatory cytokine expression in the adipose tissue and consequently improves systemic insulin resistance. Inversely, adoptive transfer of CD8+ T cells to healthy obese CD8-deficient mice increases pro-inflammatory macrophage infiltration together with a degradation of glucose tolerance and insulin sensitivity. Thus, CD8+ T cells have a crucial role in the initiation and propagation of inflammation in obese adipose tissue, by recruiting and activating macrophages to adipose tissue. However, lack of metabolic effect on lymphocyte-deficient mice (*Rag–/–*) after CD8+ T cells transfer (Winer et al. 2009) suggests that the impact of CD8+ cells on inflammatory cascade requires that of other lymphocytes.

11.2.3.2 CD4+ T Regulatory Lymphocytes

CD4+ Treg cells are a subset of T lymphocytes identified as CD4+/FoxP3+/CD25+ cells. Whereas in adipose tissue of lean mice Treg cells represent the major population of T lymphocytes, in both DIO and genetic model mice, abundance of these cells in the abdominal fat is significantly decreased (Feuerer et al. 2009). A protective role of Treg cells in the pathogenesis of insulin resistance was firstly highlighted by T cell depletion experiments (using anti-CD3 antibody treatment), leading to an accumulation of Treg cells associated with an improvement of the inflammatory state and a protection of obese mice from the development of insulin resistance (Winer et al. 2009; Ilan et al. 2010). By contrast, selective depletion of Treg cells in lean mice leads to opposite phenotype (Feuerer et al. 2009). Likewise, in db/db mouse model, Treg cells depletion using anti-CD25 antibody aggravates inflammation in adipose tissue and systemic insulin resistance, while adoptive transfer of Treg cells reduces CD8+ effector T cells infiltration and improves insulin sensitivity (Eller et al. 2011).

In humans, protective function of Treg cells in adipose tissue remains to be elucidate as, to date, much conflicting data on the abundance of Tregs cells in human visceral adipose tissue of obese patients compared to lean controls have been reported (Feuerer et al. 2009; Eller et al. 2011; Deiuliis et al. 2011; Zeyda et al. 2011).

11.2.3.3 CD4+ T Helper Lymphocytes

CD4+ cells recognize polypeptides presented by class II MHC molecules on the surfaces of APC, such as macrophages and dendritic cells. Involvement of these cells in obesity-associated insulin resistance has been explored by CD4+ cells repopulation of lymphocyte-deficient *Rag1–/–* mice. Although reconstitution of CD4+ cells improves glucose tolerance and insulin sensitivity, these experiments does not provide a clear result since these mice also display a reduction of body weight and adipocyte size (Winer et al. 2009).

Stimulated naive CD4+ cells differentiate into effectors T helper Th1, Th2, or Th17 cells. Th1 cells secrete pro-inflammatory cytokines including INFγ and promote cell-mediating immune responses, whereas Th2 anti-inflammatory cells produce IL-4 and IL-3 and contribute to the humoral immune response. While both Th1 and Th2 cells are equally present in adipose tissue of lean mice, HFD-induced obesity conducts to a severe accumulation of Th1 lymphocytes without any change in Th2 cell number (Winer et al. 2009). This increase of Th1 cells could contribute to the higher production of INFγ observed in adipose tissue of obese mice compared to lean mice (Kintscher et al. 2008). Th17 cells are also increased in adipose tissue of obese mice. Whereas in lean adipose tissue INFγ-secreting Th1 cells are predominant, a switch from a Th1 toward a Th17 response in obese HFD-fed and *ob/ob* mice has been recently described and could be partly due to a small subset of dendritic cells identified as $Cd11c^{hight}$ $F4/80^{low}$ (Bertola et al. 2012). Similarly to mice, our group showed that accumulated T lymphocytes from human adipose tissue exhibit a Th1 profile, characterized by higher expression of INFγ, TNFα, and CCL5/RANTES than blood T lymphocytes, whereas no expression of IL-4, the Th2-related cytokine, is detected (Duffaut et al. 2009b). A high expression of IL-17 by isolated CD3+ cells has been also observed in overweight and obese patients (Bertola et al. 2012), indicating that both Th1 and Th17 cells may account for obesity-induced inflammatory state.

11.3 Macrophages

11.3.1 Accumulation in Adipose Tissue

Among immune cells, the role of macrophages in adipose tissue homeostasis has been the most extensively studied. Infiltration of F4/80+ CD11b+ macrophages has been described in both murine and human adipose tissue (Hotamisligil 2006). Transplantation experiment of labeled bone marrow cells in mice demonstrated that almost all infiltrated macrophages derive from bone marrow (Cho et al. 2007). They are released as immature monocytes into circulation before their extravasation into target tissues where they differentiate into resident macrophages. Adipose tissue expansion leads to hypoxia, adipocyte cell death, increased chemokine secretion, and dysregulation of free fatty acid release which independently contribute to adipose tissue macrophage infiltration (Sun et al. 2011). In obese mice models, macrophages can represent more than one third of stromal adipose cell population (Weisberg et al. 2003; Xu et al. 2003). They are typically located in crown-like structures around necrotic adipocytes (Cinti et al. 2005). This location may be linked to their function as phagocytes to eliminate dysfunctional hypertrophic adipocytes. Moreover, as accumulation of macrophages in adipose tissue strongly correlates with insulin resistance in both obese rodents and humans (Xu et al. 2003; Bertola et al. 2009; Cancello et al. 2005), paracrine effects of macrophages in adipose tissue inflammation process have been strongly suspected.

11.3.2 Phenotypic Switch

Macrophages have remarkable plasticity that allows them to efficiently respond to environmental signals. Their phenotype and physiology can be altered by both innate and adaptive immunity. Indeed, depending on their environment, pro or anti-inflammatory, macrophages are able to adopt a distinct M1 or M2 activation state. M1 or "classically activated" macrophages are induced by pro-inflammatory cytokines, such as lipopolysaccharide (LPS) or INFγ, and exhibit a high inflammatory and bactericidal potential. However, cytokines such as IL-4 and IL-3 induce M2 or "alternatively activated" macrophages which are characterized by their antiparasitic functions and their involvement in tissue repair and remodeling. Several studies performed on mice demonstrated a phenotypic switch of macrophages, from M2 in adipose tissue of lean mice to M1 in obese mice, explaining at least in part the inflammatory state observed in adipose tissue of obese mice. In accordance with this observation, in DIO and ob/ob mice, depletion of resident macrophages by liposome clodronate injection improves insulin sensitivity and glucose tolerance (Feng et al. 2011; Lee et al. 2011). Thus, these data demonstrate the major role of macrophages in the inflammatory state during murine obesity and their contribution to associated metabolic disorders.

11.3.3 Macrophages in Human Adipose Tissue

In human adipose tissue, macrophages appear differently modulated and exhibit a less polarized phenotype. Indeed, our group showed that native population of macrophages isolated from adipose tissue expresses both M1- and M2-related markers (Bourlier et al. 2008). This mixed M1/M2 phenotype is considered as a hallmark of a chronic inflammatory process. Moreover, compared to mice, fat mass expansion is surprisingly rather associated with an accumulation of M2 macrophages and a concomitant reduction of M1 marker expression. In agreement, Zeyda et al. described an accumulation of macrophages exhibiting a M2 similar phenotype but, importantly, they are able to produce pro-inflammatory mediators (Zeyda et al. 2007). Thus, macrophages of human adipose tissue are not strictly polarized in either M1 or M2 as those of murine adipose tissue and, while they mainly exhibit an anti-inflammatory phenotype, they are nevertheless able to produce inflammatory mediators and therefore contribute to the inflammatory state. Human adipose tissue macrophages also express the lymphatic vessel endothelial hyaluronan receptor 1 (LYVE-1) (Bourlier et al. 2008). LYVE-1 positive macrophages have been involved in tumor growth and wound healing as well as in mouse adipose tissue angiogenesis (Cho et al. 2007; Schledzewski et al. 2006), suggesting that, in addition to their involvement in chronic inflammatory process during adipose tissue expansion, macrophages could have a key role in adipose tissue angiogenesis.

11.4 Macrophages and Angiogenesis

11.4.1 Postnatal and Adult Adipose Tissue Development

Cho et al. demonstrated that, in highly hypoxic region of murine adipose tissue that appears during the growth of the epididymal adult fat depot, macrophages identified as LYVE-1+ accumulate next to the vascular network and participate to its formation by inducing angiogenesis. Indeed, specific depletion of the resident macrophages by clodronate liposome injection leads to a decreased vascular network (Cho et al. 2007). The angiogenic effect of adipose tissue macrophages involves vascular endothelial growth factor (VEGF) as well as matrix metalloproteinase (MMP) 9 productions. Their accumulation within the growing adipose tissue appears to be tightly controlled by the production of the chemoattractant stromal-derived factor 1 alpha (SDF1). These findings have been confirmed and extended to the development of murine adipose tissue in the postnatal period. Indeed, LYVE-1-positive macrophages, further characterized as VEGFR1 and TIE2 positive cells (markers of angiogenic macrophages (Qian and Pollard 2010)), were identified in the primitive epididymal adipose tissue 4 days after birth (Han et al. 2011). At this stage, macrophage depletion using clodronate approaches leads to a reduction of angiogenesis, associated with a reduced expression of basic fibroblast growth factor (bFGF), MMP-9, TNFα, and transforming growth factor beta (TGFβ) in the stroma-vascular fraction of adipose tissue (Han et al. 2011).

A similar typical location of macrophages next to the vascular network has been recently observed in murine peritumoral adipose tissue (Wagner et al. 2012). Peritumoral adipose tissue, which is highly vascularized and exhibits a higher angiogenic capacity, is also characterized by a dense infiltration of macrophages mainly expressing markers of M2 polarization. Wagner et al. demonstrated that infiltrated macrophages in the peritumoral adipose tissue may play a key role in angiogenesis stimulation, creating a propitious microenvironment for tumor progression (Wagner et al. 2012).

These studies are in accordance with our results showing that CD14+ macrophages that accumulated in human subcutaneous adipose tissue of lean to overweight patients express LYVE-1 in a BMI-dependent manner. Moreover, they produce specifically the MMP-9. Finally, human adipose tissue macrophage-secreted factors stimulate in vitro the migration and organization of the human adipose tissue endothelial cells (characterized as CD34+/CD31+ cells), but also the angiogenic capacities of the human adipose tissue progenitor cells (identified as CD34+/CD31- cells) (Bourlier et al. 2008). Thus, human adipose tissue macrophages exert pro-angiogenic effects and may contribute to the extension of the vascular network during adipose tissue growth (Fig. 11.1).

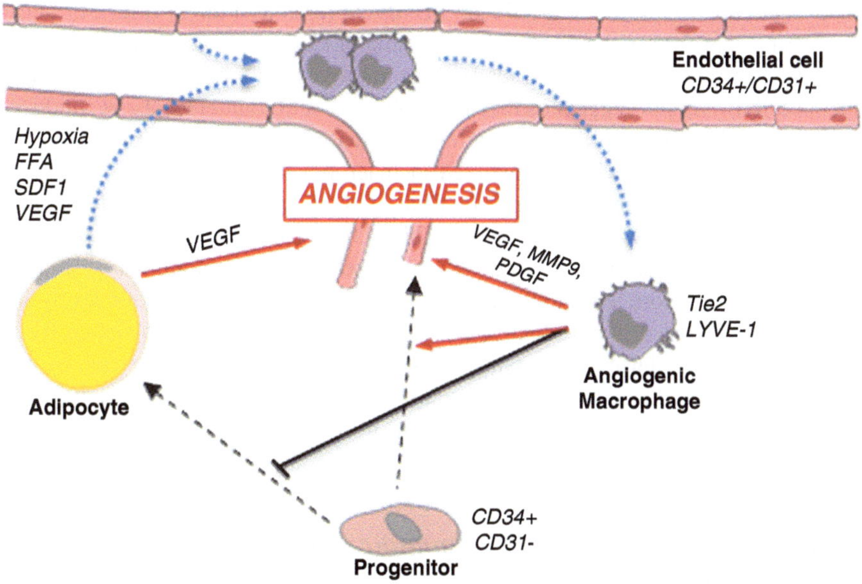

Fig. 11.1 Regulation of angiogenesis by adipose tissue macrophages. Adipose tissue macrophage accumulation, stimulated by various signals including hypoxia, free fatty acid (FFA), stromal-derived factor 1 alpha (SDF1), and vascular endothelial growth factor (VEGF), is a key player in the angiogenic process associated with adipose tissue (AT) growth. Pro-angiogenic macrophages, identified as LYVE1+ and Tie2+, may stimulate angiogenesis through effects on endothelial cells and/or progenitor cells

11.4.2 Obesity

Several angiogenic and endothelial cell growth factors are produced by adipose tissue. Among them, VEGF-A is considered as the major actor of angiogenesis and its level is positively associated with visceral fat accumulation in human obese subjects (Miyazawa-Hoshimoto et al. 2003). As macrophages produce VEGF-A, the increased levels of VEGF-A with visceral fat accumulation could be due to the higher accumulation of macrophages in visceral compared to subcutaneous adipose tissue in obese patients (Duffaut et al. 2009b; Cancello et al. 2006). However, VEGF-A is not only produced by macrophages but also by mature adipocytes (Ye 2011) and endothelial cells. Mice lacking SIRT1 (histone deacetylase sirtuin 1) and exhibiting a defective angiogenesis (Potente et al. 2007) also display a strong reduction of macrophages accumulation and subsequent decrease of cytokine production in adipose tissue (Xu et al. 2012). However, despite a strong reduced expression of several angiogenic factors including PDGF (platelet-derived growth factor), HGF (hepatocyte growth factor), and TGFβ, no modification of VEGF-A expression is observed suggesting that, at least in this model of adipose tissue, VEGF-A production is not related to adipose tissue macrophages. Macrophage accumulation in obese adipose tissue may serve as an angiogenic stimulator especially by increasing production of PDGF. Indeed, Pang et al. demonstrated that deletion of adipose tissue macrophages leads to a reduced PDGF expression in obese mice but not in lean (Pang et al. 2008). Furthermore, expression of PDGF by macrophages is stimulated by hypoxic conditions and leads to an improvement tube formation of endothelial cells in culture, suggesting that macrophages could counteract obesity-induced hypoxia and improve capillary formation by at least the production of PDGF.

VEGF has been previously involved in macrophages recruitment (Yang et al. 2004). In agreement with this capacity, Elias et al. recently demonstrated that mice over-expressing VEGF in adipose tissue exhibit an increase of M2 anti-inflammatory macrophages infiltration and, inversely, a decrease of pro-inflammatory macrophages. This therefore leads to a reduction of pro-inflammatory cytokine levels and to an improvement of whole body insulin sensitivity and glucose tolerance (Elias et al. 2012). Thus, during fat expansion, the increase of VEGF-A secretion by adipocytes could facilitate angiogenesis via the recruitment of M2 macrophages in adipose tissue, which consequently also limits the inflammatory process (Fig. 11.2).

11.4.3 *Involvement of Other Immune Cells in Adipose Tissue Angiogenesis*

Mast cells have also been involved in adipose tissue angiogenesis during adipose tissue expansion. Indeed, DIO mice lacking functional mast cells display reduced body weight, improved glucose tolerance, and an associated reduction of vascular density (Liu et al. 2009). Moreover, adoptive transfer of wild-type or cytokine-deficient bone marrow-derived mast cells demonstrated that mast cells also participate to adipose tissue angiogenesis by IL-6 and INFγ production (Liu et al. 2009). To date, the role of other immune cells, such as lymphocytes, on adipose tissue angiogenesis has not been yet explored, but it is tempting to speculate that lymphocytes might play an indirect role via their control of macrophage phenotype as well as a direct effect via their productions. Indeed, Th17 cells have been involved in the control of tumor angiogenesis (Liu et al. 2009).

11.5 Beneficial or Detrimental Effect of Angiogenesis?

Besides controlling adipose tissue angiogenesis, macrophage secretions are also able to reduce the proliferation of human adipose tissue progenitor cells in vitro (Constant et al. 2006; Lacasa et al. 2007), to impair adipogenesis (Zaragosi et al. 2010) and to promote the expression of a myofibroblast phenotype by adipose tissue progenitor cells, suggesting their involvement in fibrosis (Bourlier et al. 2012). Dynamic changes within fat mass, according to adipose tissue location and adiposity degree, may control the net impact of adipose tissue macrophages. Indeed, we showed that, according to the location of human adipose tissue (subcutaneous vs.

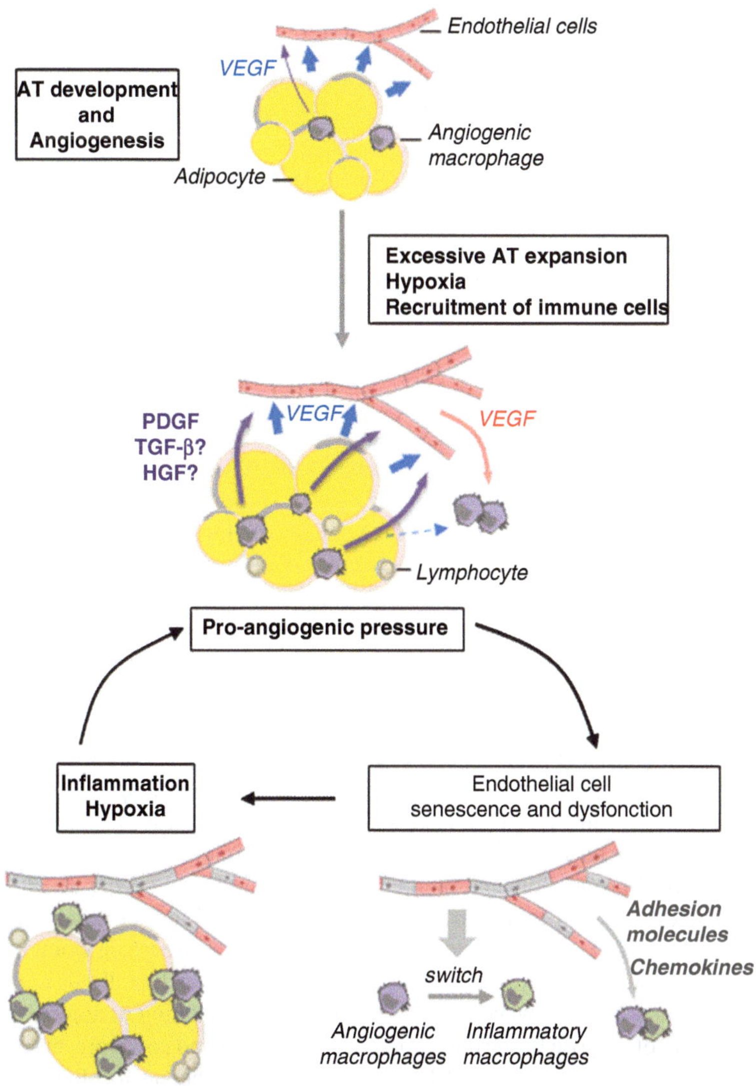

Fig. 11.2 Vicious circle of angiogenesis and inflammation. During excessive adipose tissue expansion, angiogenic function of adipose tissue cells including adipocytes, macrophages, and endothelial cells is strongly stimulated to counteract adipose tissue hypoxia. However, such pro-angiogenic pressure might, in the long run, induce senescence and impaired function of endothelial cells, leading to a defective vascular network. The arising hypoxia and immune cells recruitment may increase, in turn, adipose tissue angiogenesis in a vicious circle manner

visceral) and the adiposity degree of patients, human native adipose tissue macrophages exhibit distinct phenotypes. Macrophages isolated from human visceral adipose tissue show an increased expression of the angiogenic genes VEGF-A, LYVE-1, and IL-6 whereas macrophages isolated from human subcutaneous adipose tissue exhibit an extracellular matrix remodeling-like phenotype. Since adipokines but also hypoxia regulate human adipose tissue macrophage phenotypes, the distinct adipose tissue microenvironment according to the adipose tissue location may strongly impact the function, i.e., pro-fibrotic and/or pro-angiogenic, of adipose tissue macrophages (Bourlier et al. 2012).

Consistent with the pro-angiogenic phenotype of visceral macrophages, we showed that vascular network and endothelial cells number is higher in omental vs. subcutaneous adipose tissue of obese patients (Villaret et al. 2010).

However, a higher expression of hypoxia-related gene including VEGF-A, glucose transporter Glut-1, and hypoxia inducible factor 1 alpha is also observed in adipocytes as well as macrophages from visceral human adipose tissue, suggesting an impaired function of the dense vascular network of visceral human adipose tissue. One might consider that the extension of vascular network may be a phenomenon disconnected from the maturation and functionality of the vessels. Indeed in tumors, angiogenesis is often associated with abnormal architecture and function of the neocapillaries. Moreover, strong angiogenic stimulation per se may lead to accelerated aging of endothelial cells also called senescence. Indeed, as for other cell types, the proliferative life span of endothelial cells is limited, and after a given number of replication cycles, they undergo senescence characterized by a stop in cell cycle, expression of a pro-inflammatory phenotype, and defects in endothelial cell function including permeability and vascular tone regulation (El Assar et al. 2012). In agreement, we showed that endothelial cells from visceral adipose tissue of obese patients exhibit a senescent-like phenotype characterized by an increased expression of chemokines and adhesion molecules as well as cellular senescence markers (Villaret et al. 2010). Moreover, in vitro treatment of human adipose tissue native endothelial cells with VEGF-A is associated with an increased endothelial cell senescence. These results suggest that, via the promotion of endothelial cell senescence, the local pro-angiogenic microenvironment might contribute to an impairment of endothelial cell functions, leading to hypoxia and further accumulation of immune cells (Fig. 11.2). Ultimately, reduction of the adipose tissue storage capacity will favor ectopic fat deposits and insulin resistance. Such hypothesis remains to be clearly established. However, it may explain why stimulation of angiogenesis in the early state of adipose tissue development is beneficial and could prevent obesity-associated disorders (Elias et al. 2012; Sun et al. 2012), whereas its blockade in the late stage, such as in already severe obese and in diabetic mice (ob/ob), also improves systemic metabolic health (Sun et al. 2012; Brakenhielm et al. 2004; Dallabrida and Rupnick 2002).

11.6 Conclusion

It is now well established that immune cells play a key role in the link between obesity and associated pathologies, such as type 2 diabetes. Although the sequential and causal roles remain to be clearly elucidate, approaches performed in mice models (mostly in DIO mice) suggest that imbalanced energy homeostasis is associated with a primary accumulation of B and T lymphocytes within the expanding adipose tissue, followed by macrophages. Immune cell accumulation in adipose tissue contributes to the obesity-associated inflammatory state, promoting then insulin resistance through direct interactions between inflammatory cytokines and factors and insulin signaling pathways. Data obtained from human studies showed that established obesity is associated with a chronic unresolved inflammatory state, characterized by the accumulation of non-strictly polarized macrophages and Th-17 lymphocytes, involved in the local control of angiogenesis and adipogenesis. Their involvement in these two processes might lead to defects in both endothelial cell functions, by promoting endothelial cells senescence, and adipose tissue storage capacity, by limiting the normal turnover of mature adipocytes and the appearance of new functional adipocytes. Finally, such effects may indirectly contribute to ectopic fat depot in liver, muscle, and pancreas and therefore may lead to systemic insulin resistance and type 2 diabetes. Thus, although under normal conditions angiogenesis controlled by adipose tissue macrophages is required for proper adipose tissue function and growth, under chronic inflammation due to chronic energy overload, such a phenomenon might become pathological. Additional data are needed to clearly establish the role of immune cells on adipose tissue angiogenesis and function to determine the effects of anti-inflammatory therapies in obesity and type 2 diabetes.

Acknowledgment The work was supported by fundings from the Fondation pour la recherche médicale, Astra-Zenica, and INSERM.

References

Bastard JP, Jardel C, Bruckert E, Blondy P, Capeau J, Laville M, et al. Elevated levels of interleukin 6 are reduced in serum and subcutaneous adipose tissue of obese women after weight loss. J Clin Endocrinol Metab. 2000a;85(9):3338–42.

Bastard JP, Jardel C, Bruckert E, Vidal H, Hainque B. Variations in plasma soluble tumour necrosis factor receptors after diet-induced weight loss in obesity. Diabetes Obes Metab. 2000b;2(5):323–5.

Bendelac A, Savage PB, Teyton L. The biology of NKT cells. Annu Rev Immunol. 2007; 25:297–336.

Bertola A, Deveaux V, Bonnafous S, Rousseau D, Anty R, Wakkach A, et al. Elevated expression of osteopontin may be related to adipose tissue macrophage accumulation and liver steatosis in morbid obesity. Diabetes. 2009;58(1):125–33.

Bertola A, Ciucci T, Rousseau D, Bourlier V, Duffaut C, Bonnafous S, et al. Identification of adipose tissue dendritic cells correlated with obesity-associated insulin-resistance and inducing Th17 responses in mice and patients. Diabetes. 2012;61:2238–47.

Bourlier V, Zakaroff-Girard A, Miranville A, De Barros S, Maumus M, Sengenes C, et al. Remodeling phenotype of human subcutaneous adipose tissue macrophages. Circulation. 2008;117(6):806–15.

Bourlier V, Sengenes C, Zakaroff-Girard A, Decaunes P, Wdziekonski B, Galitzky J, et al. TGFbeta family members are key mediators in the induction of myofibroblast phenotype of human adipose tissue progenitor cells by macrophages. PLoS One. 2012;7(2):e31274.

Brakenhielm E, Cao R, Gao B, Angelin B, Cannon B, Parini P, et al. Angiogenesis inhibitor, TNP-470, prevents diet-induced and genetic obesity in mice. Circ Res. 2004;94(12):1579–88.

Cancello R, Henegar C, Viguerie N, Taleb S, Poitou C, Rouault C, et al. Reduction of macrophage infiltration and chemoattractant gene expression changes in white adipose tissue of morbidly obese subjects after surgery-induced weight loss. Diabetes. 2005;54(8):2277–86.

Cancello R, Tordjman J, Poitou C, Guilhem G, Bouillot JL, Hugol D, et al. Increased infiltration of macrophages in omental adipose tissue is associated with marked hepatic lesions in morbid human obesity. Diabetes. 2006;55(6):1554–61.

Caspar-Bauguil S, Cousin B, Bour S, Casteilla L, Penicaud L, Carpene C. Adipose tissue lymphocytes: types and roles. J Physiol Biochem. 2009;65(4):423–36.

Cho CH, Koh YJ, Han J, Sung HK, Jong Lee H, Morisada T, et al. Angiogenic role of LYVE-1-positive macrophages in adipose tissue. Circ Res. 2007;100(4):e47–57.

Cinti S, Mitchell G, Barbatelli G, Murano I, Ceresi E, Faloia E, et al. Adipocyte death defines macrophage localization and function in adipose tissue of obese mice and humans. J Lipid Res. 2005;46(11):2347–55.

Clement K, Viguerie N, Poitou C, Carette C, Pelloux V, Curat CA, et al. Weight loss regulates inflammation-related genes in white adipose tissue of obese subjects. FASEB J. 2004;18(14):1657–69.

Constant VA, Gagnon A, Landry A, Sorisky A. Macrophage-conditioned medium inhibits the differentiation of 3T3-L1 and human abdominal preadipocytes. Diabetologia. 2006;49(6):1402–11.

Dallabrida SM, Rupnick MA. Vascular endothelium in tissue remodeling: implications for heart failure. Cold Spring Harb Symp Quant Biol. 2002;67:417–27.

Deiuliis J, Shah Z, Shah N, Needleman B, Mikami D, Narula V, et al. Visceral adipose inflammation in obesity is associated with critical alterations in tregulatory cell numbers. PLoS One. 2011;6(1):e16376.

Duffaut C, Galitzky J, Lafontan M, Bouloumie A. Unexpected trafficking of immune cells within the adipose tissue during the onset of obesity. Biochem Biophys Res Commun. 2009a;384(4):482–5.

Duffaut C, Zakaroff-Girard A, Bourlier V, Decaunes P, Maumus M, Chiotasso P, et al. Interplay between human adipocytes and T lymphocytes in obesity: CCL20 as an adipochemokine and T lymphocytes as lipogenic modulators. Arterioscler Thromb Vasc Biol. 2009b;29(10): 1608–14.

El Assar M, Angulo J, Vallejo S, Peiro C, Sanchez-Ferrer CF, Rodriguez-Manas L. Mechanisms involved in the aging-induced vascular dysfunction. Front Physiol. 2012;3:132.

Elias I, Franckhauser S, Ferre T, Vila L, Tafuro S, Munoz S, et al. Adipose tissue overexpression of vascular endothelial growth factor protects against diet-induced obesity and insulin resistance. Diabetes. 2012;61:1801–13.

Eller K, Kirsch A, Wolf AM, Sopper S, Tagwerker A, Stanzl U, et al. Potential role of regulatory T cells in reversing obesity-linked insulin resistance and diabetic nephropathy. Diabetes. 2011;60(11):2954–62.

Feng B, Jiao P, Nie Y, Kim T, Jun D, van Rooijen N, et al. Clodronate liposomes improve metabolic profile and reduce visceral adipose macrophage content in diet-induced obese mice. PLoS One. 2011;6(9):e24358.

Feuerer M, Herrero L, Cipolletta D, Naaz A, Wong J, Nayer A, et al. Lean, but not obese, fat is enriched for a unique population of regulatory T cells that affect metabolic parameters. Nat Med. 2009;15(8):930–9.

Han J, Lee JE, Jin J, Lim JS, Oh N, Kim K, et al. The spatiotemporal development of adipose tissue. Development. 2011;138(22):5027–37.

Heilbronn LK, Noakes M, Clifton PM. Energy restriction and weight loss on very-low-fat diets reduce C-reactive protein concentrations in obese, healthy women. Arterioscler Thromb Vasc Biol. 2001;21(6):968–70.

Hotamisligil GS. Inflammation and metabolic disorders. Nature. 2006;444(7121):860–7.

Ilan Y, Maron R, Tukpah AM, Maioli TU, Murugaiyan G, Yang K, et al. Induction of regulatory T cells decreases adipose inflammation and alleviates insulin resistance in ob/ob mice. Proc Natl Acad Sci U S A. 2010;107(21):9765–70.

Ji Y, Sun S, Xu A, Bhargava P, Yang L, Lam KS, et al. Activation of natural killer T cells promotes M2 Macrophage polarization in adipose tissue and improves systemic glucose tolerance via interleukin-4 (IL-4)/STAT6 protein signaling axis in obesity. J Biol Chem. 2012;287(17): 13561–71.

Kintscher U, Hartge M, Hess K, Foryst-Ludwig A, Clemenz M, Wabitsch M, et al. T-lymphocyte infiltration in visceral adipose tissue: a primary event in adipose tissue inflammation and the development of obesity-mediated insulin resistance. Arterioscler Thromb Vasc Biol. 2008;28(7):1304–10.

Kotas ME, Lee HY, Gillum MP, Annicelli C, Guigni BA, Shulman GI, et al. Impact of CD1d deficiency on metabolism. PLoS One. 2011;6(9):e25478.

Lacasa D, Taleb S, Keophiphath M, Miranville A, Clement K. Macrophage-secreted factors impair human adipogenesis: involvement of proinflammatory state in preadipocytes. Endocrinology. 2007;148(2):868–77.

Lee YS, Li P, Huh JY, Hwang IJ, Lu M, Kim JI, et al. Inflammation is necessary for long-term but not short-term high-fat diet-induced insulin resistance. Diabetes. 2011;60(10):2474–83.

Liu J, Divoux A, Sun J, Zhang J, Clement K, Glickman JN, et al. Genetic deficiency and pharmacological stabilization of mast cells reduce diet-induced obesity and diabetes in mice. Nat Med. 2009;15(8):940–5.

Lynch L, O'Shea D, Winter DC, Geoghegan J, Doherty DG, O'Farrelly C. Invariant NKT cells and CD1d(+) cells amass in human omentum and are depleted in patients with cancer and obesity. Eur J Immunol. 2009;39(7):1893–901.

Mantell BS, Stefanovic-Racic M, Yang X, Dedousis N, Sipula IJ, O'Doherty RM. Mice lacking NKT cells but with a complete complement of CD8+ T-cells are not protected against the metabolic abnormalities of diet-induced obesity. PLoS One. 2011;6(6):e19831.

McDonnell ME, Ganley-Leal LM, Mehta A, Bigornia SJ, Mott M, Rehman Q, et al. B lymphocytes in human subcutaneous adipose crown-like structures. Obesity (Silver Spring). 2012;20:1372–8.

Miyazawa-Hoshimoto S, Takahashi K, Bujo H, Hashimoto N, Saito Y. Elevated serum vascular endothelial growth factor is associated with visceral fat accumulation in human obese subjects. Diabetologia. 2003;46(11):1483–8.

Nishimura S, Manabe I, Nagasaki M, Eto K, Yamashita H, Ohsugi M, et al. CD8+ effector T cells contribute to macrophage recruitment and adipose tissue inflammation in obesity. Nat Med. 2009;15(8):914–20.

Pang C, Gao Z, Yin J, Zhang J, Jia W, Ye J. Macrophage infiltration into adipose tissue may promote angiogenesis for adipose tissue remodeling in obesity. Am J Physiol Endocrinol Metab. 2008;295(2):E313–22.

Potente M, Ghaeni L, Baldessari D, Mostoslavsky R, Rossig L, Dequiedt F, et al. SIRT1 controls endothelial angiogenic functions during vascular growth. Genes Dev. 2007;21(20):2644–58.

Qian BZ, Pollard JW. Macrophage diversity enhances tumor progression and metastasis. Cell. 2010;141(1):39–51.

Rausch ME, Weisberg S, Vardhana P, Tortoriello DV. Obesity in C57BL/6J mice is characterized by adipose tissue hypoxia and cytotoxic T-cell infiltration. Int J Obes (Lond). 2008;32(3):451–63.

Schledzewski K, Falkowski M, Moldenhauer G, Metharom P, Kzhyshkowska J, Ganss R, et al. Lymphatic endothelium-specific hyaluronan receptor LYVE-1 is expressed by stabilin-1+, F4/80+, CD11b+ macrophages in malignant tumours and wound healing tissue in vivo and in bone marrow cultures in vitro: implications for the assessment of lymphangiogenesis. J Pathol. 2006;209(1):67–77.

Skurk T, Alberti-Huber C, Herder C, Hauner H. Relationship between adipocyte size and adipokine expression and secretion. J Clin Endocrinol Metab. 2007;92(3):1023–33.

Sun K, Kusminski CM, Scherer PE. Adipose tissue remodeling and obesity. J Clin Invest. 2011;121(6):2094–101.

Sun K, Asterholm IW, Kusminski CM, Bueno AC, Wang ZV, Pollard JW, et al. Dichotomous effects of VEGF-A on adipose tissue dysfunction. Proc Natl Acad Sci U S A. 2012;109(15): 5874–9.

Villaret A, Galitzky J, Decaunes P, Esteve D, Marques MA, Sengenes C, et al. Adipose tissue endothelial cells from obese human subjects: differences among depots in angiogenic, metabolic, and inflammatory gene expression and cellular senescence. Diabetes. 2010;59(11):2755–63.

Wagner M, Bjerkvig R, Wiig H, Melero-Martin JM, Lin RZ, Klagsbrun M, et al. Inflamed tumor-associated adipose tissue is a depot for macrophages that stimulate tumor growth and angiogenesis. Angiogenesis. 2012;15:481–95.

Weisberg SP, McCann D, Desai M, Rosenbaum M, Leibel RL, Ferrante Jr AW. Obesity is associated with macrophage accumulation in adipose tissue. J Clin Invest. 2003;112(12):1796–808.

Winer S, Chan Y, Paltser G, Truong D, Tsui H, Bahrami J, et al. Normalization of obesity-associated insulin resistance through immunotherapy. Nat Med. 2009;15(8):921–9.

Winer DA, Winer S, Shen L, Wadia PP, Yantha J, Paltser G, et al. B cells promote insulin resistance through modulation of T cells and production of pathogenic IgG antibodies. Nat Med. 2011;17(5):610–7.

Wu H, Ghosh S, Perrard XD, Feng L, Garcia GE, Perrard JL, et al. T-cell accumulation and regulated on activation, normal T cell expressed and secreted upregulation in adipose tissue in obesity. Circulation. 2007;115(8):1029–38.

Xu H, Barnes GT, Yang Q, Tan G, Yang D, Chou CJ, et al. Chronic inflammation in fat plays a crucial role in the development of obesity-related insulin resistance. J Clin Invest. 2003;112(12):1821–30.

Xu F, Burk D, Gao Z, Yin J, Zhang X, Weng J, et al. Angiogenic deficiency and adipose tissue dysfunction are associated with macrophage malfunction in SIRT1(−)/(−) mice. Endocrinology. 2012;153(4):1706–16.

Yang ZF, Poon RT, Luo Y, Cheung CK, Ho DW, Lo CM, et al. Up-regulation of vascular endothelial growth factor (VEGF) in small-for-size liver grafts enhances macrophage activities through VEGF receptor 2-dependent pathway. J Immunol. 2004;173(4):2507–15.

Ye J. Adipose tissue vascularization: its role in chronic inflammation. Curr Diab Rep. 2011;11(3):203–10.

Zaragosi LE, Wdziekonski B, Villageois P, Keophiphath M, Maumus M, Tchkonia T, et al. Activin a plays a critical role in proliferation and differentiation of human adipose progenitors. Diabetes. 2010;59(10):2513–21.

Zeyda M, Farmer D, Todoric J, Aszmann O, Speiser M, Gyori G, et al. Human adipose tissue macrophages are of an anti-inflammatory phenotype but capable of excessive pro-inflammatory mediator production. Int J Obes (Lond). 2007;31(9):1420–8.

Zeyda M, Huber J, Prager G, Stulnig TM. Inflammation correlates with markers of T-cell subsets including regulatory T cells in adipose tissue from obese patients. Obesity (Silver Spring). 2011;19(4):743–8.

Chapter 12
Adipose Tissue Hypoxia in Regulation of Angiogenesis and Obesity

Zoi Michailidou and Jonathan R. Seckl

Abstract Obesity is strongly associated with co-morbidities such as diabetes, hypertension, atherosclerotic cardiovascular disease and stroke, osteoarthritis, depression and certain cancers, notably of the breast, colon, oesophagus, pancreas, endometrium, kidney and gall bladder. For most of these morbidities metabolic abnormalities originating in pathologically expanded adipose tissues play a key etiological role. In obesity, rapid adipose expansion occurs. In many this is linked with local hypoxia, inadequate vascularization and consequent fibrosis. However, some individuals are resilient and less adversely affected by excess fat accumulation. The notion of 'healthy' adipose expandability is being proposed and documented in some animal models and in subgroup of obese individuals. Such healthy adipose expansion requires the fine coordination of adipogenesis and angiogenesis (vascular remodelling) to support expansion by providing the oxygen and nutrients necessary for the adipocyte survival and function. If this balance fails adipose tissue fails to respond to any metabolically challenging situation. Here we focus on one of the mechanisms, adipose tissue hypoxia, involved in regulating adiposity and angiogenesis during obesity.

Keywords Hypoxia • Adipose • Angiogenesis • Vasculature • Obesity • Hypoxia-inducible factor • Fibrosis • Macrophages • Vascular-endothelial growth factor

---

Z. Michailidou (✉)
Medical Research Council Centre for Inflammation Research, University of Edinburgh, Queen's Medical Research Institute, 47 Little France Crescent, Edinburgh EH16 4TJ, Scotland
e-mail: v1zmicha@staffmail.ed.ac.uk

J.R. Seckl (✉)
University/British Heart Foundation Centre for Cardiovascular Science, University of Edinburgh, Queen's Medical Research Institute, 47 Little France Crescent, Edinburgh EH16 4TJ, Scotland
e-mail: J.Seckl@ed.ac.uk

Y. Cao (ed.), *Angiogenesis in Adipose Tissue*, DOI 10.1007/978-1-4614-8069-3_12,

Abbreviations

ASC	Adipose stromal cells
AT	Adipose tissue
BAT	Brown adipose tissue
DIO	Diet-induced obesity
FIH	Factor inhibiting HIF
11β-HSD1	11β-Hydroxysteroid dehydrogenase type 1
HIF	Hypoxia-inducible factor
KO	Knockout
PDGF	Platelet-derived growth factor
PHD	Prolyl hydroxylase domain
POMC	Proopiomelanocortin
WAT	White adipose tissue

12.1 Obesity and Adipose Tissue Hypoxia

In obesity, adipose tissue becomes markedly expanded with both accumulation of lipid energy stores in existing adipocytes, which can massively expand, and the recruitment of new adipose cells from precursors in the stromovascular component (preadipocytes, progenitor cells). This change in tissue size, unparalleled in non-neoplastic tissues, may challenge the ability of the vasculature to supply oxygen for cellular respiration of adipocytes and other cells and to remove their metabolic waste products. The concept that reduced oxygenation (hypoxia) in adipose tissue with obesity is one of the mechanisms leading to structural and metabolic dysfunction, and in particular insulin resistance, has been increasingly documented in a number of studies published in the last 5 years. These have mainly employed diet-induced obesity (DIO) or leptin/leptin-signalling-deficient obesity models to show that with adipocyte enlargement and lipid accumulation in obesity, there is a reduction in oxygen diffusion and cells are exposed to low oxygen tension.

12.1.1 The Oxygen Sensing Pathway: HIF as a Central Regulator

In healthy tissues the oxygen tension is 20–70 mmHg (2.5–9 % oxygen). The diffusion distance of oxygen in tissues is approximately 120 μm (Helminger et al. 1997). Therefore for an enlarged adipocyte, which may be 150–200 μm in diameter (Bluher et al. 2004), cellular size may be expected to limit the oxygen diffusion, especially to distal areas remote from the vascular supply in adipose tissue, thus creating a hypoxic state. However, cells have mechanisms to sense oxygen tension.

The key mediators of cellular adaptation to oxygen lack are the hypoxia-inducible factors, HIF-1α and HIF-2α. These are basic helix-loop-helix transcriptions factors and consist of a constitutively expressed β subunit (HIF-1β or ARNT) and an oxygen-dependent HIF-α subunit. Under hypoxic conditions HIF-1α hydroxylation is inhibited and HIF-1α is stabilized (Maxwell et al. 1999; Ivan et al. 2001; Jaakkola et al. 2001). In mammalian cells, three HIF prolyl hydroxylases have been identified; PHD1, PHD2 and PHD3 (also called oxygen sensors). All three hydroxylate two proline residues (pro-402 and pro-564) in the oxygen-dependent degradation domain (ODDD) of HIF-α (Epstein et al. 2001). This allows binding to the von Hippel-Lindau (VHL) E3 ubiquitin ligase complex and proteosomal degradation of HIF-α (Jaakkola et al. 2001; Ivan et al. 2001). A fourth hydroxylase, factor inhibiting HIF (FIH), hydroxylates a conserved asparagine, thus blocking binding to the transcriptional activator p300 and subsequent transcription of hypoxia-inducible genes (Mahon et al. 2001; Lando et al. 2002). Mutations in human VHL, PHDs and FIH genes increase HIF activity (Girgis et al. 2012). Intriguingly, apart from their strong involvement in regulating oxygen transport, glucose metabolism and angiogenesis under cellular oxygen depletion, HIFs also orchestrate fibrogenesis by inducing the expression of extracellular matrix (ECM)-modifying factors (i.e. TIMP-1, PAI-1), connective tissue growth factor and lysyl oxidase genes (Higgins et al. 2004; Higgins et al. 2007; Tzouvelekis et al. 2007; Halberg et al. 2009). Scar formation may further restrict blood supply and oxygen availability, spawning a pathological cycle. It is this triangulation between hypoxia, fibrosis and angiogenesis that may underpin the likelihood of pathogenesis as adipose tissues expand under chronic caloric excess.

12.1.2 The Adipose Tissue Hypoxia Hypothesis; a Central Cause of Tissue Dysfunction in Obesity?

Excessive expansion of adipose tissue could lead to hypoxia. Indeed there is a dramatic reduction of 70 % (from 47.9 to 15.2 mmHg) in interstitial PO_2 in epididymal adipose in obese mouse models (Ye et al. 2007). Poor adipose oxygenation has been also observed in obese humans (Kabon et al. 2004; Pasarica et al. 2009), who show 15 % lower pO2 than normal weight individuals (55 ± 9 vs. 47 ± 11 mmHg) in abdominal subcutaneous fat (Pasarica et al. 2009). Hypoxia has been proposed to mediate macrophage infiltration, impaired insulin signalling and fibrosis with obesity, suggesting is via increased HIF-1α expression (Trayhurn and Wood 2004; Hosogai et al. 2007; Ye et al. 2007; Rausch et al. 2008; Yin et al. 2009; Pasarica et al. 2009; Regazzetti et al. 2009; Halberg et al. 2009). In vitro there is compelling evidence that hypoxia has profound effects on adipocyte functions, inhibiting adipogenesis, stimulating proangiogenic responses such as vascular-endothelial growth factor (VEGF), ANGPLT4 and leptin secretion and activating pro-fibrotic (collagens, matrix metalloproteinases) gene expression (Tajima et al. 2001; Grosfeld et al. 2002; Lolmede et al. 2003; Shimba et al. 2004; Lin et al. 2006; Hosogai et al. 2007; Ye et al. 2007; Halberg et al. 2009; Mazzati et al. 2012). All these studies hypothesized that excessive fat expansion leads to hypoxia and activation of HIF-1α and this in turn triggers macrophages to infiltrate the tissue. Thus proposing that (a) induction of HIF-1α is the direct effect of adipose hypoxia and (b) that this HIF-1α increase causes metabolic dysfunction, especially insulin resistance. It is important to note that, apart from tissue hypoxia *per se* HIF-1α can be regulated (stabilized) by other pathway(s), notably by inflammatory cytokines (Haddad and Land 2001; Yung et al. 2003) and insulin (Feldser et al. 1999; Treins et al. 2002; He et al. 2011) which are both hallmarks of a dysfunctional adipose in obesity, making the cause-and-effect and the time of these dynamic cellular processes difficult to dissect.

12.1.3 Modulation of Adipose HIF-System in Mice

The next key question is whether or not HIF-1α induction per se leads to adipose tissue pathology and its metabolic consequences. Recently, this has been addressed directly by manipulating HIF in adipose tissue. This has provided interesting findings but with some critical inconsistencies between studies that complicate understanding. A summary of the mouse models used to date can be found in Table 12.1. The first study published used an adipose-specific HIF-1α transgenic over-expression approach. This study reported that stabilization of HIF-1α increases body weight under basal conditions and with DIO, with hypertrophic subcutaneous adipocytes. The transgene caused glucose intolerance and macrophage infiltration (Halberg et al. 2009), suggesting that increased HIF-1α in adipose tissue in obese mouse models might play a pathogenic role. A key finding was that activation of HIF-1α failed to induce the expected proangiogenic response in adipose tissue, overturning the common view that hypoxia induces angiogenesis through HIF-1α. In contrast, HIF-1α over-expression led to collagen deposition, up-regulation of a number of pro-fibrotic genes and subsequent fibrosis by directly interacting with the major collagen I and III crosslinking molecule lysyl oxidase (Halberg et al. 2009). A series of subsequent studies used adipose-specific HIF knockout models to investigate if deletion of HIF has a protective effect in DIO. Indeed, both HIF-1α and HIF-1β (ARNT) deletion in adipocytes (using the ap2 [fabp4] promoter-driven cre to produce adipose knock-down of 'floxed' gene targets) protected mice from DIO (Jiang et al. 2011; Lee et al. 2011; Krishnan et al. 2012). All models displayed reduced fat mass and adipocyte size, increased adipose mitochondrial biogenesis, increased energy expenditure, with improved glucose tolerance and insulin sensitivity (Jiang et al. 2011; Lee et al. 2011; Krishnan et al. 2012). In contrast, another model of adipose tissue HIF-1α suppression, expression of a dominant negative human HIF-1α (also under the ap2 promoter), produced the opposite phenotype of obesity susceptibility, with all its attendant downstream metabolic abnormalities, mainly due to loss of thermogenic capacity in brown adipose tissue (BAT) (Zhang et al. 2010a, b). Adipose-VHL deletion (allowing unfettered activation of HIF) led to

Table 12.1 Summary of the phenotype of transgenic models used to dissect the role of HIF in obesity

Model	Target	Gene levels	Tissue target	AT phenotype	Metabolic phenotype	AT vasculature phenotype	Other	Proposed mechanism	Reference
HIF-Tg (HIF-1α over-expression)	Dominant active human HIF-1α (ODD removed) under ap2 promoter	Increased HIF half life from 5 to 60 min. variable levels between fat pads. Higher in scAT	Adipose	Bigger scAT adipocytes, Mφ infiltration, fibrosis/ collagen deposition	Increased BW, Glucose intolerance	Unchanged/ not affected	NA	Direct regulation of HIF-1α on LOX-mediated fibrotic pathway	Halberg et al. (2009)
dnHIF-1α-Tg (deletion mutant)	Dominant negative (dn) human HIF-1α over-expression under ap2 promoter	Increased HIF-1α dn levels. Higher in BAT	Adipose	Increased total fat mass (WAT&BAT), adipocyte size and Mφ infiltration	Increased BW, reduced energy expenditure and thermo-genesis, impaired glucose and insulin tolerance	Decreased angio-genesis (VEGF, CD31) in BAT	Lower core body temperature, decreased mitochondrial biogenesis (WAT&BAT)	Loss of thermogenic capacity in BAT	Zhang et al. (2010a, b)
					Reduced energy expenditure	Not affected in WAT			

(continued)

Table 12.1 (continued)

Model	Target	Gene levels	Tissue target	AT phenotype	Metabolic phenotype	AT vasculature phenotype	Other	Proposed mechanism	Reference
HIF-1α$^{\Delta adipo}$	Ap2 Cre/lox	Reduced HIF-1α (~88 % in adipocytes)	Adipose	Reduced fat mass and adipocyte size	Reduced BW, improved glucose tolerance and higher insulin sensitivity. Higher serum adiponectin levels	NA	Increase adipogenesis/ glucose metabolism genes. Decreased inflammation, fibrosis-related genes	Decreased SOCS3 improves insulin signalling. Unclear if it is HIF-1 direct or indirect target	Jiang et al. (2011)
ARNT$^{\Delta adipo}$	Ap2 Cre/lox	Reduced ARNT (50 % in WAT&BAT, absent in adipocytes)			Reduced serum TG & FFA				
HIF-1α iC KO	Tamoxifen-inducible Ap2cre/ lox or	Reduced HIF-1α	Adipose	Reduced visceral fat mass and adipocyte size	Reduced BW, increased energy expenditure	NA	Increased mitochondrial biogenesis	Suppression of β-oxidation through suppression of sirtuin-2	Krishnan et al. (2012)
HIF-1α BATcKO	UCP-1 cre/lox		BAT	Not affected					

FH1BKO (ARNTKO)	Ap2cre/lox	Reduced HIF-1β (ARNT) by 90 % in adipocytes- In vitro Inhibition of both HIF-α	Adipose	Reduced visceral and sc fat mass, lipid accumulation, smaller adipocytes, similar degree of hypoxia, firosis and inflammation	Reduced BW (24 %)-less BW gain after HFD, better glucose tolerance (only in older mice). Increased energy expenditure	Unaffected vessel density Reduced vascular perme-ability (reduced VEGF)	Resist hypoxia-stimulated reduction of basal and maximal mitochondrial respiration	Decreased glucose transporters Decreased glucose uptake Decreased basal and insulin-stimulated lipogenesis	Lee et al. (2011)
POMC/HIF-1βLox/Lox	POMCcre/lox	Reduced HIF-1β by 90 % in POMC neurons	POMC neurons (hypo-thalamus and pituitary)	Increased fat mass and adipocyte size after HF challenge	Increased food intake, exacerbated DIO	NA	Glucose directly upregulates HIF-2α in hypothalamus via suppres-sion of PHDs hydroxylation activity	Direct effects of HIFs on food intake. Delivery of HIF1α/HIFβ or HIF2α/HIFβ in the hypothalamus resisted DIO New role of HIFS in glucose sensing	Zhang et al. (2011)
Ap2crennVHL	Ap2cre/lox	VHL loss. Lethality E14.5–18.5	Adipose and embry-onic CNS	NA	NA	NA	Brain haemor-rhage, endothelial proliferation and capillary leakage	NA	Zhang et al. (2012)
Heterozygotes	Ap2cre/lox	VHL levels 50 %			Normal				

NA; not applicable, has not been studied or reported

brain haemorrhage and lethality, due to 'ectopic' ap2-cre expression in the CNS during development (Zhang et al. 2012). With all the limitations of the transgenic mouse models used and especially the use of ap2 (Fabp4)-cre as an 'adipose'-specific promoter (which may also delete 'floxed' gene targets in activated macrophages and cells of the developing CNS), it can be tentatively inferred that inactivation of HIF-1α in adipocytes may have beneficial effects in DIO, particularly upon glucose homeostasis and fat mass regulation. Intriguingly, a role of HIF in the central (brain) regulation of metabolism has been highlighted. Firstly, the FIH KO (HIF activation) mouse showed a hyper-metabolic phenotype with increased food intake, energy expenditure, high ventilation rate, reduced fat mass accumulation and improved insulin sensitivity, both basally and with DIO (Zhang et al. 2010a, b). Most of the hyper-metabolic phenotype features were recapitulated in a neuron-specific FIH deletion (Zhang et al. 2010a, b). Though from this model it is not robust to conclude a direct role of HIF in the CNS, since FIH was deleted from all neurons and FIH not only regulates HIF-α but other proteins too (Coleman et al. 2007; Ferguson et al. 2007; Cockman et al. 2009; Wilkins et al. 2009), Zhang et al. (2012) more-specifically deleted HIF-1β in neurons expressing proopiomelanocortin (POMC), a key neuropeptide expressed in hypothalamic (arcuate nucleus) neurons involved in suppression of food intake. The resultant mice were hyperphagic, had increased fat mass accumulation and exacerbated downstream metabolic abnormalities when fed a high fat diet. Finally, lentiviral delivery of HIF-α directly to the hypothalamus protects against DIO (Zhang et al. 2010a, b). Additionally, this study revealed a novel role of HIF in directly acting as a glucose sensor in the brain (Zhang et al. 2010a, b). Taken together, these studies clearly show an important role of the HIF system in energy balance and adipose function, but also reveal a complexity of loci involved and interrelationships that are far from resolved. Even the proposed canonical relationship between adipose tissue hypoxia and local induction of HIF-α to restore vascular supply is inconsistent. The interplay between HIF in adipocytes and brain in the regulation of body fat requires further understanding.

12.2 Adipose Tissue Hypoxia and Angiogenesis. DO HIFs MATTER?

Healthy adipose tissue growth requires a coordinated expansion of its vascular network (Hausman and Richardson 2004). Adipose tissue is normally well-vascularized and almost all adipocytes are directly supported by capillaries that provide the oxygen, nutrients, hormones and growth factors required for function and, presumably, healthy expansion under short-to-medium term caloric excess (Cao 2010). In obesity, adipose mass increases but there is often an inadequate vascular network, as detailed above. The mechanisms behind the fine balance between adipogenesis and angiogenesis and its failure in some obesity are not well understood. Additionally, adipocytes are organized in a network of ECM that provides mechanical support and mediates remodelling of the tissue during expansion or reduction of fat mass (Sun et al. 2011). During accelerated fat expansion in obesity, ECM components are up-regulated leading to collagen deposition, fibrosis and subsequent reduced plasticity and metabolic dysfunction in adipose (Kahn et al. 2009; O'Hara et al. 2009). The issue of adipose hypoxia and angiogenesis is a 'chicken and egg' question. Is it hypoxia that drives angiogenesis as an adaptive mechanism which may be inadequate for the metabolic needs of obese adipose cells, or is it failure of vascular remodelling during expansion that leads to hypoxia and fibrosis, or both?

12.2.1 Adipose Depot-Specific Responses to Hypoxia-Induced Angiogenesis

Adipose tissue from distinct anatomical sites (subcutaneous, visceral, BAT) exhibits different degrees of hypoxia under basal or obesogenic conditions. We and others have shown that visceral adipose tissue, the intraabdominal depots which drain via the hepatic portal vein directly to the liver and are conventionally associated with metabolic disease risk, is more hypoxic than subcutaneous (peripheral) adipose tissues, which are believed to reflect 'safer' more metabolically benign stores for excess calories (Michailidou et al. 2012; Farb et al. 2012). The inhomogeneity of adipose tissue oxygenation even under basal conditions suggests differences in vascularization. Indeed human subcutaneous adipose has higher capillary density compared to visceral adipose (Gealekman et al. 2011; Villaret et al. 2010). Moreover, visceral adipose expression of proinflammatory, oxidative stress-related, hypoxia-induced genes is much higher (O'Rourke et al. 2009, 2011; Farb et al. 2012), suggesting that the visceral adipose microenvironment may be intrinsically more 'toxic' to surrounding microvessels (Farb et al. 2012) and that individuals prone to visceral fat accumulation are at higher risk of hypoxia-induced metabolic dysfunction. WAT and BAT synthesize and release numerous proangiogenic factors and adipokines that stimulate neovascularization and adipose/vasculature remodelling (Cao 2007). One of the key families of angiogenic factors, VEGF, plays a crucial role in the regulation of angiogenesis and vascular remodelling in many tissues under physiological and pathological conditions (Dvorak et al. 1995; Ferrara et al. 1996; Carmeliet 2000; Carmeliet and Collen 2000). However, despite its apparently considerable proangiogenic ability, many rapidly expanding WAT depots develop hypoxia. On the other hand, BAT is particularly highly vascularized, a feature that supports its main role in thermogenesis, a function that requires a high rate of blood perfusion to supply oxygen and convey the release of heat to the body (Cao 2007). However, even brown adipose becomes hypoxic during thermogenesis (Foster and Frydman 1978). Xue et al. (2009), using cold as a stimulator of thermogenesis, showed that adipose tissues (white and brown) developed hypoxia (increased HIF-2α expression and intense hypoxyprobe staining) and exhibited a marked increase in vascularization. What is interesting is that the neovascularization was not hypoxia-dependent. Rather it was attributed to cold directly inducing the sympathetic innervation of adipose tissues and thus stimulating uncoupling protein (UCP-1) and up-regulating VEGF expression via PGC-1α (Xue et al. 2009). These authors further showed that in UCP-1 KO mice there was still active angiogenesis but no hypoxia in adipose tissues during cold exposure, indicating angiogenesis was not caused by hypoxia via the classical HIF-mediated induction of VEGF gene expression (Xue et al. 2009). Yet again adipose tissue responses to hypoxic challenge, in this case their angiogenic responses, appear at least in part independent of classical HIF-α.

12.2.2 Is Adipose Tissue Angiogenesis HIF-Dependent?

Given the canonical pathway of hypoxia-stimulated angiogenesis acting via activation of HIF-1α, it is intriguing that most studies to date (Table 12.1) provide limited information regarding the effects of HIF on adipose vasculature in obesity. Three studies have looked at the adipose vasculature under adipose-specific HIF manipulation and showed unaffected vessel density in WATs (Halberg et al. 2009; Zhang et al. 2010a, b; Lee et al. 2011). Adipose HIF-β ablation (and thus inactivation of HIF-α) led to reduced VEGF-A expression and vascular permeability (measured by extravasation of Eva's Blue dye) (Lee et al. 2011). On the other hand, the effects of a dominant negative HIF-1α reduced VEGF-A and angiogenesis was only apparent in brown adipose (Zhang et al. 2010a, b). These were all subtle effects on the adipose vasculature and certainly not consistent throughout the models used. It is clear that there is a need for the design of more specific studies that target the adipose vasculature to dissect whether or not HIF plays a direct role in stimulating adipose angiogenesis. From limited data so far it is suggested that angiogenesis during adipose expansion can develop with or without HIF activation.

Recent studies have focused on investigating if inducing angiogenesis in adipose tissue has any beneficial effects in terms of reducing hypoxia and metabolic abnormalities. Indeed, transgenic over-expression of VEGF-A in adipose tissue (either using the adiponectin or ap2 promoters) leads to higher vessel density, reduced hypoxic areas, with lower HIF-1α levels, reduced fibrosis and ECM accumulation, reduced local and systemic inflammation, improved glucose tolerance and insulin sensitivity and more efficient lipid clearance (Sun et al. 2012; Elias et al. 2012). Furthermore, adipose-specific VEGF-A over-expression led to up-regulation of UCP-1 and PGC-1α, and thus improvement of thermogenic capacity (Sun et al. 2012; Elias et al. 2012). These studies showed that by improving the angiogenic potential adipose hypoxia could be ameliorated during DIO. Interestingly Sun et al. (2012) suggested that the benefits of manipulating the adipose tissue angiogenic potential are context- and time-dependent, with induction of proangiogenic responses being advantageous at the early stages of adipose expansion in DIO (Sun et al. 2012). In a recent study we took a different approach to understand the interaction between hypoxia and angiogenesis in adipose. Glucocorticoids (cortisol, corticosterone) are key metabolic-control hormones that increase glucose and lipid levels in blood via actions in liver, pancreas, muscle and, particularly, adipose tissues. Chronic excess circulating glucocorticoids (Cushing's syndrome) is a classical cause of abdominal and truncal obesity and metabolic disease (insulin resistance, type 2 diabetes mellitus, dyslipidaemia, atherosclerosis). Glucocorticoids are also potently anti-angiogenic, inhibiting VEGF-A (Nauk et al. 1998; Leonard et al. 2005). Glucocorticoid action on target cells is determined by the density of intracellular (nuclear) receptors for the steroids and by pre-receptor metabolism, notably by 11β-hydroxysteroid dehydrogenases (11β-HSDs). 11β-HSD1 is the isoform expressed in metabolic tissues, notably adipose tissue. This catalyses the intracellular regeneration of active cortisol and corticosterone from inert 11-keto forms (cortisone, 11-dehydrocorticosterone). We used a well-established model of 'healthy' adipose tissue that resists metabolic dysfunction when challenged with high fat diet, the 11β-HSD1-deficient mouse. This primarily lacks local cellular glucocorticoid reamplification in cells including adipocytes (Morton et al. 2004; Wamil et al. 2011). Crucially, this model of intra-adipose glucocorticoid deficiency allows 'healthy' fat expansion and is highly permissive of adipose angiogenesis (Michailidou et al. 2012)—as it is of angiogenesis in other hypoxic vascular territories such as the infarcted myocardium (Small et al. 2005). The maintenance of adipose vasculature during DIO in the 11β-HSD1-deficient mouse prevented severe adipose hypoxia, reduced adipose HIF-1α levels and attenuated collagen deposition and fibrosis. Moreover, the adipose stroma cells had basally higher levels of the key angiogenic factors, VEGF-A and angiopoetin-like-protein-4, and hypoxic stimulation led to a profound up-regulation of these factors in 11β-HSD1-deficient cells (Michailidou et al. 2012). These changes plausibly explain the healthier metabolic profile of 11β-HSD1-deficient adipose tissues during DIO. Overall, the evidence so far is limited to support a direct and consistent link between HIF-1α-mediated angiogenesis in adipose tissue during fat expansion at least when HIF is manipulated in adipocytes. A HIF-independent regulation of VEGF-A and downstream angiogenic responses have been shown in other tissues (muscle and brain) and suggested to be mediated through the PPAR-γ co-activator PGC-1α (Arany et al. 2008; Ndubuizu et al. 2010). However, what remains to be seen is if the effect of hypoxia and HIF manipulation specifically in non-adipocyte cells during DIO will provide a different picture.

12.2.3 Hypoxia-Dependent AT Vascular Remodelling in Non-adipocyte Cells

The scope of this chapter was to mainly focus on the adipocyte as the central cell affected/contributing to hypoxia and angiogenesis during obesity. Human and mouse derived adipose stromal cells (ASC) have higher proangiogenic gene expression (Rahman et al. 2004; Villaret et al. 2010). Moreover hypoxia treatment of ASCs further induces VEGF-A expression and stimulates endothelial cell growth and tube-like formation in co-culture experiments (Rahman et al. 2004; Cao et al. 2005; Maumus et al. 2008; Amos et al. 2011).

In obesity, adipose tissues are in a state of chronic inflammation. The recruitment of bone-marrow-derived cells (a heterogeneous population containing endothelial, pericyte and hematopoietic lineage progenitors) (Ahan and Brown 2009) might also contribute to vessel formation and maturation. Apart from the key player of proangiogenic responses and vessel formation, the endothelial cell, there is a large body of evidence that macrophages are active players during vessel formation and that hypoxia induces a proangiogenic profile of macrophage gene expression (Lewis et al. 2007; Ye et al. 2007; Bourlier et al. 2012). Adipose tissue macrophages are the major source of the angiogenic growth factor, platelet-derived growth factor (PDGF), that stimulates endothelial tube-like formation. Its expression is highly regulated by hypoxia (Pang et al. 2008). Loss of VEGF in myeloid cells causes excessive collagen deposition and severe fibrosis suggesting that inflammatory cell-derived VEGF is required for vascular remodelling (Stockman et al. 2010).

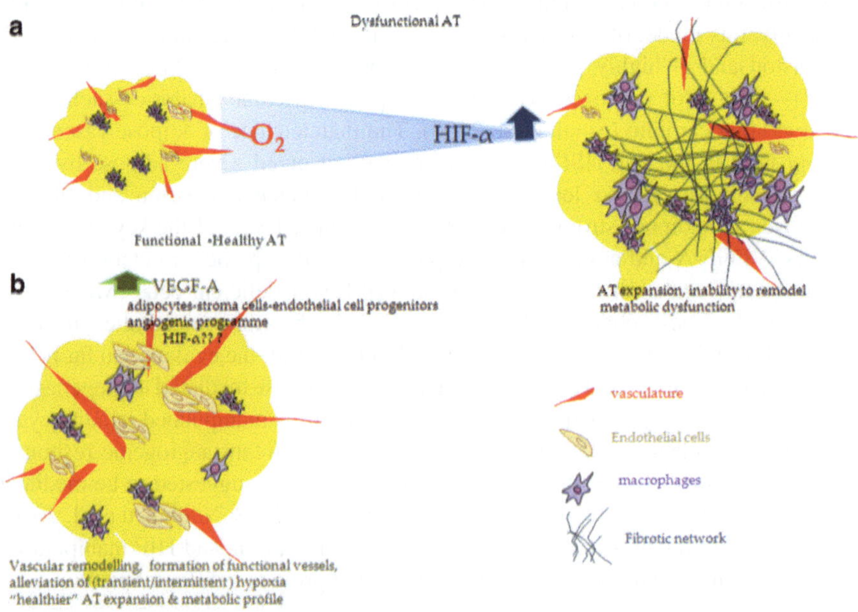

Fig. 12.1 A unified model of the potential role of adipose hypoxia on the vasculature during fat mass expansion. From *left* to *right* (**a**): the hypothesis of hypoxia as a direct cause of macrophage infiltration, adipose fibrosis and failure of vascular remodelling during adipose expansion. Excessive rapid fat mass accumulation limits oxygen availability (and activates HIF-1α); this triggers macrophages to infiltrate the tissue initiating along with hypertrophied adipocytes a pro-fibrotic signalling cascade leading to severe fibrosis and metabolically dysfunctional adipose. From *left down* (**b**): a hypothesis of healthy fat expandability due to the ability of adipose tissue to adapt and cope during growth. This requires the central angiogenic factor VEGF-A for the initiation of the angiogenic 'machinery' and functional vessel formation which is key to adipose/vasculature remodelling and requires the fine orchestration of the adipocyte and cells from the stroma (macrophages, endothelial cells and progenitors). In this model AT is alleviated from hypoxia, inflammation and severe fibrosis thus leading to a healthier metabolic profile. The direct role of the HIF-system in this model requires further investigation

Interestingly, macrophages from distinct anatomical fat depots show different phenotypes. Macrophages from subcutaneous adipose tissue acquire a pro-fibrotic profile with increased TGFβ expression, whereas visceral adipose-derived macrophages switch to a proangiogenic profile during obesity and hypoxia stimulation (Bourlier et al. 2012). Loss of HIF-1α in myeloid cells causes a depletion in the ATP cellular pool, impairment of myeloid cell motility and infiltration in models of experimental arthritis and TPA-induced ear inflammation (Cramer et al. 2003). What remains to be seen is whether macrophage HIF-loss in an obesogenic setting affects vascular remodelling during adipose expansion. Thus the complexity of the adipose microenviroment is apparent. The effects of non-adipocyte cell populations (macrophages, endothelial cells) in adipose depots and their responses to changes in oxygen tension during fat expansion and the crosstalk with adipocytes require further exploration. A working model of how these cells and their functions could be interlinked is presented in Fig. 12.1.

12.3 Conclusions

Accelerated fat mass expansion during obesity requires the fine orchestration of adipose tissue components including different cell types, the ECM network, vessel formation and oxygen level maintenance. Fine tuning of oxygen levels during expansion are fundamental for the maintenance of a functional adipose and downstream whole body metabolic health. Different cell types respond differently to changes in oxygen tension. We are just beginning to understand some of the complex events that take place during rapid adipose tissue expansion.

References

Amos PJ et al. Hypoxic culture and in vivo inflammatory environments affect the assumption of pericyte characteristics by human adipose and bone marrow progenitor cells. Am J Physiol Cell Physiol. 2011;301:C1378–88.

Ahan GO, Brown JM. Influence of bone marrow-derived hematopoietic cells on the tumor response to radiotherapy: experimental models and clinical perspectives. Cell Cycle. 2009;8:970–6.

Arany Z et al. HIF-independent regulation of VEGF and angiogenesis by the transcriptional coactivator PGC-1alpha. Nature. 2008;451:1008–12.

Bluher M et al. Role of insulin action and cell size on protein expression patterns in adipocytes. J Biol Chem. 2004;279:31902–9.

Bourlier V et al. TGFbeta family members are key mediators in the induction of myofibroblast phenotype of human adipose tissue progenitor cells by macrophages. PLoS One. 2012;7:e31274.

Cao Y et al. Human adipose tissue-derived stem cells differentiate into endothelial cells in vitro and improve postnatal neovascularization in vivo. Biochem Biophys Res Commun. 2005; 332(2):370–9.

Cao Y. Angiogenesis modulates adipogenesis and obesity. J Clin Invest. 2007;117:2362–8.

Cao Y. Adipose tissue angiogenesis as a therapeutic target for obesity and metabolic diseases. Nat Rev Drug Discov. 2010;9:107–15.

Carmeliet P. Mechanisms of angiogenesis and arteriogenesis. Nat Med. 2000;6:389–95.

Carmeliet P, Collen D. Molecular basis of angiogenesis. Role of VEGF and VE-cadherin. Ann N Y Acad Sci. 2000;902:249–62.
Cockman M et al. Proteomics-based identification of novel factor inhibiting hypoxia-inducible factor (FIH) substrates indicates widespread asparaginyl hydroxylation of ankyrin repeat domain-containing proteins. Mol Cell Proteomics. 2009;8:535–46.
Coleman ML et al. Asparaginyl hydroxylation of the Notch ankyrin repeat domain by factor inhibiting hypoxia-inducible factor. J Biol Chem. 2007;282:24027–38.
Cramer T et al. HIF-1alpha is essential for myeloid cell-mediated inflammation. Cell. 2003;113:419.
Dvorak HF et al. Vascular permeability factor/vascular endothelial growth factor, microvascular hyperpermeability, and angiogenesis. Am J Pathol. 1995;146:1029–39.
Elias I et al. Adipose tissue overexpression of vascular endothelial growth factor protects against diet-induced obesity and insulin resistance. Diabetes. 2012;61:1801–13.
Epstein A,C et al. C elegans EGLN-9 and mammalian homologs define a family of dioxygenases that regulate HIF by prolyl hydroxylation. Cell. 2001;107:43–54.
Farb MG et al. Artiriolar function in visceral adipose tissue is impaired in human obesity. Arterioscler Thromb Vasc Biol. 2012;32:467–73.
Feldser D et al. Reciprocal positive regulation of hypoxia inducible factor 1alpha and insulin growth factor 2. Cancer Res. 1999;59:3915–8.
Ferguson JE et al. ASB4 is a hydroxylation substrate of FIH and promotes vascular differentiation via an oxygen-dependent mechanism. Mol Cell Biol. 2007;27:6407–19.
Ferrara N et al. Heterozygous embryonic lethality induced by targeted inactivation of the VEGF gene. Nature. 1996;380:439–42.
Foster DO, Frydman ML. Brown adipose tissue: the dominant site of nonshivering thermogenesis in the rat. Experientia Suppl. 1978;32:147–52.
Gealekman O et al. Depot-specific differences and insufficient subcutaneous adipose tissue angiogenesis in human obesity. Circulation. 2011;123:186–94.
Girgis CM et al. Novel links between HIFs, type 2 diabetes, and metabolic syndrome. Trends Endocrinol Metab. 2012;23:372–80.
Grosfeld A et al. Hypoxia increases leptin expression in human PAZ6 adipose cells. Diabetologia. 2002;45:527–30.
Haddad J,J, Land S,C. A non-hypoxic, ROS sensitive pathway mediates TNF-alpha-dependent regulation of HIF-alpha. FEBS Lett. 2001;505:269–74.
Halberg N et al. Hypoxia-inducible factor 1alpha induces fibrosis and insulin resistance in white adipose tissue. Mol Cell Biol. 2009;29:4467–83.
Hausman GJ, Richardson RL. Adipose tissue angiogenesis. J Anim Sci. 2004;82:925–34.
He Q, Gao Z, Yin J, Zhang J, Yun Z, Ye J. Regulation of HIF-1alpha: activity in adipose tissue by obesity-associated factors: adipogenesis, insulin and hypoxia. Am J Physiol Endocrinol Metab. 2011;300:E877–85.
Helminger G et al. Interstitial pH and pO2 gradients in solid tumors in vivo: high-resolution measurements reveal a lack of correlation. Nat Med. 1997;2:177–82.
Higgins DF et al. Hypoxic induction of Ctgf is directly mediated by Hif-1. Am J Physiol Renal Physiol. 2004;287:F1223–32.
Higgins DF et al. Hypoxia promotes fibrogenesis in vivo via HIF-1 stimulation of epithelial-to-mesenchymal transition. J Clin Invest. 2007;117:3810–20.
Hosogai N et al. Adipose tissue hypoxia in obesity and its impact on adipocytokine dysregulation. Diabetes. 2007;56:901–11.
Ivan M et al. HIFalpha targeted for VHL-mediated destruction by proline hydroxylation: implications for O2 sensing. Science. 2001;292:464–8.
Jaakkola P et al. Targeting of HIF-alpha to the von Hippel-Lindau ubiquitylation complex by O2-regulated prolyl hydroxylation. Science. 2001;292:468–72.
Jiang C et al. Disruption of hypoxia-inducible factor 1 in adipocytes improves insulin sensitivity and decreases adiposity in high-fat diet-fed mice. Diabetes. 2011;60:2484–95.
Kabon B et al. Obesity decreases perioperative tissue oxygenation. Anesthesiology. 2004; 100:274–80.

Kahn T et al. Metabolic dysregulation and adipsoe fibrosis: role of collagen VI. Mol Cell Biol. 2009;29:1575–91.

Krishnan J et al. Dietary obesity-associated Hif1α activation in adipocytes restricts fatty acid oxidation and energy expenditure via suppression of the Sirt2-NAD+ system. Genes Dev. 2012;26:259–70.

Lando D et al. FIH-1 is an asparaginyl hydroxylase enzyme that regulates the transcriptional activity of hypoxia inducible factor. Genes Dev. 2002;16:1466–71.

Lee KY et al. The differential role of Hif1β/Arnt and the hypoxic response in adipose function, fibrosis, and inflammation. Cell Metab. 2011;14:491–503.

Lewis CE, De Palma M, Naldini L. Tie2-expressing monocytes and tumor angiogenesis: regulation by hypoxia and angiopoietin-2. Cancer Res. 2007;67:8429–32.

Lin Q et al. Differentiation arrest by hypoxia. J Biol Chem. 2006;281:30678–83.

Lolmede K et al. Effects of hypoxia on the expression of proangiogenic factors in differentiated 3T3-F442A adipocytes. Int J Obes Relat Metab Disord. 2003;27:1187–95.

Leonard M,O, Godson C, Brady H,R, Taylor C,T. Potentiation of glucocorticoid activity in hypoxia through the glucocorticoid receptor. J Immunol. 2005;174:2250–7.

Mahon PC, Hirota K, Semenza GL. FIH-1: a novel protein that interacts with HIF-1alpha and VHL to mediate repression of HIF-1 transcriptional activity. Genes Dev. 2001;15:2675–86.

Maumus M et al. Evidence of in situ proliferation of adult adipose tissue-derived progenitor cells: influence of fat mass microenvironment and growth. J Clin Endocrinol Metab. 2008; 93:4098–106.

Maxwell P,H et al. The tumour suppressor protein VHL targets hypoxia inducible factors to oxygen-dependent proteolysis. Nature. 1999;399:271–5.

Mazzati D et al. A microarray analysis of the hypoxia-induced modulation of gene expression in human adipocytes. Arch Physiol Biochem. 2012;118:112–20.

Michailidou Z et al. Increased angiogenesis protects against adipose hypoxia and fibrosis in metabolic disease-resistant 11β-hydroxysteroid dehydrogenase type 1 (HSD1)-deficient mice. J Biol Chem. 2012;287:4188–97.

Morton N,M et al. Novel adipose tissue mediated resistance to diet-induced visceral obesity in 11β-hydroxysteroid dehydrogenase type 1 deficient mice. Diabetes. 2004;53:931–8.

Nauk M, Karakioulakis G, Perruchoud A,P, Papakostantinou E, Roth M. Corticosteroids inhibit the expression of vascular endothelial growth factor gene in human vascular smooth cells. Eur J Pharmacol. 1998;341:309–15.

Ndubuizu O,I, Tsipis C,P, Li A, LaManna J,C. Hypoxia inducible factor (HIF-1)-independent microvascular angiogenesis in the aged rat brain. Brain Res. 2010;1366:101–9.

O'Hara A et al. Microarray analysis identifies matrix metalloproteinases (MMPSs) as key genes whose expression is upregulated in human adipocytes by macrophage-conditioned medium. Plufgers Arch. 2009;458:1103–14.

O'Rourke RW et al. Hypoxia-induced inflammatory cytokine secretion in human tissue stromovascular cells. Diabetologia. 2009;54:1480–90.

O'Rourke RW et al. Depot-specific differences in inflammatory mediators and a role of NK cells and INF-γ in inflammation in human adipose tissue. Int J Obes (Lond). 2011;33:978–90.

Pang C et al. Macrophage infiltration into adipose tissue may promote angiogenesis for adipose tissue remodelling in obesity. Am J Physiol Endocrinol Metab. 2008;295:E313–22.

Pasarica M et al. Reduced adipose tissue oxygenation in human obesity: evidence of rarefaction, macrophage chemotaxis and inflammation without an angiogenic response. Diabetes. 2009;58:718–25.

Rahman J et al. Secretion of angiogenic and antiapoptotic factors by human adipose stromal cells. Circulation. 2004;109:1292–8.

Rausch M,E, Weisberg S, Vardhana P, Tortoriello DV. Obesity in C57BL/6J mice is characterized by adipose tissue hypoxia and cytotoxic T-cell infiltration. Int J Obes (Lond). 2008; 32:451–63.

Regazzetti C et al. Hypoxia decreases insulin signaling pathways in adipocytes. Diabetes. 2009;58:95–103.

Shimba S, Wada T, Hara S, Tezuka M. EPAS1 promotes adipose differentiation in 3T3-L1 cells. J Biol Chem. 2004;279:40946–53.

Small G,R et al. Preventing local regeneration of glucocorticoids by 11beta-hydroxysteroid dehydrogenase type 1 enhances angiogenesis. Proc Natl Acad Sci U S A. 2005;102:12165–70.

Stockman C et al. Loss of myeloid cell-derived vascular endothelia growth factor accelerates fibrosis. Proc Natl Acad Sci U S A. 2010;107:4329–34.

Sun K, Kusminski CM, Scherer PE. Adipose tissue remodeling and obesity. J Clin Invest. 2011;121:2094–101.

Sun K et al. Dichotomous effects of VEGF-A on adipose tissue dysfunction. Proc Natl Acad Sci U S A. 2012;109:5874–9.

Tajima R et al. Hypoxic enhancement of type IV collagen secretion accelerates adipose conversion of 3T3-L1 fibroblasts. Biochim Biophys Acta. 2001;1540:179–87.

Trayhurn P, Wood IS. Adipokines: inflammation and the pleiotropic role of white adipose tissue. Br J Nutr. 2004;92:347–55.

Treins C, Giorgetti-Peraldi S, Murdaca J, Semenza G,L, Van Obberghen E. Insulin stimulates hypoxia inducible factor 1 through phosphatidylinositol 3-kinase/target of rapamacyn-dependent signalling pathway. J Biol Chem. 2002;277:27975–81.

Tzouvelekis A et al. Comparative expression profiling in pulmonary fibrosis suggests a role of hypoxia-inducible factor-1alpha in disease pathogenesis. Am J Respir Crit Care Med. 2007; 176(11):1108–19.

Villaret A et al. Adipose tissue endothelial cells from obese human subjects: differences among depots in angiogenic, metabolic and inflammatory gene expression and cellular senescence. Diabetes. 2010;59:2755–63.

Wamil M et al. Novel fat depot-specific mechanisms underlie resistance to visceral obesity and inflammation in 11 β-hydroxysteroid dehydrogenase type 1-deficient mice. Diabetes. 2011; 60(4):1158–67.

Wilkins SE et al. Differences in hydroxylation and binding of Notch and HIF- 1alpha demonstrate substrate selectivity for factor inhibiting HIF-1 (FIH-1). Int J Biochem Cell Biol. 2009;41: 1563–71.

Xue Y et al. Hypoxia-independent angiogenesis in adipose tissues during cold acclimation. Cell Metab. 2009;9:99–109.

Ye J, Gao Z, Yin J, He Q. Hypoxia is a potential risk factor for chronic inflammation and adiponectin reduction in adipose tissue of ob/ob and dietary obese mice. Am J Physiol Endocrinol Metab. 2007;293:E1118–28.

Yin J et al. Role of hypoxia in obesity-induced disorders of glucose and lipid metabolism in obesity. Am J Physiol Endocrinol Metab. 2009;296:E333–42.

Yung Y,J, Isaac J,S, Lee S, Trepel J, Neckers L. IL-1beta mediated up-regulation of HIF-1alpha via an NFkappaB/COX-2 pathway identifies identifies HIF-1 as a critical link between inflammation and oncogenesis. FASEB J. 2003;17:2115–7.

Yun Z, Maecker HL, Johnson RS, Giaccia AJ. Inhibition of PPAR gamma 2 gene expression by the HIF-1-regulated gene DEC1/Stra13: a mechanism for regulation of adipogenesis by hypoxia. Dev Cell. 2002;3:331–41.

Zhang N et al. The asparaginyl hydroxylase factor inhibiting HIF-1alpha is an essential regulator of metabolism. Cell Metab. 2010a;11:364–78.

Zhang J et al. Hypoxia-inducible factor 1 activation from adipose protein 2-cre mediated knockout of von hippel-lindau gene leads to embryonic lethality. Clin Exp Pharmacol Physiol. 2012;39:145–50.

Zhang H et al. Hypoxia-inducible factor directs POMC gene to mediate hypothalamic glucose sensing and energy balance regulation. PLoS Biol. 2011;9:e1001112.

Zhang X et al. Adipose tissue specific inhibition of hypoxia-inducible factor 1α induces obesity and glucose intolerance by impeding energy expenditure in mice. J Biol Chem. 2010b; 285:32869–77.

Part VI
Mouse Models to Study Adipose Angiogenesis

Chapter 13
Adipose Angiogenesis Models in Animals and Methodology

Jennifer Honek, Sharon Lim, and Yihai Cao

Abstract Adipose tissue angiogenesis plays a crucial role in the organism as blood vessels are required to supply nutrients and oxygen to the adipose tissue and to remove metabolic waste products. Both adipose depots, the lipid-storing white adipose tissue (WAT) and the metabolically active brown adipose tissue (BAT) are highly vascularized. In this chapter, we describe the morphology and physiological implications of WAT and BAT as well as the role of adipose tissue angiogenesis during development and its importance regarding the expansion of adipose tissue under physiological conditions or in the development of obesity and obesity-related metabolic disorders. We also evaluate potential therapeutic implications of WAT to brown-like adipose tissue transformation, approaches to use adipose tissue transplantation for therapeutic purposes, as well as different genetic models to study adipose angiogenesis.

Keywords Adipose tissue • Angiogenesis • Animal models • Obesity • Immunohistochemistry • Indirect calorimetry • Non-shivering thermogenesis • CAM assay • Zebrafish

---

Jennifer Honek and Sharon Lim contributed equally to this work.

J. Honek • S. Lim
Department of Microbiology, Tumor and Cell Biology, Karolinska Institute, 171 77 Stockholm, Sweden

Y. Cao, M.D., Ph.D. (✉)
Department of Microbiology, Tumor and Cell Biology (MTC), Karolinska Institute, 171 77 Stockholm, Sweden

Department of Medicine and Health Sciences, Linköping University, 581 85 Linköping, Sweden
e-mail: yihai.cao@ki.se

Y. Cao (ed.), *Angiogenesis in Adipose Tissue*, DOI 10.1007/978-1-4614-8069-3_13,

13.1 Angiogenesis and Vasculogenesis in Developing Embryos

13.1.1 Role of Adipose Tissue in Fetal Development

Observations on the intimate interactions between the growth of blood vessels and adipose tissue were initially documented in the 1870s. The Swiss scientist Toldt reported the spatial and temporal role of adipose tissue and vascularization. The spatio-temporal relation of adipocyte and vascularization begins in the fetus and persists through adulthood (Hausman et al. 1982). Studies on the development of microvasculature in adipose tissues demonstrate that angiogenesis precedes adipogenesis. Several studies have demonstrated the close relation between well-defined vessels which further support adipose tissue growth (Scroyen et al. 2010; Christiaens and Lijnen 2010; Hausman and Richardson 2004).

13.1.2 Distribution of Adipose Tissue in Fetus and Adulthood

The appearance of adipose tissue occurs around mid-gestation; both size and amount of adipose depots continue to increase throughout late gestation (Symonds et al. 2012). Adipose tissue can be categorized into two types, white adipose tissue (WAT) and brown adipose tissue (BAT). WAT is the predominant fat depot in the body where each adipocyte has a large unilocular lipid droplet which stores energy in the form of triglycerides. Its counterpart, the BAT, is essential for thermogenic activity in the newborn. BAT is mainly essential for energy metabolism and therefore has significantly higher density of blood vessels to provide efficient blood perfusion to supply nutrients, oxygen, and the removal of waste products (Hausman et al. 1984). It has been reported that the distribution and amount of BAT in a newborn is influenced by the maternal diet. The growth of BAT during gestation is largely dependent on the maternal glucose source and the amount of BAT significantly decreases after birth in newborn. The development and accumulation of different types of adipose tissues during neonatal development can highly influence the occurrence of several metabolic diseases in later life such as impairment of glucose regulation, dyslipidemia, fatty liver disease, and low grade inflammation (Cali and Caprio 2008). Unlike WAT, which mainly stores excess fat in the form of triglycerides, BAT contains a unique uncoupling protein 1 (UCP1) in the mitochondria which dissipates heat by non-shivering thermogenesis (NST). It has been reported that the BAT is able to produce 300 W/kg of heat, significantly higher than other tissues in the body (1 W/kg) (Symonds et al. 2012). BAT has been well-recognized to be metabolically active in small rodents; however, the metabolic activity of BAT in adult human has only recently been observed (Cypess et al. 2009; van Marken Lichtenbelt et al. 2009; Virtanen et al. 2009). On average an adult human has approximately 40–50 g of BAT in the supraclavicular region and "full activation" of the entire BAT would be able to burn up to 4 kg of WAT per year (van Marken Lichtenbelt et al. 2009). The distribution of BAT in rodents is mainly located in the interscapular region and retains UCP1 expression throughout the adult life. In the past decades, it has come to consensus that another category of adipose tissue exist, i.e., the brown-like adipose tissue (BRITE) which acquires both characteristics of the BAT and the WAT. The ability to stimulate or activate the transition of WAT to BRITE has increasingly gained interest due to the metabolic activity of BAT (see Sect. 13.3). If these BRITE depots are able to burn as much energy as the BAT, this could become a possible therapeutic approach for the treatment of obesity and obesity-related metabolic diseases.

13.1.3 Spatio-Temporal Relation Between Angiogenesis and Adipose Tissue

Angiogenesis is the growth of new capillaries from preexisting vessels and this process is critical for many physiological processes including tissue growth, expansion, and repair. Angiogenesis is known to be a crucial player in adipogenesis (Cao 2007; Lijnen 2008). Apart from adipocytes, the adipose tissue consists non-adipocytes structures including the stromal vascular cells, fibroblastic connective tissues, infiltrating inflammatory cells (leukocytes, monocytes, macrophages), mast cells, preadipocytes, pericytes, endothelial cells, and stem cells. Over the years, the adipose tissue has been recognized as one of the major endocrine organs that secretes a myriad of adipokines that stimulates or inhibits angiogenesis, e.g., tumor necrosis factor-alpha (TNF-α), angiopoietins (Ang), vascular endothelial growth factor-A (VEGF-A), fibroblast growth factor (FGF), hepatocyte growth factor (HGF), leptin, adiponectin, peroxisome proliferator-activated receptor-gamma (PPAR-γ), adiponectin, and visfatin (Fig. 13.1). VEGF is the most well-characterized endothelial cell growth factor that stimulates vasculogenesis and angiogenesis. VEGF family consists of VEGF-A, VEGF-B, VEGF-C, VEGF-D, and placental growth factor (PlGF). The VEGF gene encodes VEGF-$A_{121\text{-}206}$ by alternative splicing. All the different isoforms of VEGF binds to common tyrosine kinase receptors, VEGF receptor-1, -2, and -3 (VEGFR-1/Flt-1, VEGFR-2/Flk-1/KDR, VEGFR-3). VEGF-A is the major signaling factor which binds to both VEGFR-1/Flt-1 and VEGFR-2/Flk-1/KDR on the cell surface thus triggering downstream signaling cascade stimulating endothelial cell proliferation, survival, and cell migration (Fig. 13.2). Although VEGFR-1 has tenfold higher binding affinity to VEGF-A as compared to VEGFR-2, VEGFR-2 has a much stronger VEGF-A-induced kinase activity. VEGF-A is essential for the initiation and formation of immature vessels in embryonic vessels development. The expression of VEGF is influenced by hypoxia and insulin, growth factors, and other cytokines. Adipokine-derived cytokines such as leptin and insulin play an intrinsic role in regulating glucose homeostasis.

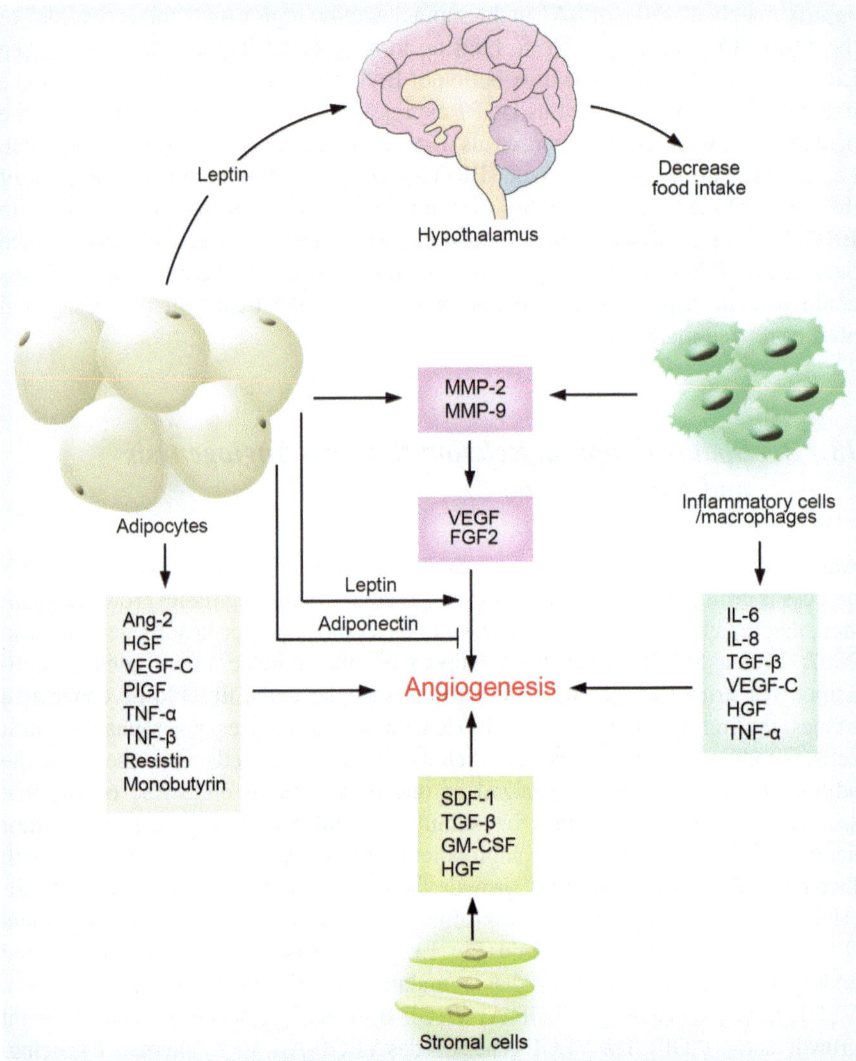

Fig. 13.1 Adipokines that modulate angiogenesis. Adipocytes and non-adipocyte cell population in the adipose tissue including stromal cells, inflammatory cells that produce a multiple angiogenic factors, and inhibitors that regulate adipose angiogenesis

It has been shown that insulin, a cytokine secreted specifically by the adipocytes, also influenced the expression of VEGF. On the other hand, leptin and glucocorticoid do not affect VEGF expression (Mick et al. 2002; Zhang et al. 1997). Leptin has been described to play a key role in controlling appetite, energy expenditure, sympathetic nervous system activation, and thermogenesis.

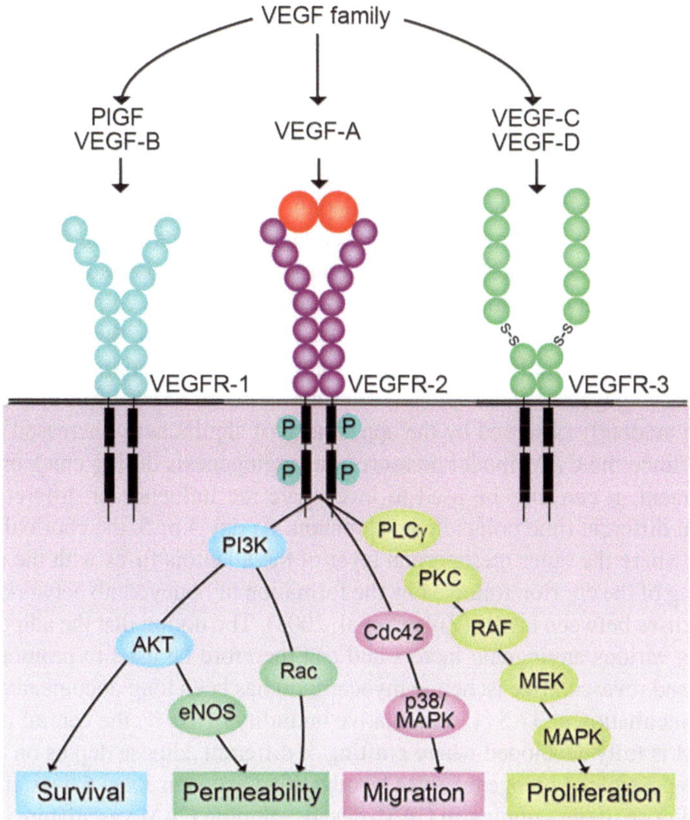

Fig. 13.2 Vascular endothelial growth factor (VEGF) family. VEGF family of proteins consist of 5 members, VEGF-A, PlGF, VEGF-B, VEGF-C, and VEGF-D which bind to their respective extracellular immunoglobulin of three transmembrane tyrosine kinase receptors (VEGFR-1, -2 and -3). Upon binding of VEGF-A to VEGFR-2, downstream signaling cascade is triggered, resulting in endothelial cell proliferation, survival, and cell migration

13.1.4 Models to Study Angiogenesis

There are different established models that can be used to study adipose tissue in stimulating angiogenesis under physiological conditions; these models include chick chorioallantoic membrane (CAM), zebrafish, rabbit cornea, and mouse models, or under pathophysiological conditions such as wound healing and ischemia model (Castellot et al. 1982; Silverman et al. 1988). Several in vitro studies have shown that adipocytes development is intrinsically related to capillary development.

13.1.4.1 CAM Assay

CAM is a well-recognized assay widely used to analyze pro- and anti-angiogenic potential of growth factors. CAM is a highly vascularized tissue of the avian embryo. The usage of the CAM assays has evolved over the years, from the use of filter disk, gelated material such as matrigel or sodium alginate, gelatin sponge to the collagen onplant model. In this chapter, we will only discuss the use of CAM collagen onplant model. Three-dimensional (3D) collagen implants on the CAM are more robust which enables better analysis, imaging, in situ analysis, and quantification of cellular and biochemical mechanisms induced by the factors compared to its counterparts. This method was established by Dr Folkman, where they investigated the effect of purified basic FGF or FGF2, or tumor cells by incorporating into the collagen gels (Nguyen et al. 1994). The vascular network formed from the induction could be evidently observed by the appearance of significantly increased vascularization. Since the CAM model measures the angiogenesis during chick embryonic development, it can thus be used to investigate the influence of different growth factors at different time points in development. At day 4 or 5, the chorioallantois is formed, where the outer mesodermal layer of the allantois fuses with the mesodermal lining of the chorion followed by the formation of honeycomb network of capillary plexuses between layers (Zijlstra et al. 2004). The notion that the adipose tissue produces various angiogenic factors and can therefore be used to promote wound healing and revascularize ischemic myocardium has been long documented. After 8 days of incubation at 37.5 °C and relative humidity of 70 %, the central portion of the CAM is fully developed where grafting of different adipose depots on the CAM becomes possible. Angiogenesis can be quantified between 3 and 5 days after layering of adipose tissue samples on CAM. The developing CAM vasculature is responsive to both pro-angiogenic and anti-angiogenic stimuli between days 8 and 12 and, therefore, making it possible to assess the effect of angiogenic stimulators or inhibitors on the different adipose depots layered on the CAM (Ledoux et al. 2008). The CAM assay once established is easy to use and less expensive compared to other models. However, similar to other models, there are a few restrictions with the CAM assay. The main limitations of CAM assay is that implantation may stimulate nonspecific inflammatory responses which induce secondary vasoproliferation thereby making the quantification of vessels difficult. To avoid inflammatory response, implantation could be made in the early stages in CAM development when the host's immune system is naïve (Ribatti et al. 2000).

13.1.4.2 Cornea Implantation Assay

The key advantage of the cornea implant assay is the avascular nature of the cornea which allows direct and easy quantification and assessment of neovascularization induced by the implants. However, this cornea implantation model requires skillful microsurgery techniques under the stereomicroscope. Different fat depots, such as the omental and the subcutaneous fat, can be implanted into the avascular rabbit or mouse cornea to investigate the role of different fat depots in inducing blood vessel formation. Briefly, different adipose tissue depots dissected from rabbits can be submerged in sterile saline at room temperature (RT) until implantation or micropellets containing angiogenic factors or adipokines should be thoroughly mixed with sucralfate (heparin-binding slow-release substance) and hydron in 99.5 % ethanol before smearing evenly on the nylon mesh (Silverman et al. 1988; Cao et al. 2011; Kadereit et al. 2008). After drying the embedded pellet for 1 h at RT, the pellet can be stored or directly implanted into the cornea. An intrastromal linear keratotomy surgical incision is made in the cornea with a surgical blade and a micropellet with angiogenic factors or adipose tissue depots is then inserted into the end of the pocket with a tissue needle. Corneal angiogenic response can be readily observed 5–7 days after implantation. The limitation of this assay is the possible death of animals due to inaccurate dosages of anesthesia used, or the unsuccessful operation due to lack of experience.

13.1.4.3 Zebrafish Models to Study Adipose Angiogenesis

Most animals use WAT as a reserve for fat storage. *C. elegans* store fat in their intestinal epithelial cells while most vertebrates store fat in the mesoderm-derived WAT (Wood 1988). The zebrafish (*Danio rerio*), a small tropical fish, is an important vertebrate model organism in research due to its relatively short generation time as compared to other laboratory animals. Similar to mammals, the primary function of adipocytes in zebrafish is to store lipids and secrete cytokines. Other cell types in zebrafish include myocytes, hepatoyctes, and chrondrocytes. Zebrafish is a suitable model for studying embryonic development, tissue/organ regeneration, and remodeling due to the short generation time and the ease of manipulation (Jensen et al. 2009, 2011). With the morpholino technology and the transparent nature of zebrafish, it is possible to investigate the interplay of blood vessels and adipocytes during development using different concentrations of VEGF morpholinos to compromise the vascular development and to use transgenic reporter strains expressing fluorescent protein such as GFP to study the distribution of adipocytes. The embryonic period of the zebrafish ranges between 0 and 3 days postfertilization (dpf), they reach the juvenile stage approximately 30 dpf, and are fully developed into mature adults by approximately 3 months of age. In 2010, Imrie et al. used the zebrafish model to study the WAT development in the relation to development time and size (Imrie and Sadler 2010). They demonstrated that adipocytes in WAT in zebrafish store lipid in a unilocular droplet and produce adipokines similar to mammalian adipocytes. WAT is not found in embryos or young larvae and the appearance of WAT in the zebrafish is first found in the pancreas around 8–15 days postfertilization (dpf), followed by appearance of WAT in the visceral, subcutaneous, and cranial regions. Adipocytes in the zebrafish can be visualized by Nile red staining which identifies lipid droplets (Flynn et al. 2009; Asano et al. 1997). Since establishing of hypoxia-induced models in mammals is a challenging task, the zebrafish provides a wonderful opportunity to investigate the effect of hypoxia in adipose tissue. An angiogenic effect in the zebrafish in response to hypoxia can be easily established by altering the nitrogen to oxygen ratio in a specifically designed aquarium (Lee et al. 2009).

13.1.4.4 Mouse Model to Study Angiogenesis and Adipose Development

The mouse is an ideal model organism to study angiogenesis and adipose development due to the genetic similarity between mice and humans. In addition, mice are prolific in reproduction bearing an average of 5–10 pups per little. They are relatively small in size and most inbred stains are docile and easy to handle for experimentations. In this chapter, we will describe the use of the cold acclimation model in mouse to study angiogenesis and to investigate transition of WAT to acquiring the BRITE phenotype (Xue et al. 2009). Vascular growth is tightly correlated with the VEGF expression level. In the cold model for instance, acclimation of mice in cold temperature for 1 h, rapidly and transiently increases the VEGF expression in BAT (Asano et al. 1997). Exposure of mice to cold, for as short as 1 week, is able to transform WAT into acquisition of the BRITE phenotype. This WAT to BAT transition in adipose tissue upon cold exposure is accompanied by increased VEGF expression as well as increased vascular density. Here, we will briefly describe the methodology to perform the cold acclimation experiment. One group of mice should be adapted at 18 °C for at least 1 week before exposure to cold at 4 °C, and the other group of mice should be acclimated at 22 °C before transferring them to their thermoneutral temperature of 30 °C. After 1 week of acclimation, transfer mice adapted at 18–4 °C, and mice adapted at 22–30 °C. Mice can be kept at 4 °C or 30 °C for 4–5 weeks depending on the experiment setup (Xue et al. 2009; Lim et al. 2012). According to Xue et al., adaptation of mice for 1 week is sufficient to induce upregulation of pro-angiogenic genes and downregulation of angiogenic inhibitors (Xue et al. 2009). The body composition, basal metabolic rate (BMR), and oxygen consumption can be measured using the magnetic resonance imaging and indirect calorimeter. The maximum capacity of BAT activity in mice can be measured by the stimulation of norepinephrine (NE) before measurement in the indirect calorimeter (Petrovic et al. 2008). At the end of the metabolic measurements, the different adipose depots can be dissected and further stained with an endothelial-specific pan marker, an anti-CD31 antibody, and/or an anti-perilipin antibody, a membrane marker for adipose tissue to investigate the number of blood vessels per adipocyte. This protocol enables one to investigate the role of angiogenesis in relation to transition of WAT to BRITE phenotype (see Sect. 13.3). The major drawbacks of this assay are the time needed to set up cold and warm facilities in the animal facility, and obtaining animal ethical permits to conduct the experiment. Also, it might be difficult to perform experiments in certain transgenic mice strains especially those that are sensitive to cold exposure. The adaptation time of these mice might be longer and larger number of mice should be used in case of unexpected death of mice during exposure to cold.

The experimental details for the induction of WAT to BRITE transformation as well as methodological details regarding the analysis of harvested tissue are described in Sect. 13.3.

13.2 Expansion of Adipose Tissue in Obese Model

Unlike most other tissues in the human body, the adipose tissue undergoes expansion and regression throughout lifetime. Adipose tissue expansion—similar to tumor growth—is dependent on angiogenesis, the sprouting of new blood vessels from preexisting vessels (Rupnick et al. 2002). Neovascularization of the expanding adipose tissue is required for a number of reasons which will be elaborated in the following.

Each adipocyte is surrounded by a blood vessel (Fig. 13.3). This spatial proximity makes it easy to understand that there is a close interplay between adipocytes and endothelial cells. Adipose tissue vasculature supplies adipocytes with oxygen and nutrients but also plays an important role in the removal of metabolic waste products. This is especially important in the metabolically highly active BAT which is characterized by an immensely dense network of blood vessels as seen in no other organ. Furthermore, vessels supply the adipose tissue with cytokines and growth factors that are enriched in the plasma. These factors are required to maintain physiological function of the adipocytes and to sustain their growth and survival. They are produced by a variety of cell types present in the adipose tissue. This heterogeneous

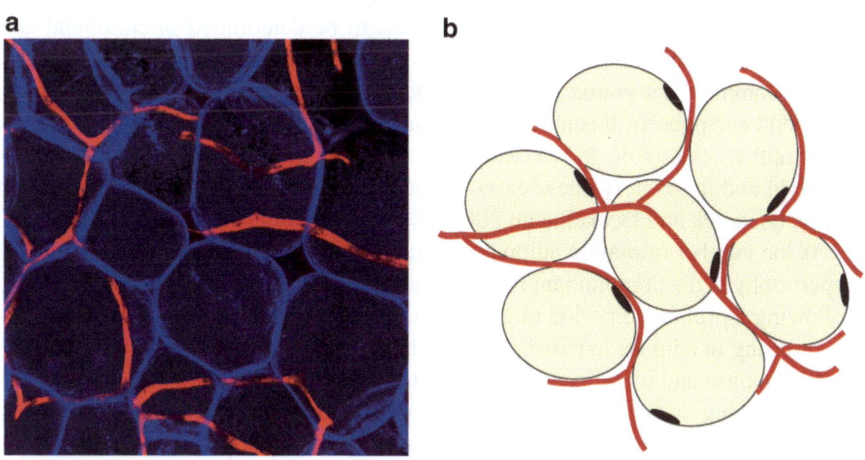

Fig. 13.3 Adipose tissue vasculature. (**a**) Adipose vasculature in murine white adipose tissue is visualized by whole mount staining with the endothelial cell marker anti-CD31 antibody (*red*). Adipocytes are visualized using an anti-perilipin antibody (*blue*). Scale bar: 50 μm. (**b**) In this schematic representation of adipose tissue vasculature, each fat cell (*yellow*) is surrounded by a blood vessel (*red*)

tissue consists of 50–85 % mature adipocytes and other of cells, including stromal cells, pericytes, and inflammatory cells, including neutrophils and monocytes (Maury and Brichard 2010; Spalding et al. 2008). These cells can easily invade the adipose tissue via the vasculature. Interestingly, in obese individuals, an augmented number of inflammatory cells in the adipose tissue have been described, underlining their importance in adipose tissue growth and expansion. Apart from growth factors produced by inflammatory cells, activated endothelial cells themselves produce a number of cytokines and growth factors which—via paracrine signaling—promote adipose tissue growth (Cao 2007; Chen et al. 2010).

Another important function of adipose tissue vasculature includes the supply of the adipose tissue with circulating stem cells. These cells can differentiate into pre-adipocytes, adipocytes, or endothelial cells thereby facilitating adipose tissue expansion.

Taken together, these vascular functions are crucial for sustainability and also expansion of adipose tissue. Consequently, in order to cope with the metabolic demand of expanding adipose tissue, vascular remodeling and an angiogenic switch are required for neovascularization (Lijnen 2008; Chen et al. 2010).

13.2.1 Adipose Hyperplasia and Hypertrophy

In response to persistent excess consumption of energy, the adipose tissue adapts by enlargement and expansion. This process occurs in obesity and follows two basic mechanisms: adipose tissue hyperplasia (i.e., increase in cell number) and hypertrophy (i.e., increase in cell size) (Fig. 13.4) (Jo et al. 2009).

However, the number of adipocytes seems to be determined in childhood and remains fairly constant during later life (Maury and Brichard 2010; Spalding et al. 2008). Commonly, a young infant possesses five to six billion adipocytes. During childhood and puberty, this number increases and reaches around 30 billion fat cells in a healthy, lean adult. In overweight individuals, this number can be increased threefold and in severely obese cases, adipocyte numbers can reach up to 250–300 billion (Bendich and Deckelbaum 2005).

As the number of mature adipocytes remains relatively stable during adulthood, hypertrophy is the predominant mechanism for adipose depots expansion in adults. Following a prolonged period of excess energy uptake, adipocytes accumulate lipids resulting in adipose hypertrophy. This increased fat storage results in a cellular stress response and often leads to an induction of pro-inflammatory signaling pathways involving nuclear factor-kappaB (NFκB) signaling, which in turn results in a dysregulation of adipokine production (Maury and Brichard 2010; Spalding et al. 2008; Jernas et al. 2006; Skurk et al. 2007). Monocyte chemotactic protein-1 (MCP-1) is an example of an adipokine secreted in large amounts by hypertrophic adipocytes. Binding to specific chemokine receptors expressed on inflammatory cells, it acts in a chemoattractant manner and facilitates monocyte and macrophage infiltration into the adipose tissue resulting in an inflammatory state. In human obesity, bone

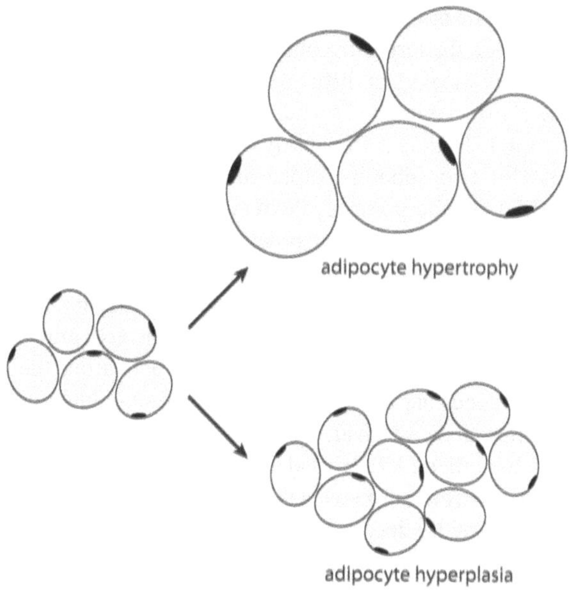

Fig. 13.4 Adipocyte hypertrophy versus hyperplasia. Upon excess uptake of energy, more triglycerides are stored within each fat cell. Adipocytes expand and increase in size. If a critical adipocyte size is reached and excess caloric intake persists over a long period of time, mitotic stimuli induce proliferation of adipocyte precursor cells that can differentiate into mature adipocytes thereby increasing the number of fat cells

marrow-derived macrophages constitute a large and important population of inflammatory cells in the adipose tissue (Maury and Brichard 2010; Weisberg et al. 2003; Xu et al. 2003).

As opposed to adipocyte hypertrophy that is the predominant mechanism of adipose tissue expansion during adulthood, adipocyte hyperplasia is mostly restricted to early stages of life. Hyperplasia mostly occurs upon mitotic activity of adipocyte precursor cells. It has been believed until recently that fat cell hyperplasia does not occur at all in the adult human. However, this opinion has been proven wrong and more recent findings show that hyperplasia in some cases can also occur during adulthood. In any case, hyperplasia is preceded by hypertrophy. Upon prolonged surplus energy balance, lipids are accumulated in adipocytes. Once the hypertrophic cells have reached a critical maximum size, precursor cells experience mitotic stimuli resulting in their proliferation (Butterwith 1997; Faust et al. 1978). To date, this is believed to only occur after tremendously long periods of extreme overeating and therefore it can only be observed in very few individuals.

13.2.2 Role of Hypoxia

Increased adipocyte size as seen in hypertrophic adipocytes, requires and expansion of the vasculature in order to maintain oxygen perfusion and nutrient supply. Consequently, hypoxia, a decreased oxygen partial pressure, is frequently observed locally in the adipose tissue of obese individuals and in rapidly expanding adipose tissue. The normal diffusion distance of oxygen in tissues is limited to 100–200 μm. However, the large size of hypertrophic adipocytes (150–200 μm) can challenge the capacity of oxygen diffusion. In the obese individuals, WAT has been described as poorly oxygenated (Trayhurn et al. 2008). The expression of the hypoxia-inducible factor-1 (HIF-1) is frequently used to assess whether hypoxia is present. HIF-1 consists of two subunits, alpha and beta. In the presence of O_2, the alpha-subunit (HIF-1α) undergoes a cycle of synthesis and proteasomal degradation. The degradation and destabilization process is mediated by HIF hydroxylases that are only active under normoxic conditions, i.e., in the presence of oxygen. Under hypoxic conditions, however, these hydroxylases are inactive resulting in a stabilization of HIF-1α. Thereby, HIF-1 can be translocated into the nucleus and bind to corresponding hypoxia response elements activating their transcription. Among these target genes are a number of angiogenesis-related genes and genes involved in cell proliferation, survival, apoptosis, and also metabolism (Trayhurn et al. 2008). Among others, VEGF is also regulated by HIF-1α.

Apart from adipocytes that respond to hypoxic conditions, macrophages are also affected by the locally lowered oxygen partial pressure. They respond by increased expression of interleukin 6 (IL6), migratory inhibitory factor (MIF), TNF-α, and VEGF (Trayhurn et al. 2008). It can therefore be concluded that hypoxia in the adipose tissue plays a role in inducing—among other processes—angiogenesis which is required for the expansion of adipose tissue and thereby facilitates the development of obesity.

13.3 White Adipose Tissue-to-BRITE Transformation

As described above, WATs and BATs are two distinct adipose depose existing in the body. Whereas WAT is storing excess energy in the form of lipids and triglycerides, BAT has a high metabolic capacity that—when fully activated—can contribute significantly to weight reduction and improve the metabolic status of individuals. It has been described that 63 g of fully metabolically active BAT can burn 4 kg of WAT each year in an adult human (Virtanen et al. 2009). Therefore, increasing the amount of BAT in the body seems to be a promising approach in order to identify a potential novel treatment option for obesity and obesity-related metabolic disorders. In an animal model, mice that are exposed to cold temperatures experience a transformation of their subcutaneous WAT into the so-called BRITE phenotype that is similar to adipocytes in BAT. This process is induced by an activation of the sympathetic nervous system by the cold. As a consequence, a dramatic increase in adipose vasculature can be observed in both WAT and BAT (Fig. 13.5). Furthermore, expression of UCP1—otherwise exclusive to BAT—is induced in the BRITE adipocytes leading to UCP-1-dependent activation of NST as a mechanism to adapt to the cold environment and compensate for the low ambient temperature.

The method that will be described in the following to induce cold-mediated activation of BAT in mice can be conducted with mice of different genetic backgrounds

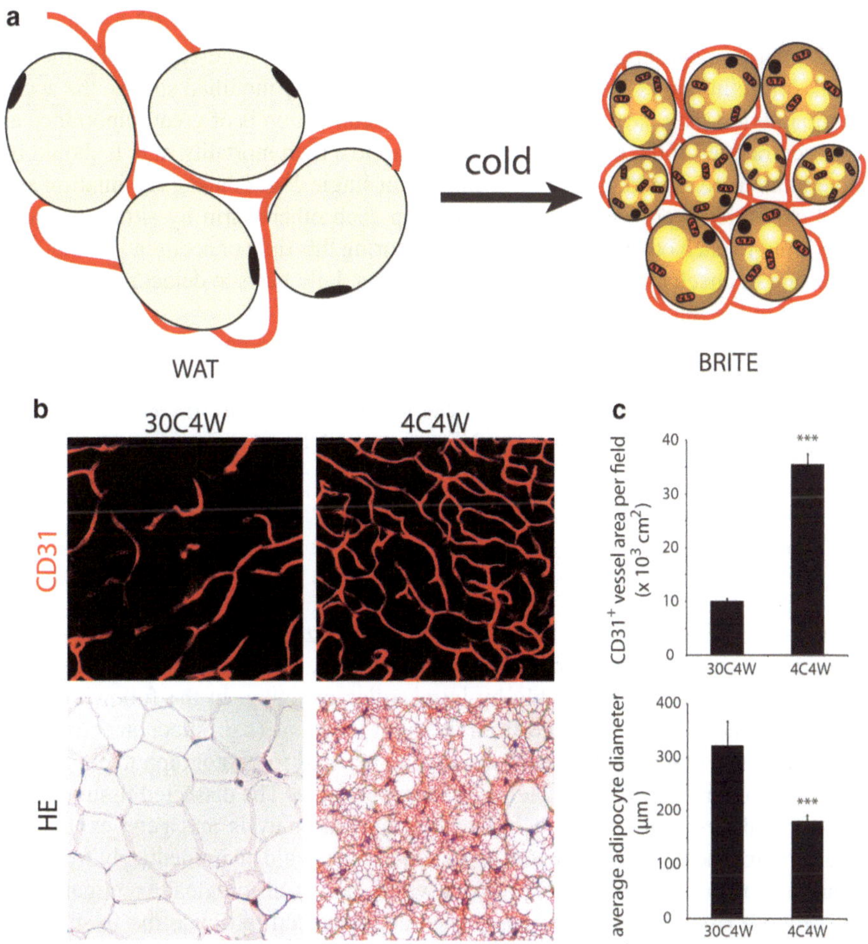

Fig. 13.5 Transformation of WAT to BRITE. (**a**) Cold acclimation induces angiogenesis in the adipose tissue and leads to a transformation of white adipocytes to smaller brown-like adipocytes which feature a multilocular lipid droplet distribution and show increased mitochondria content. (**b**) Staining with anti-CD31 antibody visualizes endothelial cells. Quantification shows a significant increase in blood vessel density at 4 °C compared to 30 °C ($p<0.001$). Hematoxylin and Eosin staining shows that adipocyte diameter decreases upon cold exposure ($p<0.001$). It is also visible that lipid droplets decrease in size and show a multilocular distribution after cold exposure. Scale bar: 50 μm

and can be combined with the administration of drugs or other chemical compounds to investigate their impact of adipose vasculature. Interestingly, the angiogenic effect of the cold as well as the BRITE transformation can already be observed after 1 week of cold exposure. However, longer exposure will produce more prominent effects. Using this model requires the permission of a local ethical committee.

In this model, the mice will be divided into two experimental groups: one group will be maintained at their thermoneutral temperature of 30 °C. At this temperature, mice have the lowest metabolic activity. The other group of mice will be exposed to 4 °C after an appropriate acclimation time at 18 °C. For wild-type (WT) mice, 1 week of acclimation is sufficient; for certain genetically modified strains, the acclimation period may have to be extended. This acclimation is of great importance as a direct transfer of the mice to 4 °C might lead to a high mortality rate. It should be noted that mice should optimally be housed in single cages during acclimation and cold exposure to prevent that the mice keep each other warm by sitting in close proximity to each other. It is very important during this time of acclimation and cold exposure to check the condition of the mice on a daily basis to detect potential discomfort of the mice immediately.

At the endpoint of the experiment, the mice from each temperature can be divided into three groups. One part of the mice can be used to collect tissue for subsequent histological analysis. A second group can be used to measure BMR and the third group can be used for assessment of NE-induced NST.

13.3.1 Histology

At the endpoint of the experiment, mice are euthanized by a lethal dose of carbon dioxide (CO_2). If blood samples are to be collected, this should be done immediately after euthanasia, preferably by intracardiac puncture. In the following step, adipose tissues can be dissected from different locations (e.g., subcutaneous WAT, visceral WAT, interscapular BAT). It is recommended to photograph the dissected tissues in order to document potential changes in color. The dissected tissue can be used for different purposes, including histological analysis and gene expression analysis (for the latter purpose, the dissected tissue should be immediately frozen in liquid nitrogen). However, this protocol will focus on histological examination of the adipose tissue. For this purpose, it is recommended to divide the freshly dissected adipose tissue into two portions: one portion will be transferred to a cryomold, immersed with Tissue-Tek solution and then transferred to dry ice. This tissue will be stored at −80 °C and can be used for cryosection immunohistostaining. The other portion of the tissue will be transferred to 4 % weight per volume (wt/vol) paraformaldehyde (PFA) and fixed for 24 h at 4 °C. After fixation, the tissue can be transferred to phosphate buffered saline (PBS) and stored at 4 °C until further use. It can then be used directly for whole mount immunohistochemistry or it can be embedded in paraffin and sectioned. All three methods will be described in the following.

13.3.2 Whole-Mount Immunohistochemistry

In order to obtain high-quality staining, it is recommended to use fresh adipose tissue.

1. Transfer the dissected and fixed tissue into Petri dishes filled with 1× PBS and cut into thin slices. Optimally, the slices are of a size of 5×5 mm. Cutting is performed using a scalpel blade. In order to obtain sections of equal thickness, the cuts should be even and little pressure is to be exerted on the actual tissue which is carefully held with a pair of forceps during the cutting process. It is recommended to always ensure that the scalpel blade is sharp to minimize damage to the tissue.
2. Transfer sections into a labeled six-well plate with about 1–1.5 mL 1× PBS per well. The tissue should be covered completely in PBS at this step. The sections are incubated for 1 h at RT in order to remove potential remaining PFA. Please not that this as well as all following washing, blocking, and incubation steps are performed on a rocking board.
3. To digest the tissue, incubate the sections with Proteinase K (20 Tris–HCl buffer (pH 7.4)) at RT for 5 min.
4. In the next step, permeabilize the tissue by incubating in about 1.5 mL methanol per well for 30 min at RT. Due to the toxicity of methanol, this step is to be performed in a fume hood.
5. To remove remaining methanol, wash the tissue 3 times with 1× PBS for 1 h.
6. For blocking of unspecific binding sites, adipose tissue sections are then incubated with 3 % (wt/vol) blocking buffer for 12–24 h at 4 °C.
7. On day 2 of the staining procedure, transfer the tissues into PBST and wash thoroughly.
8. Prepare appropriate dilutions of one or several primary antibodies. Typically, to stain blood vessels, the endothelial cell marker monoclonal rat anti-mouse CD31 antibody is used at a dilution of 1:200 in PBST. If other antibodies are used, consider the manufacturer's instructions or evaluate the optimal dilutions empirically. Antibody incubation is performed at 4 °C for 12–24 h. It is crucial that a sufficient amount of antibody is added to the tissue to ensure coverage of the sections. This step can be done using 48-well or 96-well plates in order to limit the amount of required antibody.
9. On day 3 wash the tissue with PBST for 1.5 h at 4 °C and then incubate with 3 % (wt/vol) blocking buffer for 1.5 h at 4 °C.
10. Incubate the tissue with appropriate dilutions of secondary antibody in PBST for 2 h at RT.
11. Incubate the tissue in 0.15 % (wt/vol) blocking buffer for 1 h at RT.
12. Wash the sections over night at 4 °C with PBST.
13. The next day, sections can be transferred to microscope glass slides, carefully mounted in a sufficient amount of Vectashield mounting medium and covered with a coverslip. The sections can then be stored at −20 °C for up to 4 weeks until microscopic analysis.
14. Use a confocal microscope (e.g., Nikon D-eclipse C1 and accompanying EZ-C1 3.90 software or an equivalent system) to acquire images of 5 μm thickness at ×4, ×10, ×20, or ×40 magnifications. To obtain three dimensional (3D) images, scan 8–10 layers. Analyze the images in a qualitative and quantitative manner using Adobe Photoshop software.

13.3.3 *Paraffin-Embedded Adipose Tissues*

Using a microtome, prepare sections of 5 μm thickness of paraffin-embedded tissue samples. Carefully transfer the sections into a water bath (40–45 °C) to allow the tissue and paraffin to flatten out completely and collect the sections with a Superfrost Plus glass slide. Let the sections air-dry and then remove excess paraffin by keeping the slides in an oven at 60 °C for at least 2 h. Before proceeding with Immunohistochemistry or H&E staining, let the slides adapt to RT for about 15 min.

13.3.3.1 H&E Staining on Paraffin-Embedded Adipose Tissue Sections

1. Deparaffinize the samples in a vertical staining jar in Xylen or any equivalent Tissue Clear solution for 2×5 min.
2. Rehydrate slides with 99.7 % (vol/vol) ethanol 2×5 min.
3. Rehydrate slides with 95 % (vol/vol) ethanol 2×5 min.
4. Rehydrate slides with 70 % (vol/vol) ethanol 2×5 min.
5. Wash slides with dH_2O for 5 min.
6. Stain tissue sections with fresh hematoxylin for 2–5 min.
7. Place the jar containing the slides under running tap water for 10 min to remove excess hematoxylin.
8. Stain tissue sections with fresh eosin for 1–2 min.
9. Dehydrate the stained tissue sections with 70 % (vol/vol) ethanol for 2×5 min.
10. Dehydrate the stained tissue sections with 95 % (vol/vol) ethanol for 2×5 min.
11. Dehydrate the stained tissue sections with 99.7 % (vol/vol) ethanol for 2×5 min.
12. Place the slides horizontally on a working bench and let the dry for about 15 min.
13. Mount slides with 1–2 drops of Pertex mounting medium and cover the slides with microscope coverslips. Prevent the formation of air bubbles by working very carefully and store the slides in a storage box until imaging.
14. Imaging is performed using a bright-field microscope. Acquire images at ×4, ×10, ×20, or ×40 magnifications and quantify adipocyte size using Adobe Photoshop or ImageJ software.

13.3.3.2 Immunohistochemistry on Paraffin-Embedded Adipose Tissue Sections

1. Deparaffinize the samples in a vertical staining jar in Xylen or any equivalent Tissue Clear solution for 2×5 min.
2. Rehydrate slides with 99.7 % volume per volume (vol/vol) ethanol 2×5 min.
3. Rehydrate slides with 95 % (vol/vol) ethanol 2×5 min.
4. Rehydrate slides with 70 % (vol/vol) ethanol 2×5 min.
5. Wash slides with dH_2O for 5 min.

6. Boil the slides at ~100 °C in a microwave in antigen unmasking solution (1:100 in dH_2O) for 10 min.
7. Allow the slides to cool down for about 20 min and then wash them in 1× PBS for 2×5 min.
8. Encircle the tissue sections on the slides using a PAP hydrophobic pen. Thereby, the amount of reagent required for the staining process will be limited to the tissue area. Carefully remove excess liquid from the slides using a paper towel. Avoid touching the tissue sections during this step as it might destroy tissue integrity.
9. If using a biotinylated primary antibody, block endogenous biotin using an Avidin-Biotin blocking kit. Otherwise proceed to step 14.
10. Apply one drop of avidin-blocking solution on each tissue section and incubate for 15 min.
11. Rinse slides in 1× PBS for 2×5 min.
12. Apply one drop of biotin-blocking solution on each tissue section and incubate for 15 min.
13. Rinse slides in 1× PBS for 2×5 min.
14. If using a horse radish peroxidase (HRP)-linked secondary antibody, block endogenous peroxidase activity by incubation in 3 % H_2O_2 for 10 min and then rinse in 1× PBS for 2×5 min. Otherwise proceed to step 15 directly.
15. Block unspecific binding sites by incubating the slides with 1× PBS containing 4 % (vol/vol) normal goat serum.
16. Incubate tissue sections with primary antibody at an appropriate dilution in 1× PBS containing 4 % (vol/vol) normal goat serum in a humidified chamber for 12–24 h at 4 °C. One antibody or several antibodies in combination can be used.
17. On day 2, wash slides 3 times in 1× PBS for 5 min each.
18. Prepare appropriate dilutions of secondary antibodies in 1× PBS containing 4 % (vol/vol) normal goat serum. If using fluorochrome-linked secondary antibodies, protect tissue slides from light in order to prevent bleaching.
19. Wash slides 3 times in 1× PBS for 5 min each.

    For fluorescence detection of signals, proceed to step 20, for chromogenic detection proceed to step 21.
20. Fluorescence detection of the signals, mount slides in using 1–2 drops of Vectashield mounting medium and cover with a microscope coverslip. If desired, Vectashield mounting medium containing DAPI can be used to stain cell nuclei. Slides can be stored in the dark at –20 °C for several weeks until microscopic analysis.
21. Chromogenic detection: use DAB substrates to develop positive signals. Stop development at appropriate time points by transferring the slides into dH_2O. If required, counterstain nuclei with hematoxylin for 30 s and wash the slides in dH_2O to remove excess hematoxylin. Dehydrate samples by rinsing with 70 % (vol/vol) and 99.7 % (vol/vol) ethanol for 5 min each. Let the sections air-dry and mount with Pertex. Cover the sections carefully with coverslips and place in a storage box until microscopy analysis.

For analysis of the staining, imaging is performed using either a fluorescence microscope or a bright field microscope if using the chromogenic method.

22. Acquire images at ×4, ×10, ×20, or ×40 magnifications and quantify the obtained signals using Adobe Photoshop software.

13.3.3.3 Immunohistochemistry on Cryosections

Prepare sections of the Tissue Tek-embedded adipose tissue using a microtome. Adjust the cryostat to −30 °C prior to sectioning. This temperature is lower compared to other tissues to maintain adipose tissue integrity. Carefully prepare sections of 15 μm thickness and transfer them to Superfrost Plus microscope slides. The prepared sections can be stored at −80 °C for about 2 years until further use.

For immunohistochemical staining of the cryosections, the steps in the protocol below should be performed:

1. Transfer the tissue slides from −80 °C to RT and allow them to adapt for about 30 min.
2. Fix the tissue with 100 % acetone for 10 min in a vertical staining jar.
3. Wash the tissue slides in 1× PBS for 5 min. Repeat this step 3 times.
4. Encircle the tissue sections on the slides using a PAP hydrophobic pen. Thereby, the amount of reagent required for the staining process will be limited to the tissue area. Carefully remove excess liquid from the slides using a paper towel. Avoid touching the tissue sections during this step as it might destroy tissue integrity.
5. Block unspecific binding sites by incubating the slides with blocking buffer (1× PBS containing 4 % (vol/vol) normal goat serum).
6. Incubate tissue sections with primary antibody at an appropriate dilution in blocking buffer in a humidified chamber for 12–24 h at 4 °C. One antibody or several antibodies in combination can be used.
7. On day 2, wash slides 3 times in 1× PBS for 5 min each.
8. Incubate sections with appropriate dilutions of secondary antibodies in blocking buffer at RT for 45 min in the humidified incubation chamber. If using fluorochrome-linked secondary antibodies, protect tissue slides from light in order to prevent bleaching.
9. Wash slides 3 times in 1× PBS for 5 min each.
10. Carefully remove excess PBS using a paper towel.
11. Mount slides using 1–2 drops of Vectashield mounting medium and cover with a microscope coverslip. Slides can be stored in the dark at −20 °C for several weeks until microscopic analysis.
12. Imaging is performed using a fluorescence microscope with different excitation wavelengths. Acquire images at ×4, ×10, ×20, or ×40 magnifications and quantify the obtained signals using Adobe Photoshop software.

13.3.4 Measurement of Basal Metabolic Rate and NE-Induced NST

Another factor in assessing the capacity of WAT to transform to a BRITE phenotype is measuring the murine metabolism, especially the maximum extent of NST which is based on UCP1 expression. Metabolic rate can be measured using an indirect calorimeter. Optimally, calorimeters are equipped with a time that maintains a regular 12 h light/dark cycle corresponding to the circadian rhythm of the mice. The procedure will only be explained generally in this paragraph. However, detailed protocol is available (Lim et al. 2012).

To metabolize carbohydrates, fats, or proteins, oxygen (O_2) is consumed while CO_2 and heat are released. In indirect calorimetry, energy expenditure (i.e., BMR and NE-induced NST) are assessed by determining a respiratory quotient (RQ) based on CO_2 production and O_2 consumption (Ferrannini 1988). For this purpose, the air inside the chamber is calibrated in the beginning of each measurement so it contains a defined O_2 and CO_2 saturation. During the measurement, the air inside the indirect calorimeter is sampled in fixed and regular time intervals and the levels of the respiratory gases are measured (Ferrannini 1988). Typically, the metabolic chambers are equipped with a dehumidifying filter system to remove humidity from the air that is generated during respiration of the animals. This is an important factor as humidity can alter the outcome of the measurement which is especially problematic during BMR measurement where the animals remain in the chambers for 24 h and a significant amount of humidity can be produced over time.

During the course of basal metabolic measurement, mice should be provided sufficient food and water. Ideally, mice are not interrupted during the measurement and remain inside the closed calorimeters in order not to alter the concentration of the respiratory gases by opening the chamber. Many calorimeters are also equipped with a temperature regulating system so that BMR can be assessed, e.g., at RT or at the thermoneutral temperature. At the end of the measurement, mice can be transferred back to their home cages.

Measurement of NE-induced NST requires more effort compared to BMR. Since this measurement should only assess the energy consumption due to NST, it is required to anesthetize the animals prior to starting the experiment in order to prevent unreliable results due to skeletal muscle activity. Anesthesia is commonly done using Pentobarbital (40 mg kg^{-1}) which is administered by intraperitoneal injection. Place the mice in a calm and warm environment once they are anesthetized, and then transfer them into the calorimeters.

In the beginning of the measurement, BMR should be measured for about 30 min to obtain a stable baseline. Ensure that the mice are fully asleep and the inject 1 mg kg^{-1} NE subcutaneously in the dorsal back. Already after a few minutes, a response to the NE injection can be observed as a significant increase of the RQ above the baseline level. Following injection of this adrenergic agonist, brown fat and UCP1 activities are stimulated and reach their maximum capacity. Thereby, the response to the NE injection can be correlated to the extent of BAT and also BRITE tissue present in the mouse. After approximately 45 min to 1 h the response reaches a plateau and starts to decline. At this point, the measurement can be terminated and mice can either be placed in a warming cabinet to recover from anesthesia or they can be sacrificed according to the respective regulations provided by the local ethical committee.

13.4 Transplantation of Adipose Tissue

In general, there are two different approaches in transplanting adipose tissue, fat grafting, and transplantation of adipose-derived stem cells (ADSCs). In the first method, the adipose tissue is harvested from the donor site using a cannula. The sample is sedimented by centrifugation and injected into the recipient site of the same individual. This method is frequently used in reconstructive or plastic surgery and has been first described in 1893 (Neuber 1893). Generally, it has been proven beneficial to transplant small grafts instead of larger amounts of fat in order to maximize graft survival and revascularization (Wilson et al. 2011).

For ADSC therapy, harvesting can be performed by liposuction. After washing and collagenase digestion, the stromal vascular fraction (SVF) containing the stem cells can be separated from the mature adipocyte partition by centrifugation. While mature adipocytes will be collected in the upmost layer due to their high lipid content and thereby low density, the SVF will be sedimented in the pellet. After separating the mature adipocytes from the SVF, ASDCs can be implanted into the recipient site. It is critical to use a sufficient amount of starting lipoaspirate (at least 250 mL) in order to guarantee an appropriate number of ADSCs in the sample. Typically, 250 mL of aspirate contain 10^7–10^8 ADSCs (Tabit et al. 2012).

The biological basis underlying the idea of transplanting adipose tissue or ADSCs lies in their inherent properties. ADSCs are mesenchymal cells that are relatively similar compared to bone marrow-derived cells (mesenchymal stem cells) (Wagner et al. 2005; Lee et al. 2004; Hong et al. 2010). They are multipotent cells which can be differentiated into multiple different cell types including adipocytes, smooth muscle cells, cardiomyocytes, and endothelial cells (Hong et al. 2010; Zuk et al. 2001; Tran and Kahn 2010; Planat-Benard et al. 2004). Another benefit of using ADSCs is their ability to secrete remarkable amounts of both angiogenic factors—thereby promoting revascularization of, for example, ischemic areas—and also anti-apoptotic factors that prevent cell death.

Currently, there are a number of studies underway that investigate the potential of autologous fat grafting in relation to facial reconstruction, reconstruction of the breast after mastectomy as well as for the treatment of lipidystrophies (http://www.clinicaltrials.gov/. U.S. National Institutes of Health). Additionally, many preclinical studies evaluate the potential of ADSC-enriched grafting focus on stroke, spinal cord trauma, multiple sclerosis, type 1 diabetes, ischemia, and burn wound treatment (Tabit et al. 2012; Tran and Kahn 2010; Ryu et al. 2009).

One potential drawback that has to be kept in mind when discussing the potential of ADSCs in reconstructive or regenerative research is that individuals that have a history of tumor formation might not be suitable for this method. This is due to the stem cell character of ADSCs. To date, it seems that only ouvert cancer but not resting tumor cells experience a growth enhancement upon ADSC delivery (Tabit et al. 2012; Zimmerlin et al. 2011). Even though results obtained from preclinical studies in animal models are inconclusive, the extent of the potential benefit compared to the risk for the patient need to be evaluated (Tabit et al. 2012).

13.4.1 Transplantation of Adipose Tissue for Regenerative Medicine

The potential of grafting adipose tissue after an ischemic myocardial infarction to induce revascularization of the infarcted area has been recognized for decades. However, we are beginning to understand more and more the mechanisms behind this process. Myocardial infarction (MI)—also known as heart attack—is a rather frequent cardiac event and accounts for over 10 % of deaths worldwide. In western developed countries, it is the leading cause of death and can be associated frequently with sedentary lifestyle and unhealthy nutrition based on a high amount of carbohydrates and fat. Further risk factors are diabetes, hypertension, dyslipidemia and also genetic factors. The most commonly described symptom of MI is chest pain that often gradually develops. Chest pain can be due to ischemia of the cardiac muscle (angina pectoris). More advanced symptoms include dyspnea (shortness of breath) due to a limited cardiac output that develops in the left ventricle and can lead to pulmonary edema. However, in some cases MI remains unnoticed and is only detected postmortem during autopsy (Mazo et al. 2010).

The cardiac event in most cases is initiated by rupturing of an atherosclerotic plaque in the coronary artery, triggering a clotting cascade that can lead to complete arterial occlusion. The impaired cardiac blood flow can further lead to ischemia. Necrotic as well as apoptotic cell death causes myocardial cells to die leaving behind nonfunctional scar tissue. This damage to the heart is permanent and myocardial scarring can even deteriorate the health condition by causing arrhythmias or the development of lethal aneurysms. These subsequent conditions can increase coronary ischemia and further aggravate the cardiac function.

In order to reduce the extent of permanent damage to the heart upon MI, immediate medical care is crucial. These interventions have become more and more successful over the past years and could ensure post-MI survival of an increasing number of patients. However, in most cases, the cell death of cardiomyocytes—that usually peaks at the cardiac event—persists after the MI even though to a smaller extent. Also, the number of chronically infarcted patients increases (Mazo et al. 2010). Even though many approaches have been undertaken to restore the function of the injured heart, none of the attempts has reached groundbreaking results yet and chronic infarction often results eventually in heart failure (Mazo et al. 2010). At that stage, a heart transplant is the only option to ensure patient survival; however, the number of organs available is limited and it is difficult to find a matching donor heart.

In order to encompass restoration of the injured heart, stem cell therapy has been established over a decade ago to (1) regenerate the myocardial mass, (2) recreate a functional vascular network, and (3) return the ventricle to its appropriate geometry (Mazo et al. 2010). Even though this attempt is far from being able to offer a cure for the infarcted heart, promising approaches have been described. In this context, adipose tissue grafting and transplantation of adipose tissue-derived mesenchymal stem cells have become a major focus.

As opposed to grafting of autologous fat alone, it has been proven beneficial to transplant adipose tissue grafts enriched with ADSCs. Thereby, neovascularization is facilitated and even cardiac remodeling has been described in rodent MI models (Tabit et al. 2012; Valina et al. 2007; Strem and Hedrick 2005; Hwangbo et al. 2010).

13.4.2 Transplantation of Adipose Tissue for Reconstructive Medicine

Adipose tissue or ADSC transplantation for the use in plastic surgery has been established for over hundred years (Neuber 1893; Tran and Kahn 2010). A pioneer in this field was the German doctor Franz Neuber who transplanted a fat graft from the upper arm of a tuberculosis patient to his face to fill facial depressions (Neuber 1893; Wilson et al. 2011; Tran and Kahn 2010). However, early grafting attempts performed by him as well as others often had poor resorption rates. The size of the graft influences the survival rate of the transplanted fat by affecting the graft's ability to establish blood supply in the recipient site (Peer 1955).

13.4.3 Transplantation of Adipose Tissue for Physiological Research

Several attempts of transplanting adipose tissue (WAT and BAT) aim at a better understanding regarding adipose tissue development. One of the pioneering experiments in this area was performed in the late 1930s. Fibroblast-like preadipocytes were transplanted from the gonadal white adipose depot of the donor to a recipient rat. These cells developed into fat pads that responded appropriately upon food deprivation or feeding (Tran and Kahn 2010; Hausberger 1938, 1955). Similar studies have further highlighted the role of vasculature in the development of adipose tissue as well as the importance of the host environment in determining the phenotype of the transplanted cells (Tran and Kahn 2010; Iyama et al. 1979; Enser and Ashwell 1983). The latter finding has been observed both with adipocytes from WAT and BAT (Tran and Kahn 2010; Enser and Ashwell 1983; Roberts et al. 1986).

13.4.4 Transplantation of Adipose Tissue for the Development of Novel Treatment Options for Metabolic Disorders

From a therapeutic aspect—especially regarding the increasing number of overweight and obese individuals worldwide—the concept of adipose tissue grafting to open new avenues in the treatment of obesity and related metabolic disorders seems very tempting. However, not only combatting the adverse effects of excess fat storage but also treating the opposite phenomenon—inability to store lipids within adipocytes (lipodystrophy)—is an important issue as it is also associated with metabolic disruptions (Tran and Kahn 2010).

In regard to develop novel methods for combatting the metabolic syndrome, transplantation of adipose tissue might be a promising option keeping in mind the beneficial effect of BAT on energy expenditure (Tran and Kahn 2010; Dellagiacoma et al. 1992; Smahel 1989). Although some approaches in transplantation of BAT have been undertaken, the tissue grafts cannot be maintained within the host and undergo necrotic cell death. Interestingly, it has also been shown that the host environment has a strong influence on the transplanted tissue in terms of, for example, fatty acid composition (Tran and Kahn 2010; Ryu et al. 2009).

13.4.5 Conclusion

The field of adipose tissue or ADSC transplantation has the potential to open up novel therapeutic approaches in the treatment of obesity and obesity-related metabolic diseases as well as in reconstructive and regenerative medicine. However, in particular the attempts to improve metabolic syndrome need to be improved. To date, the techniques used in preclinical studies are not optimal and often fail to lead to consistent long-term improvements. Above all, survival of the graft is a major obstacle that needs to be overcome. But also the strategies currently used in ADSC transplantation leave room for improvement. Apart from the complications regarding survival of the graft, it is also crucial to consider the safety of the patient. Potential effects of adipose tissue or ADSC grafting on the whole body need to be evaluated thoroughly.

13.4.6 Genetic Models

During the past century, the mouse has gradually become the model organism of choice to increase our knowledge to translate in vivo experiments to simulate human diseases in research. The genome of the mouse shows extensive synteny and the physiology of the mouse is highly similar to humans; therefore, the mouse has over the years developed into becoming a suitable model to be used to study different diseases in humans. Currently, there are thousands of inbred mouse strains with genetically engineered mutants that are prone to develop different diseases that can be used to study human diseases. The rapid advancement in technology to create different genetically modified mice using techniques such as overexpression, knock-out and knock-in, tissue- and time-specific deletion of target genes using the *Cre-loxP* system, gene-trap approaches allows us to now investigate the role and function of genes in vivo (Davey and MacLean 2006). Despite the advancement in creating transgenic mice and the high synteny between mice and human, many confounding factors still have to be considered while choosing a suitable genetically modified mouse strain to translate to the human endocrine system.

13.4.7 Mouse Models to Study Angiogenesis in Obesity and Diabetes

The expansion and regression of adipose tissue is highly coupled to angiogenesis. The metabolic demand in the adipose tissue is also dependent on the adipose vasculature. Obesity has over the years become a major concern in the modern society. The incidence of many obesity-related medical problems including type II diabetes, cardiovascular diseases, and hypertension has been escalating tremendously and has raised the health care expenditures and therefore requires immediate attention and intervention (Cao 2010). Genetic- and diet-induced mouse models are excellent models and provide immense opportunities to investigate the modulation of angiogenesis during adipogenesis and obesity. At present, there are several mouse models available to study monogenic obesity and diabetes including: (1) Mice with spontaneous mutation for obesity, the leptin-deficient mice (Lep^{ob}, commonly known as *ob*/*ob*), with structurally defective leptin which does not bind to the leptin receptor (ObR). These mice do not have leptin activity and exhibit obesity, hyperphagia, glucose intolerance, and elevated plasma insulin at the age of 4 weeks of age. (2) Mice with spontaneous mutation for diabetes ($lepr^{db}$ or *db*/*db*), which develop obesity with the onset of 3–4 weeks of age. These mice have dysfunctional endothelial and vascular function and could be used to study metabolic syndrome and type 2 diabetes in relation to blood vessel growth and functionality. (3) Diet-induced obese (DIO) mice have elevated blood glucose and impaired glucose tolerance suitable for studying prediabetic type 2 diabetes and obesity. These mouse models are ideal for the investigation of the spatial and functional relations between adipocytes and vasculature including blood vessels density, structure, and remodeling during onset of disease, and treatment of obesity and diabetes.

13.4.8 Different Transgenic Mouse Model to Study Embryonic Angiogenesis

There are two major families of endothelial-specific receptor tyrosine kinases that play a critical role in mediating signaling that regulates growth, survival, proliferation, and differentiation of endothelial cells; the VEGFRs and Tie family. The Tie receptor family consists of two members, Tie-1 and Tie-2; and four ligands, Angiopoietin 1-4 (Ang1-4). Angiopoietin 1-4 bind to Tie-2 while Tie-1 is an orphan tyrosine receptor. Tie-1 is an essential endothelial receptor tyrosine kinase playing crucial role in the development and sustenance of the vascular system. Several transgenic mouse models of abnormal embryonic vascular development have been developed over the years. Mice lacking VEGF, FLT-1, FLK-1, and VE-cadherin result in the formation of primitive capillary plexus (Carmeliet et al. 1996; Fong et al. 1995; Shalaby et al. 1995). Mice lacking Tie-1, Tie-2, Ang-1 Braf, VHL, and VCAM-1 exhibit defects in expansion and remodeling during embryonic development. Since these genes are essential in embryonic development, deletion of these genes results in embryonic lethality. Therefore Conditional gene targeting enables inactivation or modification of genes in specific cell types and also in spatio-temporal manner (Kuhn and Schwenk 1997). Conditional recombination serves as an excellent way to circumvent lethality at early embryonic stages. Conditional recombination technique uses the loxP sequences which flank both ends of the target gene. The site-specific DNA recombinase Cre recognizes the loxP and excises the DNA segments that are flanked by loxP sites. The target gene can be knocked out by the presence of Cre recombinase which activates the loxP recombination. To achieve a tissue-specific knockout of the gene, mice with Cre recombinase gene with tissue-specific promoter can be crossed with desired loxP mice resulting in loxP recombination occurring in a specific tissue where the promoter is active (Houdebine 2002; Mallo 2006).

13.4.9 FOXC2-TM to Study Vascular Remodeling, Patterning, and Maturation

In this section, we will discuss the use of a transgenic mouse strain, FOXC2-transgenic mice (FOXC2-TM) to study vascular remodeling including vessel maturation, patterning, and stability. FOXC2 belongs to a member of the forkhead/winged helix transcription factor family. FOXC2 has been reported to be implicated in many human diseases such as lymphedema-distichiasis, type 2 diabetes and

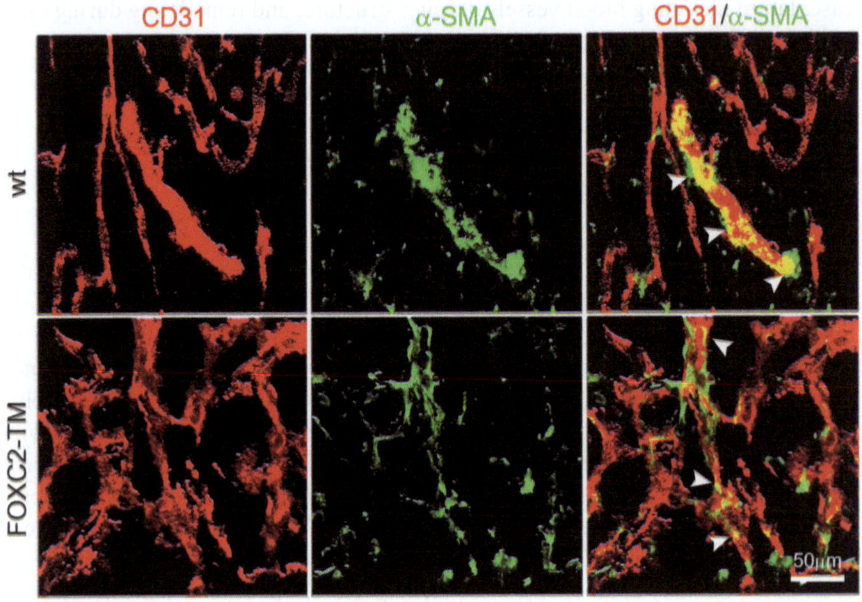

Fig. 13.6 Vascular remodeling in FOXC2-TM. The WAT depots from WT and FOXC2-TM were stained with an endothelial cell marker anti-CD31 antibody (*red*), and vascular smooth muscle marker, an anti-SMA antibody (*green*). *Arrowheads* point to double-positive areas

interstitial nephritis. Cederberg et al. have shown that FOXC2 expression in adipocytes could lead to the conversion of WAT to acquire a more BRITE-like phenotype (Cederberg et al. 2001). In addition, expression of FOXC2 in adipocytes resulted in increase of oxygen consumption (up to 4 times higher compared to WT mice), and increased metabolism in mice. Interestingly, necropsy observation of the inguinal WAT (iWAT) of the FOXC2-TM showed a red diffused phenotype resembling that of BAT suggesting the possible phenotypic change from WAT to BAT (Xue et al. 105). Combination of genetic models and immunohistochemistry techniques serves as an important platform for the analysis of blood vasculature density, patterning remodeling, and different microvasculature networks (Xue et al. 2010). Immunohistochemistry staining using anti-CD31 antibody allows visualization of the density and structural networks of blood vessels. To further study vascular remodeling and maturation of blood vessels, immunohistochemistry stainings using alpha smooth muscle actin (α-SMA) antibody and proteoglycan sulfate (NG2) antibody allow the investigation of the distribution of smooth muscle cells and pericytes on the blood vasculature (Fig. 13.6). The blood vasculature in FOXC2-TM iWAT is highly dense, irregular, and comprises highly disorganized vascular plexuses resembling honeycomb structures as compared to vasculature in the WT mice. The blood vessel density in the BAT is up to 5 times higher compared to the blood vessels in WAT. Here, we observed that the blood vessel density in WAT of FOXC2-TM is comparable to that of BAT in WT mice. In WT mice, vascular smooth muscle cells

(VSMCs) and pericytes in WAT are associated with large arterial vessels and are very rarely found on microvessels. In the WAT and BAT of FOXC2-TM, the percentage of αSMA-positive VSMCs is significantly increased and distributed on the microvessels as compared to WT mice demonstrating the role of FOXC2 in regulating the distribution of VSMCs in adipose maturation. On the other hand, the NG2-positive pericyte coverage in FOXC2-TM was not only found on microvessels but was also redistributed to large arterial vessels. Vascular remodeling through the redistribution of pericytes and VSMCs on the blood vessels could be essential to regulate and support the high metabolic demand in the FOXC2 expressing adipose tissue. Affymetrix microarray gene chip and PCR analyses revealed that the expression level of angiogenic genes such as vascular VEGF, platelet-derived growth factor (PDGF), Ang-2, transforming growth factor-beta (TGF-β) were significantly upregulated in the adipose tissues of FOXC2-TM. Among these angiogenic factors, Ang-2 was one of the most upregulated genes; a nearly sixfold increase of Ang-2 was detected in the WAT of FOXC2-TM as compared to WT mice. Ang-2 promoter was subsequently fused with the luciferase reporter system, cloned and transfected into 3T3 differentiated preadipocytes in the absence and presence of FOXC2. The Ang-2 activity was increased in a dose-dependent manner upon addition of FOXC2, suggesting that Ang-2 could be an essential mediator of FOXC2-induced angiogenic phenotype. Ang-2 is essential for vascular remodeling, patterning, and maturation (Huang et al. 2009; Maisonpierre et al. 1997). Ang-1 and Ang-2 bind to tyrosine kinase Tie-2 thereby triggering downstream activation to initiate blood vessel remodeling and maturation. An Ang-2-specific inhibitor L1-10 was used to investigate if the angiogenic phenotype in the WAT and BAT of FOXC2-TM could be reversed. Treatment of FOXC2-TM with L1-10 diminished the reddish phenotype in the WAT, and normalized the dense, irregular, and highly disorganized vascular plexuses to a highly organized blood vasculature as observed in WT mice. In addition, α-SMA-positive VSMCs were redistributed to the large arterial vessels similar to that of WT mice. Blockade with L1-10 decreased the total number of VSMC-coated vascular area and increased the association to large arterial vessel instead of microvessel. This showed that Ang-2 was responsible for FOXC2-induced vascular maturation and redistribution in the adipose tissue. Alteration of blood vasculature and increased formation of premature new vessels in FOXC2-TM could potentially affect wound-healing; therefore, a full skin wound-healing experiment was performed to investigate the role of FOXC2 in vascular functionality. The wound-healing capacity of FOXC2-TM was indeed impaired and it took approximately 3 days longer for wounds to heal in FOXC2-TM as compared to WT mice.

13.4.10 Conclusion and Outlook

Here we described an example to use the transgenic mouse model FOXC2-TM accompanied with immunohistochemistry techniques to explore the role of blood vessel patterning, distribution, and maturation. To further study the functionality of blood vessels in the adipose tissues, in terms of perfusion and leakiness, tetramethylrhodamine dextran of different molecular weight could be used to investigate either perfusion (2,000 kDa) or leakiness of vessels (70 kDa). Dextran can be injected into the tail vein of mice and is allowed to circulate before sacrificing the mice. Further immunohistochemistry staining with different markers such as anti-CD31-specific antibody, a blood vessel pan-marker, could be used to investigate the area of perfused vessels and area of leakiness (Xue et al. 2010; Hedlund et al. 2009). The vascular permeability can be assayed by monitoring extravasation of injected dextran. There is no single model that is ideal to elucidate the development of angiogenesis in adipose tissue, a combination of one or more models would be necessary to encapsulate the human diseases. However, over the years, genetically engineered animal model have gradually shown to be a better and clean system to evaluate the progress and development of human diseases. With the availability of more and more transgenic mice strains, it is possible to choose suitable genetically modified mice to mimic the human disease thereby making the research work more relevant to the clinical settings.

References

Asano A, Morimatsu M, Nikami H, Yoshida T, Saito M. Adrenergic activation of vascular endothelial growth factor mRNA expression in rat brown adipose tissue: implication in cold-induced angiogenesis. Biochem J. 1997;328(Pt 1):179–83.

Bendich A, Deckelbaum RJ. Preventive nutrition: the comprehensive guide for health professionals, vol. 1. Totowa, NJ: Humana Press; 2005. p. 351.

Butterwith SC. Regulators of adipocyte precursor cells. Poult Sci. 1997;76(1):118–23.

Cali AM, Caprio S. Prediabetes and type 2 diabetes in youth: an emerging epidemic disease? Curr Opin Endocrinol Diabetes Obes. 2008;15(2):123–7.

Cao Y. Angiogenesis modulates adipogenesis and obesity. J Clin Invest. 2007;117(9):2362–8.

Cao Y. Adipose tissue angiogenesis as a therapeutic target for obesity and metabolic diseases. Nat Rev Drug Discov. 2010;9(2):107–15.

Cao R, Lim S, Ji H, Zhang Y, Yang Y, Honek J, et al. Mouse corneal lymphangiogenesis model. Nat Protoc. 2011;6(6):817–26.

Carmeliet P, Ferreira V, Breier G, Pollefeyt S, Kieckens L, Gertsenstein M, et al. Abnormal blood vessel development and lethality in embryos lacking a single VEGF allele. Nature. 1996;380(6573):435–9.

Castellot Jr JJ, Karnovsky MJ, Spiegelman BM. Differentiation-dependent stimulation of neovascularization and endothelial cell chemotaxis by 3T3 adipocytes. Proc Natl Acad Sci U S A. 1982;79(18):5597–601.

Cederberg A, Gronning LM, Ahren B, Tasken K, Carlsson P, Enerback S. FOXC2 is a winged helix gene that counteracts obesity, hypertriglyceridemia, and diet-induced insulin resistance. Cell. 2001;106(5):563–73.

Chen L, Cao ZL, Han F, Gao ZC, He QY. Chronic intermittent hypoxia from pedo-stage decreases glucose transporter 4 expression in adipose tissue and causes insulin resistance. Chin Med J (Engl). 2010;123(4):463–70.

Christiaens V, Lijnen HR. Angiogenesis and development of adipose tissue. Mol Cell Endocrinol. 2010;318(1–2):2–9.

Cypess AM, Lehman S, Williams G, Tal I, Rodman D, Goldfine AB, et al. Identification and importance of brown adipose tissue in adult humans. N Engl J Med. 2009;360(15):1509–17.
Davey RA, MacLean HE. Current and future approaches using genetically modified mice in endocrine research. Am J Physiol Endocrinol Metab. 2006;291(3):E429–38.
Dellagiacoma G, Sbarbati A, Rossi M, Zancanaro C, Benati D, Merigo F, et al. Brown adipose tissue: magnetic resonance imaging and ultrastructural studies after transplantation in syngeneic rats. Transplant Proc. 1992;24(6):2986.
Enser M, Ashwell M. Fatty acid composition of triglycerides from adipose tissue transplanted between obese and lean mice. Lipids. 1983;18(11):776–80.
Faust IM, Johnson PR, Stern JS, Hirsch J. Diet-induced adipocyte number increase in adult rats: a new model of obesity. Am J Physiol. 1978;235(3):E279–86.
Ferrannini E. The theoretical bases of indirect calorimetry: a review. Metabolism. 1988;37(3):287–301.
Flynn III EJ, Trent CM, Rawls JF. Ontogeny and nutritional control of adipogenesis in zebrafish (Danio rerio). J Lipid Res. 2009;50(8):1641–52.
Fong GH, Rossant J, Gertsenstein M, Breitman ML. Role of the Flt-1 receptor tyrosine kinase in regulating the assembly of vascular endothelium. Nature. 1995;376(6535):66–70.
Hausberger FX. Quantitative studies on the development of transplanted fat tissue stores in rats [German]. Virchows Arch Pathol Anat Physiol Klin Med. 1938;302:640–56.
Hausberger FX. Quantitative studies on the development of autotransplants of immature adipose tissue of rats. Anat Rec. 1955;122(4):507–15.
Hausman GJ, Richardson RL. Adipose tissue angiogenesis. J Anim Sci. 2004;82(3):925–34.
Hausman GJ, Kasser TR, Martin RJ. The effect of maternal diabetes and fasting on fetal adipose tissue histochemistry in the pig. J Anim Sci. 1982;55(6):1343–50.
Hausman GJ, Novakofski JE, Martin RJ, Thomas GB. The development of adipocytes in primary stromal-vascular culture of fetal pig adipose tissue. Cell Tissue Res. 1984;236(2):459–64.
Hedlund EM, Hosaka K, Zhong Z, Cao R, Cao Y. Malignant cell-derived PlGF promotes normalization and remodeling of the tumor vasculature. Proc Natl Acad Sci U S A. 2009;106(41):17505–10.
Hong SJ, Traktuev DO, March KL. Therapeutic potential of adipose-derived stem cells in vascular growth and tissue repair. Curr Opin Organ Transplant. 2010;15(1):86–91.
Houdebine LM. Animal transgenesis: recent data and perspectives. Biochimie. 2002;84(11):1137–41.
Huang J, Bae JO, Tsai JP, Kadenhe-Chiweshe A, Papa J, Lee A, et al. Angiopoietin-1/Tie-2 activation contributes to vascular survival and tumor growth during VEGF blockade. Int J Oncol. 2009;34(1):79–87.
Hwangbo S, Kim J, Her S, Cho H, Lee J. Therapeutic potential of human adipose stem cells in a rat myocardial infarction model. Yonsei Med J. 2010;51(1):69–76.
Imrie D, Sadler KC. White adipose tissue development in zebrafish is regulated by both developmental time and fish size. Dev Dyn. 2010;239(11):3013–23.
Iyama K, Ohzono K, Usuku G. Electron microscopical studies on the genesis of white adipocytes: differentiation of immature pericytes into adipocytes in transplanted preadipose tissue. Virchows Arch B Cell Pathol Incl Mol Pathol. 1979;31(2):143–55.
Jensen LD, Cao R, Cao Y. In vivo angiogenesis and lymphangiogenesis models. Curr Mol Med. 2009;9(8):982–91.
Jensen LD, Rouhi P, Cao Z, Lanne T, Wahlberg E, Cao Y. Zebrafish models to study hypoxia-induced pathological angiogenesis in malignant and nonmalignant diseases. Birth Defects Res C Embryo Today. 2011;93(2):182–93.
Jernas M, Palming J, Sjoholm K, Jennische E, Svensson PA, Gabrielsson BG, et al. Separation of human adipocytes by size: hypertrophic fat cells display distinct gene expression. FASEB J. 2006;20(9):1540–2.
Jo J, Gavrilova O, Pack S, Jou W, Mullen S, Sumner AE, et al. Hypertrophy and/or hyperplasia: dynamics of adipose tissue growth. PLoS Comput Biol. 2009;5(3):e1000324.

Kadereit B, Kumar P, Wang WJ, Miranda D, Snapp EL, Severina N, et al. Evolutionarily conserved gene family important for fat storage. Proc Natl Acad Sci U S A. 2008;105(1):94–9.

Kuhn R, Schwenk F. Advances in gene targeting methods. Curr Opin Immunol. 1997;9(2):183–8.

Ledoux S, Queguiner I, Msika S, Calderari S, Rufat P, Gasc JM, et al. Angiogenesis associated with visceral and subcutaneous adipose tissue in severe human obesity. Diabetes. 2008;57(12):3247–57.

Lee RH, Kim B, Choi I, Kim H, Choi HS, Suh K, et al. Characterization and expression analysis of mesenchymal stem cells from human bone marrow and adipose tissue. Cell Physiol Biochem. 2004;14(4–6):311–24.

Lee SL, Rouhi P, Dahl Jensen L, Zhang D, Ji H, Hauptmann G, et al. Hypoxia-induced pathological angiogenesis mediates tumor cell dissemination, invasion, and metastasis in a zebrafish tumor model. Proc Natl Acad Sci U S A. 2009;106(46):19485–90.

Lijnen HR. Angiogenesis and obesity. Cardiovasc Res. 2008;78(2):286–93.

Lim S, Honek J, Xue Y, Seki T, Cao Z, Andersson P, et al. Cold-induced activation of brown adipose tissue and adipose angiogenesis in mice. Nat Protoc. 2012;7(3):606–15.

Maisonpierre PC, Suri C, Jones PF, Bartunkova S, Wiegand SJ, Radziejewski C, et al. Angiopoietin-2, a natural antagonist for Tie2 that disrupts in vivo angiogenesis. Science. 1997;277(5322):55–60.

Mallo M. Controlled gene activation and inactivation in the mouse. Front Biosci. 2006;11:313–27.

Maury E, Brichard SM. Adipokine dysregulation, adipose tissue inflammation and metabolic syndrome. Mol Cell Endocrinol. 2010;314(1):1–16.

Mazo M, Pelacho B, Prosper F. Stem cell therapy for chronic myocardial infarction. J Cardiovasc Transl Res. 2010;3(2):79–88.

Mick GJ, Wang X, McCormick K. White adipocyte vascular endothelial growth factor: regulation by insulin. Endocrinology. 2002;143(3):948–53.

Neuber F. Fetttransplantation. Bericht uber die Verhandlungen der Deutschen Gesellschaft fur Chirurgie [German]. Zentralbl Chir. 1893;22:66.

Nguyen M, Shing Y, Folkman J. Quantitation of angiogenesis and antiangiogenesis in the chick embryo chorioallantoic membrane. Microvasc Res. 1994;47(1):31–40.

Peer LA. Cell survival theory versus replacement theory. Plast Reconstr Surg (1946). 1955;16(3):161–8.

Petrovic N, Shabalina IG, Timmons JA, Cannon B, Nedergaard J. Thermogenically competent nonadrenergic recruitment in brown preadipocytes by a PPARgamma agonist. Am J Physiol Endocrinol Metab. 2008;295(2):E287–96.

Planat-Benard V, Silvestre JS, Cousin B, Andre M, Nibbelink M, Tamarat R, et al. Plasticity of human adipose lineage cells toward endothelial cells: physiological and therapeutic perspectives. Circulation. 2004;109(5):656–63.

Ribatti D, Vacca A, Roncali L, Dammacco F. The chick embryo chorioallantoic membrane as a model for in vivo research on anti-angiogenesis. Curr Pharm Biotechnol. 2000;1(1):73–82.

Roberts JL, Ashwell M, Enser M. Brown adipose tissue triacylglycerol fatty acids of obese and lean mice: in situ and in transplants. Lipids. 1986;21(3):195–201.

Rupnick MA, Panigrahy D, Zhang CY, Dallabrida SM, Lowell BB, Langer R, et al. Adipose tissue mass can be regulated through the vasculature. Proc Natl Acad Sci U S A. 2002;99(16):10730–5.

Ryu HH, Lim JH, Byeon YE, Park JR, Seo MS, Lee YW, et al. Functional recovery and neural differentiation after transplantation of allogenic adipose-derived stem cells in a canine model of acute spinal cord injury. J Vet Sci. 2009;10(4):273–84.

Scroyen I, Jacobs F, Cosemans L, De Geest B, Lijnen HR. Blood vessel density in de novo formed adipose tissue is decreased upon overexpression of TIMP-1. Obesity (Silver Spring). 2010;18(3):638–40.

Shalaby F, Rossant J, Yamaguchi TP, Gertsenstein M, Wu XF, Breitman ML, et al. Failure of blood-island formation and vasculogenesis in Flk-1-deficient mice. Nature. 1995;376(6535):62–6.

Silverman KJ, Lund DP, Zetter BR, Lainey LL, Shahood JA, Freiman DG, et al. Angiogenic activity of adipose tissue. Biochem Biophys Res Commun. 1988;153(1):347–52.

Skurk T, Alberti-Huber C, Herder C, Hauner H. Relationship between adipocyte size and adipokine expression and secretion. J Clin Endocrinol Metab. 2007;92(3):1023–33.

Smahel J. Experimental implantation of adipose tissue fragments. Br J Plast Surg. 1989;42(2):207–11.

Spalding KL, Arner E, Westermark PO, Bernard S, Buchholz BA, Bergmann O, et al. Dynamics of fat cell turnover in humans. Nature. 2008;453(7196):783–7.

Strem BM, Hedrick MH. The growing importance of fat in regenerative medicine. Trends Biotechnol. 2005;23(2):64–6.

Symonds ME, Pope M, Sharkey D, Budge H. Adipose tissue and fetal programming. Diabetologia. 2012;55(6):1597–606.

Tabit CJ, Slack GC, Fan K, Wan DC, Bradley JP. Fat grafting versus adipose-derived stem cell therapy: distinguishing indications, techniques, and outcomes. Aesthetic Plast Surg. 2012;36:704–13.

Tran TT, Kahn CR. Transplantation of adipose tissue and stem cells: role in metabolism and disease. Nat Rev Endocrinol. 2010;6(4):195–213.

Trayhurn P, Wang B, Wood IS. Hypoxia in adipose tissue: a basis for the dysregulation of tissue function in obesity? Br J Nutr. 2008;100(2):227–35.

Valina C, Pinkernell K, Song YH, Bai X, Sadat S, Campeau RJ, et al. Intracoronary administration of autologous adipose tissue-derived stem cells improves left ventricular function, perfusion, and remodelling after acute myocardial infarction. Eur Heart J. 2007;28(21):2667–77.

van Marken Lichtenbelt WD, Vanhommerig JW, Smulders NM, Drossaerts JM, Kemerink GJ, Bouvy ND, et al. Cold-activated brown adipose tissue in healthy men. N Engl J Med. 2009;360(15):1500–8.

Virtanen KA, Lidell ME, Orava J, Heglind M, Westergren R, Niemi T, et al. Functional brown adipose tissue in healthy adults. N Engl J Med. 2009;360(15):1518–25.

Wagner W, Wein F, Seckinger A, Frankhauser M, Wirkner U, Krause U, et al. Comparative characteristics of mesenchymal stem cells from human bone marrow, adipose tissue, and umbilical cord blood. Exp Hematol. 2005;33(11):1402–16.

Weisberg SP, McCann D, Desai M, Rosenbaum M, Leibel RL, Ferrante Jr AW. Obesity is associated with macrophage accumulation in adipose tissue. J Clin Invest. 2003;112(12):1796–808.

Wilson A, Butler PE, Seifalian AM. Adipose-derived stem cells for clinical applications: a review. Cell Prolif. 2011;44(1):86–98.

Wood WB. Determination of pattern and fate in early embryos of Caenorhabditis elegans. Dev Biol. 1988;5:57–78.

Xu H, Barnes GT, Yang Q, Tan G, Yang D, Chou CJ, et al. Chronic inflammation in fat plays a crucial role in the development of obesity-related insulin resistance. J Clin Invest. 2003;112(12):1821–30.

Xue Y, Cao R, Nilsson D, Chen S, Westergren R, Hedlund EM, et al. FOXC2 controls Ang-2 expression and modulates angiogenesis, vascular patterning, remodeling, and functions in adipose tissue. Proc Natl Acad Sci U S A. 105;105(29):10167–72.

Xue Y, Petrovic N, Cao R, Larsson O, Lim S, Chen S, et al. Hypoxia-independent angiogenesis in adipose tissues during cold acclimation. Cell Metab. 2009;9(1):99–109.

Xue Y, Lim S, Brakenhielm E, Cao Y. Adipose angiogenesis: quantitative methods to study microvessel growth, regression and remodeling in vivo. Nat Protoc. 2010;5(5):912–20.

Zhang Y, Olbort M, Schwarzer K, Nuesslein-Hildesheim B, Nicolson M, Murphy E, et al. The leptin receptor mediates apparent autocrine regulation of leptin gene expression. Biochem Biophys Res Commun. 1997;240(2):492–5.

Zijlstra A, Aimes RT, Zhu D, Regazzoni K, Kupriyanova T, Seandel M, et al. Collagenolysis-dependent angiogenesis mediated by matrix metalloproteinase-13 (collagenase-3). J Biol Chem. 2004;279(26):27633–45.

Zimmerlin L, Donnenberg AD, Rubin JP, Basse P, Landreneau RJ, Donnenberg VS. Regenerative therapy and cancer: in vitro and in vivo studies of the interaction between adipose-derived stem cells and breast cancer cells from clinical isolates. Tissue Eng Part A. 2011;17(1–2):93–106.

Zuk PA, Zhu M, Mizuno H, Huang J, Futrell JW, Katz AJ, et al. Multilineage cells from human adipose tissue: implications for cell-based therapies. Tissue Eng. 2001;7(2):211–28.

Chapter 14
Mouse Genetic Models in Studying Adipose Angiogenesis

Ilse Scroyen and H. Roger Lijnen

Abstract Development of obesity is associated with modulation of adipose tissue structure, involving adipogenesis, angiogenesis, and extracellular matrix remodeling. These processes require proteolytic activity provided mainly by the fibrinolytic, matrix metalloproteinase, and ADAM/ADAMTS systems. Adipogenesis is tightly associated with angiogenesis, as shown by the findings that adipose tissue explants trigger blood vessel formation, whereas in turn adipose tissue endothelial cells promote preadipocyte differentiation. Modulation of angiogenesis may have the potential to impair adipose tissue development.

One of the most commonly used approaches to study protein function in vivo is the use of genetically engineered animals, with either transgenic or knock-out models. Therefore, murine genetic models have been very useful tools to study mechanisms of adipose tissue development.

Keywords Fibrinolytic system • Matrix metalloproteinase system • Angiogenic • Anti-angiogenic compounds • Mouse models

14.1 Introduction

Development of obesity is associated with substantial modulation of adipose tissue structure, involving adipogenesis, angiogenesis, and extracellular matrix remodeling. As obesity is also linked with pathological angiogenesis and increasingly more anti-angiogenic compounds are characterized in the cancer field, modulation of angiogenesis may be a promising novel therapeutic approach (Cao 2010; Lijnen 2008).

I. Scroyen • H.R. Lijnen (✉)
Center for Molecular and Vascular Biology, KU Leuven, Campus Gasthuisberg, O&N 1, Herestraat 49, Box 911, 3000 Leuven, Belgium
e-mail: ilse.scroyen@med.kuleuven.be; roger.lijnen@med.kuleuven.be

Y. Cao (ed.), *Angiogenesis in Adipose Tissue*, DOI 10.1007/978-1-4614-8069-3_14,

To supply growing adipose tissue with nutrients and oxygen, the vasculature responds by increasing the number and/or size of blood vessels (capillaries). In early stage development of adipose tissue, adipogenesis is tightly associated with angiogenesis. Expansion of white adipose tissue (WAT) can be supported by either neovascularization for new adipocyte differentiation (hyperplasia) or dilatation and remodeling of existing capillaries for adipocyte expansion (hypertrophy).

The angiogenic activity in adipose tissue is often presented by the blood vessel density that is defined as the number of capillaries for a given surface. However, it does not truly reflect angiogenic activity. The number of cells that can be supported by a blood vessel in tumors varies, influencing in turn the vascular density (Hlatky et al. 2002). In addition, a study on capillary fenestrations in adipose tissue showed that the microvessel density was lower in genetically obese *ob/ob* mice as compared with wild-type (WT) controls, as a result of increased adipocyte size in these obese mice (Cao et al. 2001). To take this into account, blood vessel density in adipose tissue can be normalized to the adipocyte density.

One of the most commonly used approaches to study protein function in vivo is the use of genetically engineered animals, with either gain-of-function (transgenic) or loss-of-function (knock-out) models. This strategy provides powerful tools to discover the importance of individual proteins in a variety of physiological as well as pathological conditions. Additionally, obesity mouse models have been developed which can be divided in three major groups. The first one includes mice that acquire obesity after dietary manipulation. Such a nutritionally induced obesity model in rodents represents a well-characterized tool for the study of adipose tissue-related angiogenesis. Indeed, many mouse strains become obese after consuming fatty diets with adjusted calories over time (West and York 1998) and therefore provide excellent models to study different stages of adipose tissue development, which are highly analogous to human obesity. However, there are mouse strain-specific differences in responses to a high fat diet (HFD) (Kanasaki and Koya 2011). For instance, C57BL6/J and AKR mice are sensitive to HFD-induced obesity and diabetes whereas some other strains are to a certain degree obesity- and diabetes-resistant (Kanasaki and Koya 2011). The second group includes genetically modified mice, which spontaneously become obese. Different mouse models of genetically determined obesity are available (Kanasaki and Koya 2011). The *ob/ob* mice lack the gene *leptin* whereas the *db/db* mice lack the leptin receptor. Leptin was discovered in 1994, and it was shown that this hormone, the product of the *obese* (*ob*) gene, is produced in white adipose tissue and serves as the peripheral signal of nutritional status directed to the brain (Zhang et al. 1994). Therefore, *ob/ob* mice exhibit uncontrollable food intake, obesity, type 2 diabetes, and insulin resistance with hyperinsulinemia. Additionally, the db stands for "diabetes" and these mice have a point mutation in the leptin receptor gene, resulting in defective leptin signaling (Chen et al. 1996). Consequently they suffer from persistent hyperphagia and obesity associated with hyperleptinemia, insulin resistance, and increased insulin levels. In addition, mice with combined leptin and LDL-receptor deficiency feature most of the confounding metabolic syndrome components including a more severe hypercholesterolemia and hypertriglyceridemia, as in obese people (Hasty et al. 2001). A third in vivo model monitors de novo adipogenesis: athymic NUDE mice, injected in the back with 3T3-F442A preadipocytes, develop fat pads at the site of injection (Mandrup et al. 1997; Neels et al. 2004).

This section will review studies on adipose tissue angiogenesis in genetically modified mice, modulating proteolysis as well as pro- and anti-angiogenic factors.

14.2 Proteolysis in Growing Adipose Tissue

There is growing evidence that proteolysis is required for several aspects of adipose tissue expansion. Besides hypertrophy of adipocytes, proteolysis is also required for cell migration during the development of blood vessels and peripheral nerves as well as for the migration of macrophages in the growing adipose tissue. The proteolytic activity required for adipogenesis and angiogenesis is mainly mediated by the fibrinolytic (plasminogen/plasmin) and matrix metalloproteinase (MMP) systems. Also proteins of the ADAM (a disintegrin and metalloproteinase) and ADAMTS (ADAM with thrombospondin (TSP) motif) families may be implicated. These proteinases are collectively able to cleave a wide variety of substrates. In addition, MMPs and plasmin are important for the bioavailability of several growth factors and cytokines implicated in obesity, with major roles in angiogenesis.

14.2.1 The Fibrinolytic System

The fibrinolytic system comprises an inactive proenzyme plasminogen, which can be converted to the active enzyme plasmin. The latter degrades fibrin into soluble fibrin-degradation products. Two immunologically distinct plasminogen activators have been identified: tissue-type plasminogen activator (t-PA) and urokinase-type plasminogen activator (u-PA). u-PA binds to a specific cellular receptor (u-PAR or CD87) resulting in enhanced activation of cell-bound plasminogen. Inhibition of the fibrinolytic system occurs either at the level of plasmin by α2-antiplasmin, or at the level of the plasminogen activators, mainly by plasminogen activator inhibitor-1 (PAI-1) (Lijnen 2009; Lijnen and Collen 1995). A relationship between blood fibrinolytic activity and body fatness was already reported in the early sixties (Shaw and Machaughton 1963), but the interactions between individual fibrinolytic parameters and other systems, such as angiogenesis, appear more complex than anticipated.

Several nutritionally induced obesity models in transgenic mice have been used to study the role of the fibrinolytic system in the development of obesity (Table 14.1). t-PA-deficient mice, kept on HFD, had a higher body weight and adipose tissue mass than WT controls (Morange et al. 2002). There was an increase in the number of endothelial cells (ECs) and stroma cells in the fat tissues, suggesting that targeted inactivation of t-PA increases angiogenesis in the adipose tissue, which may promote adipose tissue formation. In contrast, deficiency in u-PA had no effect on

Table 14.1 Role of the fibrinolytic system in adipose tissue development and obesity studied in genetic mouse models

	Effect on adipose tissue	Effect on adipose angiogenesis	Reference
Deficiency			
t-PA	Increased	Increased	Morange et al. (2002)
u-PA	No effect	No data	Morange et al. (2002)
Plasminogen	Decreased	No data	Hoover-Plow et al. (2002)
α2-antiplasmin	No effect	No data	Lijnen (2007)
PAI-1	No effect	No data, increased	Lijnen (2005), Scroyen et al. (2007)
	Decreased	No data	Ma et al. (2004), Schafer et al. (2001), De Taeye et al. (2006)
PAI-2	Decreased	No effect	Lijnen et al. (2007)
Overexpression			
PAI-1	Decreased	Decreased	Lijnen et al. (2003a)
De novo			
local PAI-1 expression	Increased	No effect	Scroyen et al. (2009a)
systemic PAI-1 expression	Decreased	Decreased	Scroyen et al. (2009a)

nutritionally induced obesity (Morange et al. 2002), although it was previously shown that overexpression of u-PA in the brain resulted in reduced body weight and size (Miskin and Masos 1997). Mice deficient in plasminogen ($Plg^{-/-}$), the substrate for both plasminogen activators, showed reduced fat accumulation associated with reduced differentiation of stromal cells in culture. This study suggested that differences in the stromal cell population are responsible for the reduced adipose tissue accumulation in the $Plg^{-/-}$ mice, and that the plasminogen system plays an important role in adipose tissue accumulation (Hoover-Plow et al. 2002). No data are available about the effect on adipose tissue angiogenesis, but in a mouse cornea model a significantly decreased angiogenic response of new vessel formation to both vascular endothelial growth factor (VEGF) and basic fibroblast growth factor (bFGF) was observed upon plasminogen deficiency (Oh et al. 2003). Deficiency of α2-antiplasmin had no significant effect on adipose tissue development in mice (Lijnen 2007).

PAI-1 is strongly expressed in murine and in human adipose tissue and its role is extensively studied in adipose tissue development (Alessi et al. 1997; Samad and Loskutoff 1996; Shimomura et al. 1996). In human adipose tissue, its expression is positively correlated with BMI (Alessi et al. 2000). In addition, increased levels of PAI-1 are also observed in patients with the insulin resistance syndrome (Juhan-Vague et al. 1991). In obese women, plasma PAI-1 levels during body weight reduction and body weight regain are positively correlated with the amount of body fat (Mavri et al. 1999). The role of PAI-1 in adipose tissue development at present still remains controversial. Some studies with transgenic mice reported that PAI-1 deficiency has no effect on adipose tissue development (Lijnen 2005; Scroyen et al. 2007), whereas others reported reduced development of adipose tissue (Ma et al. 2004). In contrast, some studies indicated that high PAI-1 levels impair obesity (Lijnen et al. 2003a; Morange et al. 2000). Disruption of the PAI-1 gene in genetically obese and diabetic *ob/ob* mice also resulted in reduced adiposity and improvement of the metabolic profile (Schafer et al. 2001). A recent study indicated that only the complete absence of PAI-1 protected against the development of diet-induced obesity, whereas the absence of bone marrow-derived PAI-1 protected against expansion of the visceral fat mass (De Taeye et al. 2006). The discrepancies between these studies may be explained in part by the fact that nutritionally induced models depend on the composition and timing of the diet, and on the age and genetic background of the mice used. It can also not be excluded that modifier genes in a specific genetic background affect the outcome of such obesity studies. There is also a difference between genetically determined obesity models (mainly *ob/ob* or *db/db* mice lacking leptin or its receptor) and nutritionally induced models in WT mice with intact genome. Leptin provides a local angiogenic signal and improves the efficiency of lipid release from fat stores to maintain energy homeostasis (Sierra-Honigmann et al. 1998).

To further elucidate the role of PAI-1 in the development of obesity, transgenic (Tg) mice were generated with overexpression of murine PAI-1 under control of the adipocyte-specific promoter aP2. High circulating PAI-1 levels and reduced fibrinolytic activity in adipose tissue resulted in a reduction of nutritionally induced obesity (Lijnen et al. 2003a). Besides adipocyte hypotrophy, a significantly lower ratio of stroma cells/adipocytes was observed in the Tg mice. Analysis of blood vessels did not reveal significant differences, except for a somewhat smaller average vessel size but higher density in adipose tissue of Tg mice on HFD as compared with controls (Lijnen et al. 2003a). Overexpression of PAI-1 thus seems to modify the cellularity of adipose tissue, however without significantly affecting angiogenesis. In order to further substantiate a potential functional role of PAI-1 in the development of obesity, a de novo adipogenesis model, based on the injection of 3T3-F442A preadipocytes in the back of NUDE mice, was used (Mandrup et al. 1997; Neels et al. 2004). Injection of preadipocytes expressing mPAI-1 resulted in larger fat pads with a higher adipocyte density without affecting angiogenesis, whereas a systemic overexpression in this model revealed smaller adipocytes as well as blood vessel size (Scroyen et al. 2009a). Taken together these studies support a pro-angiogenic effect of PAI-1 in adipose tissue at low concentration and an anti-angiogenic effect at high concentration. Indeed, systemic PAI-1 overexpression results in smaller blood vessels (Scroyen et al. 2009a), whereas a larger blood vessel size is observed in subcutaneous (SC) adipose tissue of PAI-1-deficient mice as compared with WT mice (Scroyen et al. 2007). In addition, also PAI-1 transgenic mice on high fat diet have smaller blood vessels than controls and develop less adipose tissue (Lijnen et al. 2003a). Overall, these data are compatible with a bell-shaped dose-dependent effect of PAI-1 on adipogenesis and angiogenesis; the optimal level of PAI-1 to promote or inhibit adipose tissue development remains, however, to be established. Thus, pharmacological inhibition of PAI-1 could be beneficial in diseases associated with expansion of adipose tissue. Therefore, PAI-1 inhibitors will be needed that target active circulating PAI-1 as well as PAI-1 stabilized by binding to vitronectin. Interestingly, comparative evaluation of the reactivity of PAI-1 inhibiting monoclonal antibodies toward mouse wild-type and glycosylation-deficient PAI-1 mutants revealed a clear glycosylation-dependent reactivity profile that corresponds well with negative in vivo results in an endotoxemia model (Van De Craen et al. 2012). Therefore, PAI-1 inhibitors for in vivo application should be selected on their adequate cross-reactivity and cross-inhibition with the biologically active target.

PAI-2 is an intracellular serpin that efficiently inhibits t-PA and u-PA, albeit at a somewhat lower rate than PAI-1 (Kruithof et al. 1987). Generation of PAI-2-deficient mice has revealed that it is not required for normal development or survival (Dougherty et al. 1999) and has made it possible to study its specific role in biological processes, like obesity. This indicated that PAI-2 is expressed in murine adipose tissues and that its deficiency results in impaired adipose tissue development in mice fed an HFD (Lijnen et al. 2007). The mechanism is independent of the antifibrinolytic activity of PAI-2. Blood vessel density, normalized to adipocyte number, was comparable in SC fat of WT and PAI-$2^{-/-}$ mice on HFD, but was lower in gonadal (GON) fat of PAI-$2^{-/-}$ mice, whereas blood vessel size was overall comparable (Lijnen et al. 2007). It is therefore unlikely that PAI-2 deficiency significantly affects angiogenesis in adipose tissue.

Overall, studies using a nutritionally induced obesity model in genetically modulated mice support a role of the fibrinolytic system in adipogenesis, as well as in adipose tissue-related angiogenesis.

14.2.2 The Matrix Metalloproteinase System

The MMPs belong to a family of about 26 neutral endopeptidases, of which 24 are found in mammals. Together, they are able to cleave all of the ECM components as well as several non-ECM proteins, such as adhesion molecules, cytokines, protease inhibitors, and other (pro)MMPs. Generally, MMPs are expressed at low levels but are rapidly induced at times of active tissue remodeling. Most MMPs are secreted as inactive pro-enzymes and require proteolytic processing to become active. MMPs have the ability to regulate cell behavior in several ways and are thereby involved in a large variety of physiological and pathological processes (Sternlicht and Werb 2001). MMP activity is modulated through interactions with tissue inhibitors of MMPs (TIMPs). Four TIMPs have been characterized that are able to inhibit the activities of all known MMPs. Consequently, the net MMP activity in tissues is locally determined by the balance between the levels of activated MMPs and TIMPs (Gomez et al. 1997). Today, the majority of MMP knock-out mice generated show no major physiological alterations. This lack of severe phenotypes is probably due to enzymatic redundancy, enzymatic compensation, and/or adaptive development. Indeed, in these animal models, related MMPs could enzymatically compensate for the loss of one specific MMP. Therefore, almost all described MMP deficiencies in mice only show significant defects in tissues in response to environmental challenges such as wounding, infection, or inflammation.

Several lines of evidence suggest a potential role of MMPs in the development of adipose tissue, and the first indications came from in vitro observations. Conditioned medium of rat adipocytes contains an MMP-2 (gelatinase A) like gelatinolytic activity, which may play a role in their organization in multicellular clusters (Brown et al. 1997). In 2001, the presence of MMP-2 in adipose tissue was documented and increased expression of MMP-2 was observed in adipose tissue of mice with nutritionally induced obesity as well as in genetically obese mice (Lijnen et al. 2001, 2002a). MMP-2 and MMP-9 expression and secretion have also been demonstrated in human adipose tissue and are modulated during adipocyte differentiation (Bouloumie et al. 2001). Furthermore, MMP-2 levels increase and TIMP-1 levels decrease during adipocyte differentiation (Johnson et al. 1994). To gain further insight into the involvement of the MMPs in the development of adipose tissue, the expression of MMPs and TIMPs was monitored in lean and diet-induced obese mice (Maquoi et al. 2002). With the exception of MMP-8, all MMP and TIMP transcripts were detected in both SC and GON fat depots. In addition, this study revealed upregulation with obesity of mRNA levels of some MMPs (MMP-3, -11, -12, -13, -14) and downregulation of others (MMP-7, -9, -16, -24). Most of these modulations were specific to the GON fat, supporting the concept that the different fat depots are not identical. Similar experiments in two genetic models of obesity (*ob/ob* and *db/db* mice) compared with a diet-induced obesity model (AKR mice) revealed that mRNA levels for MMP-2, -3, -12, -14, and -19 are strongly induced in obese adipose tissue as compared with lean tissues (Chavey et al. 2003).

TIMP-1, which is synthesized by most types of connective tissue cells as well as macrophages, acts against all members of the collagenase, stromelysin, and gelatinase classes. Analysis of mRNA expression in adipose tissue of lean and obese mice revealed significant upregulation of TIMP-1 with obesity. In contrast, TIMP-4 was downregulated with obesity, whereas TIMP-2 and TIMP-3 expression levels were not significantly modulated, at least in gonadal adipose tissue (Maquoi et al. 2002; Chavey et al. 2003).

In the past decades, efforts were made to identify crucial MMPs involved in adipose tissue development, using different MMP- and TIMP-deficient mice in a nutritionally induced obesity model (Table 14.2). Inactivation of the stromelysin-1 (MMP-3) gene in mice leads to enhanced development of adipose tissue when fed an HFD (Maquoi et al. 2003). The higher body weight of MMP-$3^{-/-}$ mice resulted essentially from a specific increase of their adiposity, characterized by hypertrophic adipocytes in the SC and GON fat pads. Also a higher blood vessel density was observed in the adipose tissue of MMP-$3^{-/-}$ mice, suggesting that MMP-3 affects adipose tissue-related angiogenesis. A regulatory role of MMP-3 has also been suggested in adipogenesis during mammary gland involution in mice; MMP-$3^{-/-}$ mice showed accelerated differentiation and hypertrophy of adipocytes (Alexander et al. 2001). These data thus suggest an inhibitory effect of MMP-3 on adipocyte metabolism and differentiation. Similar results were seen upon inactivation of the stromelysin-3 (MMP-11) gene. Indeed, MMP-11 deficiency promoted adipose tissue

Table 14.2 Role of the matrix metalloproteinase system in adipose tissue development and obesity studied in genetic mouse models

	Effect on adipose tissue	Effect on adipose angiogenesis	Reference
Deficiency			
MMP-3	Increased	Increased	Maquoi et al. (2003)
MMP-11	Increased	No data	Lijnen et al. (2002b)
MMP-19	Increased	No data	Pendas et al. (2004)
MMP-10	No effect	No data	Lijnen et al. (2009a)
MMP-2	Decreased	No effect	Van Hul and Lijnen (2008)
MMP-9	No effect	No effect	Van Hul et al. (2010)
TIMP-1	Decreased	No effect	Lijnen et al. (2003b)
TIMP-2	Increased	No effect	Stradecki and Jaworski (2011), Jaworski et al. (2011)
Overexpression			
TIMP-1	No effect	Decreased	Demeulemeester et al. (2006), Scroyen et al. (2010)
De novo			
local hTIPM-1 expression	No effect	No effect	Scroyen et al. (2010)
systemic hTIMP-1 expression	No effect	Decreased	Scroyen et al. (2010)

development and resulted in adipocyte hypertrophy (Lijnen et al. 2002b). Enhanced adipose tissue growth associated with adipocyte hypertrophy was also seen in MMP-19-deficient mice upon HFD feeding (Pendas et al. 2004). In contrast, MMP-10-deficient mice on HFD did not differ markedly from their WT counterparts (Lijnen et al. 2009a).

Since the discovery of the expression of gelatinases in adipose tissue, elucidating their functional role in adipose tissue development became very attempting (Van Hul and Lijnen 2008, 2011a, b; Van Hul et al. 2010, 2012a). MMP-2-deficient (MMP-2$^{-/-}$) mice when kept on an HFD, but not MMP-9-deficient mice, showed significantly reduced obesity associated with adipocyte hypotrophy, without an effect on angiogenesis (Van Hul and Lijnen 2008; Van Hul et al. 2010). The potential to reduce adipose tissue development by pharmacological inhibition of gelatinases was also shown. WT mice fed an HFD supplemented with Tolylsam (a relative gelatinase-specific inhibitor, efficiently inhibiting MMP-2, -9, and -12) showed similar effects on adipose tissue as the MMP-2$^{-/-}$ mice. An interesting observation was that mice treated with this inhibitor had smaller blood vessels and lower normalized blood vessel densities as compared with their controls (Van Hul and Lijnen 2008). It is conceivable that MMP-2 deficiency as such does not affect angiogenesis in fat pads, whereas a combined inhibition of MMP-2, -9, and -12 does. This could imply that the effect of Tolylsam on adipose tissue is dual: it may slow down fat mass accumulation by inhibiting neovascularization of the adipose tissue, thereby depriving adipocytes of sufficient access to nutrition and circulating triglycerides, and by inhibiting adipocyte hyperplasia.

TIMP-1-deficient (TIMP-$1^{-/-}$) mice on HFD gained less weight than their WT counterparts and developed less adipose tissue (Lijnen et al. 2003b). A significantly higher blood vessel density and size in SC and GON adipose tissue of TIMP-$1^{-/-}$ mice as compared to control mice was observed. However, after normalization for adipocyte density, this difference was no longer observed (Lijnen et al. 2003b). Plasma leptin levels were significantly elevated in WT as compared with TIMP-$1^{-/-}$ mice on HFD. Leptin acts as satiety factor and increases energy expenditure, while its secretion is strongly correlated with body fat mass and adipocyte size (Friedman and Halaas 1998). This suggests an effect of TIMP-1 deficiency on leptin secretion, or may rather be due to the lower body fat mass in the TIMP-$1^{-/-}$ mice. To further substantiate a role of TIMP-1 in nutritionally induced obesity, the effect of TIMP-1 overexpression by adenoviral gene transfer in mice was studied on adipogenesis and adipose tissue development (Demeulemeester et al. 2006). Long-term expression of highly elevated levels of human TIMP-1 (hTIMP-1) was associated with reduced MMP activity in plasma, as well as in adipose tissue. However, there was no significant effect on body weight or fat mass when the mice were kept on HFD. Similar observations were made after overexpression of hTIMP-1 in mice with lipectomy of the SC adipose tissue, kept on an HFD for 20 weeks (Demeulemeester et al. 2006). This is somewhat surprising since TIMP-1 deficiency resulted in impaired adipose tissue development. However, it is possible that physiologic TIMP-1 concentrations in mice are sufficient to promote adipogenesis and adipose tissue development, whereas overexpression has no further effect, and deficiency results in impairment. Interestingly, in both in vivo models, blood vessels in the adipose tissues were significantly smaller after hTIMP-1 gene transfer than in control mice (Demeulemeester et al. 2006). Similarly, systemic hTIMP-1 overexpression in an in vivo de novo adipogenesis model had no significant effect on in vivo adipogenesis, although blood vessel density in de novo formed fat pads was decreased (Scroyen et al. 2010). In contrast, local hTIMP-1 overexpression did not significantly affect early stages of adipogenesis as evaluated from in vivo de novo adipose tissue formation (Scroyen et al. 2009b).

TIMP-2 exerts MMP-dependent (MMP inhibition and pro-MMP-2 activation) and MMP-independent functions. To determine the contribution of TIMP-2 to the hypothalamic regulation of feeding, body mass and food consumption were monitored in TIMP-2-deficient (TIMP-$2^{-/-}$) mice kept on a standard chow or HFD (Stradecki and Jaworski 2011). Before the onset of obesity, TIMP-$2^{-/-}$ mice were hyperphagic, leptin-resistant, and gained more weight than WT mice. Furthermore, proteolysis was increased in the arcuate nucleus of these mice, suggesting a role for TIMP-2 in hypothalamic control of feeding and energy homeostasis (Stradecki and Jaworski 2011). Moreover, male but not female HFD-fed TIMP-$2^{-/-}$ mice developed insulin resistance. TIMP-2 expression is downregulated and MMP-14 activity is increased in male, but not female, adipose tissue of WT mice, suggesting that dysregulated MMP-14/TIMP-2 expression may contribute to the observed sex-specific diet-induced insulin resistance (Jaworski et al. 2011). Although, TIMP-2 exerts anti-angiogenic effects (Stetler-Stevenson and Seo 2005), the latter study suggested that TIMP-2 deficiency resulted in increased vascularization upon HFD feeding in male, but not in female mice, but no genotype differences were observed (Jaworski et al. 2011).

Overall, the available data suggest the potential to impair adipose tissue development by MMP inhibition and demonstrate that the MMP/TIMP balance plays a pivotal role in adipose tissue angiogenesis. Identification of MMPs that play key roles in adipose tissue development and generation of more specific inhibitors will be required to further explore their potential to affect obesity.

14.2.3 ADAM and ADAMTS Families

ADAMTS-1 and ADAMTS-8 can inhibit vascular endothelial growth factor (VEGF)-induced angiogenesis and suppress fibroblast growth factor-2 (FGF-2)-induced vascularization (Vazquez et al. 1999). Both mediate a greater anti-angiogenic response than either Thrombospondin-1 (TSP-1) or endostatin (see below), with ADAMTS-1 showing a greater inhibitory capacity than ADAMTS-8. The anti-angiogenic activity of ADAMTS-1 and -8 is mediated through their TSP motifs. Interestingly, ADAMTS-1 deficient mice are exceptionally lean, their volume of epididymal fat is significantly smaller than in WT mice, and the formation of capillaries in the adrenal gland is drastically impaired (Luque et al. 2003). ADAM-17, ADAMTS-1, and ADAMTS-8 mRNA are detected in both SC and GON adipose tissue of lean mice. In SC adipose tissue of obese mice (HFD), the expression of ADAM-17 is enhanced and that of ADAMTS-1 reduced, whereas in GON adipose tissue expression of ADAMTS-8 is reduced. In lean and obese mice, expression of ADAM-17, ADAMTS-1, and ADAMTS-8 is higher in the stromal-vascular cell fraction than in mature adipocytes. During differentiation of murine 3T3-F442A preadipocytes, expression of ADAM-17 and ADAMTS-1 remained virtually unaltered, whereas that of ADAMTS-8 decreased as adipocytes matured (Voros et al. 2003). Aggrecan, as well as the two aggrecanases ADAMTS-4 and ADAMTS-5 mRNAs, are expressed in SC and GON adipose tissues of mice (Voros et al. 2006). In mice with nutritionally induced obesity as well as in lean controls, aggrecan mRNA expression is downregulated, whereas ADAMTS-4 and ADAMTS-5 are upregulated with time. In mice with genetically determined obesity (*ob/ob*), ADAMTS-5 mRNA is upregulated in both SC and GON adipose tissues, as compared to WT mice. Thus, aggrecan levels are high at the early stages of adipose tissue development in mice, whereas its production decreases and its degradation increases during development of obesity. A functional role of aggrecan in promoting early stages of adipogenesis is supported by the findings that it stimulates the in vitro differentiation of 3T3-F442A preadipocytes and the de novo in vivo accumulation of fat in Matrigel plaques injected into WT mice (Voros et al. 2006). Proteoglycans in the ECM of adipose tissue, such as aggrecan, may thus contribute to the regulation of lipid uptake and obesity in mice.

14.3 Pro- and Anti-angiogenic Factors

Many components affecting angiogenesis have been identified in adipose tissues (Table 14.3), including platelet-derived growth factor (PlGF), neuropelin 1 (Np-1), FGF-2, angiopoietin 2 (Ang-2), leptin, thrombospondin 1 (TSP-1), osteonectin, adiponectin, resistin, tissue factor (TF), TNF-α, hepatocyte growth factor (HGF), etc. The angiogenic potential of the main pro- or anti-angiogenic components is sometimes context-dependent and variable in different fat pads.

14.3.1 Pro-angiogenic Factors

VEGF is a major contributor to angiogenesis, increasing the number of capillaries in a given network. VEGF-A (17–23 kDa) is a major angiogenic factor that stimulates proliferation and migration of ECs (Carmeliet et al. 1996). Three forms of VEGF-A are produced in the mouse as a result of alternative splicing (VEGF-A121, VEGF-A165, and VEGF-A189). Several studies indicate that VEGF-A stimulates both physiological and pathological angiogenesis by signaling through VEGF

Table 14.3 Role of pro- and anti-angiogenic components in adipose tissue development and obesity studied in genetic mouse models

	Effect on adipose tissue	Effect on adipose angiogenesis	Reference
Deficiency			
VEGF-D	No effect	No effect	Lijnen et al. (2009b)
PlGF	Decreased	Decreased	Lijnen et al. (2006)
FGF-2	Increased	No data	Xiao et al. (2010)
leptin	Increased	Decreased	Cao et al. (2001), Voros et al. (2005)
SPARC	Increased	No data	Bradshaw et al. (2003), Nie et al. (2011)
TSP-1	No effect	No data	Li et al. (2011), Voros and Lijnen (2006)
TSP-2	Increased	No data	Shitaye et al. (2010)
	Decreased	No effect	Van Hul et al. (2011)
CD36	Decreased	No effect	Christiaens et al. (2012), Kennedy et al. (2011)
Over-expression			
VEGF	No effect	Increased	Elias and Franckhauser (2012)
Ang-1	Decreased	No data	Dallabrida et al. (2003)
De novo			
Local HGF knockdown	No effect	Decreased	Bell et al. (2008)
Local HGF expression	No effect	Increased	Bell et al. (2008)

Receptor-2 (VEGFR-2) in a strict dose-dependent manner. Loss of a single allele causes embryonic vascular defects, while reduction of VEGF-A levels by only 25 % impairs spinal cord perfusion, resulting in motor neuron degeneration reminiscent of amyotrophic lateral sclerosis (Carmeliet et al. 1996; Ferrara et al. 1996; Oosthuyse et al. 2001). VEGF-B (21 kDa) is 43 % identical to VEGF-A165; it also promotes angiogenesis and is implicated in ECM degradation via regulation of plasminogen activation (Olofsson et al. 1998). VEGF-C (23 kDa) displays 30 % homology with VEGF-A165 and plays an important role both in angiogenesis and lymph angiogenesis (Joukov et al. 1996; Karkkainen et al. 2004). VEGF-D (22 kDa) is 48 % identical to VEGF-C and also promotes the growth of lymhatic vessels (Stacker et al. 2001).

PlGF, a 25 kDa homologue of VEGF-A (53 % sequence identity with VEGF-A165), enhances angiogenesis, but only in pathological conditions. Loss of PlGF impairs angiogenesis in the ischemic retina, limb, and heart, in wounded skin and in tumors, without affecting physiological angiogenesis (Carmeliet et al. 2001).

The members of the VEGF family bind to transmembrane tyrosine kinase receptors [VEGFR-1 (206 kDa), VEGFR-2 (218 kDa), and VEGFR-3 (150 kDa)]. VEGF-A interacts with both VEGFR-1 and VEGFR-2, whereas VEGF-B and PlGF bind to VEGFR-1. VEGF-C and VEGF-D activate VEGFR-3, but VEGF-C can also bind to VEGFR-2. VEGFR-1 and VEGFR-2 mediate angiogenesis, whereas VEGFR-3 is involved mainly in lymph angiogenesis (Veikkola et al. 2001). VEGF-A165, PlGF, and VEGF-B also bind to another transmembrane receptor, Np-1. Inactivation of the Np-1 gene in mice causes disturbances in development of the vascular and nervous system (Kawasaki et al. 1999).

mRNA levels of VEGF-A isoforms, VEGF-B, VEGF-C, and its receptors, are not markedly modulated by obesity (Voros et al. 2005). Reciprocal regulation of adipogenesis and angiogenesis suggests that blockade of VEGF signaling may inhibit in vivo adipose tissue formation (Fukumura et al. 2003). In addition, with in vitro adipose tissue-imaging techniques, it was demonstrated that angiogenesis requires a close interplay between differentiating adipocytes, stromal cells, and blood cells (Nishimura et al. 2007). The use of an anti-VEGF antibody indeed inhibited not only angiogenesis, but also the formation of adipo/angiogenic cell clusters, indicating that coupling of adipogenesis and angiogenesis is essential for differentiation of adipocytes, thus confirming that VEGF is a key mediator (Nishimura et al. 2007). A recent study showed that overexpression of VEGF in adipose tissue induces vessel formation in both WAT and brown AT (BAT), with beneficial effects (Elias et al. 2012). This study highlighted the crucial role of vascularization in promoting size and thermogenesis in BAT as a means to counteract the development of HFD-induced obesity, without an increase in WAT mass. Moreover, adipose tissue from animals overexpressing VEGF present increased recruitment of M2 anti-inflammatory macrophages, avoiding insulin resistance (Elias et al. 2012). On the other hand, using VEGF-D-deficient mice, no important role of VEGF-D in (lymph) angiogenesis or in adipose tissue development was observed (Lijnen et al. 2009b). In this study, the composition of SC and GON adipose tissues in terms of size and density of adipocytes or blood vessels was also comparable between VEGF-D-deficient and WT control mice (Lijnen et al. 2009b).

To establish a functional role in obesity, PlGF-deficient (PlGF$^{-/-}$) and WT mice with the same genetic background were kept on HFD for 15 weeks. PlGF$^{-/-}$ mice had a significant lower body weight and less total SC plus GON adipose tissue. Blood vessel size was lower in GON adipose tissue of PlGF$^{-/-}$ mice and blood vessel density, normalized to adipocyte number, was significantly lower in their SC adipose tissue. These differences were not observed for PlGF$^{-/-}$ and WT mice kept on normal chow, in line with the emerging concept that PlGF deficiency has little or no effect on angiogenesis under normal (feeding) conditions, but is associated with impaired angiogenesis under stress conditions (HFD) during the early stages of adipose tissue formation (Lijnen et al. 2006). Neovascularization in new adipose tissue originates by sprouting from larger host-derived blood vessels that run parallel to peripheral nerves, but endothelial progenitor cells do not play an important role in this process (Neels et al. 2004). Adipogenesis and de novo fat pad formation can be impaired by inhibition of PPAR-γ or VEGFR-2 (Fukumura et al. 2003) and by administration of a PlGF neutralizing monoclonal antibody (Lijnen et al. 2006).

The FGF family with its prototype members FGF-1 (acidic FGF) and FGF-2 (basic FGF) consists of at least 22 members. In general, FGFs stimulate a variety of cellular functions by binding to cell surface FGF-receptors in the presence of heparin proteoglycans. FGF-2 (25 kDa) is a potent stimulator of differentiation, migration, and proliferation of ECs and enhances adipocyte differentiation in vivo (Kawaguchi et al. 1998). During angiogenesis, FGF-2 stimulates the synthesis of proteinases such as collagenase and u-PA, and of integrins to form new capillary cord structures (Okamura et al. 1991; Tienari et al. 1991). Several studies suggested both pro- and anti-adipogenic effects of FGF-2 (Kawaguchi et al. 1998; Kimura et al. 2002; Xiao et al. 2010). Recently, it was shown that disruption of FGF-2 gene activates the adipogenic program in mesenchymal marrow stromal stem cells (Xiao et al. 2010).

Another signaling system contributing to maintenance, growth, and stabilization of blood vessels involves the tyrosine kinase (T) with Ig (I) and epidermal (E) growth factor (EGF) homology domains (TIE)-1 and -2 receptors (140–145 kDa). TIE-2 binds the angiopoietins (Ang-1 and Ang-2), whereas the ligands of TIE-1 are not known yet. Unlike Ang-2, which activates TIE-2 on some cells, but blocks it on others (Davis et al. 2003), Ang-1 consistently activates TIE-2 (Davis et al. 1996). The role of Ang-1 in vascularization is pleiotropic and context-dependent. Ang-1 tightens vessels by affecting junctional molecules (Thurston et al. 2000), by promoting the interaction between ECs and mural cells, and by recruiting pericytes (Carlson et al. 2001). In addition, Ang-1 mRNA levels inversely correlate with the rates of change in body weight. It regulates adipose tissue growth, as shown by the finding that obese mice injected with ang1/pcDNA have reduced rates of weight gain and fat pad weights, regardless of the route of plasmid administration (Dallabrida et al. 2003). Ang-2 may stimulate vessel growth by loosening endothelial/peri-EC interactions and degrading the ECM (Ahmad et al. 2001). TIE-1 and TIE-2 are expressed in adipose tissues, but their role in adipose tissue associated angiogenesis has not been clearly established.

HGF is an angiogenic growth factor in models of hindlimb ischemia and myocardial infarction (Aoki et al. 2000; Van Belle et al. 1998). Furthermore, HGF is a potent mitogenic and angiogenic factor produced in human adipose tissue. Recently, a central role for HGF in adipose tissue angiogenesis was proposed (Bell et al. 2008). In this study, however, knockdown of HGF had no effect on differentiation of 3T3-F442A preadipocytes to mature adipocytes in vivo. In developing fat pads under the skin of HGF overexpressing transgenic mice, TIE-1 and PECAM-1 mRNA were increased correlating with immunohistochemical evidence of EC migration in the developing fat pad (Bell et al. 2008).

Leptin (16 kDa), a satiety hormone produced by the adipose tissue, promotes migration of ECs. Interaction of leptin with its receptor on ECs leads to activation of the Stat3 pathway and enhancement of its DNA-binding activity (Sierra-Honigmann et al. 1998). Besides a direct pro-angiogenic activity, leptin upregulates VEGF expression via activation of the Jak/Stat3 signaling pathway (Suganami et al. 2004). Similar to VEGF-A, leptin induces the formation of fenestrated capillaries, as confirmed by the absence of fenestrations in leptin-deficient *ob/ob* mice (Cao et al. 2001). Leptin has a synergistic stimulatory effect on angiogenesis induced by VEGF or FGF-2 (Cao et al. 2001). Furthermore, regulation of mRNA expression of some pro- and anti-angiogenic factors is considerably different between the nutritionally and genetically (*ob/ob*) induced diet models. Leptin deficiency results in faster adipose tissue formation and a higher degree of obesity as compared with the HFD-fed WT animals (Voros et al. 2005). These differences may be explained by the enhanced adipose tissue formation or by the lack of leptin itself, which affects the expression of several angiogenic components.

14.3.2 Anti-angiogenic Factors

Adipose tissue also produces several anti-angiogenic components, but their effect on adipose vessel growth and remodeling is not well understood. They may restrict further vessel growth when the growth rate of adipose tissue stabilizes.

Osteonectin, also known as secreted protein acidic and rich in cysteine (SPARC) or basement-membrane protein 40 (BM-40), binds to VEGF-A, impairs VEGFR-1 activation, and inhibits FGF-2, resulting in inhibition of EC proliferation. Osteonectin is produced by the adipose tissue and its expression is strongly elevated in adipocytes of obese mice (Tartare-Deckert et al. 2001). SPARC-deficient mice on HFD develop larger fat pads as compared with WT mice (Bradshaw et al. 2003). It acts on the production and remodeling of adipose tissue, as well as in the regulation of preadipocyte differentiation (Nie et al. 2011).

Thrombospondins [TSP-1 (145 kDa) and TSP-2 (145 kDa)] are components of the ECM in remodeling tissues and, like other matricellular proteins, bind to matrix proteins and cell-surface receptors, including proteoglycans, non-integrin, and integrin receptors. Despite its pleiotropic biological role, mice deficient in TSP-1 are viable and exhibit only subtle abnormalities in development, although they develop pneumonia and show delayed organization and neovascularization of skin wounds (Agah et al. 2002). TSP-1 and -2 are anti-angiogenic, as they inhibit the proliferation and migration of ECs by interacting with CD36, expressed on the surface of these cells (Silverstein and Febbraio 2007). Binding of TSP-1 to CD36 induces the expression of FAS ligand (FasL), which activates its specific, ubiquitous receptor, Fas. This in turn leads to the activation of caspases and apoptosis of the cell (Armstrong and Bornstein 2003). TSP-1 deficiency does not affect the development of HFD-induced obesity (Li et al. 2011; Voros and Lijnen 2006). However, TSP-1 deficiency reduces macrophage accumulation in adipose tissue and protects against obesity-related inflammation and insulin function (Li et al. 2011). Evaluation of weight gain over time in TSP-2$^{-/-}$ and control mice fed normal chow or HFD indicated that TSP-2-deficient female mice are 30 % heavier than wild-type controls due to an increase in non-visceral adipose tissue (Shitaye et al. 2010). In contrast, a similar study using TSP-2$^{-/-}$ and WT littermate mice revealed that TSP-2$^{-/-}$ mice kept on HFD had a significantly lower total body weight throughout the experimental period, without an important functional role in adipose tissue-related angiogenesis (Van Hul et al. 2012b).

In a nutritionally induced obesity model, total body weight, as well as SC and GON adipose tissue mass, was significantly lower in CD36-deficient mice as compared to WT littermates (Christiaens et al. 2012; Kennedy et al. 2011). However, CD36 deficiency had no effect on adipose tissue angiogenesis (Christiaens et al. 2012). In addition, de novo fat pad formation in NUDE mice following injection of preadipocytes was significantly reduced upon CD36 silencing as compared with control preadipocyte injection. This reduction was associated with marked adipocyte hypotrophy, without affecting angiogenesis (Christiaens et al. 2012).

Overall these endogenous molecules, playing a role in the control of angiogenesis, may constitute novel therapeutic approaches for treatment of obesity.

14.4 Conclusions

Modulation of angiogenesis may have the potential to impair adipose tissue development and obesity. Since many anti-angiogenic agents are characterized in the cancer field, the evaluation of such components in obesity models in vivo has become a real option. Nevertheless, since the development of adipose tissue is a complex and multifactorial process, it is unlikely that a single protease or angiogenesis inhibitor will allow reduction of obesity without any side effect. It should be kept in mind that both proteolytic systems and angiogenic components are not only involved in adipogenesis and adipose tissue development, but are also critical to many other biological processes.

References

Agah A, Kyriakides TR, Lawler J, Bornstein P. The lack of thrombospondin-1 (TSP1) dictates the course of wound healing in double-TSP1/TSP2-null mice. Am J Pathol. 2002;161:831–9.

Ahmad SA, Liu W, Jung YD, et al. The effects of angiopoietin-1 and -2 on tumor growth and angiogenesis in human colon cancer. Cancer Res. 2001;61:1255–9.

Alessi MC, Peiretti F, Morange P, Henry M, Nalbone G, Juhan-Vague I. Production of plasminogen activator inhibitor 1 by human adipose tissue: possible link between visceral fat accumulation and vascular disease. Diabetes. 1997;46:860–7.

Alessi MC, Bastelica D, Morange P, et al. Plasminogen activator inhibitor 1, transforming growth factor-beta1, and BMI are closely associated in human adipose tissue during morbid obesity. Diabetes. 2000;49:1374–80.

Alexander CM, Selvarajan S, Mudgett J, Werb Z. Stromelysin-1 regulates adipogenesis during mammary gland involution. J Cell Biol. 2001;152:693–703.

Aoki M, Morishita R, Taniyama Y, et al. Angiogenesis induced by hepatocyte growth factor in non-infarcted myocardium and infarcted myocardium: up-regulation of essential transcription factor for angiogenesis, ets. Gene Ther. 2000;7:417–27.

Armstrong LC, Bornstein P. Thrombospondins 1 and 2 function as inhibitors of angiogenesis. Matrix Biol. 2003;22:63–71.

Bell LN, Cai L, Johnstone BH, Traktuev DO, March KL, Considine RV. A central role for hepatocyte growth factor in adipose tissue angiogenesis. Am J Physiol Endocrinol Metab. 2008;294:E336–44.

Bouloumie A, Sengenes C, Portolan G, Galitzky J, Lafontan M. Adipocyte produces matrix metalloproteinases 2 and 9: involvement in adipose differentiation. Diabetes. 2001;50:2080–6.

Bradshaw AD, Graves DC, Motamed K, Sage EH. SPARC-null mice exhibit increased adiposity without significant differences in overall body weight. Proc Natl Acad Sci U S A. 2003;100:6045–50.

Brown LM, Fox HL, Hazen SA, LaNoue KF, Rannels SR, Lynch CJ. Role of the matrixin MMP-2 in multicellular organization of adipocytes cultured in basement membrane components. Am J Physiol. 1997;272:C937–49.

Cao Y. Adipose tissue angiogenesis as a therapeutic target for obesity and metabolic diseases. Nat Rev Drug Discov. 2010;9:107–15.

Cao R, Brakenhielm E, Wahlestedt C, Thyberg J, Cao Y. Leptin induces vascular permeability and synergistically stimulates angiogenesis with FGF-2 and VEGF. Proc Natl Acad Sci U S A. 2001;98:6390–5.

Carlson TR, Feng Y, Maisonpierre PC, Mrksich M, Morla AO. Direct cell adhesion to the angiopoietins mediated by integrins. J Biol Chem. 2001;276:26516–25.

Carmeliet P, Ferreira V, Breier G, et al. Abnormal blood vessel development and lethality in embryos lacking a single VEGF allele. Nature. 1996;380:435–9.

Carmeliet P, Moons L, Luttun A, et al. Synergism between vascular endothelial growth factor and placental growth factor contributes to angiogenesis and plasma extravasation in pathological conditions. Nat Med. 2001;7:575–83.

Chavey C, Mari B, Monthouel MN, et al. Matrix metalloproteinases are differentially expressed in adipose tissue during obesity and modulate adipocyte differentiation. J Biol Chem. 2003;278:11888–96.

Chen H, Charlat O, Tartaglia LA, et al. Evidence that the diabetes gene encodes the leptin receptor: identification of a mutation in the leptin receptor gene in db/db mice. Cell. 1996;84:491–5.

Christiaens V, Van Hul M, Lijnen HR, Scroyen I. CD36 promotes adipocyte differentiation and adipogenesis. Biochim Biophys Acta. 2012;1820(7):949–56.

Dallabrida SM, Zurakowski D, Shih SC, et al. Adipose tissue growth and regression are regulated by angiopoietin-1. Biochem Biophys Res Commun. 2003;311:563–71.

Davis S, Aldrich TH, Jones PF, et al. Isolation of angiopoietin-1, a ligand for the TIE2 receptor, by secretion-trap expression cloning. Cell. 1996;87:1161–9.

Davis S, Papadopoulos N, Aldrich TH, et al. Angiopoietins have distinct modular domains essential for receptor binding, dimerization and superclustering. Nat Struct Biol. 2003;10:38–44.

De Taeye BM, Novitskaya T, Gleaves L, Covington JW, Vaughan DE. Bone marrow plasminogen activator inhibitor-1 influences the development of obesity. J Biol Chem. 2006;281: 32796–805.

Demeulemeester D, Scroyen I, Voros G, et al. Overexpression of tissue inhibitor of matrix metalloproteinases-1 (TIMP-1) in mice does not affect adipogenesis or adipose tissue development. Thromb Haemost. 2006;95:1019–24.

Dougherty KM, Pearson JM, Yang AY, Westrick RJ, Baker MS, Ginsburg D. The plasminogen activator inhibitor-2 gene is not required for normal murine development or survival. Proc Natl Acad Sci U S A. 1999;96:686–91.

Elias I, Franckhauser S, Ferre T, et al. Adipose tissue overexpression of vascular endothelial growth factor protects against diet-induced obesity and insulin resistance. Diabetes. 2012;61(7):1801–13.

Ferrara N, Carver-Moore K, Chen H, et al. Heterozygous embryonic lethality induced by targeted inactivation of the VEGF gene. Nature. 1996;380:439–42.

Friedman JM, Halaas JL. Leptin and the regulation of body weight in mammals. Nature. 1998;395:763–70.

Fukumura D, Ushiyama A, Duda DG, et al. Paracrine regulation of angiogenesis and adipocyte differentiation during in vivo adipogenesis. Circ Res. 2003;93:e88–97.

Gomez DE, Alonso DF, Yoshiji H, Thorgeirsson UP. Tissue inhibitors of metalloproteinases: structure, regulation and biological functions. Eur J Cell Biol. 1997;74:111–22.

Hasty AH, Shimano H, Osuga J, et al. Severe hypercholesterolemia, hypertriglyceridemia, and atherosclerosis in mice lacking both leptin and the low density lipoprotein receptor. J Biol Chem. 2001;276:37402–8.

Hlatky L, Hahnfeldt P, Folkman J. Clinical application of antiangiogenic therapy: microvessel density, what it does and doesn't tell us. J Natl Cancer Inst. 2002;94:883–93.

Hoover-Plow J, Ellis J, Yuen L. In vivo plasminogen deficiency reduces fat accumulation. Thromb Haemost. 2002;87:1011–9.

Jaworski DM, Sideleva O, Stradecki HM, et al. Sexually dimorphic diet-induced insulin resistance in obese tissue inhibitor of metalloproteinase-2 (TIMP-2)-deficient mice. Endocrinology. 2011;152:1300–13.

Johnson MD, Kim HR, Chesler L, Tsao-Wu G, Bouck N, Polverini PJ. Inhibition of angiogenesis by tissue inhibitor of metalloproteinase. J Cell Physiol. 1994;160:194–202.

Joukov V, Pajusola K, Kaipainen A, et al. A novel vascular endothelial growth factor, VEGF-C, is a ligand for the Flt4 (VEGFR-3) and KDR (VEGFR-2) receptor tyrosine kinases. EMBO J. 1996;15:290–8.

Juhan-Vague I, Alessi MC, Vague P. Increased plasma plasminogen activator inhibitor 1 levels. A possible link between insulin resistance and atherothrombosis. Diabetologia. 1991;34: 457–62.

Kanasaki K, Koya D. Biology of obesity: lessons from animal models of obesity. J Biomed Biotechnol. 2011;2011:197636.

Karkkainen MJ, Haiko P, Sainio K, et al. Vascular endothelial growth factor C is required for sprouting of the first lymphatic vessels from embryonic veins. Nat Immunol. 2004;5:74–80.

Kawaguchi N, Toriyama K, Nicodemou-Lena E, Inou K, Torii S, Kitagawa Y. De novo adipogenesis in mice at the site of injection of basement membrane and basic fibroblast growth factor. Proc Natl Acad Sci U S A. 1998;95:1062–6.

Kawasaki T, Kitsukawa T, Bekku Y, et al. A requirement for neuropilin-1 in embryonic vessel formation. Development. 1999;126:4895–902.

Kennedy DJ, Kuchibhotla S, Westfall KM, Silverstein RL, Morton RE, Febbraio M. A CD36-dependent pathway enhances macrophage and adipose tissue inflammation and impairs insulin signalling. Cardiovasc Res. 2011;89:604–13.

Kimura Y, Ozeki M, Inamoto T, Tabata Y. Time course of de novo adipogenesis in matrigel by gelatin microspheres incorporating basic fibroblast growth factor. Tissue Eng. 2002;8:603–13.

Kruithof EK, Tran-Thang C, Gudinchet A, et al. Fibrinolysis in pregnancy: a study of plasminogen activator inhibitors. Blood. 1987;69:460–6.

Li Y, Tong X, Rumala C, Clemons K, Wang S. Thrombospondin1 deficiency reduces obesity-associated inflammation and improves insulin sensitivity in a diet-induced obese mouse model. PLoS One. 2011;6:e26656.

Lijnen HR. Effect of plasminogen activator inhibitor-1 deficiency on nutritionally-induced obesity in mice. Thromb Haemost. 2005;93:816–9.

Lijnen HR. Deficiency of alpha2-antiplasmin does not affect murine adipose tissue development. J Thromb Haemost. 2007;5:420–1.

Lijnen HR. Angiogenesis and obesity. Cardiovasc Res. 2008;78:286–93.

Lijnen HR. Role of fibrinolysis in obesity and thrombosis. Thromb Res. 2009;123 Suppl 4:S46–9.

Lijnen HR, Collen D. Mechanisms of physiological fibrinolysis. Baillieres Clin Haematol. 1995;8:277–90.

Lijnen HR, Maquoi E, Holvoet P, et al. Adipose tissue expression of gelatinases in mouse models of obesity. Thromb Haemost. 2001;85:1111–6.

Lijnen HR, Maquoi E, Demeulemeester D, Van Hoef B, Collen D. Modulation of fibrinolytic and gelatinolytic activity during adipose tissue development in a mouse model of nutritionally induced obesity. Thromb Haemost. 2002a;88:345–53.

Lijnen HR, Van HB, Frederix L, Rio MC, Collen D. Adipocyte hypertrophy in stromelysin-3 deficient mice with nutritionally induced obesity. Thromb Haemost. 2002b;87:530–5.

Lijnen HR, Maquoi E, Morange P, et al. Nutritionally induced obesity is attenuated in transgenic mice overexpressing plasminogen activator inhibitor-1. Arterioscler Thromb Vasc Biol. 2003a;23:78–84.

Lijnen HR, Demeulemeester D, Van Hoef B, Collen D, Maquoi E. Deficiency of tissue inhibitor of matrix metalloproteinase-1 (TIMP-1) impairs nutritionally induced obesity in mice. Thromb Haemost. 2003b;89:249–55.

Lijnen HR, Christiaens V, Scroyen I, et al. Impaired adipose tissue development in mice with inactivation of placental growth factor function. Diabetes. 2006;55:2698–704.

Lijnen HR, Frederix L, Scroyen I. Deficiency of plasminogen activator inhibitor-2 impairs nutritionally induced murine adipose tissue development. J Thromb Haemost. 2007;5:2259–65.

Lijnen HR, Van Hoef B, Rodriguez JA, Paramo JA. Stromelysin-2 (MMP-10) deficiency does not affect adipose tissue formation in a mouse model of nutritionally induced obesity. Biochem Biophys Res Commun. 2009a;389:378–81.

Lijnen HR, Frederix L, Van Hoef B, Dewerchin M. Deficiency of vascular endothelial growth factor-D does not affect murine adipose tissue development. Biochem Biophys Res Commun. 2009b;378:255–8.

Luque A, Carpizo DR, Iruela-Arispe ML. ADAMTS1/METH1 inhibits endothelial cell proliferation by direct binding and sequestration of VEGF165. J Biol Chem. 2003;278:23656–65.

Ma LJ, Mao SL, Taylor KL, et al. Prevention of obesity and insulin resistance in mice lacking plasminogen activator inhibitor 1. Diabetes. 2004;53:336–46.

Mandrup S, Loftus TM, MacDougald OA, Kuhajda FP, Lane MD. Obese gene expression at in vivo levels by fat pads derived from s.c. implanted 3T3-F442A preadipocytes. Proc Natl Acad Sci U S A. 1997;94:4300–5.

Maquoi E, Munaut C, Colige A, Collen D, Lijnen HR. Modulation of adipose tissue expression of murine matrix metalloproteinases and their tissue inhibitors with obesity. Diabetes. 2002;51:1093–101.

Maquoi E, Demeulemeester D, Voros G, Collen D, Lijnen HR. Enhanced nutritionally induced adipose tissue development in mice with stromelysin-1 gene inactivation. Thromb Haemost. 2003;89:696–704.

Mavri A, Stegnar M, Krebs M, Sentocnik JT, Geiger M, Binder BR. Impact of adipose tissue on plasma plasminogen activator inhibitor-1 in dieting obese women. Arterioscler Thromb Vasc Biol. 1999;19:1582–7.
Miskin R, Masos T. Transgenic mice overexpressing urokinase-type plasminogen activator in the brain exhibit reduced food consumption, body weight and size, and increased longevity. J Gerontol A Biol Sci Med Sci. 1997;52:B118–24.
Morange PE, Lijnen HR, Alessi MC, Kopp F, Collen D, Juhan-Vague I. Influence of PAI-1 on adipose tissue growth and metabolic parameters in a murine model of diet-induced obesity. Arterioscler Thromb Vasc Biol. 2000;20:1150–4.
Morange PE, Bastelica D, Bonzi MF, et al. Influence of t-pA and u-PA on adipose tissue development in a murine model of diet-induced obesity. Thromb Haemost. 2002;87:306–10.
Neels JG, Thinnes T, Loskutoff DJ. Angiogenesis in an in vivo model of adipose tissue development. FASEB J. 2004;18:983–5.
Nie J, Bradshaw AD, Delany AM, Sage EH. Inactivation of SPARC enhances high-fat diet-induced obesity in mice. Connect Tissue Res. 2011;52:99–108.
Nishimura S, Manabe I, Nagasaki M, et al. Adipogenesis in obesity requires close interplay between differentiating adipocytes, stromal cells, and blood vessels. Diabetes. 2007;56: 1517–26.
Oh CW, Hoover-Plow J, Plow EF. The role of plasminogen in angiogenesis in vivo. J Thromb Haemost. 2003;1:1683–7.
Okamura K, Sato Y, Matsuda T, et al. Endogenous basic fibroblast growth factor-dependent induction of collagenase and interleukin-6 in tumor necrosis factor-treated human microvascular endothelial cells. J Biol Chem. 1991;266:19162–5.
Olofsson B, Korpelainen E, Pepper MS, et al. Vascular endothelial growth factor B (VEGF-B) binds to VEGF receptor-1 and regulates plasminogen activator activity in endothelial cells. Proc Natl Acad Sci U S A. 1998;95:11709–14.
Oosthuyse B, Moons L, Storkebaum E, et al. Deletion of the hypoxia-response element in the vascular endothelial growth factor promoter causes motor neuron degeneration. Nat Genet. 2001;28:131–8.
Pendas AM, Folgueras AR, Llano E, et al. Diet-induced obesity and reduced skin cancer susceptibility in matrix metalloproteinase 19-deficient mice. Mol Cell Biol. 2004;24:5304–13.
Samad F, Loskutoff DJ. Tissue distribution and regulation of plasminogen activator inhibitor-1 in obese mice. Mol Med. 1996;2:568–82.
Schafer K, Fujisawa K, Konstantinides S, Loskutoff DJ. Disruption of the plasminogen activator inhibitor 1 gene reduces the adiposity and improves the metabolic profile of genetically obese and diabetic ob/ob mice. FASEB J. 2001;15:1840–2.
Scroyen I, Christiaens V, Lijnen HR. No functional role of plasminogen activator inhibitor-1 in murine adipogenesis or adipocyte differentiation. J Thromb Haemost. 2007;5:139–45.
Scroyen I, Jacobs F, Cosemans L, De Geest B, Lijnen HR. Effect of plasminogen activator inhibitor-1 on adipogenesis in vivo. Thromb Haemost. 2009a;101:388–93.
Scroyen I, Cosemans L, Lijnen HR. Effect of tissue inhibitor of matrix metalloproteinases-1 on in vitro and in vivo adipocyte differentiation. Thromb Res. 2009b;124:578–83.
Scroyen I, Jacobs F, Cosemans L, De Geest B, Lijnen HR. Blood vessel density in de novo formed adipose tissue is decreased upon overexpression of TIMP-1. Obesity (Silver Spring). 2010;18:638–40.
Shaw DA, Machaughton D. Relationship between blood fibrinoytic activity and body fatness. Lancet. 1963;1:352–4.
Shimomura I, Funahashi T, Takahashi M, et al. Enhanced expression of PAI-1 in visceral fat: possible contributor to vascular disease in obesity. Nat Med. 1996;2:800–3.
Shitaye HS, Terkhorn SP, Combs JA, Hankenson KD. Thrombospondin-2 is an endogenous adipocyte inhibitor. Matrix Biol. 2010;29:549–56.
Sierra-Honigmann MR, Nath AK, Murakami C, et al. Biological action of leptin as an angiogenic factor. Science. 1998;281:1683–6.

Silverstein RL, Febbraio M. CD36-TSP-HRGP interactions in the regulation of angiogenesis. Curr Pharm Des. 2007;13:3559–67.

Stacker SA, Caesar C, Baldwin ME, et al. VEGF-D promotes the metastatic spread of tumor cells via the lymphatics. Nat Med. 2001;7:186–91.

Sternlicht MD, Werb Z. How matrix metalloproteinases regulate cell behavior. Annu Rev Cell Dev Biol. 2001;17:463–516.

Stetler-Stevenson WG, Seo DW. TIMP-2: an endogenous inhibitor of angiogenesis. Trends Mol Med. 2005;11:97–103.

Stradecki HM, Jaworski DM. Hyperphagia and leptin resistance in tissue inhibitor of metalloproteinase-2 deficient mice. J Neuroendocrinol. 2011;23:269–81.

Suganami E, Takagi H, Ohashi H, et al. Leptin stimulates ischemia-induced retinal neovascularization: possible role of vascular endothelial growth factor expressed in retinal endothelial cells. Diabetes. 2004;53:2443–8.

Tartare-Deckert S, Chavey C, Monthouel MN, Gautier N, Van Obberghen E. The matricellular protein SPARC/osteonectin as a newly identified factor up-regulated in obesity. J Biol Chem. 2001;276:22231–7.

Thurston G, Rudge JS, Ioffe E, et al. Angiopoietin-1 protects the adult vasculature against plasma leakage. Nat Med. 2000;6:460–3.

Tienari J, Alanko T, Lehtonen E, Saksela O. The expression and localization of urokinase-type plasminogen activator and its type 1 inhibitor are regulated by retinoic acid and fibroblast growth factor in human teratocarcinoma cells. Cell Regul. 1991;2:285–97.

Van Belle E, Witzenbichler B, Chen D, et al. Potentiated angiogenic effect of scatter factor/hepatocyte growth factor via induction of vascular endothelial growth factor: the case for paracrine amplification of angiogenesis. Circulation. 1998;97:381–90.

Van De Craen B, Scroyen I, Vranckx C, et al. Maximal PAI-1 inhibition in vivo requires neutralizing antibodies that recognize and inhibit glycosylated PAI-1. Thromb Res. 2012;129: e126–33.

Van Hul M, Lijnen HR. A functional role of gelatinase A in the development of nutritionally induced obesity in mice. J Thromb Haemost. 2008;6:1198–206.

Van Hul M, Lijnen HR. Effect of weight loss on gelatinase levels in obese mice. Clin Exp Pharmacol Physiol. 2011a;38:647–9.

Van Hul M, Lijnen HR. Matrix metalloproteinase inhibition impairs murine adipose tissue development independently of leptin. Endocr J. 2011b;58:101–7.

Van Hul M, Piccard H, Lijnen HR. Gelatinase B (MMP-9) deficiency does not affect murine adipose tissue development. Thromb Haemost. 2010;104:165–71.

Van Hul M, Lupu F, Dresselaers T, Buyse J, Lijnen HR. Matrix metalloproteinase inhibition affects adipose tissue mass in obese mice. Clin Exp Pharmacol Physiol. 2012a;39(6):544–50.

Van Hul M, Frederix L, Lijnen HR. Role of thrombospondin-2 in murine adipose tissue angiogenesis and development. Obesity (Silver Spring). 2012b;20(9):1757–62.

Vazquez F, Hastings G, Ortega MA, et al. METH-1, a human ortholog of ADAMTS-1, and METH-2 are members of a new family of proteins with angio-inhibitory activity. J Biol Chem. 1999;274:23349–57.

Veikkola T, Jussila L, Makinen T, et al. Signalling via vascular endothelial growth factor receptor-3 is sufficient for lymphangiogenesis in transgenic mice. EMBO J. 2001;20:1223–31.

Voros G, Lijnen HR. Deficiency of thrombospondin-1 in mice does not affect adipose tissue development. J Thromb Haemost. 2006;4:277–8.

Voros G, Maquoi E, Collen D, Lijnen HR. Differential expression of plasminogen activator inhibitor-1, tumor necrosis factor-alpha, TNF-alpha converting enzyme and ADAMTS family members in murine fat territories. Biochim Biophys Acta. 2003;1625:36–42.

Voros G, Maquoi E, Demeulemeester D, Clerx N, Collen D, Lijnen HR. Modulation of angiogenesis during adipose tissue development in murine models of obesity. Endocrinology. 2005;146:4545–54.

Voros G, Sandy JD, Collen D, Lijnen HR. Expression of aggrecan(ases) during murine preadipocyte differentiation and adipose tissue development. Biochim Biophys Acta. 2006;1760:1837–44.

West DB, York B. Dietary fat, genetic predisposition, and obesity: lessons from animal models. Am J Clin Nutr. 1998;67:505S–12.

Xiao L, Sobue T, Esliger A, et al. Disruption of the Fgf2 gene activates the adipogenic and suppresses the osteogenic program in mesenchymal marrow stromal stem cells. Bone. 2010;47:360–70.

Zhang Y, Proenca R, Maffei M, Barone M, Leopold L, Friedman JM. Positional cloning of the mouse obese gene and its human homologue. Nature. 1994;372:425–32.

Part VII
Adipose Tissue Facilitates Tumor Angiogenesis and Growth

Chapter 15
Adipose Tissue-Derived Progenitor Cells and Cancer

Chieh Tseng and Mikhail G. Kolonin

Abstract Obesity, a condition of white adipose tissue (WAT) overgrowth and perturbation of system metabolic homeostasis, is a significant risk factor for cancer mortality. The mechanisms underlying the obesity-cancer link remain poorly understood. Elevated systemic circulation of pathogenic adipokines and chronic low-grade inflammation associated with obesity have indicated a potential tumor-promoting role of endocrine WAT signaling. In addition, possible effects of WAT-derived cell populations localized at the tumor site have been uncovered. Recent studies have shown that adipose stromal cell (ASC), the progenitors of adipocytes, can traffic to tumors. As a constituent of tumor microenvironment, ASC may engage in paracrine signaling and thereby promote cancer progression. Here, we discuss tumor-promoting functions of ASC and a potential of targeting these cells as an approach to thwarting cancer progression.

Keywords Adipocyte • Adipose • Stromal • Progenitor cell • Adipokine • Vasculature • Vessel • Endothelial • Cancer • Tumor

15.1 Introduction

The pandemic of obesity has become a global health crisis (Basen-Engquist and Chang 2011; Khandekar et al. 2011). Obesity is defined as the body mass index (BMI = weight/height^2) of $\geq$30 kg/m^2 (Basen-Engquist and Chang 2011).

C. Tseng • M.G. Kolonin, Ph.D. (✉)
Institute of Molecular Medicine, Center for Stem Cell and Regenerative Medicine, The University of Texas Health Science Center, 1825 Pressler St., Rm 630G, Houston, TX 77030, USA

The Brown Foundation Institute of Molecular Medicine, The University of Texas Health Science Center, 1825 Pressler St., Rm 630G, Houston, TX 77030, USA
e-mail: Mikhail.G.Kolonin@uth.tmc.edu

Y. Cao (ed.), *Angiogenesis in Adipose Tissue*, DOI 10.1007/978-1-4614-8069-3_15,

Accumulating epidemiological evidence demonstrates a strong link between obesity and several cancers (Calle and Kaaks 2004; Renehan et al. 2008; Basen-Engquist and Chang 2011), including colorectal, esophageal, kidney, and pancreatic cancer, as well as other cancer types with weaker association (Li et al. 2009; Basen-Engquist and Chang 2011; Dalamaga et al. 2012). Given the significant impact of obesity on cancer risk and mortality, increasing efforts have been devoted to obesity-cancer research. Irrespective of its underlying causes (genetic predisposition, environmental and psychosocial factors), obesity is manifested as a surplus in energy deposited and accumulated in white adipose tissue (WAT) (Cinti 2005; Basen-Engquist and Chang 2011; Cawthorn et al. 2012). The amount of WAT within the body has significant impact on homeostasis, as evidenced by obesity comorbidities such as the metabolic syndrome, type 2 diabetes, and cardiovascular diseases (Khandekar et al. 2011). Cancer often develops or/and progresses when normal physiological homeostasis is perturbed and inflammation-driven tissue remodeling fuels the genetic cell abnormalities. Therefore, to better appreciate the underlying mechanisms of obesity-cancer link, it is fundamental to dissect the possible contributions of WAT to pathological tissue reorganization. The switch from viewing obesity as solely an aberration in endocrine signaling modulating energy intake/expenditure and the metabolic repercussions to recognizing potential additional paracrine effects of cells trafficking from WAT (Zhang et al. 2010) has opened new perspectives. Here, we review recent research on the pathophysiological relationship between WAT and cancer and discuss its potential clinical implications.

15.2 Composition and Biological Function of Adipose Tissue

Histologically, adipose tissue appears as a type of loose connective tissue which can be divided into two functionally and anatomically distinctive organs: WAT and brown adipose tissue (BAT) (Gesta et al. 2007). BAT is composed of adipocytes rich in mitochondria and multilocular lipid droplets (Virtanen et al. 2009). BAT expresses high levels of uncoupling protein-1(UCP-1) important for basal and adaptive thermogenesis (Virtanen et al. 2009). BAT utilizes lipid storage mainly for heat production through dissipating proton electron gradient established by oxidative phosphorylation (Cinti 2005). BAT is found in periadrenal, cervical, axillary regions of human system in young, but decreases significantly in adulthood (Gesta et al. 2007). In contrast, WAT acts as long-term fuel repository, storing excessive calories as high energy fuel: triglycerides (Frayn et al. 2003). Lipid droplets in WAT are prevalently unilocular (Trayhurn 2007). WAT accumulation is a physiological adaptation to excess food intake that ensures availability of energy during nutritional deprivation (Cinti 2005). Furthermore, adipose tissue also provides mechanical support and thermal insulation of vital organs (Frayn et al. 2003), as reflected by distribution of WAT throughout major intra-abdominal and subcutaneous depots (Gesta et al. 2007). In addition to the distinct partition of BAT and WAT in some depots (recognizable by their color and gene expression profile) (Cinti 2005), several

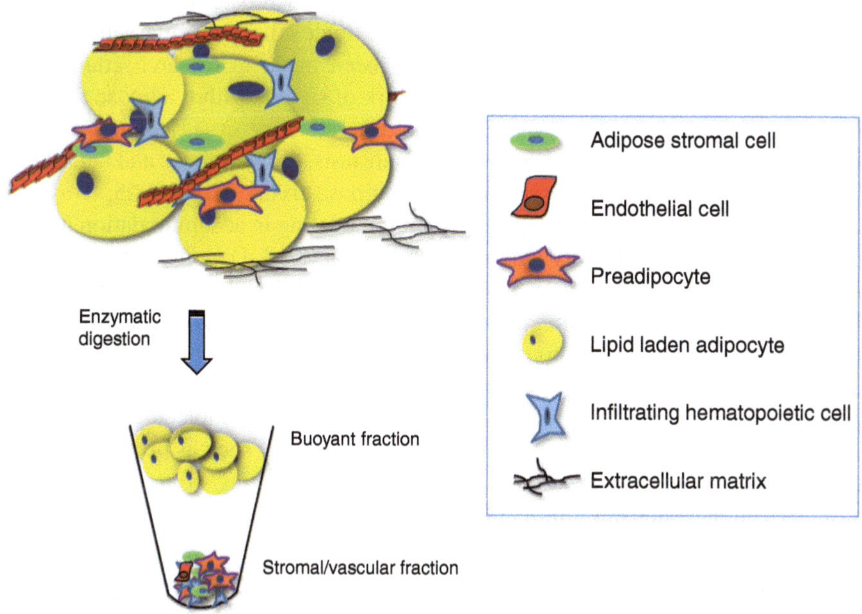

Fig. 15.1 Cell populations of white adipose tissue (WAT). Cellular components of WAT can be separated by enzymatic digestion into a buoyant fraction and a stromal/vascular fraction. The buoyant fraction is enriched with lipid-laden pre-adipocytes and mature adipocytes, whereas the pelletable SVF is constituted by EC, perivascular ASC, and infiltrating leukocytes

groups reported that mixed BAT and WAT are found interspersed within the system (Cinti 2005, 2011; Virtanen et al. 2009). Furthermore, induction of WAT transition to brown-like adipose tissue, composed of "brite" (beige) adipocytes, takes place in response to beta-adrenergic stimulation (Lim et al. 2012; Vitali et al. 2012). The importance and functional aspects of beige adipose tissue are being investigated (Wu et al. 2012).

Structurally, WAT is a heterogeneous organ composed of a number of distinct cell populations (Fig. 15.1). Enzymatic digestion separates WAT into a buoyant fraction and a pelletable stromal/vascular fraction (SVF) (Daquinag et al. 2011b; Gimble et al. 2011; Cawthorn et al. 2012). The buoyant fraction contains lipid-laden pre-adipocytes and mature adipocytes, whereas the SVF is a heterogeneous mixture of cell populations: adipose progenitor cells/stromal cells (ASC), adipose endothelial cells (EC), and infiltrating hematopoietic cells (Daquinag et al. 2011b). WAT expansion relies not only on adipocyte hypertrophy but also on proliferation of ASC that, upon differentiation, generate new adipocytes (Cinti 2011). Expansion of the SVF ensures sufficient blood flow and modulates compensatory tissue remodeling associated with adipocyte death, which is elevated in obesity (Eto et al. 2011). Blood flow in adipose tissue is highly labile and is altered in response to physiological stimuli such as fasting, postprandial, and exercise (Frayn et al. 2003). Overall, WAT is a relatively hypoxic tissue receiving about 5 % of total cardiac output

(Blogowski et al. 2012). Expansion of WAT further exacerbates the hypoxic condition, in particular in morbid obesity featuring extreme adipocyte enlargement.

Over the past decade, great strides have been made in research attaining to unravel the intricate and tightly knitted network of signals within adipose tissue. It is now well appreciated that WAT is the largest endocrine organ (Cinti 2005). To accommodate one's nutritional state and homeostasis, gut hormones and neural factors converge with adipokines secreted by adipose tissue (Cinti 2005; Trayhurn 2007). WAT is an organ undergoing constant remodeling in accommodation to fluctuations in energy availability (Mariman and Wang 2010; Daquinag et al. 2011b). Innervation and vascularization have been investigated as key factors regulating WAT mass and function (Smith and Minson 2012). The sympathetic and parasympathetic innervation of WAT controls adipocyte lipolysis and lipogenesis (Trayhurn 2007). In response to energy excess, the expansion of WAT triggers signaling changes, many of which are mediated by secreted paracrine molecules. These signals feed back upon the cellular components within WAT, such as progenitor cells, to further stimulate adipogenesis, as well as the accompanying stromatogenesis and vascularization (Yang and Smith 2007).

15.3 WAT and Cancer: The Role of Adipokines

Cytokines and other factors secreted by WAT, termed adipokines, are bioactive systemically, which makes WAT the largest endocrine organ (Calle and Kaaks 2004). At present, there are more than 50 adipokines identified, and under normal physiological condition, adipokines integrate into communication network between WAT and other organs, modulating events ranging from ECM remodeling and immune response to energy balance and metabolism (Khandekar et al. 2011). For example, leptin is an adipocyte-secreted hormone that regulates energy intake and metabolism through binding to receptors in hypothalamus, which stimulates anorexigenic peptides aiming at reduction in energy intake (Khandekar et al. 2011). Adiponectin is another adipokine controlling energy homeostasis and metabolism, expression of which by adipocytes is inversely correlated with increased adiposity (Calle and Kaaks 2004; Kelesidis et al. 2006). Adiponectin intracellular signaling, mainly through AMP-activated protein kinase (AMPK), is critical for modulating insulin sensitivity, vessel function, and immunity (Kelesidis et al. 2006; Dalamaga et al. 2012). Increased circulation of leptin and decreased circulation of adiponectin are observed in obese cancer patients (Grossmann et al. 2010).

In obesity, accumulated adipose tissue leads to aberration in secretion of various adipokines, contributing to obese-related pathological conditions including cancer (Calle and Kaaks 2004).

Some adipokines are secreted by cells of the SVF rather than by adipocytes. Aberrations in cytokines such as interleukin-6 (IL-6) and tumor necrosis factor alpha (TNF-α), secreted by WAT-infiltrating monocytes/macrophages, account for chronic low-grade inflammatory state setting in obesity (Park et al. 2010).

As these inflammatory adipokines are known to control tumor cell proliferation, survival, and invasion, they may link cancer with obesity. Indeed, both IL-6 and TNF-α have been associated with higher cancer risk in colorectal cancer (Fenton and Birmingham 2010; Calle and Kaaks 2004). Other adipokines, including plasminogen activator inhibitor-1 (PAI-1), insulin-like growth factors (IGFs), epidermal growth factor (EGF), transforming growth factor-beta (TGF-β) and leptin, may function as tropic factors, for instance by supporting malignant cell proliferation and survival (Khandekar et al. 2011). IL-6 and TGF-β can also support tumor vasculature through activating pro-angiogenic signaling mediated by vascular endothelial growth factor (VEGF), hepatocyte growth factor (HGF), and fibroblast growth factors (FGF-1 and FGF-2) (Mueller and Fusenig 2004; Rehman et al. 2004). In addition, obesity-induced perturbation of insulin sensitivity signals through adipokines contributes to impaired insulin action, altering secretion and responsiveness of corresponding organs to dietary nutrients. The resultant hyperinsulinemia act as potent growth factor and also enhances the expression of other tumor-promoting factors such as IGFs (Calle and Kaaks 2004).

The importance of WAT-derived endocrine factors in promoting cancer progression has been well supported by experimental models (Calle and Kaaks 2004; Khandekar et al. 2011; Nieman et al. 2011). However, based on the evidence of ASC trafficking in obesity (Zhang et al. 2010; Bellows et al. 2011a; Klopp et al. 2012) and the presence of adipocytes within tumors (Zyromski et al. 2009; Nieman et al. 2011), we have proposed that paracrine adipokines regulation may also take place.

15.4 Adipose Stromal Progenitor Cells and Their Potential Pathological Functions

The heterogeneous ASC population of WAT is activated under ischemic condition and contributes to adipose tissue remodeling and expansion (Zhang et al. 2010). ASC possess traits first described for the bone marrow mesenchymal stem cells (MSC), initially characterized as fibroblast colony-forming units (CFU-F) (Bianco et al. 2008). These features include the capacity to differentiate into mesodermal lineages: osteoblasts, chondrocytes, and adipocytes and more recently extended to non-mesodermal origin (Gimble et al. 2011; Cawthorn et al. 2012). With the ease of isolation and inherent multipotency characteristic of ASC, WAT emerged as an attractive cell source for regenerative medicine and basic research (Cawthorn et al. 2012). To date, the immunophenotypic signature of ASC remains controversial, largely due to the heterogeneity of this population. To isolate ASC, our group and others have utilized their plastic adherence properties and the CD34-positive CD45-negative CD31-negative (CD34+CD45−CD31−) immunophenotype (Traktuev et al. 2008; Zhang et al. 2009). More recently, by combinatorial phage display approaches, a new decorin isoform has been identified as a potential marker to distinguish and enrich for ASC with progenitor capacity (Daquinag et al. 2011a).

The ASC pool is expanded in obesity (Maumus et al. 2008; Joe et al. 2009) and serves as a major component of SVF (Zhang et al. 2010). The main role of ASC is serving as progenitors of adipocytes. However, other important functions have been assigned to these cells, some of which may have pathological implications. One of the key roles of ASC is their involvement in vascularization. Development of functional blood vessels relies on orchestrated assembly of pericytes, EC, and perivascular matrix (Song et al. 2005; von Tell et al. 2006). Through direct association with EC, pericytes lend support for formation of mature vasculature (Gerhardt and Semb 2008). Pericyte depletion in transgenic mouse models (Song et al. 2005) or blockage of pericytes recruitment by inhibitors results in increased vessel leakage and impairment in patency (Cooke et al. 2012). The localization of ASC in contact with EC has revealed them as pericytes/adventitial cells (Tang et al. 2008; Traktuev et al. 2008; Corselli 2011). The paracrine effect of perivascular ASC through secretory factors such as VEGF, HGF, and TGF-β is believed to be critical for modulating EC proliferation, differentiation, and survival (von Tell et al. 2006; Gerhardt and Semb 2008).

Another function of ASC is the control of tissue stiffness. It is well recognized that extracellular matrix (ECM) composition regulates cell survival, proliferation, migration, and differentiation (Levental et al. 2009). Therefore, the dynamics of ECM is crucial for directing cell behavior (Mariman and Wang 2010). As in other organs, ECM in WAT is composed mainly of collagens and other fibrillar proteins, such as laminin and fibronectin (Frantz et al. 2010; Mariman and Wang 2010). Collagen, the most abundant fibrous ECM protein, is known to be produced by adipocytes, ASC, and other SVF cells (Mariman and Wang 2010). With increased adiposity, the ECM is remodeled with a shift from fibrillar to laminar ECM in adaptation to support the increased lipid content resulting from adipocyte hypertrophy (Henegar et al. 2008; Mariman and Wang 2010). Several structural components involved in ECM remodeling and inflammation-related genes are found to be upregulated in obesity, contributing to its increased fibrosis (Henegar et al. 2008). Like bone marrow-derived MSC, ASC produce large amounts of fibronectin and collagen (Kolonin et al. 2012). Recent studies revealed that ASC mediate ECM remodeling and enhance ECM rigidity (Chandler et al. 2012). It has been shown that cells display higher invasiveness with increased invadopodia and divide much more rapidly on rigid than on compliant ECM (Alexander et al. 2008; Ulrich et al. 2009). Therefore, the ASC-enforced changes in ECM rigidity may regulate cell migratory ability and proliferation.

Finally, MSC have potent immunomodulatory properties (Nauta and Fibbe 2007). It has been shown that ASC are immunosuppressive and capable of attenuating T-cell function (Puissant et al. 2005). In response to injury-associated factors, ASC can participate in resolving scar tissue and increasing capillary density (Eto et al. 2011). Clinical studies in myocardial infarction models demonstrated improved cardiac function following human ASC engraftment (Cai et al. 2009).

15.5 Evidence for ASC Mobilization in Cancer

Cancer progression is a multi-step and dynamic process involving heterotypic interactions between distinct cell populations (Hanahan and Weinberg 2011). The establishment of functional tumor microenvironment relies on recruitment of cells locally from adjacent tissues and systemically from distant organs via cell mobilization. A number of circulating cell populations with progenitor properties have been described: hematopoietic and endothelial progenitor cells, mature EC, monocytic fibroblast progenitors (also known as fibrocytes), and stromal cells (Pitchford et al. 2009; Kolonin et al. 2012). Origins of these cell populations have been explored in animal models, and bone marrow has been demonstrated as a source of mobilized progenitor cells in circulation. However, evidence of progenitor cell mobilization from other organs has also begun to emerge (Aicher et al. 2007). We have proposed WAT as a possible alternative source of circulating progenitor cells (Kolonin and Simmons 2009; Zhang et al. 2010). Consistent with this possibility, increased circulation of progenitor cells of not only hematopoietic, but also of mesenchymal origin has been detected in obesity (Bellows et al. 2011b). We hypothesized that the recruited cells may become involved in supporting blood vessel formation (Kolonin et al. 2012), thus complementing the function previously assigned to circulating EC progenitors (Laird et al. 2008) and possibly the trafficking of progenitors elevated in obesity may explain the obesity paradox: the improved postoperative healing in obese patients (Curtis et al. 2005; Mullen et al. 2009). Interestingly, in our studies, systemic circulation of MSC with the ASC immunophenotype (CD34-positive) was found to be further increased in cancer (Bellows et al. 2011a). This suggested mobilization of ASC from WAT in response to cancer signals has led to the hypothesis that in cancer settings, instead of mending damaged benign tissues, ASC recruited by tumors could affect cancer progression.

15.6 Toward Mechanisms Controlling ASC Trafficking

Chronic inflammation is increasingly recognized as a substantial constituent of cancer initiation and progression (Hanahan and Weinberg 2011). In the event of obesity, the delicate homeostatic balance is often tipped as a result of expanding WAT with inadequate oxygenation, creating hypoxic environment within the organ (Bertolini et al. 2012). Expression of many chemokines is induced by hypoxia, which subsequently promotes recruitment of macrophages, mast cells, neutrophils, lymphocytes, and other leukocyte populations to WAT (Balkwill 2004). The chemokine gradient of stromal cell-derived factor (SDF-1) and its receptor (CXCR4) appears to be key in this process (Balkwill 2012). The accumulated WAT and increased infiltrated inflammatory cells secrete and directly increase numerous inflammatory factors, growth molecules, and bioavailability of steroid hormones (Sirin and Kolonin 2013).

Like other mesenchymal progenitors, ASC demonstrate tropism to inflammatory sites. In response to ischemic conditions, CXCR4 expression is augmented in ASC (Thangarajah et al. 2009), which may contribute to their migration. In addition, chemokine receptors CCR1, CCR4, CCR7, and CXCR5 are found to be expressed on ASC and bone marrow-derived MSC, with many induced by inflammation stimuli (Von Luttichau et al. 2005; Ponte et al. 2007; Klopp et al. 2012). Our studies suggest that expression of CXCR1 and CXCR2 on ASC may explain their migration in response to IL-8 and CXCL1 secreted by endometrial tumor cells (Klopp et al. 2012). Additional studies are required to elucidate the complex signaling networks involved in regulating ASC migration in cancer.

15.7 The Proposed Roles of ASC in Cancer

The possible functional role of circulating progenitor cells from WAT in the context of cancer has been evaluated in mouse models. In the pioneering study, mice expressing green fluorescence protein (GFP) have been used to demonstrate the positive effect of recruited ASC on tumor growth (Zhang et al. 2009). It has been shown that ASC incorporate into tumor vasculature likely by providing support to vasculature integrity, as well as contributing to ECM remodeling, hence enabling increased proliferation and survival of malignant cells (Zhang et al. 2009; Klopp et al. 2012). While a number of independent studies have confirmed this observation (Kidd et al. 2012; Klopp et al. 2012), the effect of ASC on tumor growth appears to depend on the tumor model used (Ra et al. 2011). The current, largely hypothetical, mechanisms through which ASC could contribute to tumor growth are summarized in Fig. 15.2.

Studies from several groups have suggested that ASC may serve as cancer-associated fibroblasts and contribute to "reactive stroma" (Kolonin et al. 2012). Studies on fibroblasts from cancerous and noncancerous tissues revealed remarkable differences (Olumi et al. 1999; Mueller and Fusenig 2004). In normal physiological condition, fibroblasts exhibit low metabolic activity with minimal proliferative capacity (Xouri and Christian 2010). However, cancer-associated fibroblasts (CAF) acquire pronounced contractile features and express higher amount of chemokines, matrix metalloproteinases (MMPs), and ECM components (Mueller and Fusenig 2004). Similar to myofibroblasts, CAF express increased level of α-Smooth Muscle Actin (α-SMA), desmin, fibroblast activation protein (FAP), and intermediate filament Vimentin (Xouri and Christian 2010; Kolonin et al. 2012). The active contribution of CAF to cancer progression has been attributed to them providing supportive structural foundation, serving as additional oncogenic signals by secreting and modulating the ECM components that foster cancer cell proliferation and invasion (Karnoub et al. 2007). In mouse models, potent tumor-promoting properties have been demonstrated for CAF when those were mixed with cancer cells or initiated nontumorigenic epithelial cells (Olumi et al. 1999; Karnoub et al. 2007). Furthermore, CAF appear to confer chemoresistance to

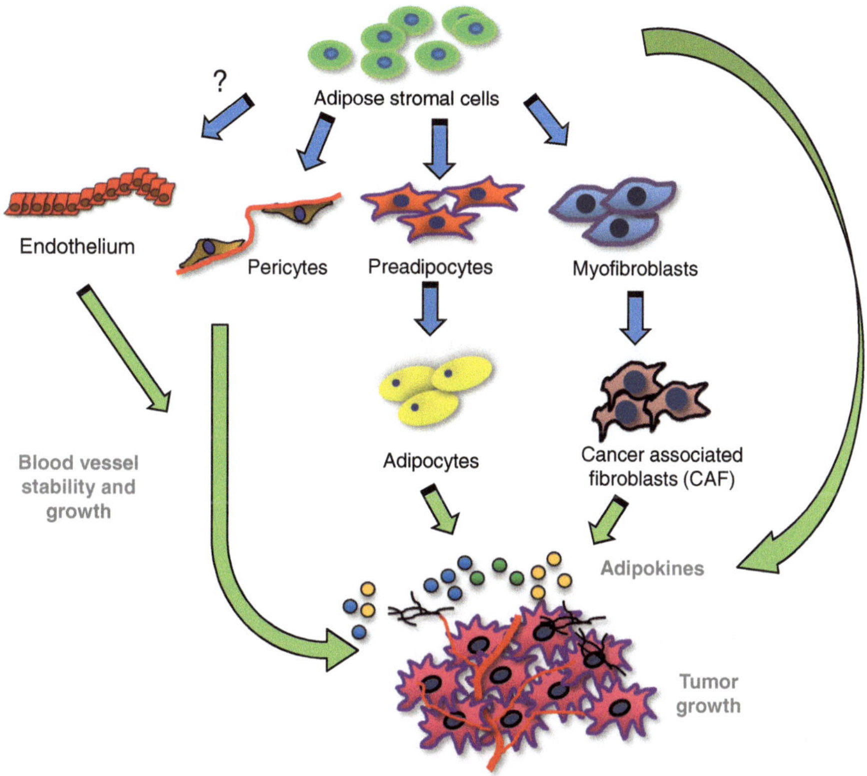

Fig. 15.2 Proposed tumor-promoting functions of ASC-derived cells. ASC contribute to different cell lineages in response to signaling cues. The endothelial and pericytes enhance tumor vasculature development through direct incorporation into lumen of tumor vessel or/and provide growth factors. The adipocytes and CAF secret pro-inflammatory cytokines and growth factors and modulate tumor ECM. The collective growth-stimulating and anti-apoptotic effect of these cells promote tumor development

tumors through decreasing uptake of chemotherapeutic drugs (Loeffler et al. 2006) and inducing nuclear factor-κB (NF-κB)-regulated anti-apoptotic cascades (Loeffler et al. 2006). Although the ontogeny of CAF remains controversial, as shown in Fig. 15.3, they may originate from several cell types, including normal tissue-resident fibroblasts and infiltrating monocytes (Cirri and Chiarugi 2011; Kolonin et al. 2012). In response to tumor-secreted soluble factors such as TGFβ (Kalluri and Weinberg 2009), tissue-resident cells may also undergo the epithelial-mesenchymal transition (EMT) (Kalluri and Zeisberg 2006; Kalluri and Weinberg 2009; Kolonin et al. 2012) or endothelial-mesenchymal transition (Cirri and Chiarugi 2011) and contribute to the CAF pool. Based on studies by us and others, tumor myofibroblasts origination from ASC has been proposed (Zhang et al. 2009; Kidd et al. 2012; Klopp et al. 2012; Kolonin et al. 2012). As further evidence of ASC serving as CAF, upon tumor homing ASC acquire myofibroblast phenotype

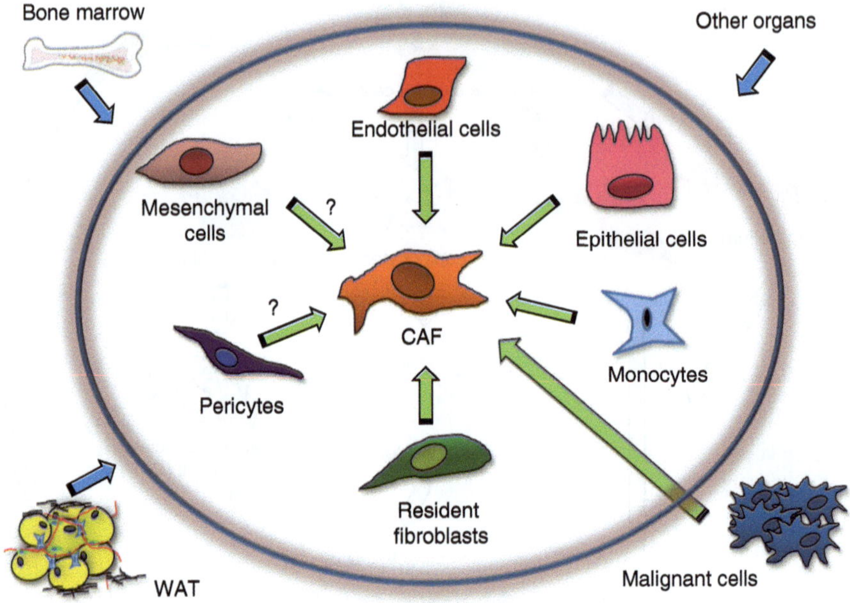

Fig. 15.3 Possible origins of cancer-associated fibroblasts. CAF may be derived from distinct cells recruited from different organs (bone marrow, WAT, and others), as well as from tumor cells by trans-differentiation processes: epithelial-mesenchymal transition and endothelial-mesenchymal transition

with increased contractility, as well as higher expression of α-SMA and ECM proteins that modulate tumor rigidity (Chandler et al. 2012; Kidd et al. 2012). While animal model data continues to build, the extent of ASC contribution to the CAF pool in different types/stages of cancer in patients remains unknown.

15.8 ASC and Tumor Vasculature

Tumor growth and progression relies on its vasculature that enables efficient oxygenation, nutrient transport, removal of waste, and recruitment of cells shaping tumor microenvironment (Hanahan and Weinberg 2011). Blood vessels are composed of endothelial cells as well as smooth muscle cells and pericytes wrapping around the endothelium and adding stability to the vasculature (Gerhardt and Semb 2008). However, tumor vascular trees are often tortuous, flawed, and leaky with abnormalities in the arrangement of EC, pericytes, and the basement membrane (Jain 2005). As a result of compromised perfusion, delivery of drugs to tumor cells is often jeopardized (Jain 2005).

The ability of WAT-derived cells to stimulate tumor neovascularization has been supported by a number of recent studies (Zhang et al. 2009; Kidd et al. 2012; Klopp et al. 2012). While adipose EC can directly incorporate into the lumen of tumor vasculature (Zhang et al. 2009), ASC acquire perivascular localization in tumors. Circulation of EC is increased in cancer (Asahara et al. 2011); however, the origin and identification of endothelial progenitor cells remain elusive (Timmermans et al. 2009). WAT has been proposed as one of their sources (Zhang et al. 2009; Kidd et al. 2012; Klopp et al. 2012). Recent studies demonstrate association of obesity with increased circulation of progenitors with ASC properties, suggesting that WAT may serve a source of cells contributing to pericytes in tumor vasculature. ASC secrete angiogenic factors such as HGF, VEGF, and fibroblast growth factor known to promote blood vessel formation (Traktuev et al. 2008), which is expected to enhance survival, proliferation, and sprouting of EC in tumors. The apparent vasculogenic role of ASC is possibly their most important cancer-promoting attribute. Other inherent characteristics of ASC, such as immunomodulation and adipocyte differentiation, may also contribute to cancer progression.

15.9 ASC Targeting as an Approach to Cancer Therapy

As discussed above, the interplay of ASC with tumor components appears to be critical for cancer progression. The paracrine ASC signaling is likely to be an important mechanistic contributor to obesity effect on cancer. Accumulating data indicate an essential role of ASC in modulating various aspects of cancer progression, including metastasis (Kidd et al. 2012; Klopp et al. 2012). As shown in Fig. 15.2, ASC may contribute to tumor ECM, stroma, and mediate diverse effects such as providing growth stimulatory signals, as well as activate immunomodulatory and pro-angiogenic cascades (Zhang et al. 2010; Eto et al. 2011). The dependence of tumor on its stromal microenvironment makes targeting stroma a potentially attractive therapy (Mueller and Fusenig 2004). Direct targeting of stromal cell recruitment factors, such as PDGFRα (Anderberg et al. 2009) and PDGFRβ (Cooke et al. 2012), results in reduced tumor growth. Consistently, genetic ablation of PDGFRβ or use of tyrosine kinase inhibitors results in smaller tumors.

Because ASC appear to serve as CAF (Fig. 15.3), some of these effects could be due to ASC inhibition. Currently, there are no tools to specifically deplete ASC and their development is much anticipated. Specific tissue-homing peptide coupled to pro-apoptotic peptide such as $(KLAKLAK)_2$ have been used to target WAT endothelium (Kolonin et al. 2004). Similar approaches, based on ASC markers being uncovered (Daquinag et al. 2011a), could potentially be developed for targeted depletion of ASC in WAT or/and in tumors. As chemotaxis appears to be involved in ASC migration (Klopp et al. 2012), their modulation by pharmacological agents could be employed to disrupt ASC tumor recruitment. Inhibition of key chemokine gradients has been shown to inhibit cancer progression (Phillips et al. 2003; Kucia et al. 2004; Acharyya et al. 2012). While inhibitory effects have been attributed to suppressed migration of cancer cells, complementary incapacitation of tumor microenvironment components, such as ASC, has not been ruled out. In accordance with the proposed involvement of CXCL1/IL-8 in ASC migration (Klopp et al. 2012), interfering with the corresponding chemokine axis by specific inhibitors might be sought to inactivate ASC trafficking. However, as most known chemokine gradients are implicated in trafficking of several cell populations, their inhibition is not expected to be ASC-specific. In addition, a cautionary note is that the concept of ASC targeting may represent a double-edged sword for cancer patients. As turns out to be the case for vascular disrupting agents, crippling tumor vasculature potentially makes chemotherapy less efficient (Jain 1997, 2005) and a similar scenario is expected for ASC targeting. Another arising concern is increased cancer aggressiveness potentially resulting from tumor vasculature inhibition (Kerbel 2008; Kerbel et al. 2008). Moreover, increased metastatic dissemination has been associated with pericyte depletion in experimental mouse models (Gerhardt and Semb 2008; Cooke et al. 2012), therefore specific caution will need to be taken for advanced cancer cases.

While depletion of ASC may have therapeutic benefits, another reason why this cell population may be relevant in cancer medicine relates to their inherent tumor-homing capacity. In this regard, the potential of ASC to serve as vehicles for agent delivery (Kucerova et al. 2007) could be exploited for therapeutic tumor targeting. The use of genetically modified bone marrow MSC expressing anti-tumor interleukins has been explored (Studeny et al. 2004; Loebinger et al. 2009) and similar approaches have shown successful for ASC (Kucerova et al. 2007). Delivery of cytotoxic agents with ASC to tumors could facilitate treatment efficacy and reduce its off-site toxicity. Because the dynamic function of ASC may be both beneficial and potentially pro-tumorigenic, special caution should be implemented with clinical use of ASC (Bertolini et al. 2012).

15.10 Conclusions and Future Directions

The studies discussed in this chapter indicate a previously overseen role of WAT in cancer development and progression. In addition to serving as a source of endocrine adipokines, WAT is a source of ASC, the migratory cells whose paracrine contribution to the tumor microenvironment appears to be important. Despite the progress in understanding of the mechanisms linking obesity and cancer via ASC trafficking, challenges and difficulties remain. This is largely due to the lack of definitive cell surface markers for the cell populations activated in cancer, as well as their plasticity and heterogeneity. At this point, the spectrum of circulating progenitor cells, their tissue origins, and their functional roles in cancer progression remain largely unclear. To overcome the current limitation, development of standard criteria for enumeration of ASC among other cell populations in WAT, peripheral circulation, and tumors is imperative. Despite these uncertainties, the accumulating data indicate that adipose progenitor cells are capable of infiltrating tumor, and promoting cancer progression, which may partially account for the association of obesity with increased cancer progression.

References

Acharyya S, Oskarsson T, Vanharanta S, Malladi S, Kim J, Morris PG, et al. A CXCL1 paracrine network links cancer chemoresistance and metastasis. Cell. 2012;150:165–78.

Aicher A, Rentsch M, Sasaki K, Ellwart JW, Fändrich F, et al. Nonbone marrow-derived circulating progenitor cells contribute to postnatal neovascularization following tissue ischemia. Circ Res. 2007;100:581–9.

Alexander NR, Branch KM, Parekh A, Clark ES, Iwueke IC, Guelcher SA, et al. Extracellular matrix rigidity promotes invadopodia activity. Curr Biol. 2008;18:1295–9.

Anderberg C, Li H, Fredriksson L, Andrae J, Betsholtz C, Li X, et al. Paracrine signaling by platelet-derived growth factor-CC promotes tumor growth by recruitment of cancer-associated fibroblasts. Cancer Res. 2009;69:369–78.

Asahara T, Kawamoto A, Masuda H. Concise review: circulating endothelial progenitor cells for vascular medicine. Stem Cells. 2011;29:1650–5.

Balkwill F. Cancer and the chemokine network. Nat Rev Cancer. 2004;4:540–50.

Balkwill FR. The chemokine system and cancer. J Pathol. 2012;226:148–57.

Basen-Engquist K, Chang M. Obesity and cancer risk: recent review and evidence. Curr Oncol Rep. 2011;13:71–6.

Bellows CF, Zhang Y, Chen J, Frazier ML, Kolonin MG. Circulation of progenitor cells in obese and lean colorectal cancer patients. Cancer Epidemiol Biomarkers Prev. 2011a;20:2461–8.

Bellows CF, Zhang Y, Simmons PJ, Khalsa AS, Kolonin MG. Influence of BMI on level of circulating progenitor cells. Obesity (Silver Spring). 2011b;19:1722–6.

Bertolini F, Lohsiriwat V, Petit JY, Kolonin MG. Adipose tissue cells, lipotransfer and cancer: a challenge for scientists, oncologists and surgeons. Biochim Biophys Acta. 2012;1826: 209–14.

Bianco P, Robey PG, Simmons PJ. Mesenchymal stem cells: revisiting history, concepts, and assays. Cell Stem Cell. 2008;2:313–9.

Blogowski W, Ratajczak MZ, Zyzniewska-Banaszak E, Dolegowska B, Starzynska T. Adipose tissue as a potential source of hematopoietic stem/progenitor cells. Obesity (Silver Spring). 2012;20:923–31.

Cai L, Johnstone BH, Cook TG, Tan J, Fishbein MC, Chen PS, et al. IFATS collection: human adipose tissue-derived stem cells induce angiogenesis and nerve sprouting following myocardial infarction, in conjunction with potent preservation of cardiac function. Stem Cells. 2009;27:230–7.

Calle EE, Kaaks R. Overweight, obesity and cancer: epidemiological evidence and proposed mechanisms. Nat Rev Cancer. 2004;4:579–91.

Cawthorn WP, Scheller EL, Macdougald OA. Adipose tissue stem cells: the great WAT hope. Trends Endocrinol Metab. 2012;23(6):270–7.

Chandler EM, Seo BR, Califano JP, Andresen Eguiluz RC, Lee JS, Yoon CJ, et al. Implanted adipose progenitor cells as physicochemical regulators of breast cancer. Proc Natl Acad Sci U S A. 2012;109:9786–91.

Cinti S. The adipose organ. Prostaglandins Leukot Essent Fatty Acids. 2005;73:9–15.

Cinti S. Between brown and white: novel aspects of adipocyte differentiation. Ann Med. 2011;43:104–15.

Cirri P, Chiarugi P. Cancer associated fibroblasts: the dark side of the coin. Am J Cancer Res. 2011;1:482–97.

Cooke VG, LeBleu VS, Keskin D, Khan Z, O'Connell JT, Teng Y, et al. Pericyte depletion results in hypoxia-associated epithelial-to-mesenchymal transition and metastasis mediated by met signaling pathway. Cancer Cell. 2012;21:66–81.

Corselli M, Chen CW, Sun B, Yap S, Rubin JP, Peault B. The tunica adventitia of human arteries and veins as a source of mesenchymal stem cells. Stem Cells Dev. 2011;21:1299–308.

Curtis JP, Selter JG, Wang Y, Rathore SS, Jovin IS, Jadbabaie F, et al. The obesity paradox: body mass index and outcomes in patients with heart failure. Arch Intern Med. 2005;165:55–61.

Dalamaga M, Diakopoulos KN, Mantzoros CS. The role of adiponectin in cancer: a review of current evidence. Endocr Rev. 2012;33(4):547–94.
Daquinag AC, Zhang Y, Amaya-Manzanares F, Simmons PJ, Kolonin MG. An isoform of decorin is a resistin receptor on the surface of adipose progenitor cells. Cell Stem Cell. 2011a;9:74–86.
Daquinag AC, Zhang Y, Kolonin MG. Vascular targeting of adipose tissue as an anti-obesity approach. Trends Pharmacol Sci. 2011b;32:300–7.
Eto H, Suga H, Inoue K, Aoi N, Kato H, Araki J, et al. Adipose injury-associated factors mitigate hypoxia in ischemic tissues through activation of adipose-derived stem/progenitor/stromal cells and induction of angiogenesis. Am J Pathol. 2011;178:2322–32.
Fenton JI, Birmingham JM. Adipokine regulation of colon cancer: adiponectin attenuates interleukin-6-induced colon carcinoma cell proliferation via STAT-3. Mol Carcinog. 2010;49:700–9.
Frantz C, Stewart KM, Weaver VM. The extracellular matrix at a glance. J Cell Sci. 2010;123:4195–200.
Frayn KN, Karpe F, Fielding BA, Macdonald IA, Coppack SW. Integrative physiology of human adipose tissue. Int J Obes Relat Metab Disord. 2003;27:875–88.
Gerhardt H, Semb H. Pericytes: gatekeepers in tumour cell metastasis? J Mol Med (Berl). 2008;86:135–44.
Gesta S, Tseng YH, Kahn CR. Developmental origin of fat: tracking obesity to its source. Cell. 2007;131:242–56.
Gimble JM, Bunnell BA, Chiu ES, Guilak F. Concise review: Adipose-derived stromal vascular fraction cells and stem cells: let's not get lost in translation. Stem Cells. 2011;29:749–54.
Grossmann ME, Ray A, Nkhata KJ, Malakhov DA, Rogozina OP, Dogan S, et al. Obesity and breast cancer: status of leptin and adiponectin in pathological processes. Cancer Metastasis Rev. 2010;29:641–53.
Hanahan D, Weinberg RA. Hallmarks of cancer: the next generation. Cell. 2011;144:646–74.
Henegar C, Tordjman J, Achard V, Lacasa D, Cremer I, Guerre-Millo M, et al. Adipose tissue transcriptomic signature highlights the pathological relevance of extracellular matrix in human obesity. Genome Biol. 2008;9:R14.
Jain RK. The Eugene M. Landis Award Lecture 1996. Delivery of molecular and cellular medicine to solid tumors. Microcirculation. 1997;4:1–23.
Jain RK. Normalization of tumor vasculature: an emerging concept in antiangiogenic therapy. Science. 2005;307:58–62.
Joe AW, Yi L, Even Y, Vogl AW, Rossi FM. Depot-specific differences in adipogenic progenitor abundance and proliferative response to high-fat diet. Stem Cells. 2009;27:2563–70.
Kalluri R, Weinberg RA. The basics of epithelial-mesenchymal transition. J Clin Invest. 2009;119:1420–8.
Kalluri R, Zeisberg M. Fibroblasts in cancer. Nat Rev Cancer. 2006;6:392–401.
Karnoub AE, Dash AB, Vo AP, Sullivan A, Brooks MW, Bell GW, et al. Mesenchymal stem cells within tumour stroma promote breast cancer metastasis. Nature. 2007;449:557–63.
Kelesidis I, Kelesidis T, Mantzoros CS. Adiponectin and cancer: a systematic review. Br J Cancer. 2006;94:1221–5.
Kerbel RS. Tumor angiogenesis. N Engl J Med. 2008;358:2039–49.
Kerbel RS, Benezra R, Lyden DC, Hattori K, Heissig B, Nolan DJ, et al. Endothelial progenitor cells are cellular hubs essential for neoangiogenesis of certain aggressive adenocarcinomas and metastatic transition but not adenomas. Proc Natl Acad Sci U S A. 2008;105:E54–5.
Khandekar MJ, Cohen P, Spiegelman BM. Molecular mechanisms of cancer development in obesity. Nat Rev Cancer. 2011;11:886–95.
Kidd S, Spaeth E, Watson K, Burks J, Lu H, Klopp A, et al. Origins of the tumor microenvironment: quantitative assessment of adipose-derived and bone marrow-derived stroma. PLoS One. 2012;7:e30563.

Klopp AH, Zhang Y, Solley T, Amaya-Manzanares F, Marini F, Andreeff M, et al. Omental adipose tissue-derived stromal cells promote vascularization and growth of endometrial tumors. Clin Cancer Res. 2012;18:771–82.

Kolonin MG, Simmons PJ. Combinatorial stem cell mobilization. Nat Biotechnol. 2009;27:252–3.

Kolonin MG, Saha PK, Chan L, Pasqualini R, Arap W. Reversal of obesity by targeted ablation of adipose tissue. Nat Med. 2004;10:625–32.

Kolonin MG, Evans KW, Mani SA, Gomer RH. Alternative origins of stroma in normal organs and disease. Stem Cell Res. 2012;8:312–23.

Kucerova L, Altanerova V, Matuskova M, Tyciakova S, Altaner C. Adipose tissue-derived human mesenchymal stem cells mediated prodrug cancer gene therapy. Cancer Res. 2007;67: 6304–13.

Kucia M, Jankowski K, Reca R, Wysoczynski M, Bandura L, Allendorf DJ, et al. CXCR4-SDF-1 signalling, locomotion, chemotaxis and adhesion. J Mol Histol. 2004;35:233–45.

Laird DJ, von Andrian UH, Wagers AJ. Stem cell trafficking in tissue development, growth, and disease. Cell. 2008;132:612–30.

Levental KR, Yu H, Kass L, Lakins JN, Egeblad M, Erler JT, et al. Matrix crosslinking forces tumor progression by enhancing integrin signaling. Cell. 2009;139:891–906.

Li D, Morris JS, Liu J, Hassan MM, Day RS, Bondy ML, et al. Body mass index and risk, age of onset, and survival in patients with pancreatic cancer. JAMA. 2009;301:2553–62.

Lim S, Honek J, Xue Y, Seki T, Cao Z, Andersson P, et al. Cold-induced activation of brown adipose tissue and adipose angiogenesis in mice. Nat Protoc. 2012;7:606–15.

Loebinger MR, Eddaoudi A, Davies D, Janes SM. Mesenchymal stem cell delivery of TRAIL can eliminate metastatic cancer. Cancer Res. 2009;69:4134–42.

Loeffler M, Kruger JA, Niethammer AG, Reisfeld RA. Targeting tumor-associated fibroblasts improves cancer chemotherapy by increasing intratumoral drug uptake. J Clin Invest. 2006;116:1955–62.

Mariman EC, Wang P. Adipocyte extracellular matrix composition, dynamics and role in obesity. Cell Mol Life Sci. 2010;67:1277–92.

Maumus M, Sengenes C, Decaunes P, Zakaroff-Girard A, Bourlier V, Lafontan M, et al. Evidence of in situ proliferation of adult adipose tissue-derived progenitor cells: influence of fat mass microenvironment and growth. J Clin Endocrinol Metab. 2008;93:4098–106.

Mueller MM, Fusenig NE. Friends or foes—bipolar effects of the tumour stroma in cancer. Nat Rev Cancer. 2004;4:839–49.

Mullen JT, Moorman DW, Davenport DL. The obesity paradox: body mass index and outcomes in patients undergoing nonbariatric general surgery. Ann Surg. 2009;250:166–72.

Nauta AJ, Fibbe WE. Immunomodulatory properties of mesenchymal stromal cells. Blood. 2007;110:3499–506.

Nieman KM, Kenny HA, Penicka CV, Ladanyi A, Buell-Gutbrod R, Zillhardt MR, et al. Adipocytes promote ovarian cancer metastasis and provide energy for rapid tumor growth. Nat Med. 2011;17:1498–503.

Olumi AF, Grossfeld GD, Hayward SW, Carroll PR, Tlsty TD, Cunha GR. Carcinoma-associated fibroblasts direct tumor progression of initiated human prostatic epithelium. Cancer Res. 1999;59:5002–11.

Park EJ, Lee JH, Yu GY, He G, Ali SR, Holzer RG, et al. Dietary and genetic obesity promote liver inflammation and tumorigenesis by enhancing IL-6 and TNF expression. Cell. 2010;140: 197–208.

Phillips RJ, Burdick MD, Lutz M, Belperio JA, Keane MP, Strieter RM. The stromal derived factor-1/CXCL12-CXC chemokine receptor 4 biological axis in non-small cell lung cancer metastases. Am J Respir Crit Care Med. 2003;167:1676–86.

Pitchford SC, Furze RC, Jones CP, Wengner AM, Rankin SM. Differential mobilization of subsets of progenitor cells from the bone marrow. Cell Stem Cell. 2009;4:62–72.

Ponte AL, Marais E, Gallay N, Langonne A, Delorme B, Herault O, et al. The in vitro migration capacity of human bone marrow mesenchymal stem cells: comparison of chemokine and growth factor chemotactic activities. Stem Cells. 2007;25:1737–45.

Puissant B, Barreau C, Bourin P, Clavel C, Corre J, Bousquet C, et al. Immunomodulatory effect of human adipose tissue-derived adult stem cells: comparison with bone marrow mesenchymal stem cells. Br J Haematol. 2005;129:118–29.

Ra JC, Shin IS, Kim SH, Kang SK, Kang BC, Lee HY, et al. Safety of intravenous infusion of human adipose tissue-derived mesenchymal stem cells in animals and humans. Stem Cells Dev. 2011;20:1297–308.

Rehman J, Traktuev D, Li J, Merfeld-Clauss S, Temm-Grove CJ, Bovenkerk JE, et al. Secretion of angiogenic and antiapoptotic factors by human adipose stromal cells. Circulation. 2004;109:1292–8.

Renehan AG, Tyson M, Egger M, Heller RF, Zwahlen M. Body-mass index and incidence of cancer: a systematic review and meta-analysis of prospective observational studies. Lancet. 2008;371:569–78.

Sirin O, Kolonin MG. Treatment of obesity as a potential complementary approach to cancer therapy. Drug Discov Today. 2013;18(11–12):567–73.

Smith MM, Minson CT. Obesity and adipokines: effects on sympathetic overactivity. J Physiol. 2012;590:1787–801.

Song S, Ewald AJ, Stallcup W, Werb Z, Bergers G. PDGFRbeta+ perivascular progenitor cells in tumours regulate pericyte differentiation and vascular survival. Nat Cell Biol. 2005;7:870–9.

Studeny M, Marini FC, Dembinski JL, Zompetta C, Cabreira-Hansen M, Bekele BN, et al. Mesenchymal stem cells: potential precursors for tumor stroma and targeted-delivery vehicles for anticancer agents. J Natl Cancer Inst. 2004;96:1593–603.

Tang W, Zeve D, Suh JM, Bosnakovski D, Kyba M, Hammer RE, et al. White fat progenitor cells reside in the adipose vasculature. Science. 2008;322:583–6.

Thangarajah H, Vial IN, Chang E, El-Ftesi S, Januszyk M, Chang EI, et al. IFATS collection: adipose stromal cells adopt a proangiogenic phenotype under the influence of hypoxia. Stem Cells. 2009;27:266–74.

Timmermans F, Plum J, Yoder MC, Ingram DA, Vandekerckhove B, Case J. Endothelial progenitor cells: identity defined? J Cell Mol Med. 2009;13:87–102.

Traktuev DO, Merfeld-Clauss S, Li J, Kolonin M, Arap W, Pasqualini R, et al. A population of multipotent CD34-positive adipose stromal cells share pericyte and mesenchymal surface markers, reside in a periendothelial location, and stabilize endothelial networks. Circ Res. 2008;102:77–85.

Trayhurn P. Adipocyte biology. Obes Rev. 2007;8 Suppl 1:41–4.

Ulrich TA, de Juan Pardo EM, Kumar S. The mechanical rigidity of the extracellular matrix regulates the structure, motility, and proliferation of glioma cells. Cancer Res. 2009;69:4167–74.

Virtanen KA, Lidell ME, Orava J, Heglind M, Westergren R, Niemi T, et al. Functional brown adipose tissue in healthy adults. N Engl J Med. 2009;360:1518–25.

Vitali A, Murano I, Zingaretti MC, Frontini A, Ricquier D, Cinti S. The adipose organ of obesity-prone C57BL/6J mice is composed of mixed white and brown adipocytes. J Lipid Res. 2012;53:619–29.

Von Luttichau I, Notohamiprodjo M, Wechselberger A, Peters C, Henger A, Seliger C, et al. Human adult CD34- progenitor cells functionally express the chemokine receptors CCR1, CCR4, CCR7, CXCR5, and CCR10 but not CXCR4. Stem Cells Dev. 2005;14:329–36.

von Tell D, Armulik A, Betsholtz C. Pericytes and vascular stability. Exp Cell Res. 2006;312:623–9.

Wu J, Bostrom P, Sparks LM, Ye L, Choi JH, Giang AH, et al. Beige adipocytes are a distinct type of thermogenic fat cell in mouse and human. Cell. 2012;150:366–76.

Xouri G, Christian S. Origin and function of tumor stroma fibroblasts. Semin Cell Dev Biol. 2010;21:40–6.

Yang X, Smith U. Adipose tissue distribution and risk of metabolic disease: does thiazolidinedione-induced adipose tissue redistribution provide a clue to the answer? Diabetologia. 2007;50:1127–39.

Zhang Y, Daquinag A, Traktuev DO, Amaya-Manzanares F, Simmons PJ, March KL, et al. White adipose tissue cells are recruited by experimental tumors and promote cancer progression in mouse models. Cancer Res. 2009;69:5259–66.

Zhang Y, Bellows CF, Kolonin MG. Adipose tissue-derived progenitor cells and cancer. World J Stem Cells. 2010;2:103–13.

Zyromski NJ, Mathur A, Pitt HA, Wade TE, Wang S, Nakshatri P, et al. Obesity potentiates the growth and dissemination of pancreatic cancer. Surgery. 2009;146:258–63.

Chapter 16
Adipose-Derived Endothelial Precursor Cells Supporting Tumor Growth

Patrizia Mancuso, Ines Martin-Padura, and Francesco Bertolini

Abstract It has been recently found that the human white adipose tissue (WAT), similarly to neoplasia, shows in vivo a robust angiogenic switch when the growth rate exceeds a given expansion threshold. When compared to the bone marrow (BM), human WAT contains significantly more $CD45^{-}CD34^{+}$ progenitors that express high levels of angiogenesis-related genes and can generate in culture endothelial cells and tubes as efficiently as mesenchymal cells. We recently reported that human WAT $CD45^{-}CD34^{+}$ progenitors obtained from lipotransfer procedures contributed to tumor vascularization and significantly increased tumor growth and metastases in several orthotopic models of human breast cancer. In another recent study, 321 consecutive patients operated for primary breast cancer who subsequently underwent a lipotransfer procedure were compared with two matched patients with similar characteristics who did not undergo lipotransfer. In this retrospective study, the lipotransfer group exhibited a higher risk of local events compared to the controls. A second data revision after prolonged follow-up confirmed this significant difference. It seems urgent to clarify which WAT cell populations can be used safely for tissue/organ reconstruction in cancer patients and what have the potential for reactivating dormant tumor cells.

Keywords Cancer • Angiogenesis • Adipose tissue • Progenitor cell • Endothelial cells

P. Mancuso • I. Martin-Padura • F. Bertolini, M.D., Ph.D. (✉)
Laboratory of Hematology-Oncology, European Institute of Oncology,
via Ripamonti 435, 20141 Milan, Italy
e-mail: francesco.bertolini@ieo.it

Y. Cao (ed.), *Angiogenesis in Adipose Tissue*, DOI 10.1007/978-1-4614-8069-3_16,

16.1 Introduction

The role of bone marrow (BM)-derived endothelial progenitor cells (EPCs) in cancer growth has been intensively debated in the last decade. Donor-derived endothelial cells have been found, albeit in limited number, in patients who received allogeneic BM transplants. Some investigators found a crucial and quantitatively relevant role for BM-derived vessels in the early phases of cancer growth in some preclinical models of neoplasia; some others failed to find any significant role for BM-derived EPCs in cancer vessels and in cancer development in several other preclinical models (reviewed in Bertolini et al. 2009). One study has recently described that EPCs are present in tissues other than the BM, and in particular in the adipose tissue of mice (Grenier et al. 2007). Along this line, we have recently described that the human WAT is a very rich reservoir of $CD45^{-}CD34^{+}$ EPCs, and investigated their role in several in vivo models of breast cancer growth and metastases (Martin-Padura et al. 2012).

16.2 $CD34^{+}$ Cell Purification from Lipotransfer Samples

Human WAT samples were obtained from lipotransfer procedures for breast reconstruction in breast cancer patients who signed an informed consent. Most of these procedures involved WAT collection from the abdomen.

Stromal-vascular cell fractions were obtained after collagenase digestion (HBSS, Gibco) containing 2 mg/mL of collagenase type I (Sigma Aldrich) and 3.5 % bovine serum albumin (BSA; Sigma Aldrich) at 37 °C with constant shaking for 60 min. The digestion was blocked with RPMI 1640 supplemented by 20 % FBS (Euroclone), 0.3 and a cell pellet was obtained by centrifugation at 200 × *g* for 10 min at 4 °C (Lohsiriwat et al. 2011; Petit et al. 2010; Sengenès et al. 2005).

$CD34^{+}$ cells were purified from WAT samples by means of anti-CD34 microbeads (Miltenyi Biotec) according to the manufacturer's instructions. Final $CD34^{+}$ cell purity was evaluated by flow cytometry and found in each instance to be more than 95 %. Overall, 113 human WAT samples were studied in vitro, and 38 of these were used to provide $CD34^{+}$ cells that were used in 124 different individual mouse studies. $CD34^{+}$ cells obtained from WAT were investigated by flow cytometry, gene expression analysis, culture, and in vivo studies.

16.3 Flow Cytometry Studies Showed That WAT Are Very Rich Reservoirs of $CD45^{-}CD34^{+}$ Cells

By means of six-color flow cytometry and following an approach recently validated for the quantification of circulating EPCs and perivascular progenitors (Mancuso et al. 2009, 2011), we enumerated the numbers of EPCs and other subsets of progenitors in the bone marrow and in WAT of humans. In Table 16.1 the list of

Table 16.1 Monoclonal antibodies used in the study

MoAb	Fluorochrome	Clone	Supplier
CD45	APC-Cy7	2D1	BD
CD34	APC	8G12	BD
	Pe-Cy7		
CD31	Pe-Cy7	L133.1	BD
CD13	APC	WM15	BD
CD10	APC	H10a	BD
CD140b	PE	28D4	BD
CD29	PE	MAR4	BD
CD90	PE	5E10	BD
VEGFR-2	PE	89106	R&D
VEGFR-3	PE	54733	R&D
CD44	APC	BJ18	Bio-Legend
CD144	PE	TEA 1/31	Beckman-Coulter
7-AAD			Sigma
Syto16	FITC		Invitrogen

monoclonal antibodies used is reported. The nuclear staining Syto16 was used to discriminate between DNA containing cells, platelets, and cell debris, while 7-amino-actinomycin D (7-AAD) to determine the viability status of the cells. The absolute count of $CD45^-CD34^+$ cells was obtained using reference beads in Trucount tubes (BD).

WAT was found to contain a large amount of $CD45^-CD34^+$ cells that fulfill the most recent criteria for EPC identification (Hirschi et al. 2008; Yoder and Ingram 2009; Bertolini et al. 2009). These $CD45^-CD34^+$ cells included two subpopulations: $CD34^{++}CD13^+CD140b^+CD44^+CD90^{++}$ cells and $CD34^+CD31^+CD140b^-$ cells (Fig. 16.1a–j). The immunomagnetic purification procedure used in the study led to a cell population which included 79–96 % of $CD45^-CD34^{++}CD13^+CD140b^+CD44^+CD90^{++}$ cells and 2–18 % of $CD45^-CD34^+CD31^+CD105^+$ cells. $CD34^-$ cells always made up less than 5 % of the purified cell population.

Quantitative studies showed that WAT contains about 263-fold more $CD45^-CD34^+$ EPCs/mL than bone marrow obtained from disease-free patients undergoing a follow-up involving bone marrow investigation for hematologic or solid neoplasia ($n=32$). In particular, median WAT $CD45^-CD34^+CD31^-$ cells were 181,046/mL (range, 35,970–465,357) and $CD45^-CD34^+CD31^+$ cells were 76,946/mL (range, 13,982–191,287). Correlations were found between the body mass index and total $CD34^+$ cells ($r=0.608$, $P<0.001$) and between WAT donor age and total $CD34^+$ cells ($r=0.387$).

We also found $CD45^+CD34^+$ hematopoietic progenitor cells in WAT (median 6,141/mL, range <1–161,338). On average, WAT contained 87 times less $CD45^+CD34^+$ hematopoietic progenitor cells/mL than did bone marrow. The presence of hematopoietic progenitors in WAT was confirmed by culture studies, where a mean of 1.4 ± 0.2 granulocyte macrophage colony-forming units and 3.4 ± 1.8 BFU-E/10^5 seeded cells were obtained from WAT tissue.

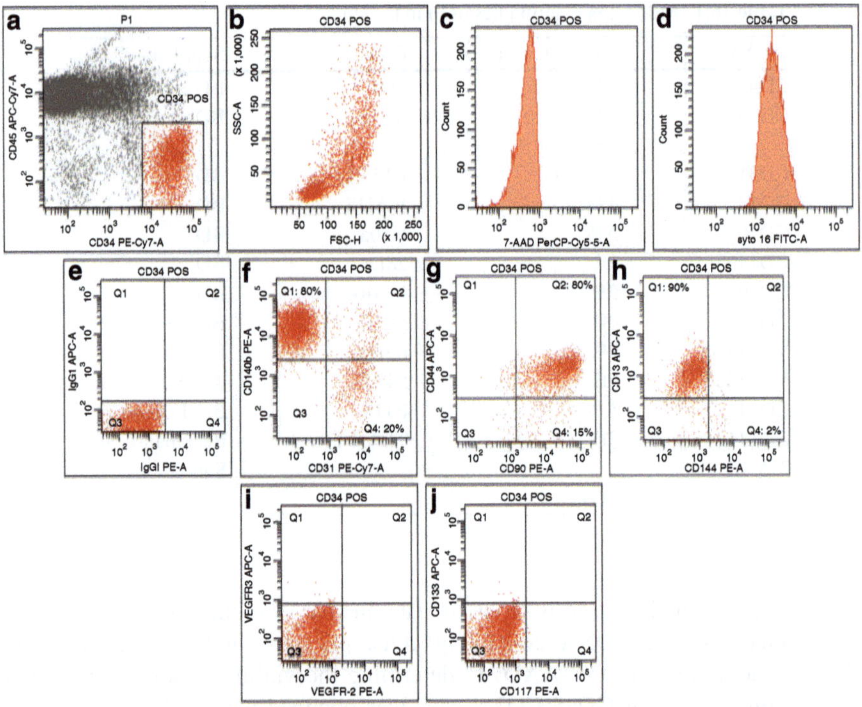

Fig. 16.1 Flow cytometry evaluation and enumeration of CD45⁻CD34⁺ cells in human WAT. Representative evaluation of CD45⁻CD34⁺ cells in human WAT tissue from lipotransfer procedures. (**a–d**) The gate used to investigate CD34⁺CD45⁻ cells, DNA (Syto16)+ viable (7-AAD−) cells. (**e**) The negative controls. (**f–j**) The expression of CD31, CD140b, CD44, CD90, CD13, CD144, VEGFR-2, VEGFR-3, CD133, and CD117 in the DNA (Syto16)+ viable (7-AAD−) CD34⁺CD45⁻ WAT cells

16.4 WAT CD34⁺ Cells Express Stemness-Related Genes and Very High Levels of Angiogenesis-Related Genes

CD34⁺ cells from human WAT were purified and their gene expression profile compared (by Affymetrix human gene 1.0 ST) with that of purified bone marrow CD34⁺ cells mobilized in the blood of healthy donors by G-CSF administration ($n=5$ per study arm). Data were confirmed by qRT-PCR (Martin-Padura et al. 2012).

RNA isolation was carried out using QIAmp RNA Blood Mini Kit (Qiagen) and cDNA was generated from 40 ng of RNA using the high-capacity cDNA reverse transcription kit (Applied Biosystems); quantitative real-time PCR (qRT-PCR) was carried out with an ABI Prism 700 platform as previously described (Rabascio et al. 2004) using primers and probes from the TaqMan Gene Expression Assay.

For microarray analysis, synthesis of labeled targets, array hybridization (Affymetrix GeneChip Gene ST 1.0 Human array; Affymetrix), staining, and scanning were carried out according to Affymetrix standard protocols, starting from 500 ng of total RNA. Duplicate microarrays were hybridized with each DNA sample. The MAS5 algorithm was used to determine the expression levels of mRNAs; the absolute analysis was carried out using default parameters and scaling factor 500. Report files were extracted for each microarray chip and performance of labeled target was evaluated on the basis of several values (e.g., scaling factor, background and noise values, percentage of present calls, and average signal value). The data were deposited at GEO (GSE31415).

When compared with bone marrow-derived CD34$^+$ cells human WAT CD34$^+$ cells expressed significantly higher levels of genes associated with angiogenesis (e.g., VEGFR-1 and -2, NRP1, TEK, VE-Cadherin, VCAM-1, and ALK1), adipogenesis (e.g., LPL, FABP4, and PPARG), and endothelial and mesenchymal (e.g., RGS5, insulin-like growth factor I, and platelet-derived growth factor receptor β) differentiation. A large majority of the panel of genes associated with stemness (e.g., SOX2, LIF, WNT3A, and Nanog) and hematopoiesis (e.g., RUNX, IL-6, and CSF-1) were expressed in WAT and bone marrow-derived CD34$^+$ cells at similar levels.

16.5 WAT CD34$^+$ Cells Generate Mature Endothelial Cells and Capillary Tubes

When cultured in vitro in appropriate endothelial-differentiation media, human WAT-derived CD34$^+$ cells generated mature endothelial cells.

In brief, cells were plated in complete EBM-2 medium (Lonza) in 12- or 24-well plates precoated with rat tail collagen I. Plates were placed in a 37 °C, 5 % CO_2 humidified incubator. The seeding density ranged from 50×10^3 to 500×10^3 cells. The presence of ECs and colonies was assessed using an inverted microscope. After 3–7 days of culture, endothelial cell colonies were identified morphologically and were subsequently picked out using cloning rings. Fibroblast contamination was avoided by depleting them from cell suspensions with the Anti-Fibroblast Microbead kit (Miltenyi). Endothelial cell surface antigen expression was assessed by flow cytometry and immunofluorescence staining of VE-Cadherin was done as previously described (Rabascio et al. 2004; Corada et al. 2001).

Endothelial capillary tubes were also generated using the appropriate culture procedure in Matrigel. Briefly, matrix solution was thawed on ice, seeded on 24-well plates, and incubated at 37 °C to solidify. ECs were harvested, resuspended in complete media, seeded at a final concentration of 5×10^4 cells per cell onto the polymerized Matrigel, and incubated at 37 °C in a tissue incubator. After 17 h, tube formation was inspected under an inverted light microscope at ×20 magnification.

16.6 The Coinjection of Human WAT CD34⁺ Cells from Lipotransfer Procedures Significantly Increases Tumor Growth and Metastases in Breast Cancer Models

To examine the in vivo involvement of $CD34^+$ WAT cells in promoting tumor growth in female NSG mice two separate studies were conducted. First of all, the role of purified $CD34^+$ cells from human WAT was investigated in bilateral and monolateral studies.

In bilateral studies 1×10^6 MDA-MB-436 or HCC1937 (triple-negative human breast cancer cell line) were coinjected with 2×10^5 human $CD34^+$ WAT cells in one of the lateral mammary fat pads of NSG mice ($N=124$), with the contralateral mammary fat pad injected with breast cancer cells alone as control (Shultz et al. 2007; Agliano et al. 2008).

In monolateral studies a total of 1×10^6 MDA-MB-436 or HCC1937 cells were injected into the right fourth mammary fat pad and 1×10^6 MDA-MB-436 or HCC1937 cells were coinjected with 2×10^5 human $CD34^+$ WAT cells into the left fourth mammary fat pad of the same mouse.

In both sets of studies, tumors were measured weekly using digital calipers and tumor volume was calculated according to the formula: $L \times W^2/2 = \text{mm}^3$.

Breast cancer cells generated tumors in the mammary fat pad. Purified human $CD34^+$ WAT cells, when injected in the mammary fat pad in the absence of tumor cells, did not generate tumors. The coinjection of breast cancer cells and unprocessed nucleated cells from human WAT significantly increased tumor growth. The coinjection of breast cancer cells and purified $CD34^+$ WAT cells increased tumor growth to a similar extent, suggesting that the large majority of the tumor-promoting activity in human WAT cells reside in the $CD34^+$ WAT cell fraction. The coinjection of $CD34^-$ WAT cells was less effective than the coinjection of $CD34^+$ WAT cells in promoting tumor growth.

Tumor growth was slightly (albeit not significantly) higher in bilateral studies compared with monolateral studies. These data suggest that human WAT $CD34^+$ cells exert most (if not all) of their tumor-promoting activity locally and not via soluble factors that are released in circulation, which would also have promoted the growth of tumors in the opposite mammary fat pad that was not coinjected with human WAT $CD34^+$ cells.

Similar results were obtained in NSG mice injected with HCC1937 breast cancer cells alone, or in combination with human $CD34^+$ WAT cells. Histology studies ruled out the possibility that the larger size of tumors in animals coinjected with $CD34^+$ WAT cells was due to the generation of adipocytes.

In this model of breast cancer, lung metastases were observed around day 70. The number of lung metastases was significantly increased in mice coinjected with breast cancer and $CD34^+$ WAT cells compared with mice injected with breast cancer cells alone or mice injected with breast cancer cells and $CD34^-$ WAT cells.

In another model of breast cancer metastasis, MDA-MB-436 breast cancer cells were injected into the mammary fat pad of NSG mice to produce orthotopic primary tumors. When the tumor size was 200–250 mm^3, that is, about 45 days after tumor implant, the tumor was surgically removed. Mice were then divided into an experimental group in which $CD34^+$ WAT cells were injected and a control group without WAT cell injection. After 2 months, mice injected with $CD34^+$ WAT cells had significantly more axillary and lung metastases, compared with mice injected with $CD34^-$ cells and controls.

Immunohistochemistry and confocal/Z-stack microscopy studies showed the presence of human $CD31^+$, $CD34^+$, $CD105^+$ endothelial vessels, and perivascular cells in the mammary fat pad and in tumors of mice coinjected with breast cancer cells and $CD34^+$ human WAT cells. Confocal microscopy confirmed the presence of a lumen and of circulating red blood cells in human $CD34^+$ and $CD31^+$ vessels in mice injected with human $CD34^+$ WAT cells. We were never able to observe this effect in our previous studies using bone marrow-derived cells; consequently, these results show an important bona fide functional difference between WAT- and bone marrow-derived progenitors.

16.7 Discussion

Our data offer new insight into the controversy about the quantitative and the catalytic role of EPCs in cancer growth. All previous studies investigating this topic enumerated the role of EPCs in mice carrying GFP+ (or otherwise genetically labeled) BM. This approach excluded the quantification of the role of WAT-derived EPCs, that in our work were found to be in numbers significantly higher than in the BM. Moreover, the present array studies indicate that WAT-derived $CD34^+$ cells express significantly higher levels of angiogenesis-related genes compared to BM-derived, mobilized $CD34^+$ cells.

To our knowledge, our study is one of the first reporting about the tumor-promoting activity of fresh human WAT-derived purified $CD34^+$ cells. As shown by our data in mice receiving breast cancer and WAT cells only in one of the two mammary fat pads and breast cancer cells alone in the other mammary fat pad, the cancer promoting activity of WAT $CD34^+$ cells is likely to be exerted through a local, rather that systemic, activity. These data nicely complement the recent observation from the Kolonin laboratory (Corada et al. 2001) that in mouse models WAT cells are mobilized and recruited by experimental tumors to promote cancer progression (Fig. 16.2).

A novel brain/adipocyte-BDNF/leptin axis has been recently proposed to play a potentially relevant role in cancer progression (Rabascio et al. 2004). WAT-derived $CD34^+$ progenitors express high levels of the receptors involved in this axis. Along a similar line, WAT $CD34^+$ cells express very high levels of FAP-alpha, a crucial suppressor of antitumor immunity (Purhonen et al. 2008).

Finally, our data suggest caution about the clinical use of lipotransfer-derived WAT cells for tissue reconstruction in patients with cancer. We have recently reported a study of 321 consecutive patients operated for a primary breast cancer between 1997 and 2008 who subsequently underwent a lipotransfer procedure for esthetical purpose, compared with two matched patients with similar characteristics who did not undergo lipotransfer (Petit et al. 2012). In this study, the lipotransfer

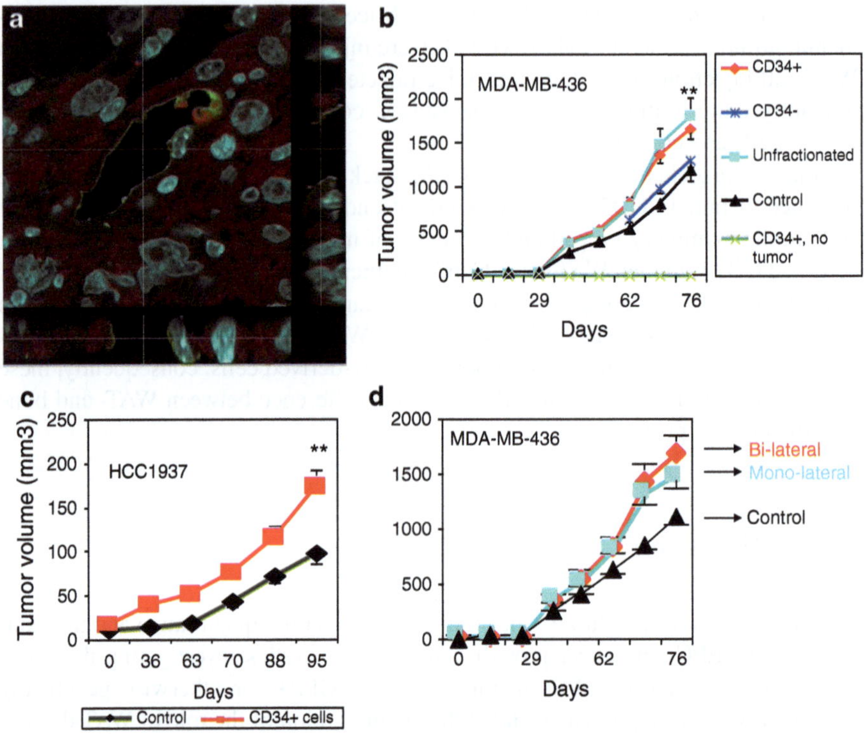

Fig. 16.2 Orthotopic models of breast cancer. (**a**) Confocal laser-scanning of human CD34 antigen distribution in a tumor section in mice injected with MDA-MB-436 tumor cells plus WAT-derived CD34⁺ human cells. For the imaging of the red cells the 561 laser line was used and the auto-fluorescence of the cells was collected. Snapshot images are orthogonal sections of the z-stacks taken at points along the vessel cavity. (**b**) Tumor growth in NSG mice injected with WAT CD34⁺ cells alone, MDA-MB-436 cells alone, MDA-MB-436 cells plus unfractionated WAT cells, and MDA-MB-436 cells plus CD34⁺ or CD34⁻ WAT cells. (**c**) Tumor growth in NSG mice injected with HCC1937 breast cancer cells alone and in NSG mice injected with the same number of breast cancer cells plus human CD34⁺ WAT cells. (**d**) Tumor growth in NSG mice injected with MDA-MB-436 breast cancer cells in mono- and bilateral studies. $^{**}P<0.005$ vs control by Mann–Whitney *U* test. From Martin-Padura et al. (2012), modified

group resulted at higher risk of local events (four events) compared to the control group (no event) when the analysis was limited to intraepithelial neoplasia. A second data revision after prolonged follow-up confirmed this significant difference (Petit et al. 2013).

The investigation of the different roles of purified populations of WAT-derived progenitors and mature cells is urgently needed to clarify what WAT cell populations can be used safely for breast reconstruction and what are at risk for supporting the growth of otherwise quiescent cancer cells still resident after surgery. It has been recently found that zoledronic acid inhibits the interaction between mesenchymal stem cells and breast cancer cells (Gallo et al. 2012). This finding suggests an interesting possible pharmaceutical strategy to be investigated in preclinical models.

References

Agliano A, Martin-Padura I, Marighetti P, et al. Human acute leukemia cells injected in NOD/LtSz-scid/IL-2Rgamma null mice generate a faster and more efficient disease compared to other NOD/scid-related strains. Int J Cancer. 2008;123:2222–7.

Bertolini F, Mancuso P, Braidotti P, Shaked Y, Kerbel RS. The multiple personality disorder phenotype(s) of circulating endothelial cells in cancer. Biochim Biophys Acta. 2009;1796: 27–32.

Corada M, Liao F, Lindgren M, et al. Monoclonal antibodies directed to different regions of vascular endothelial cadherin extracellular domain affect adhesion and clustering of the protein and modulate endothelial permeability. Blood. 2001;97:1679–84.

Gallo M, De Luca A, Lamura L, Normanno N. Zoledronic acid blocks the interaction between mesenchymal stem cells and breast cancer cells: Implications for the adjuvant therapy of breast cancer. Ann Oncol. 2012;23:597–604.

Grenier G, Scimè A, Le Grand F, et al. Resident endothelial precursors in muscle, adipose, and dermis contribute to postnatal vasculogenesis. Stem Cells. 2007;25:3101–10.

Hirschi KK, Ingram DA, Yoder MC. Assessing identity, phenotype, and fate of endothelial progenitor cells. Arterioscler Thromb Vasc Biol. 2008;28:1584–95.

Lohsiriwat V, Curigliano G, Rietjens M, Goldhirsch A, Petit YV. Autologous fat transplantation in patients with breast cancer: "silencing" or "fueling" cancer recurrence? Breast. 2011;20: 351–7.

Mancuso P, Antoniotti P, Quarna J, et al. Validation of a standardized method for enumerating circulating endothelial cells and progenitors: flow cytometry and molecular and ultrastructural analyses. Clin Cancer Res. 2009;15:267–73.

Mancuso P, Martin-Padura I, Calleri A, et al. Circulating perivascular progenitors, a target of PDGFR inhibition. Int J Cancer. 2011;129:1344–50.

Martin-Padura I, Gregato G, Marighetti P, et al. The white adipose tissue used in lipotransfer procedures is a rich reservoir of CD34+ progenitors able to promote cancer progression. Cancer Res. 2012;72(1):325–34. Epub 2011 Nov 3.

Petit JY, Clough K, Sarfati I, et al. Lipotransfer in cancer patients: from surgical technique to oncologic point of view. Plast Reconstr Surg. 2010;126:262–3.

Petit JY, Botteri E, Lohsiriwat V, et al. Locoregional recurrence risk after lipotransfer in breast cancer patients. Ann Oncol. 2012;23:582–8.

Petit JY, Rietjens M, Botteri E, et al. Evaluation of fat grafting safety in patients with intra epithelial neoplasia: a matched-cohort study. Ann Oncol. 2013;24:1479–84.

Purhonen S, Palm J, Rossi D, et al. Bone marrow-derived circulating endothelial precursors do not contribute to vascular endothelium and are not needed for tumor growth. Proc Natl Acad Sci U S A. 2008;105:6620–5.

Rabascio C, Muratori E, Mancuso P, et al. Assessing tumor angiogenesis: increased circulating VE-cadherin RNA in patients with cancer indicates viability of circulating endothelial cells. Cancer Res. 2004;15:4373–7.

Sengenès C, Lolmède K, Zakaroff-Girard A, et al. Preadipocytes in the human subcutaneous adipose tissue display distinct features from the adult mesenchymal and hematopoietic stem cells. J Cell Physiol. 2005;205:114–22.

Shultz LD, Ishikawa F, Greiner DL. Humanized mice in translational biomedical research. Nat Rev Immunol. 2007;7:118–30.

Yoder MC, Ingram DA. The definition of EPCs and other bone marrow cells contributing to neoangiogenesis and tumor growth: is there common ground for understanding the roles of numerous marrow-derived cells in the neoangiogenic process? Biochim Biophys Acta. 2009;1796: 50–4.

Part VIII
Engineering of Vascularized Adipose Tissue

Chapter 17
Successful Angiogenesis as Key to Engineered Adipose Tissue

Paul Severin Wiggenhauser and Jan-Thorsten Schantz

Abstract The transplantation of fat tissue is a common procedure in plastic surgery. But the volume of the transplanted fat tissue is strictly limited by the viability of the fat graft depending on blood supply and oxygen diffusion. Therefore there are only two possibilities for the transplantation of larger grafts: first repeated transplantation of small pieces and second transplantation of a large prevascularized fat graft that is reconnected to blood circulation.

Consequently, the same rules apply for fat tissue created with adipose tissue engineering. Thus, an approach for the engineering of a large adipose tissue graft needs a concomitant strategy for vascularization. So a successful approach needs a working interplay of stem cells, differentiation factors, scaffold architecture and vascularization strategy. Moreover concepts must emphasize the importance of the micro and macro scaffold architecture not only in reference to tissue creation but also in reference to vascularization.

In conclusion every adipose tissue engineering approach should care for a holistic concept from tissue engineering to viable graft implantation. Otherwise there is no chance for a successful clinical application.

Keywords Scaffold materials • Scaffold architecture • Adipogenic differentiation • Tissue integration • Bioactive implants • Regeneration • Vessel loop • Chamber model • COPROG • Transgene fibroblasts

---

P.S. Wiggenhauser
Klinik für Hals-Nasen-Ohrenheilkunde, Uniklinik Ulm, 89070 Ulm, Frauensteige 12

J.-T. Schantz (✉)
Klinik und Poliklinik für Plastische Chirurgie und Handchirurgie, TU München, 81675 München, Ismaninger Str 22
e-mail: jtschantz@lrz.tum.de

Y. Cao (ed.), *Angiogenesis in Adipose Tissue*, DOI 10.1007/978-1-4614-8069-3_17,

17.1 Introduction

17.1.1 Clinical Reasoning

The covering of large soft tissue defects and the correction of body contour deformities are the classical indications for surgical fat transplantation. These conditions can be found in reconstructive as well as in aesthetic surgery. Representative procedures are the reconstruction of the breast mound after the complete removal of all breast tissue in the course of breast cancer treatment. Another example is the covering of a deep soft tissue and skin defect after the excision of an invasive skin cancer or the remodelling of contour deformities due to developmental disorders as seen in funnel chest deformity.

The main goals of fat transplantation surgery are firstly to shape the body surface by expanding subcutaneous soft tissue and secondly to repair large skin and soft tissue defects with skin-adipose composite grafts.

17.1.2 Understanding Fat Transplantation

17.1.2.1 First Fat Graft by Neuber

The first report of a free fat graft can be found in 1893. During the congress of the German society of surgeons in 1893, the surgeon Neuber told his colleagues about the treatment of a 20-year-old man who suffered from a large conical infraorbital scar due to tuberculous osteitis during his childhood. Neuber filled the conical scar with a fat graft harvested from the upper limb and achieved a satisfying cosmetic result after an uncomplicated healing process (Neuber 1893).

In contrast to this case report and to other successful cases, Neuber clearly recognized the limitations of his technique and told his colleagues about the limitations of free fat transplantation. "According to my experience, fat particles that exceed the size of a bean or almond will not be integrated successfully". Instead there will be a wound healing disorder and the complete graft will become necrotic.

Remarkably, Neuber described the crucial challenge of fat transplantation right from the beginning: *the need for vascularization in large volume fat grafts*.

17.1.2.2 Feasibility of Large Volume Fat Transplantation

Basically there are two ways to facilitate fat transplantation: firstly surgeons transplant only solid fat tissue with a pre-existing and functional vascular system or secondly they inject small non-vascularized portions of fat tissue repetitively so that this portions sum up to a volume effect. This last procedure is also known as lipostructure or lipotransfer.

DIEP-Flap

The DIEP-Flap[1] represents the gold standard of free large volume transplantation using a solid vascularized fat graft (Hijjawi and Blondeel 2010).

The feasibility of this technique is based on the perforator vessels that nourish the subcutaneous skin of the belly. The perforator vessels are small vessels originating from the deep inferior epigastric artery and direct through the rectus abdominis muscle to the subcutaneous tissue. Therefore every perforator vessel supplies a specified area of subcutaneous tissue around its exit of the abdominal muscle (Baumann et al. 2010).

Consequently surgeons dissect the perforator vessels and harvest the subcutaneous fat tissue belonging to the previously dissected perforator vessels en bloc. By that means, they are able to create a solid fat graft with a totally independent but functional vascular system. Reconnection of the perforator vessel to another artery after transplantation leads to a complete recovery of circulation within the fat graft (Hijjawi and Blondeel 2010).

In summary the reconnection of the inherent vascular network of a fat graft maintains the vitality of the graft even in large volumes.

Coleman's Lipostructure

Coleman's technique on the contrary is based on the transfer of small adipose particles without a vascular network that are nourished by diffusion. Volume gain is only achieved by the repetitive injection of small amounts of non-vascularized fat tissue that sums up to a volume effect.

After the introduction of liposuction into aesthetic surgery by Klein, there was a new way to harvest fat tissue: in contrast to surgical resection of fat en bloc, Klein's technique harvested fat particles that were dissolved in tumescent solution (Klein 1988, 1990).

The pioneering work of Coleman was to optimize the liposuction process by inducing blunt cannulas and a harvesting process that cropped small blocks of fat tissue by curetting instead of traumatic suctioning. By that means, Coleman enhanced vitality of transferred particles and was able to achieve long-term stable and safe volume corrections (Coleman 1997).

But this technique only allows correction of smaller defects as it needs repetitive procedures to achieve substantial volume effects. Therefore this technique is not feasible for larger volume transfers as needed for breast reconstructive procedures.

Moreover the circumstance of non-vascularized transplantation leads to a partial resorption of the transplanted fat tissue because of missing nourishment and oxygen supply. The transplanted fat does not become until reintegrated into the host site vascular system.

---

[1] DIEP stands for deep inferior epigastric <artery> perforator.

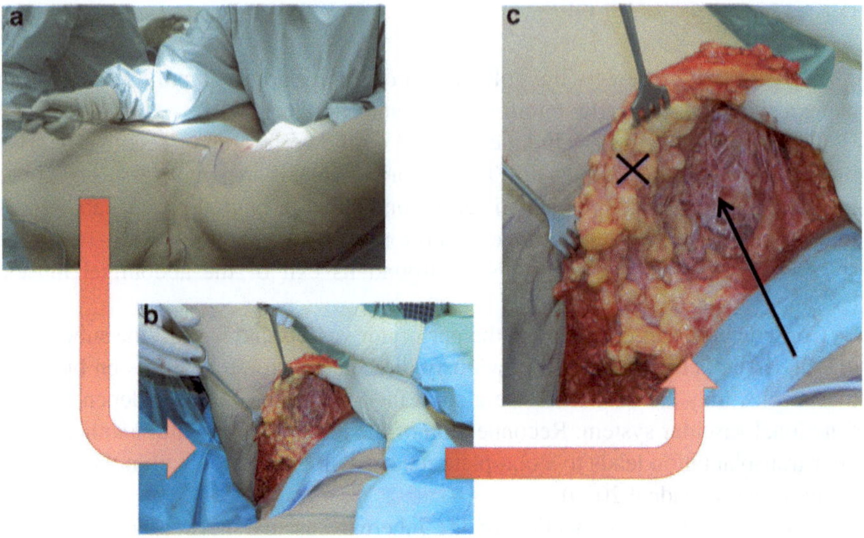

Fig. 17.1 Intraoperative photographs of subcutaneous fat tissue. (**a**) Liposuction removes mature adipocytes. (**b**) Snapshot during thighplasty revealing macroscopic anatomy of fat tissue structures. (**c**) Close-up of (**b**). Subcutaneous fat tissue (*x*) is broadly interspersed with connective tissue septa containing major blood and lymphatic vessels directing to the enclosed fat tissue next to the septum. *Arrow* points at these septa uncovered through the removal of the enclosed fat tissue by liposuction

Consequently there have been studies to treat the recipient site preoperatively to enhance the integration of the transplanted fat particles. The BRAVA technology, for example, is famous method to expand breast skin and to simultaneously induce angiogenesis. This preoperative treatment led to lower resorption rates of injected fat tissue in the course of breast augmentation procedures (Zocchi and Zuliani 2008).

In conclusion Coleman refined the technique of non-vascularized fat transplantation as described by Neuber, using an approach of repetitive transfer of small volume particles. His method, however, still faces problems of undernourishment and low perfusion after transplantation due to the missing vascularization (Fig. 17.1).

17.2 Adipose Tissue Engineering

17.2.1 History and Impact

In 1999 Patrick et al. created fat tissue de novo by tissue engineering technologies. Tissue engineering means formation of tissue using a carrier material and stem cells for the formation of tissue in vitro and bioactive agents for formation of tissue in situ.

Table 17.1 Advantages and disadvantages of stem cell sources for adipose tissue engineering

	+	–	
ESCs	Differentiation in every cell type	Teratoma No controlled differentiation Ethic problems	Wernig et al. (2003)
iPSCs	Differentiation in every cell type Easy access via skin fibroblasts	Genetic reprogramming, safety?	Takahashi et al. (2007), Takahashi and Yamanaka (2006), Tenzen et al. (2010)
MSCs	Differentiation in every mesenchymal tissue Long clinical experience	Invasive harvesting process Low stem cell ratio	Caplan and Dennis (2006), Klopsch et al. (2011), Wernig et al. (2003)
ASCs	Differentiation in every mesenchymal tissue Easy access via liposuction High stem cell ratio		Peng et al. (2008), Brown et al. (2010)

Patrick used a collagen scaffold as carrier material and adipose tissue-derived stem cells (ASCs) for the formation of fat cells in vitro. After implantation of seeded scaffolds into rats, he observed fat tissue development on explanted scaffolds during the histological work-up (Patrick et al. 1999).

Since that time, research has focused on a possible third way for fat transplantation: tissue engineered fat grafts created in the laboratory. The creation of fat tissue with tissue engineering is referred to as adipose tissue engineering.

17.2.2 Stem Cells Sources

Principally every stem cell source that can give rise to the adipose lineage can be used for adipose tissue engineering approaches: embryonic stem cells (ESCs), mesenchymal stem cells (MSCs), induced pluripotent stem cells (iPSCs) and ASCs.

Although there are many different stem cell sources, research has started to exclusively focus on ASCs. That is due to adult stem cell characteristics: safe use in vivo, easy differentiation process, high stem cell rate in source tissue and the ubiquitous and easily accessible stem cell source.

Especially the easy harvesting process made ASCs the stem cell source of choice in current research. There are two ways of isolation: from solid fat blocks after surgery or from liposuction aspirate. Both methods harvest ASCs with similar qualities and in equal ratios. But liposuction is considered as the easiest, safest and fastest way to ASCs (Bunnell et al. 2008; Fernández et al. 2010; Schreml et al. 2009; Baglioni et al. 2009) (Table 17.1).

17.2.3 *Scaffold Materials*

Suitable carrier materials, namely scaffolds, are the decisive condition of an engineering approach. There are plenty of different scaffold types and those are all different in respect to specific properties. Therefore the experimental and later clinical setting must be taken into account to choose an appropriate type of scaffold. For example, treatment procedure can be performed with injection or implantation and thus completely different scaffold qualities are necessary for the respective treatment. Otherwise scaffolds can differ in way of fabrication, costs, handling, differentiation potential, volume maintenance and degradation (Table 17.2).

17.2.4 *Differentiation Stimuli*

The differentiation of ASCs in mature adipocytes is achieved via multimodal stimulation in cell culture experiments. The application of hormonal stimuli as well as the influence of the micro milieu (pO_2, movement, cocultivation with mature adipocytes and the surround matrix in reference to material and architecture) is used for that purpose. The most common hormonal factors are: insulin, ibmx, dexamethasone, indometacine and thiazolidindione (Stacey et al. 2009; Hausman et al. 1996; von Heimburg et al. 2001; Fischbach et al. 2004; Carriere et al. 2004; Fink et al. 2004; Rodriguez Fernandez and Ben-Ze'ev 1989).

For further understanding please refer to chapters dealing with adipose stem cells and adipose tissue development (Table 17.3).

17.2.5 *Influence of Scaffold Architecture*

The influence of scaffold architecture on the regeneration result has to be considered in two different dimensions: micro and macro architecture. Micro architecture refers to the material qualities itself and their potential to resemble natural substrate for cell growth and proliferation. Macro architecture alludes to the macroscopic structure of the scaffold playing a key role for cell and vascular invasion and finally tissue integration.

17.2.5.1 Micro Architecture

The micro architecture of a scaffold determines the interactions of the scaffold itself with the stem cells. Interactions can be measured by cell adhesion, cell distribution, cell proliferation, cell differentiation and cell vitality.

Table 17.2 Most common scaffold types in adipose tissue engineering (modified from thesis of P.S. Wiggenhauser (in German))

Material	Origin	Observations	Stem cells[a]	Representative
PEG (Polyethylenglycol)	Synthesized	+ Form stability + Differentiation after pretreatment in vitro	Human MSCs	Alhadlaq et al. (2005)
PLGA (Polylactid-co-Glycolic-Acid)	Synthesized	+ Mechanic stability and volume maintenance after 6 weeks + Phenotype of mature fat cells	Human preadipocytes	Cho et al. (2005)
PCL (Polycaprolacton)	Synthesized	+ Spontaneous differentiation without further stimuli on the scaffold + Remaining stem cells between differentiated cells – Only in vitro study	Murine ESC	Kang et al. (2007)
Collagen type 1	Decellularized porcine or bovine skin	+ High porosity for cell migration + Enhancement of profliferation and differentiation + Minimal inflammatory reaction	Human ASC	Itoi et al. (2010)
HYAFF11	Derived from hyaluronic acid	+ Complete differentiation in vivo + High porosity for differentiation – Volume shrinkage	Human preadipocytes	von Heimburg et al. (2001)
Fibrin	From human plasma	+ Natural, vital adipose tissue + No inflammatory response + No Necrosis – Initial volume shrinkage	Human preadipocytes	Torio-Padron et al. (2007)
Gelatin	Porcine skin gelatin	+ Adipogenic differentiation and lipid droplet formation – Only in vitro in this setting	MSCs	Hong et al. (2005)
Matrigel	Basement membrane of rat sarcoma	+ De novo Adipose tissue formation + Used for in situ approach with bFGF – No in vitro studies	Unseeded	Kimura et al. (2002)
Alginate	Cell wall of brown algae	+ Adipose tissue formation + Vascular and tissue ingrowths + Minimal inflammatory response + Volume maintenance for 3 months of observation	Sheep preadipocytes	Halberstadt et al. (2002)

[a]Preadipocytes are a not clearly defined cell source from stromal vascular fraction. Since the definition of ASCs the term has become obsolete

Table 17.3 Overview of soluble factors used to achieve differentiation in adipose tissue engineering (taken from thesis of P.S. Wiggenhauser (in German))

Way of action	Point of interaction	Stimulus	
Signalling transduction cascades	Tyrosine kinase	Insulin	Pairault and Lasnier (1987), Harrison et al. (1985), Chen and London (1981), Hauner (1990)
		IGF	Wabitsch et al. (1995)
		PGF2	Catalioto et al. (1991)
		GH	Hauner and Loffler (1986), Nixon and Green (1984), Morikawa et al. (1982)
	cAMP/PKA	cAMP	Cook et al. (1988), Darimont et al. (1993), Kurten et al. (1988), Verhaegen et al. (1979)
		IBMX	Lequeux et al. (2009), Schimmel et al. (1980), Swierczewski et al. (1987), Kim et al. (2010)
		Arachidonic acid	Gaillard et al. (1989)
Transcriptional factors	SREB	Glucocorticoids	Hauner et al. (1989), Cook et al. (1988), Pairault and Lasnier (1987), Knight et al. (1987), Ringold et al. (1986), Chapman et al. (1985), Hauner (1990)
	RXR	Retinoic acid	Pairault and Lasnier (1987), Safonova et al. (1994), Pairault et al. (1988)
	PPAR	Thiazolidindiones	Sandouk et al. (1993), Gimble et al. (1996), Lehmann et al. (1995)
		Prostacyclin PGI2	Catalioto et al. (1991), Negrel et al. (1989), Kliewer et al. (1995)
		Indomethacin, Ibuprofen	Lehmann et al. (1997), Knight et al. (1987)
	RXR/PPAR	Free fatty acids	Takase et al. (1998), Suruga et al. (1995)
	THR	T3	Flores-Delgado et al. (1987), Darimont et al. (1993)

In the beginning researchers were looking for organic scaffold materials to achieve optimal results. For example, collagen type 1 was commonly studied, as it is the natural extracellular matrix of adipose tissue. Studies by Itoi could show an influence of adipogenic differentiation only by adding collagen type 1. Due to the limited access to natural scaffolds, investigators developed synthetic scaffolds that were intended to resemble natural extracellular matrix but with modified stability, degradation and in vivo behaviour (Itoi et al. 2010; Engler et al. 2006; Brown et al. 2010; Gleeson et al. 2010; Hausman et al. 1996).

The micro architecture determines the behaviour of stem cells on the scaffold and hereby significantly influences the regeneration result.

17.2.5.2 Macro Architecture

In addition to the crucial micro architecture of scaffolds, the macro architecture plays a key role for in vivo integration and thus the regeneration result of the scaffold.

The macroscopic structure of a scaffold, referred to as macro architecture, determines cell and vascular invasion and overall tissue distribution. The authors showed this impact of scaffold structure on the tissue engineering approach. They studied two scaffolds made out of equal materials, both Polycaprolactone-based, but with two different macro architectures. As shown in Fig. 17.2, the random pattern architecture of the first scaffold inhibited cell and vascular invasion by blocking intruding pathways. On the other hand, the well-structured second scaffold presented deep invasion due to its preformed invasion pathways, as can be seen in Fig. 17.3 (Wiggenhauser et al. 2011).

Moreover the cellular and vascular invasion determines the overall integration of the implant and the quality of the tissue regeneration result. So far, most in vivo experiments have been performed as proof of principle studies with construct smaller than 10 mL. But a large volume regeneration of clinical significance is only possible if the entire scaffold cannot be integrated within a sufficient timeframe. That is why cell and vascular invasion parameters are fundamental to clinical success.

Otherwise the quality of the regeneration result depends on a homogenous and coherent tissue formation. This can only be achieved if the scaffold enables a full and entire integration within the host organism. If parts of the scaffold cannot be accessed by invading cells and blood vessels, existing adipose tissue becomes necrotic because of missing nourishment or de novo adipose tissue formation does not take place and empty wholes with clinical dangers remain.

The macro architecture of a scaffold determines the clinical quality of the regeneration approach and must always be considered during scaffold design.

17.2.6 Essential Challenge of Vascularization

Neuber experienced the correlation of graft volume and success rate of implantation in 1893. Some decades later, Folkman et al. showed the growth arrest of tumour spheroids in cell culture experiments and gave a scientific explanation for Neubers observations: the cells within the spheroids were nourished by diffusion of oxygen

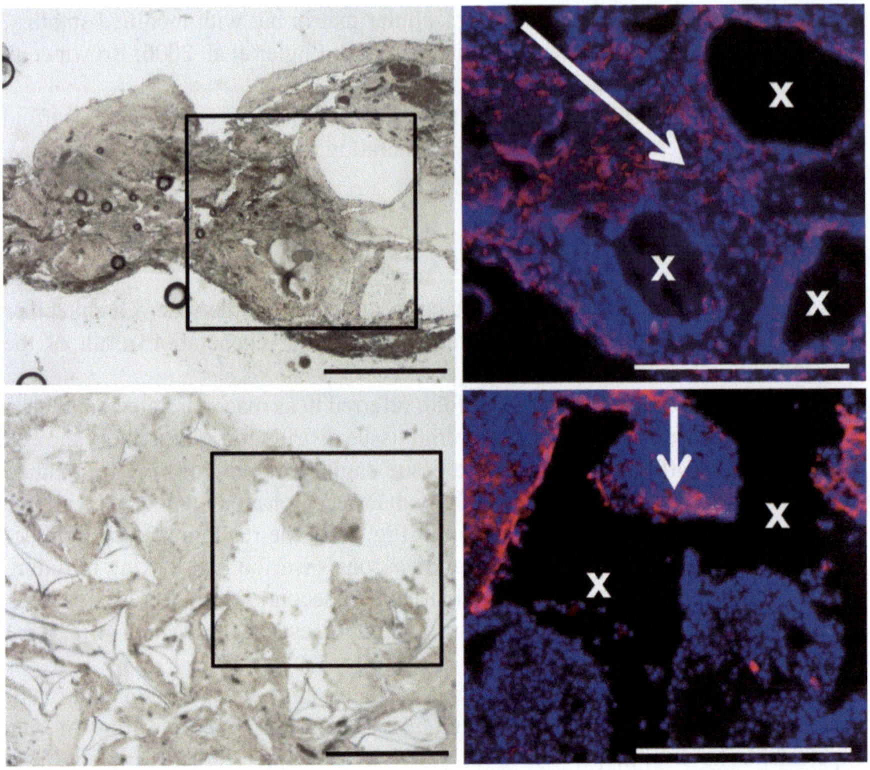

Fig. 17.2 Integration of scaffolds in the host organism. *Upper* row represents regularly structured scaffold, *lower* row random pattern scaffold. First column shows native section through scaffold. Notice lower cell density on random pattern scaffold. Second column shows immunostaining with anti-smooth muscle antibodies to show blood vessels and DAPI staining of cell nuclei. The invasion of blood vessels (*arrow*) in central areas of the scaffold was shown on the regularly structured scaffold. Random pattern scaffolds inhibited cell invasion on non-interconnected parts of the scaffold. Notice the avascular lower part of the image. (*x*) shows negative of through processing removed scaffold (bar = 500 μm) (taken from thesis of P. S. Wiggenhauser)

and nutrition and therefore maximal distance of diffusion limited the growth of cells in 3D to the size of a pea (Folkman and Hochberg 1973).

As a consequence of this experiment tissue engineers had to focus on strategies to create a vascular network within the created fat tissue. But in adipose tissue, vascularization is well advanced. Histology shows that every mature adipocyte is attached to at least one capillary. This intensive vascularization yields a special challenge for every adipose tissue engineering approach (Welsch 2006).

Therefore adipose tissue or constructs cannot be created without providing a functional and deeply branched vascular network. At present, the induction of angiogenesis is the most important part of research in adipose tissue engineering.

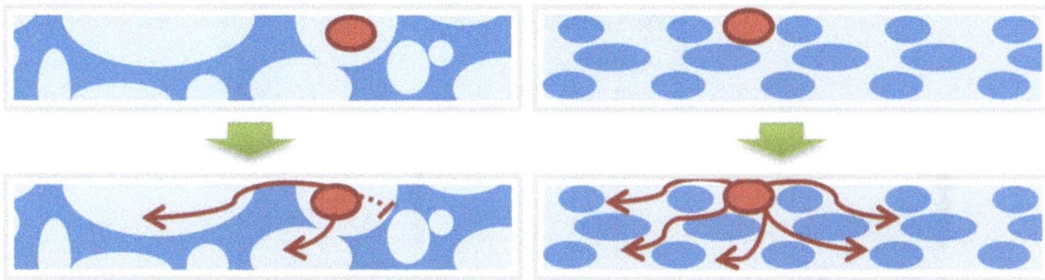

Fig. 17.3 Facilitation of cell and vascular invasion through macro architecture: Schematic drawing of random pattern scaffold (*left*) and regularly structured scaffold (*right*) and process of vascular invasion starting from a vessel loop (*red*) on cross-section view. The scaffolds are shown in *dark blue*, origination vessels in *red*. The random pattern leads to long and laborious routes for invasion and even stops invasion (*dotted line*) at non-interconnected parts. Contrarily regularly structured scaffolds facilitate straightforward directed invasion and lead cells and vessels by guiding structures (taken from thesis of P.S. Wiggenhauser)

17.2.7 In Vitro vs. In Situ

In principle, tissue engineering yields two ways of tissue creation: in vitro and in situ approaches. Despite the general difference of these two approaches, each of them confronts researchers with its unique challenge of angiogenesis. In vitro adipose tissue engineering needs an intensive and fast angiogenesis after implantation to maintain the viability of a large adipose construct in vivo. Whereas in situ approaches can employ a two-step technology: firstly caring for angiogenesis and secondly inducing fat tissue formation. In such an in vivo approach, the time needed for angiogenesis can take longer because vascularization is not needed to keep an in vitro engineered construct viable (Tanzi and Fare 2009).

This is why it is of great importance that the process of fat tissue formation and the way of angiogenesis go hand in hand in all adipose tissue engineering approaches. The time available for angiogenesis determines the portfolio of angiogenesis methods that can be used for the specific engineering approach.

17.3 Engineering Approaches and the Strategy of Vascularization

Sufficient vascularization remains a challenge in all tissue engineering approaches. There are numerous strategies to provide sufficient angiogenesis. The concepts range from using decellularized vascular matrices from animals to in situ induction of vascular formation.

The following section will give an overview of innovative and common concepts in adipose tissue engineering. A structured outline is given in Fig. 17.4.

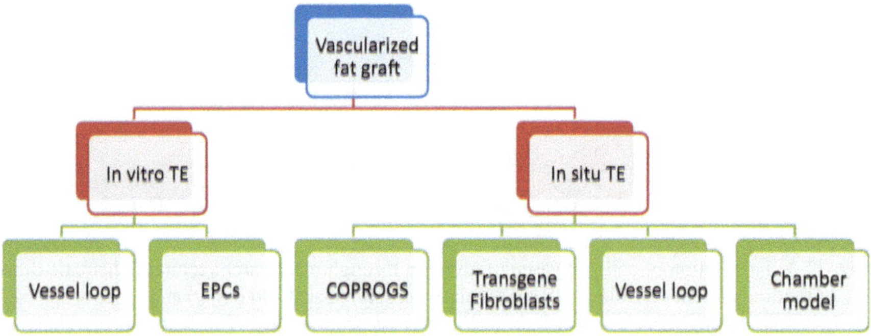

Fig. 17.4 Strategies to create vascularized adipose tissue. *TE* tissue engineering, *EPCs* endothelial progenitor cells, *COPROGS* copolymer-protected gene vectors

17.3.1 Vascularized Single Graft

The engineering of a single high volume fat graft is still the most favoured goal in adipose tissue engineering. There are two reasons for that: first the chance for a one step repair of deep and large skin and soft tissue defects and second the need for a single fat graft, if there is no remaining underlying soft tissue.

17.3.2 In Vitro Tissue Engineering

Surgeons are trained to harvest and transplant fat using flaps. As described for DIEP flaps, the common characteristic of these flaps is their vascular network that is accessible through a major artery and vein. Surgeons require these larger vessels to reconnect the transplant to the systemic circulation at the recipient site of reconstruction. Therefore tissue engineers aim for a vascularized fat construct with a major artery and vein to facilitate the transplantation and to substitute autologous tissue that has to be harvested from a donor site. By that means research wants to eliminate doner site defects including wound healing disorders, infections, hematoma and physiological weakness after dissection of muscles and fascia.

17.3.2.1 Vessel Loop

The vessel loop model is based on plastic surgical method that was developed for the preparation of skin flap. The finding that poorly vascularized tissue can be vascularized by adjacent highly vascularized tissue under hypoxic conditions is fundamental for the understanding of this method (Dolderer et al. 2007; Lokmic et al. 2007; Morrison et al. 1997).

In tissue engineering experiments, this technique was refined and reduced to a vascular bundle that can be used to vascularize constructs created in the cell culture laboratory. For that purpose, constructs have to be implanted in the near of the vascular bundle. There are several possibilities of arrangement: placement of vessels through or on top of the construct and the constellation of the vessels itself.

Tanaka et al. studied diverse constellations where a vessel bundle was embraced by a construct like a sandwich. The aim of the study was to identify the most suitable arrangement of the vessels. They compared an arteriovenous shunt-loop (that has lead to the name vessel loop), a ligated vascular bundle and an intact vascular bundle that was directing through the construct. The flow-through vascular bundle was found to be superior to the other forms of vascular constellations. There was a distinct angiogenesis with an outgrowth of vessels from the venous vasa vasorum (Tanaka et al. 2003; Hofer et al. 2003).

All in all the vessel loop technique is a widely spread technology in tissue engineering. It is an easy and functional method to vascularize engineered constructs. But still little is known about the vascularization of non-experimental volumes of clinical significance.

17.3.2.2 Endothelial Progenitors

Endothelial progenitor cells are the natural precursor cells of endothelial cells that form the capillary system and the intima of bigger vessels. Therefore they were used to build blood vessels de novo (Bleiziffer et al. 2011; Zhang et al. 2011).

Singh et al. seeded porous polycaprolactone scaffolds with endothelial progenitor cells and implanted the scaffolds in subcutaneous tissue of nude mice. In addition scaffolds were treated with heparin and VEGF. After 7 days in vivo scaffolds were explanted and examined. The researchers found that the endothelial progenitor cells had given rise to a vascular network that was connected to the host circulatory system. By changing the seeding cell density, they were able to optimize the formed vascular network. Furthermore they found out that endothelial progenitor cells could accelerate blood vessel formation in co-culture settings. Unfortunately central necrosis was found in larger volume scaffolds (Singh et al. 2011).

That is why endothelial progenitor cells are a promising alternative for de novo blood vessel formation to support the perfusion of in vitro build adipose tissue. However, endothelial progenitor cells face the same problems as other tissue engineered constructs with central necrosis in larger volumes.

17.3.3 In Situ Tissue Engineering

The concept of in situ engineering is based on the facilitation and induction of tissue regeneration in situ. The goal is to create a highly vascularized tissue that can receive stem cells for tissue regeneration in another step or that can lead to tissue regeneration by homing of desired stem cells. Therefore the major task of angiogenesis is not to maintain a transplanted construct as seen in in vitro engineering but to establish a vascular network. Thus, angiogenesis strategies have to support tissue formation and researchers have to employ strategies to induce blood vessel formation to create a vascularized bed for tissue regeneration.

17.3.3.1 COPROGS

The term COPROGS stands for copolymer-protected gene vectors. They offer a method for non-viral gene transfer and minimize the risks of viral gene transfers.

They consist of DNA encapsulated with polymers. These particles get into the cytoplasm via endocytosis and escape degradation in the endosomes through their copolymer design that destroys the endosome and releases the coding DNA into the cytoplasm. After incorporation of the DNA in the nucleus, the transfected cells start to produce proteins encoded by the COPROGS gene vector (Kolk et al. 2011).

COPROGS have already been studied in respect of skin regeneration models. In these studies they were used to induce VEGF production and thus to trigger angiogenesis within the scaffold. In one study COPROGS were loaded with DNA encoding for VEGF. Afterwards the COPROGS were attached to a collagen scaffold with a coating technology. In vitro results showed no difference in the behaviour of seeded cells in respect to cell distribution, attachment, differentiation and vitality. Furthermore in vitro testing showed a 3 weeks lasting release of VEGF. In vivo models showed a significant increase of VEGF on COPROGS treated scaffolds in comparison to untreated scaffold in a wound model. Moreover histological analysis indicated the presence of newly build blood vessels (Reckhenrich et al. 2011).

Summing up, COPROGS are still an experimental technology for angiogenesis induction. Still, their safe mode of action and their high innovative potential predestine them for further in situ adipose tissue engineering approaches.

17.3.3.2 Transgene Fibroblasts

Transgene fibroblasts are an alternative technology for transient enrichment of bioactive agents in native tissue. Here, fibroblasts are genetically modified to produce and release bioactive agents (e.g. VEGF) in the surrounding tissue to trigger angiogenesis. These agents can be used to induce vascular growth and formation of vascular networks.

Machens et al. studied the applicability of transfected fibroblasts in rat models to improve the vascularization of tissue for flap surgery. In different studies they investigated the influence of PDGF, bFGF and VEGF release on angiogenesis. Firstly, rat fibroblasts were isolated, transfected and tested for safe transfection. Afterwards a skin flap surgery was performed and transfected fibroblasts were injected in the flap tissue. After surgery, the flaps were monitored and constantly evaluated: the treated flaps showed a higher rate of survival and a higher rate of vascularization after histological analysis ex vivo in comparison to control groups (Holzbach et al. 2010; Machens et al. 1998; Spanholtz et al. 2009, 2011).

Machens et al. proposed the application of this technology in critically ischemic tissue. Every adipose tissue engineering approach focuses critical ischemia within the construct sooner or later. Therefore the use of transgene fibroblasts could be an innovative vascularization strategy for adipose tissue constructs.

17.3.3.3 Vessel Loop

The vessel loop technique has already been described in reference to in vitro tissue engineering. But it can also be applied for in situ approaches.

The most significant difference is the need to vascularize immature tissue or unseeded scaffolds. This circumstance facilitates invasion possibilities, as blood vessels are not forced to make their way through coherent cell layers. Moreover the time pressure for angiogenesis is reduced, due to the formation of new tissue and the progress of angiogenesis going hand in hand within the implanted scaffold.

The vessel loop method is an easy method to lead stem cells for de novo tissue formation through the blood stream to the site of interest and to care for a simultaneous angiogenesis.

17.3.3.4 Chamber Model

The chamber model uses a hollow plastic chamber to generate fat tissue originating from a pedicled fat flap within the chamber.

The chamber model was propagated by Morrison et al. and is a classical hybrid technology of surgery and adipose tissue engineering. Morrison used a hollow plastic chamber (solid or perforated) and inserted a pedicled fat flap inside the chamber. After 6 weeks in situ, histological analysis showed a spontaneous formation of fat tissue within the chamber. Hereby the perforated chamber showed best results. During further research Morrison refined the model and was able to go beyond the step of clinical graft volume. In the latest study he was able to show the formation of long-term stable adipose tissue with a perforated chamber filled with PLGA and an inserted fat pedicle or even only a vascular pedicle alone (Dolderer et al. 2007, 2011; Findlay et al. 2011; Hofer et al. 2003; Lokmic et al. 2007).

The chamber model currently represents the most advanced adipose tissue engineering approach using classical surgical techniques and adipose tissue engineering.

17.4 Summary

The number of engineering approaches in adipose tissue engineering is still increasing. Most of them are still in the experimental phase and have only been performed in small animal models. Only few of them have taken the step to large volume experiments, e.g. cell-assisted-lipotransfer or the chamber model. Although all of them face the problem of achieving sufficient vascularization, in situ approaches may be more promising, due to their two-step concept giving more time for the crucial angiogenesis.

References

Alhadlaq A, Tang M, Mao JJ. Engineered adipose tissue from human mesenchymal stem cells maintains predefined shape and dimension: implications in soft tissue augmentation and reconstruction. Tissue Eng. 2005;11(3–4):556–66.

Baglioni S, Francalanci M, Squecco R, Lombardi A, Cantini G, Angeli R, Gelmini S, Guasti D, Benvenuti S, Annunziato F, Bani D, Liotta F, Francini F, Perigli G, Serio M, Luconi M. Characterization of human adult stem-cell populations isolated from visceral and subcutaneous adipose tissue. FASEB J. 2009;23(10):3494–505.

Baumann DP, Lin HY, Chevray PM. Perforator number predicts fat necrosis in a prospective analysis of breast reconstruction with free TRAM, DIEP, and SIEA flaps. Plast Reconstr Surg. 2010;125(5):1335–41.

Bleiziffer O, Hammon M, Naschberger E, Lipnik K, Arkudas A, Rath S, Pryymachuk G, Beier JP, Stürzl M, Horch RE, Kneser U. Endothelial progenitor cells are integrated in newly formed capillaries and alter adjacent fibrovascular tissue after subcutaneous implantation in a fibrin matrix. J Cell Mol Med. 2011;15(11):2452–61. doi:10.1111/j.1582-4934.2010.01247.x.

Brown SA, Levi B, Lequeux C, Lequex C, Wong VW, Mojallal A, Longaker MT. Basic science review on adipose tissue for clinicians. Plast Reconstr Surg. 2010;126(6):1936–46. doi:10.1097/PRS.0b013e3181f44790.

Bunnell BA, Flaat M, Gagliardi C, Patel B, Ripoll C. Adipose-derived stem cells: isolation, expansion and differentiation. Methods. 2008;45(2):115–20.

Caplan A, Dennis J. Mesenchymal stem cells as trophic mediators. J Cell Biochem. 2006;98(5):1076–84.

Carriere A, Carmona MC, Fernandez Y, Rigoulet M, Wenger RH, Penicaud L, Casteilla L. Mitochondrial reactive oxygen species control the transcription factor CHOP-10/GADD153 and adipocyte differentiation: a mechanism for hypoxia-dependent effect. J Biol Chem. 2004;279(39):40462–9.

Catalioto RM, Gaillard D, Maclouf J, Ailhaud G, Negrel R. Autocrine control of adipose cell differentiation by prostacyclin and PGF2 alpha. Biochim Biophys Acta. 1991;1091(3):364–9.

Chapman AB, Knight DM, Ringold GM. Glucocorticoid regulation of adipocyte differentiation: hormonal triggering of the developmental program and induction of a differentiation-dependent gene. J Cell Biol. 1985;101(4):1227–35.

Chen JJ, London IM. Hemin enhances the differentiation of mouse 3T3 cells to adipocytes. Cell. 1981;26(1 Pt 1):117–22.

Cho SW, Kim SS, Rhie JW, Cho HM, Choi CY, Kim BS. Engineering of volume-stable adipose tissues. Biomaterials. 2005;26(17):3577–85.

Coleman SR. Facial recontouring with lipostructure. Clin Plast Surg. 1997;24(2):347–67.

Cook JS, Lucas JJ, Sibley E, Bolanowski MA, Christy RJ, Kelly TJ, Lane MD. Expression of the differentiation-induced gene for fatty acid-binding protein is activated by glucocorticoid and cAMP. Proc Natl Acad Sci U S A. 1988;85(9):2949–53.

Darimont C, Gaillard D, Ailhaud G, Negrel R. Terminal differentiation of mouse preadipocyte cells: adipogenic and antimitogenic role of triiodothyronine. Mol Cell Endocrinol. 1993;98(1):67–73.

Dolderer JH, Abberton KM, Thompson EW, Slavin JL, Stevens GW, Penington AJ, Morrison WA. Spontaneous large volume adipose tissue generation from a vascularized pedicled fat flap inside a chamber space. Tissue Eng. 2007;13(4):673–81. doi:10.1089/ten.2006.0212.

Dolderer JH, Dolderer JH, Doldere JH, Thompson EW, Slavin J, Trost N, Cooper-White JJ, Cao Y, O'connor AJ, Penington A, Morrison WA, Abberton KM. Long-term stability of adipose tissue generated from a vascularized pedicled fat flap inside a chamber. Plast Reconstr Surg. 2011;127(6):2283–92. doi:10.1097/PRS.0b013e3182131c3e.

Engler AJ, Sen S, Sweeney HL, Discher DE. Matrix elasticity directs stem cell lineage specification. Cell 2006;126(4):677–89. doi: S0092-8674(06)00961-5 [pii] 10.1016/j.cell.2006.06.044.

Fernández M, Acuña MJ, Reyes M, Olivares D, Hirsch S, Bunout D, de la Maza MP. Proliferation and differentiation of human adipocyte precursor cells: differences between the preperitoneal and subcutaneous compartments. J Cell Biochem. 2010;111(3):659–64. doi:10.1002/jcb.22753.

Findlay MW, Dolderer JH, Trost N, Craft RO, Cao Y, Cooper-White J, Stevens G, Morrison WA. Tissue-engineered breast reconstruction: bridging the gap toward large-volume tissue engineering in humans. Plast Reconstr Surg. 2011;128(6):1206–15. doi:10.1097/PRS.0b013e318230c5b2.

Fink T, Abildtrup L, Fogd K, Abdallah BM, Kassem M, Ebbesen P, Zachar V. Induction of adipocyte-like phenotype in human mesenchymal stem cells by hypoxia. Stem Cells. 2004;22(7):1346–55.

Fischbach C, Seufert J, Staiger H, Hacker M, Neubauer M, Gopferich A, Blunk T. Three-dimensional in vitro model of adipogenesis: comparison of culture conditions. Tissue Eng. 2004;10(1–2):215–29.

Flores-Delgado G, Marsch-Moreno M, Kuri-Harcuch W. Thyroid hormone stimulates adipocyte differentiation of 3T3 cells. Mol Cell Biochem. 1987;76(1):35–43.

Folkman J, Hochberg M. Self-regulation of growth in three dimensions. J Exp Med. 1973;138(4):745–53.

Gaillard D, Negrel R, Lagarde M, Ailhaud G. Requirement and role of arachidonic acid in the differentiation of pre-adipose cells. Biochem J. 1989;257(2):389–97.

Gimble JM, Robinson CE, Wu X, Kelly KA, Rodriguez BR, Kliewer SA, Lehmann JM, Morris DC. Peroxisome proliferator-activated receptor-gamma activation by thiazolidinediones induces adipogenesis in bone marrow stromal cells. Mol Pharmacol. 1996;50(5):1087–94.

Gleeson JP, Plunkett NA, O'Brien FJ. Addition of hydroxyapatite improves stiffness, interconnectivity and osteogenic potential of a highly porous collagen-based scaffold for bone tissue regeneration. Eur Cell Mater. 2010;20:218–30.

Halberstadt C, Austin C, Rowley J, Culberson C, Loebsack A, Wyatt S, Coleman S, Blacksten L, Burg K, Mooney D, Holder Jr W. A hydrogel material for plastic and reconstructive applications injected into the subcutaneous space of a sheep. Tissue Eng. 2002;8(2):309–19. doi:10.1089/107632702753725067.

Harrison JJ, Soudry E, Sager R. Adipocyte conversion of CHEF cells in serum-free medium. J Cell Biol. 1985;100(2):429–34.

Hauner H. Complete adipose differentiation of 3T3 L1 cells in a chemically defined medium: comparison to serum-containing culture conditions. Endocrinology. 1990;127(2):865–72.

Hauner H, Loffler G. Adipogenic factors in human serum promote the adipose conversion of 3T3-L1 fibroblasts. Int J Obes. 1986;10(4):323–30.

Hauner H, Entenmann G, Wabitsch M, Gaillard D, Ailhaud G, Negrel R, Pfeiffer EF. Promoting effect of glucocorticoids on the differentiation of human adipocyte precursor cells cultured in a chemically defined medium. J Clin Invest. 1989;84(5):1663–70.

Hausman GJ, Wright JT, Richardson RL. The influence of extracellular matrix substrata on preadipocyte development in serum-free cultures of stromal-vascular cells. J Anim Sci. 1996;74(9):2117–28.

Hijjawi JB, Blondeel PN. Advancing deep inferior epigastric artery perforator flap breast reconstruction through multidetector row computed tomography: an evolution in preoperative imaging. J Reconstr Microsurg. 2010;26(1):11–20.

Hofer SO, Knight KM, Cooper-White JJ, O'Connor AJ, Perera JM, Romeo-Meeuw R, Penington AJ, Knight KR, Morrison WA, Messina A. Increasing the volume of vascularized tissue formation in engineered constructs: an experimental study in rats. Plast Reconstr Surg. 2003;111(3):1186–92. discussion 1193–4.

Holzbach T, Vlaskou D, Neshkova I, Konerding MA, Wortler K, Mykhaylyk O, Gansbacher B, Machens HG, Plank C, Giunta RE. Non-viral VEGF(165) gene therapy—magnetofection of acoustically active magnetic lipospheres ('magnetobubbles') increases tissue survival in an oversized skin flap model. J Cell Mol Med. 2010;14(3):587–99.

Hong L, Peptan I, Clark P, Mao JJ. Ex vivo adipose tissue engineering by human marrow stromal cell seeded gelatin sponge. Ann Biomed Eng. 2005;33(4):511–7.

Itoi Y, Takatori M, Hyakusoku H, Mizuno H. Comparison of readily available scaffolds for adipose tissue engineering using adipose-derived stem cells. J Plast Reconstr Aesthet Surg. 2010;63(5):858–64.

Kang X, Xie Y, Powell HM, James Lee L, Belury MA, Lannutti JJ, Kniss DA. Adipogenesis of murine embryonic stem cells in a three-dimensional culture system using electrospun polymer scaffolds. Biomaterials. 2007;28(3):450–8.

Kim SP, Ha JM, Yun SJ, Kim EK, Chung SW, Hong KW, Kim CD, Bae SS. Transcriptional activation of peroxisome proliferator-activated receptor-gamma requires activation of both protein kinase A and Akt during adipocyte differentiation. Biochem Biophys Res Commun 2010;399(1):55–9. doi: S0006-291X(10)01340-9 [pii] 10.1016/j.bbrc.2010.07.038.

Kimura Y, Ozeki M, Inamoto T, Tabata Y. Time course of de novo adipogenesis in matrigel by gelatin microspheres incorporating basic fibroblast growth factor. Tissue Eng. 2002;8(4): 603–13. doi:10.1089/107632702760240526.

Klein JA. Anesthesia for liposuction in dermatologic surgery. J Dermatol Surg Oncol. 1988;14(10):1124–32.

Klein JA. Tumescent technique for regional anesthesia permits lidocaine doses of 35 mg/kg for liposuction. J Dermatol Surg Oncol. 1990;16(3):248–63.

Kliewer SA, Lenhard JM, Willson TM, Patel I, Morris DC, Lehmann JM. A prostaglandin J2 metabolite binds peroxisome proliferator-activated receptor gamma and promotes adipocyte differentiation. Cell. 1995;83(5):813–9.

Klopsch C, Donndorf P, Kaminski A, Ma N, Steinhoff G. Cell sources for cardiovascular tissue engineering. Chirurg. 2011;82(4):295–302. doi:10.1007/s00104-010-2030-3.

Knight DM, Chapman AB, Navre M, Drinkwater L, Bruno JJ, Ringold GM. Requirements for triggering of adipocyte differentiation by glucocorticoids and indomethacin. Mol Endocrinol. 1987;1(1):36–43.

Kolk A, Haczek C, Koch C, Vogt S, Kullmer M, Pautke C, Deppe H, Plank C. A strategy to establish a gene-activated matrix on titanium using gene vectors protected in a polylactide coating. Biomaterials. 2011;32(28):6850–9. doi:10.1016/j.biomaterials.2011.05.071.

Kurten RC, Navre M, Gaddy-Kurten D, Semenkovich CF, Ringold GM, Chan L, Richards JS. Induction of adenosine 3′,5′-monophosphate-dependent protein kinase subunits during adipogenesis in vitro. Endocrinology. 1988;123(5):2408–18.

Lehmann JM, Moore LB, Smith-Oliver TA, Wilkison WO, Willson TM, Kliewer SA. An antidiabetic thiazolidinedione is a high affinity ligand for peroxisome proliferator-activated receptor gamma (PPAR gamma). J Biol Chem. 1995;270(22):12953–6.

Lehmann JM, Lenhard JM, Oliver BB, Ringold GM, Kliewer SA. Peroxisome proliferator-activated receptors alpha and gamma are activated by indomethacin and other non-steroidal anti-inflammatory drugs. J Biol Chem. 1997;272(6):3406–10.

Lequeux C, Auxenfans C, Mojallal A, Sergent M, Damour O. Optimization of a culture medium for the differentiation of preadipocytes into adipocytes in a monolayer. Biomed Mater Eng. 2009;19(4–5):283–91.
Lokmic Z, Stillaert F, Morrison WA, Thompson EW, Mitchell GM. An arteriovenous loop in a protected space generates a permanent, highly vascular, tissue-engineered construct. FASEB J. 2007;21(2):511–22.
Machens HG, Morgan JR, Berthiaume F, Stefanovich P, Reimer R, Berger AC. Genetically modified fibroblasts induce angiogenesis in the rat epigastric island flap. Langenbecks Arch Surg. 1998;383(5):345–50.
Morikawa M, Nixon T, Green H. Growth hormone and the adipose conversion of 3T3 cells. Cell. 1982;29(3):783–9.
Morrison WA, Penington AJ, Kumta SK, Callan P. Clinical applications and technical limitations of prefabricated flaps. Plast Reconstr Surg. 1997;99(2):378–85.
Negrel R, Gaillard D, Ailhaud G. Prostacyclin as a potent effector of adipose-cell differentiation. Biochem J. 1989;257(2):399–405.
Neuber FF. Bericht uber die Verhandlungen der Deutschen Gesellschaft fur Chirurgie [German]. Zentralbl Chir. 1893;22:66.
Nixon T, Green H. Contribution of growth hormone to the adipogenic activity of serum. Endocrinology. 1984;114(2):527–32.
Pairault J, Lasnier F. Control of the adipogenic differentiation of 3T3-F442A cells by retinoic acid, dexamethasone, and insulin: a topographic analysis. J Cell Physiol. 1987;132(2):279–86.
Pairault J, Quignard-Boulange A, Dugail I, Lasnier F. Differential effects of retinoic acid upon early and late events in adipose conversion of 3T3 preadipocytes. Exp Cell Res. 1988;177(1):27–36.
Patrick Jr CW, Chauvin PB, Hobley J, Reece GP. Preadipocyte seeded PLGA scaffolds for adipose tissue engineering. Tissue Eng. 1999;5(2):139–51.
Peng L, Jia Z, Yin X, Zhang X, Liu Y, Chen P, Ma K, Zhou C. Comparative analysis of mesenchymal stem cells from bone marrow, cartilage, and adipose tissue. Stem Cells Dev. 2008;17(4): 761–73. doi:10.1089/scd.2007.0217.
Reckhenrich AK, Hopfner U, Krotz F, Zhang Z, Koch C, Kremer M, Machens HG, Plank C, Egana JT. Bioactivation of dermal scaffolds with a non-viral copolymer-protected gene vector. Biomaterials. 2011;32(7):1996–2003.
Ringold GM, Chapman AB, Knight DM. Glucocorticoid control of developmentally regulated adipose genes. J Steroid Biochem. 1986;24(1):69–75.
Rodriguez Fernandez JL, Ben-Ze'ev A. Regulation of fibronectin, integrin and cytoskeleton expression in differentiating adipocytes: inhibition by extracellular matrix and polylysine. Differentiation. 1989;42(2):65–74.
Safonova I, Darimont C, Amri EZ, Grimaldi P, Ailhaud G, Reichert U, Shroot B. Retinoids are positive effectors of adipose cell differentiation. Mol Cell Endocrinol. 1994;104(2):201–11.
Sandouk T, Reda D, Hofmann C. Antidiabetic agent pioglitazone enhances adipocyte differentiation of 3T3-F442A cells. Am J Physiol. 1993;264(6 Pt 1):C1600–8.
Schimmel RJ, Honeyman TW, McMahon KK, Serio R, Clark RB. Inhibition of cyclic AMP accumulation in hamster adipocytes with phosphatidic acid: differences and similarities with alpha adrenergic effects. J Cyclic Nucleotide Res. 1980;6(6):437–49.
Schreml S, Babilas P, Fruth S, Orso E, Schmitz G, Mueller MB, Nerlich M, Prantl L. Harvesting human adipose tissue-derived adult stem cells: resection versus liposuction. Cytotherapy. 2009;11(7):947–57.
Singh S, Wu BM, Dunn JCY. Accelerating vascularization in polycaprolactone scaffolds by endothelial progenitor cells. Tissue Eng Part A. 2011;17(13–14):1819–30. doi:10.1089/ten.TEA.2010.0708.
Spanholtz T, Maichle A, Niedworok C, Stoeckelhuber BM, Krüger S, Wedel T, Aach T, Middeler G, Hellwig-Bürgel T, Bader A, Krengel S, Müller OJ, Franz WM, Lindenmaier W, Machens

H-G. Timing and targeting of cell-based VEGF165 gene expression in ischemic tissue. J Surg Res. 2009;151(1):153–62. doi:10.1016/j.jss.2008.01.038.

Spanholtz TA, Theodorou P, Holzbach T, Wutzler S, Giunta RE, Machens HG. Vascular endothelial growth factor (VEGF165) plus basic fibroblast growth factor (bFGF) producing cells induce a mature and stable vascular network—a future therapy for ischemically challenged tissue. J Surg Res. 2011;171(1):329–38.

Stacey DH, Hanson SE, Lahvis G, Gutowski KA, Masters KS. In vitro adipogenic differentiation of preadipocytes varies with differentiation stimulus, culture dimensionality, and scaffold composition. Tissue Eng Part A. 2009;15(11):3389–99.

Suruga K, Suzuki R, Goda T, Takase S. Unsaturated fatty acids regulate gene expression of cellular retinol-binding protein, type II in rat jejunum. J Nutr. 1995;125(8):2039–44.

Swierczewski E, Pello JY, Arapinis C, Aron Y, Krishnamoorthy R. Characterization of rat preadipocytes from normal rat adipose tissue by their effector response. Biochem J. 1987;248(2):383–7.

Takahashi K, Yamanaka S. Induction of pluripotent stem cells from mouse embryonic and adult fibroblast cultures by defined factors. Cell. 2006;126(4):663–76. doi:10.1016/j.cell.2006.07.024.

Takahashi K, Tanabe K, Ohnuki M, Narita M, Ichisaka T, Tomoda K, Yamanaka S. Induction of pluripotent stem cells from adult human fibroblasts by defined factors. Cell. 2007;131(5): 861–72. doi:10.1016/j.cell.2007.11.019.

Takase S, Tanaka K, Suruga K, Kitagawa M, Igarashi M, Goda T. Dietary fatty acids are possible key determinants of cellular retinol-binding protein II gene expression. Am J Physiol. 1998;274(4 Pt 1):G626–32.

Tanaka Y, Sung KC, Tsutsumi A, Ohba S, Ueda K, Morrison WA. Tissue engineering skin flaps: which vascular carrier, arteriovenous shunt loop or arteriovenous bundle, has more potential for angiogenesis and tissue generation? Plast Reconstr Surg. 2003;112(6):1636–44.

Tanzi MC, Fare S. Adipose tissue engineering: state of the art, recent advances and innovative approaches. Expert Rev Med Devices. 2009;6(5):533–51.

Tenzen T, Zembowicz F, Cowan CA. Genome modification in human embryonic stem cells. J Cell Physiol. 2010;222(2):278–81. doi:10.1002/jcp.21948.

Torio-Padron N, Baerlecken N, Momeni A, Stark GB, Borges J. Engineering of adipose tissue by injection of human preadipocytes in fibrin. Aesthetic Plast Surg. 2007;31(3):285–93.

Verhaegen M, Correze C, Krug E, Nunez J. Cyclic AMP and lipogenesis in fat cells from thyroidectomized rats. Mol Cell Endocrinol. 1979;14(3):167–80.

von Heimburg D, Zachariah S, Low A, Pallua N. Influence of different biodegradable carriers on the in vivo behavior of human adipose precursor cells. Plast Reconstr Surg. 2001;108(2): 411–20. discussion 421–2.

Wabitsch M, Hauner H, Heinze E, Teller WM. The role of growth hormone/insulin-like growth factors in adipocyte differentiation. Metabolism. 1995;44(10 Suppl 4):45–9.

Welsch U. Sobotta Lehrbuch Histologie. 2nd ed. München: Elsevier, Urban & Fischer Verlag; 2006.

Wernig M, Scheffler B, Brüstle O. Medizinische Perspektiven der Stammzellforschung. In: Ganten D, Ruckpaul K, editors. Grundlagen der Molekularen Medizin. 2nd ed. Berlin: Springer; 2003.

Wiggenhauser PS, Müller DF, Melchels FPW, Egaña JT, Storck K, Mayer H, Leuthner P, Skodacek D, Hopfner U, Machens HG, Staudenmaier R, Schantz JT. Engineering of vascularized adipose constructs. Cell Tissue Res. 2011. doi:10.1007/s00441-011-1226-2.

Zhang Z, Ito WD, Hopfner U, Bohmert B, Kremer M, Reckhenrich AK, Harder Y, Lund N, Kruse C, Machens HG, Egana JT. The role of single cell derived vascular resident endothelial progenitor cells in the enhancement of vascularization in scaffold-based skin regeneration. Biomaterials. 2011;32(17):4109–17.

Zocchi ML, Zuliani F. Bicompartmental breast lipostructuring. Aesthetic Plast Surg. 2008;32(2):313–28. doi:10.1007/s00266-007-9089-3.

Index

A

Adipocytes
  adipose organ, plasticity of
    adipose tissue, 132, 133
    depots and adipocyte types, 133–134
    origin of adipocyte subtypes, 134
    types of, 132
  hypertrophy *vs.* hyperplasia, 275
  precursors and vasculature
    adipose tissue capillaries, 135
    embryogenesis, 135
    endothelial cells in adipose stem cells, 148–150
    human adipose stem cells and vasculature, 150, 151
    hypertrophy, 134
    lineage tracing evidence, 145–148
    pericytes (*see* Pericytes)
    preadipocytes, 135
  transformation, 176
  and vascular cells
    adipocyte-vasculature interplay, 95–96
    effects of, 94–95
    endothelial cells, 91–93
    perivascular cells, 94
    vascular remodeling and neovascularization, 91, 93

Adipokines
  adiponectin (*see* Adiponectin)
  angiogenin, 205–206
  apelin, 206
  chemerin, 87–88, 207
  endocrine organ, 85
  fat depot-specific differences, 210–211
  impact angiogenesis., 188
  insulin sensitivity, 88
  interleukin-6, 88, 203–204
  leptin (*see* Leptin)
  lipid derivatives, 209–210
  lipid metabolism, 88
  macrophages/stromal cells, 88
  metallothionein, 88
  obesity, 188, 211–215
  omentin, 207
  PAI-1, 88, 208–209
  physiological roles, 189
  resistin (*see* Resistin)
  TGB-β, 204–205
  TNFα and TGF-β, 88
  vaspin, 209
  visfatin, 87, 209
  WAT and cancer, 324–325

Adiponectin, 88
  abundant adipokine, 193
  ADIPOR1 and ADIPOR2, 194–195
  AMPK, 195
  angiogenesis effects, 195–198
  enzymatic digestion, 193
  HMW oligomeric species, 193
  multifunctional protein, 194
  N-terminal signal sequence, 193
  physiological processes, 189, 194
  pleiotropic insulin-sensitizing effects, 194
  vascular function effects, 198

Adipose-derived endothelial precursor cells. *See* Endothelial progenitor cells (EPCs)

Adipose-derived stem cells (ADSCs), 31, 82–83, 284–285

Adipose stem cells
  adipocytes, pre-adipocytes, 27
  brown adipogenesis, 23–24

Y. Cao (ed.), *Angiogenesis in Adipose Tissue*, DOI 10.1007/978-1-4614-8069-3,

Adipose stem cells (*cont.*)
brown pre-adipocyte recruitment
β-adrenergic signaling, 29
COX-2, 29–30
clinical aspects
clinical practice and limitations, 32–33
metabolic control and obesity-related disorders, 32–33
metabolocentric therapy, 34–35
MSC, 30–32
therapeutic use, 35–36
determination and differentiation phases, 20
mature adipocytes, 20
microenvironment, 25
pre-adipocyte cell lines, 20
pro-adipogenic signals
bone morphogenic proteins, 25–26
Wnt Signaling, 26–27
transcriptional mediators of adipogenesis
C/EBP family members, 28
PPARg, 28–29
white adipogenesis
adipose tissue nonresident progenitors, 22
adipose tissue-resident progenitors, 21–22
Adipose stromal cells (ASCs), 158
angiogenic potential
CD34, 159
CD31-positive cell, 161
endothelial CD31 marker, 160
endothelial-like cells, 161
endothelial-like phenotype, 158–160
fetal calf serum, 160
heart ischemia, 159
hypoxia promotes, 161
immature cells and endothelial cells, 159
matrigel plug, 159
mesenchymal to endothelial transformation, 160
paracrine activity, 161
SDF1/CXCR4 and PI3 kinase pathways, 160
shear stress, 160
subcutaneous blood flow, 159
vascular tropism, 160
VEGF uptaked acetylated LDL, 160
in cancer
cancer therapy, 331–332
mobilization, 327
role of, 328–330
trafficking, 327–328
tumor vasculature, 330–331
extracellular matrix, 326
hypoxia treatment of, 257
pathological functions, 325–326
smooth muscle phenotype, 162
Adipose tissue-derived progenitor cells. *See* Progenitor cells
Adipose tissue engineering
advantages, 355
differentiation stimuli, 356, 358
disadvantages, 355
history and impact, 354, 355
scaffold architecture
macro architecture, 359–361
micro architecture, 356, 358–359
scaffold materials, 356, 357
stem cells sources, 355
vascularization (*see* Vascularization)
Adipose tissue macrophages (ATMs), 47–48
Adipose tissue remodeling, 44–45
Adipose tissue transplantation
FOXC2-TM, 289–291
genetic models, 288
metabolic disorders, 287
mouse models
embryonic angiogenesis, 289
obesity and diabetes, 288–289
physiological research, 286–287
reconstructive medicine, 286
regenerative medicine, 285–286
ADSC. *See* Adipose-derived stem cells (ADSCs)
Age-related macular degeneration (AMD), 10
Angiogenesis
adipokines, 267, 268
adipose tissue hypoxia (*see* Hypoxia)
in developing embryos, vasculogenesis (*see* Vasculogenesis)
disease and therapy
adipose angiogenesis, 6
blood vessel formation and growth, 4
cancer therapy (*see* Cancer therapy)
cardiovascular diseases, 11
metabolic diseases, 6–7
neovascularization, 4
Notch signaling systems, 4
ophthalmological diseases, 10
tumor tissue vasculatures, 4, 5
vasculogenesis, 4
VEGF expression, 4, 5
VEGFR-2-triggered signaling pathway, 5
mouse genetic models (*see* Mouse genetic models)
Angiogenin, 205–206
Angiopoietin, 85

Angiopoietin-like protein 4 (Angptl-4), 200–201
Apelin, 206
Arcuate nucleus (ARC), 172–173
ASCs. *See* Adipose stromal cells (ASCs)
ATMs. *See* Adipose tissue macrophages (ATMs)

B
β3-adrenergic receptor (β3-AR) agonists, 171
β-adrenergic signaling, 29
Basal metabolic rate (BMR), 283–284
Basement membrane synthesis, 54–56
BAT. *See* Brown adipose tissue (BAT)
11β-hydroxysteroid dehydrogenases (11β-HSDs), 257
B lymphocytes, 233–234
Bone morphogenic proteins, 25–26
Bromodeoxyuridine (BrdU), 24
Brown adipose tissue (BAT)
  adipokines (*see* Adipokines)
  adipomyocytes, 171
  adipose stem cells
    β-adrenergic signaling, 29
    BrdU, 24
    CD137 and TMEM26, 24
    human physiology, 23
    metabolocentric therapy, 34–35
    metabolo-regulatory impact, 35
    muscle-like precursors, 23
  adult humans, 170
  angiogenic factors, 84–85
  energy metabolism, 266
  18FDG, 170
  gestation, 266
  multilocular and mitochondrial-rich adipocytes, 170
  overweight/obese, 170
  PET-CT technology, 170
  plasticity of adipose vasculatures
    anti-angiogenic agents, 90
    pro-angiogenic agents, 89
  structures and functions, 78–80
  thermogenesis, 255
  thermogenic organ, 170
  UCP1, 170
  vascular density, 79–81
  vascular function
    adipose vasculature and ADSC, 82–83
    blood vessels in, 81, 82
    infiltrating inflammatory cells and obesity, 83–84

C
Cancer-associated fibroblasts (CAF), 328–330
Cancer therapy
  angiogenic factor antagonists, 8
  antiangiogenic drug resistance, 10
  anti-VEGF drugs, 9
  ASC, 331–332
  bevacizumab, 8
  cancer cachexia, 9
  chemotherapy, 8, 9
  endogenous angiogenesis inhibitors, 8
  paraneoplastic syndrome, 9
  preclinical animal models, 7
  tumor angiogenesis, 7
  tumor vasculat, 8
CCAAT/enhancer-binding protein-α (CEBP-α), 111
Chemerin, 87–88, 207
Chorioallantoic membrane (CAM), 191, 270
Chronic inflammation, 232
Copolymer-protected gene vectors (COPROGS), 364
Cornea implantation assay, 270–271
Cyclooxygenase-2 (COX-2), 29–30

D
Diabetic macular edema (DME), 7
DIEP-Flap, 353
Dorsomedial hypothalamus (DMH)
  brown adipocyte development, 174–176
  energy expenditure, 178–181
  inguinal lipid mobilization and blood fat levels, 177–178
  interscapular BAT activity, 178, 179
  leptin-independent control, 173–174
  obesity, 173

E
Endothelial cell dysfunction, 47
Endothelial progenitor cells (EPCs)
  CD34+ cell purification, 340
  flow cytometry studies, 340–342
  WAT CD34$^+$ cells
    angiogenesis-related genes, 342–343
    breast cancer models, 344–345
    capillary tubes, 343
    mature endothelial cells, 343
    stemness-related genes, 342–343
Epididymal fat (eWAT), 135–136
Extracellular matrix (ECM), 44–45, 326, 328, 329

F

FACS. *See* Fluorescence-activated cell sorting (FACS)
Factor inhibiting HIF (FIH), 249, 254
Fasting-induced adipose factor (FIAF), 200
Fat deposition, 110, 118
Fat transplantation
  first fat graft, 352
  goals of, 352
  large volume fat transplantation
    Coleman's lipostructure, 353–354
    DIEP-Flap, 353
    intraoperative photographs, 354
Fibrinolytic system
  α2-antiplasmin, 300
  PAI-1, 299–302
  PAI-2, 302
  t-PA, 299, 300, 302
  u-PA, 299, 300, 302
Fluorescence-activated cell sorting (FACS), 22

G

Glucocorticoids, 256–257

H

Hepatocyte growth factor (HGF), 190, 307, 310
High fat diet (HFD), 301–305
High-molecular weight (HMW) oligomeric species, 193
Horse radish peroxidase (HRP), 281
Human tissue inhibitors of MMPs (TIMPs), 302–306
Hypertrophy, 20
Hypoxia
  and angiogenesis, 254–255
    11β-HSD1, 257
    depot-specific responses, 255–256
    HIF, 256–257
    in non-adipocyte cells, vascular remodelling, 257–259
    VEGF-A, 256, 257
  obesity
    in mice, adipose HIF-system, 250–254
    oxygen sensing pathway, 248–249
    tissue dysfunction in, 249–250
Hypoxia-inducible factor (HIF)
  HIF-1, 43, 44
  induction of, 250
  in mice, adipose HIF-system modulation, 250–254
  oxygen sensing pathway, 248–249
  transgenic models, 251–253

I

Immune cells, 48–49
  adipose tissue inflammation, 232
  chronic inflammation, 232
  involvement of, 240
  lymphoid cells
    B lymphocytes, 233–234
    NKT cells, 233
    T lymphocytes, 234–236
  role of macrophages (*see* Macrophages)
Immunohistochemistry
  on cryosections, 282
  on paraffin-embedded adipose tissue, 280–282
Inducible brown fat, 24
Interleukin-6 (IL-6), 203–204

L

Leptin, 86
  angiogenesis effects, 191–192
  food intake and energy metabolism, 191
  peripheral organs, 191
  subcutaneous WAT adipocytes, 191
  transmembrane-spanning receptor, 191
  vascular function effects, 192–193
Lipid derivatives, 209–210
Lymphangiogenesis
  adipose-associated immune cells, 116
  blood vasculature, 115
  cancer metastasis, 119
  inflammatory disorders, 117
  leptin, an anti-satiety signal, 116
  lymphatic endothelial cells, 113
  lymphatic endothelium, 114
  maturation and remodeling, 115–116
  migration and proliferation, 115
  molecular events, 113, 114
  obesity and metabolic disorders, 120–121
  signaling molecules, 113
  VEGF-C, 118–119
Lymphatic system
  adipolymphatic connections
    adipocyte accumulation, 111
    adipogenesis, 111
    adipogenic property of lymph, 112
    CEBP-α, 111
    chylomicrons, 110
    collagen-rich connective tissue, 109
    CREB-α participates, 112
    lipid absorption, 109, 110
    lymphatic pumping, 113
    lymphatic vessels, 110–112
    lymphedema, 111

lymph stasis promotes inflammation, 111
PAT, 113
adipose tissue, 116–117
discovery of, 104
lymphangiogenesis (*see* Lymphangiogenesis)
structure and function
collecting lymphatic vessels, 106–108
fluid homeostasis, 104
initial lymphatic vessels, 105–107
lymph nodes, 106, 108, 109
structural organization, 105
Lymphatic vessel endothelial hyaluronan receptor 1 (LYVE-1), 237, 238
Lymphoid cells
B lymphocytes, 233–234
NKT cells, 233
T lymphocytes
CD4+ T helper lymphocytes, 235–236
CD8+ T lymphocytes, 234–235
CD4+ T regulatory lymphocytes, 235

M
Macrophages, 47–48
accumulation in adipose tissue, 236
and angiogenesis
inflammatory process, 240, 241
obesity, 239–241
postnatal and adult adipose tissue development, 238–239
VEGF-A, 239, 240
in human adipose tissue, 237
phenotypic switch, 237
Matrix metalloproteinase system (MMPs), 302–306
Mesenchymal stem cells (MSCs)
adipose stem cells
adipocytes, 20, 21
adipose tissues, 23, 30–31
β-catenin targets transcription factors, 27
bone morphogenic proteins, 25, 26
pre-adipocyte lineage, 25
regenerative medicine, 31, 32
SVF, 21
Wnt signaling, 27
vascular adipose complex
adipose cells/induction, 58
bone marrow, 54, 56, 58, 62
CM-DiI-labeled, 62, 63
dermal fibroblasts, 59
Metabolic syndrome, 32, 33
MMPs. *See* Matrix metalloproteinase system (MMPs)
Monoclonal antibodies, 341
Monocyte chemotactic protein-1 (MCP-1), 274
Mouse genetic models
anti-angiogenic factors, 310–311
pro-angiogenic factors
FGF family, 309
HGF, 310
leptin, 310
PIGF, 307–309
VEGF-A, 307, 308
VEGF-B, 308
VEGF-D, 308
proteolysis
ADAM and ADAMTS families, 306
fibrinolytic system, 299–302
matrix metalloproteinase system, 302–306
MSCs. *See* Mesenchymal stem cells (MSCs)
Myeloid cells
HIF-1α loss of, 259
resistin, 199
VEGF, loss of, 259
Myocardial infarction (MI), 285

N
Natural killer T cells, 233
Neuropeptide Y (NPY)
ARC, 172–173
BAT, 170, 171
central and peripheral nervous system, 171
DMH (*see* Dorsomedial hypothalamus (DMH))
peripheral, 181
PYY and PP, 171
WAT, 170, 171
Non-shivering thermogenesis (NST)
BMR and NE-induced NST, 283–284
UCP-1-dependent activation, 276
Norepinephrine (NE), 283–284
NPY. *See* Neuropeptide Y (NPY)

O
Obesity
adipose tissue hypoxia
in mice, adipose HIF-system, 250–254
oxygen sensing pathway, 248–249
tissue dysfunction in, 248–249
adipose tissue inflammation, 232
defintion, 321
infiltrating inflammatory cells, 83–84
inflammation, 232
metabolic disorders, 232

Obesity (*cont.*)
  in obese model, adipose tissue
    adipose tissue vasculature, 273
    hyperplasia and hypertrophy, 274–275
    hypoxia, 275–276
Omentin, 207
Ophthalmological diseases, 10

P
Pancreatic polypeptide (PP), 171
Paraffin-embedded adipose tissues
  cryosections, 282
  H&E staining, 280
  immunohistochemistry on, 280–282
Peptide tyrosine-tyrosine (PYY), 171
Pericytes, 162
  brown adipocytes
    E17-18, 141
    interscapular fat depot, 141
    morphology of, 141, 142
    preadipocytes, 142–145
  white adipocoytes
    abundant pericytes, 137
    endothelial cells, 138–141
    endothelial-pericytic cell, 137, 138
    eWAT, 135, 136
    fibroblast-like cells, 136
    pericyte-like elements, 138
    pericytes preadipocytes, 138
    vasculo-adipocytic islets, 136, 138, 139
Perinodal adipose tissue (PAT), 113
Perivascular cells, 61–63
Peroxisome proliferator-activated receptor gamma (PPARg), 22
Plasminogen activator inhibitor-1 (PAI-1), 208–209, 299–302
Platelet-derived growth factor (PDGF), 239, 240, 307–309
PP. *See* Pancreatic polypeptide (PP)
PR domain containing 16 (PRDM16), 23
Pre-B cell colony-enhancing factor (PBEF), 209
Progenitor cells
  ASC in cancer
    cancer therapy, 331–332
    mobilization, 327
    role of, 328–330
    trafficking, 327–328
    tumor vasculature, 330–331
  composition and biological function, 322–324
  pathological functions, 325–326
  WAT and cancer, 324–325
Proliferative diabetic retinopathy (PDR), 7
Prolylhydrylases (PHDs), 7
Proteolysis
  ADAM and ADAMTS families, 306
  fibrinolytic system
    α2-antiplasmin, 300
    PAI-1, 299–302
    PAI-2, 302
    t-PA, 299, 300, 302
    u-PA, 299, 300, 302
  matrix metalloproteinase system, 302–306
PYY. *See* Peptide tyrosine-tyrosine (PYY)

R
Resistin, 86–87
  adipose tissue-resident macrophages, 199
  adipose tissue-specific secretory factor, 199
  Angptl-4, 200
  cellular receptors, 199
  FIZZ3, 199
  hepatic Kupffer cells, 199
  human adipose tissues, obesity, 199
  obesity and insulin resistance, 199
  pro-angiogenic factor, 200
  RELMs, 199
Resistin-like molecules (RELMs), 199

S
Stromal vascular cells
  adipose tissue development and blood vessels, 42–44
  adipose tissue inflammation, 45, 46
  ECM, 44
  functions of
    adipocyte progenitor cells, 46–47
    B cells, 48
    $CD8^+$ T cells, 48
    $CD4^+$ Th1 cells, 48
    endothelial cells, 47
    iNKT cells, 49
    macrophages, 47–48
    mast cells, 48–49
    Treg cells, 49
  MMPs, 45
  obese adipose tissue, 42
  obesity, 42
  triacylglycerol storage, 44
  visceral adipose tissue, 41–42
Stromal/vascular fraction (SVF), 21, 323
Sympathetic nervous system (SNS), 170

T
Thrombospondin 1 (TSP-1), 307, 310–311
Tissue engineering, 59
Tissue inhibitors of MMPs (TIMPs), 302–306
Tissue-type plasminogen activator (t-PA), 299, 300, 302
T lymphocytes (CD3+)
  CD8+ lymphocytes, 234–235
  CD4+ T helper lymphocytes, 235–236
  CD4+ T regulatory lymphocytes, 235
Transforming growth factor-β (TGF)-β, 88, 204–205
Transmembrane protein 26 (TMEM26), 24
Tumor necrosis factor alpha (TNFα), 88, 202–203

U
Uncoupling protein 1 (UCP1), 170
Urokinase-type plasminogen activator (u-PA), 299, 300, 302

V
Vascular adipose complex
  adipocyte development, 54–56
  basement membrane synthesis, 54–56
  cellular constituents of adipose tissue
    adipose cells/induction of MSC, 58
    ATM, 57
    endothelial cells, 59–61
    noncellular extracts, 57
    perivascular cells, 61–63
    pre-vascularized tissue, 57
    vascular support cells, 58–60
  endocrine function, 54
  heparin-binding angiogenic factors, 63, 65
  HUVECs, 63, 64
  immunohistochemical staining, 63
  implantable tissue
    adipocyte precursors, 65, 67
    histological structure, 65, 67
    host vasculature invested implants, 65, 68
    pre-vascularized adipose tissue, 65, 66
    production of, 65, 66
    vasculature, 65, 69
  mesenchymal adipocytes, 54
  non-basement membrane extracellular matrix molecules, 56–57
  standard immunostaining for CD31, 62, 63
  type IV collagen, 62, 64
Vascular endothelial cells, 59–61
Vascular endothelial growth factor (VEGF), 4, 5, 9, 84–85, 118–119, 160, 188–190, 239, 240, 256, 257, 259, 267–269, 307, 308
Vascular functions in adipose tissue
  endothelial cells, 91–93
  perivascular cells, 94
Vascularization
  challenge of, 359
  single graft, 362
  in situ tissue engineering, 363–364
    chamber model, 365
    COPROGS, 364
    transgene fibroblasts, 364–365
    vessel loop method, 365
  in vitro tissue engineering
    endothelial progenitors, 363
    vessel loop model, 362–363
  in vitro *vs.* in situ, 361
Vascular remodelling
  angiopoietin, 85
  in FOXC2-TM, 289–291
  in non-adipocyte cells, 257–259
Vascular smooth muscle cells (VSMCs), 290–291
Vascular support cells, 58–60
Vasculogenesis
  adipose tissue
    in fetal development, 266
    in fetus and adulthood, 266–267
    VEGF, 267–269
  CAM assay, 270
  cornea implantation assay, 270–271
  mouse model, 272–273
  zebrafish models, 271–272
Vaspin, 209
Vessel loop model, 362–363, 365
Visfatin, 87
VSMCs. *See* Vascular smooth muscle cells (VSMCs)

W
White adipose tissue (WAT)
  adipokines (*see* Adipokines)
  adipose stem cells
    aberrant metabolic properties., 33
    BAT, 23
    brown adipocytes, 24
    energy storage, 23, 33
    experimental animal models, 20
    human patient samples, 20
    insulin resistance, 34
    tissue inflammation, 33

White adipose tissue (WAT) (*cont.*)
angiogenic factors, 84–85
to BRITE transformation, 276–278
BMR and NE-induced NST, 283–284
histology, 278
paraffin-embedded adipose tissues (*see* Paraffin-embedded adipose tissues)
whole-mount immunohistochemistry, 278–279
and cancer, 324–325
$CD45^{-}CD34^{+}$ cells, 340–342
$CD34^{+}$ cells, tumor growth
angiogenesis-related genes, 342–343
breast cancer models, 344–345
capillary tubes, 343
mature endothelial cells, 343
stemness-related genes, 342–343
cell populations, 323
NPY
β3-AR agonists, 171
energy reservoir-store excess calories, 170
sympathetic innervation induces lipolysis, 170
unilocular adipocytes, 170
orthotopic models, 346
structures and functions, 78, 79
vascular density, 79–81
vascular function
adipose vasculature and adipose-derived stem cells, 82–83
blood vessels in adipose tissue, 81, 82
infiltrating inflammatory cells and obesity, 83–84

Z

Zebrafish models, 271–272

MIX
Papier aus verantwortungsvollen Quellen
Paper from responsible sources
FSC® C105338

If you have any concerns about our products,
you can contact us on
ProductSafety@springernature.com

In case Publisher is established outside the EU,
the EU authorized representative is:
Springer Nature Customer Service Center GmbH
Europaplatz 3, 69115 Heidelberg, Germany

Printed by Libri Plureos GmbH
in Hamburg, Germany